S0-ASW-698

Foundations of Hadronic Chemistry

Fundamental Theories of Physics

An International Book Series on The Fundamental Theories of Physics: Their Clarification, Development and Application

Volume 120

Foundations of Hadronic Chemistry

With Applications to New Clean Energies and Fuels

by

Ruggero Maria Santilli

Institute for Basic Research,
Palm Harbor, FL, U.S.A.

KLUWER ACADEMIC PUBLISHERS

DORDRECHT / BOSTON / LONDON

A C.I.P. Catalogue record for this book is available from the Library of Congress.

ISBN 1-4020-0087-1

Published by Kluwer Academic Publishers,
P.O. Box 17, 3300 AA Dordrecht, The Netherlands.

Sold and distributed in North, Central and South America
by Kluwer Academic Publishers,
101 Philip Drive, Norwell, MA 02061, U.S.A.

In all other countries, sold and distributed
by Kluwer Academic Publishers,
P.O. Box 322, 3300 AH Dordrecht, The Netherlands.

Printed on acid-free paper

Printed in the Netherlands.

This monograph is dedicated to

Professor **T. Nejat Veziroglu**,
Director,
Clean Energy Research Institute,
University of Miami, Coral Gables, Florida,
and
Editor in Chief,
International Journal of Hydrogen Energy,
Elsevier Science, Oxford, England,

because his commitment to scientific
democracy for qualified inquiries and
his impeccable editorial processing
have permitted the birth of the new
discipline presented in this monograph.

Contents

Preface

A Physicist's Perspective on the Insufficiencies and Generalizations of Quantum Chemistry

My Undergraduate and Graduate Studies in Italy on the Insufficiencies of Quantum Mechanics and Chemistry

I was first exposed to quantum chemistry during my undergraduate courses in physics at the University of Naples, Italy, in the late 1950s. My teacher was Prof. Bakunin, a well known lady chemist in Europe at that time, who escaped from Russia with her family during the advent of communism. My three exams with her (inorganic chemistry, organic chemistry, and laboratory chemistry) were, by far, the most difficult exams of my life (although I did please Prof. Bakunin during the examinations).

Besides chemistry, during my undergraduate studies I plunged into the study of physics, with particular reference to quantum mechanics and its mathematical structure. My mathematics teacher was Prof. Caccioppoli, one of the most famous Italian mathematicians of that time, who taught me the necessity of advanced mathematics for quantitative physical studies.

By reading the works of the founders of contemporary physics, it was easy for me to see the lack of final character of quantum mechanics already in these undergraduate studies.

For instance, I was impressed by Enrico Fermi (whose departure in 1954 was still felt in the physics classrooms of the late 1950s) when in p. 111 of his celebrated *Nuclear Physics* (University of Chicago Press, 1950) he states:

> *"there are doubts as to whether the usual concepts of geometry hold for such small region of space (those of nuclear forces)."*

I subsequently spent several years of my research life to construct geometries, which are more suitable to represent the complexity of the nuclear forces because physics can indeed be reduced to primitive geometrical notions.

Equally inspiring was the teaching of Emilio Segre, Victor Weisskopf and other leading scientists of the time when they showed the impossibility for quantum mechanics to achieve an exact representation of total nuclear magnetic moments of *all* nuclei, rather than that of the deuteron alone (impossibility still in full existence today) because of apparent "alterations" of the intrinsic characteristics of protons and neutrons in the transition from their condition in vacuum (which have been used for their detection until now) to the novel conditions when members of a nuclear structure.

For instance, I still remember vividly my first reading of the statement by Blatt and Weisskopf in their *Theoretical Nuclear Physics* (John Wiley & Sons, first edition, 1952) in the section on nuclear magnetic moments (page 31 of the 1963 edition):

> *"It is possible that the intrinsic magnetism of a nucleon is different when it is in close proximity to another nucleon."*

It was easy to see since my undergraduate studies the impossibility for quantum mechanics to represent the alteration of an *intrinsic* characteristic of a particle as stable as the proton, thus mandating a generalization of the theory. I subsequently spent years of my research life studying the problem and did indeed achieve more recently an exact representation of *all* total nuclear magnetic moments, thanks to the hadronic generalization of quantum mechanics at the foundation of this monograph (see later on footnotes 1 and 6 of this Preface and Vol. III of Refs. [9]).

I was also impressed by the *"lack of completion" of quantum mechanics* voiced by Einstein, Podolsky and Rosen (Phys. Rev. **47**, 777, 1935) which I was reading in epistemological books circulating in Italy at that time. In fact, I subsequently spent decades of my research life to identify an appropriate and consistent "completion" of quantum mechanics. The entire contents of this monograph, beginning with the contents of this Preface, can be considered a study on the completion of quantum mechanics and, therefore, of quantum chemistry.

I still remember articles in the Italian newspapers of my college years debating the numerous doubts expressed by Albert Einstein (then still alive, since he died in 1955) on the lack of deterministic character of quantum mechanics.

I felt the expression of these basic doubts to be the ultimate manifestation of teaching as well as of science, because it stimulates minds of all ages. I did and still feel ascientific any teaching or textbook presenting a discipline as the end of the scientific adventure in the field.

In my undergraduate studies of the history of science, I saw the evidence that *science will never admit final theories.* I therefore acquired the conviction that the *belief on the terminal character of a given discipline is purely political-nonscientific.* True scientists can indeed debate the way in which a given theory should be generalized, but questioning, obstructing, or jeopardizing qualified technical studies on broader theories is scientific dishonesty. After all, history teaches that studies on the generalization of existing theories are always productive, irrespectively of whether successful or not, while their suppression is manifestly sterile.

As is well known and lamented by various scholars, in the remaining decades of the 20-th century, physics and chemistry evolved into a form of scientific religion because of the lack of proper treatment of historical doubts recalled above. Jointly, we have passed from the momentous discoveries of the first half of the 20-th century to a virtually stagnating scientific scene. On my part, I always ignored these political trends in science and remained always faithful to the historical doubts on the final character of quantum mechanics[1].

---

[1]It is important for the reader to understand how valid said doubts remain to this day. Consider, for instance, the problem of the total nuclear magnetic moments addressed by Segre, Weisskopf, and others. The unadulterated use of quantum mechanics and the value of the intrinsic magnetic moments of protons and neutrons as measured when in vacuum, miss about 3.8% of the experimental value of the magnetic moment of the simplest possible nucleus, the deuteron, when in its ground state. The error becomes bigger in heavier nuclei, to assume very large values for heavy nuclei such as the Zirconium. Various recent papers and monographs claim the reduction of the error for the deuteron down to 1% for nonrelativistic treatments and claim essentially smaller errors for relativistic corrections. However, these claims demand the mixing of three or more energy levels of the deuteron, thus losing scientific credibility (no matter how authoritative the source) for various reasons, such as: 1) the experimental value of the magnetic moment of the deuteron is measured for its *ground state*, and definitely not for a mixture of different energy states; 2) the claims assume the joint existence of different energy levels without any emission or absorption of quanta, thus violating basic laws of quantum mechanics; 3) the reduction of the unsolved problem to quarks *multiplies* the inconsistencies, rather than reduces them, as we shall see later on in this Preface. At any rate, assuming that the above manipulations (perpetrated in the intent of preserving at whatever cost a beloved theory against basic advances) are indeed successful in representing the magnetic moment of the deuteron, can *the same manipulations work for the remaining nuclei, such as the Copper, the Gadolinium or the Zirconium?* Any scholar with a minimum of ethical standards must answer in the negative, thus confirming that, on real scientific grounds without academic politics, quantum mechanics *cannot* represent *exactly* the magnetic ,moments of *all* nuclei. Quantum mechanics certainly remains valid, but only as a first approximation of an expected more adequate theory. Numerous additional insufficiencies and inconsistencies not reported here for brevity (see papers [12]) imply beyond *credible* doubts that quantum mechanics *cannot* be exactly for the nuclear structure, thus requiring a suitable generalization.

As the reader will see, the novel hadronic mechanics and chemistry presented in this monograph see their ultimate inspiration and guidance, first, on the teaching of Lagrange, Hamilton and Jacobi, and, second, on the more recent teaching by Fermi, Segre, Weisskopf, Einstein, and numerous other founders of contemporary science (see later).

Upon the completion of my undergraduate studies in Naples, in the early 1960s I initiated my graduate studies in physics at the University of Turin, Italy, with the specific intent of dedicating my research life to the structural generalization of quantum mechanics. An outline of my early studies appears to be significant for this monograph.

The foundations of quantum mechanics are given by Lie's theory in its operator realization on a Hilbert space $\mathcal{H}$ over the field $\mathbb{C}$ of complex numbers, which is characterized by the celebrated *Heisenberg's representation* in finite and infinitesimal forms

$$A(t) = U \times A(0) \times U^\dagger = e^{i \times H \times t} \times A(0) \times e^{-i \times t \times H}, \tag{1a}$$

$$i \frac{dA}{dt} = [A, B]_{\text{operator}} = A \times H - H \times A, \tag{1b}$$

$$U = e^{i \times H \times t}, \quad U \times U^\dagger = U^\dagger \times U = I, \tag{1c}$$

with the complementary *Schrödinger representation*

$$H(r, p) \times |\psi\rangle = E \times |\psi\rangle, \tag{2a}$$

$$p_k \times |\psi\rangle = -i \times \hbar \times \partial_k |\psi\rangle, \tag{2b}$$

and classical counterpart given by the celebrated *Hamiltonian mechanics*

$$A(t) = e^{-t \times (\partial H / \partial r^k) \times (\partial / \partial p_k)} \times A(0) \times e^{t \times (\partial / \partial r^k) \times (\partial H / \partial p_k)}, \tag{3a}$$

$$\frac{dr^k}{dt} = \frac{\partial H(t, r, p)}{\partial p_k}, \quad \frac{dp_k}{dt} = -\frac{\partial H(t, r, p)}{\partial r^k}. \tag{3b}$$

$$\frac{dA}{dt} = [A, H]_{\text{classical}} = \frac{\partial A}{\partial r^k} \times \frac{\partial H}{\partial p_k} - \frac{\partial H}{\partial r^k} \times \frac{\partial A}{\partial p_k}. \tag{3c}$$

where $\times$ represents the conventional associative product.

In this way, *Lie groups* are realized by the finite time evolutions (1a) and (3a), while *Lie algebras* are realized by the brackets $[A, B]$ of the infinitesimal forms.

A dominant physical characteristic at both classical and operator levels is that the antisymmetric character of brackets $[A, B]$ implies the familiar *conservation laws of total physical quantities*, such as that of the total energy

$$i \times \frac{dH}{dt} = [H, H] = H \times H - H \times H = 0. \tag{4}$$

The starting point of my research (which I realized since the early days of my graduate studies) was the clear insufficiency of the above theory for the representation of *nonconservative systems*, trivially, because the emphasis of quantum mechanics is on *total conservation laws*, while the systems considered need *time-rate-of-variations of physical quantities*, such as the energy, linear momentum, angular momentum, *etc*[2].

In my first paper [1] of 1967 (which was part of my Ph.D. thesis, see also [2]) I submitted the following generalization (I call "lifting") of Heisenberg's representation

$$A(t) = U \times A(0) \times W^\dagger = e^{i \times H \times q \times t} \times A(0) \times e^{-i \times t \times p \times H}, \quad H = H^\dagger, \quad (5a)$$

$$\begin{aligned} i\, dA/dt = (A, B)_{\text{operator}} &= p \times A \times H - q \times H \times A = \\ &= m \times (A \times B - B \times A) + n \times (A \times B + B \times A) = \\ &= m \times [A, B] + n \times \{A, B\}, \end{aligned} \quad (5b)$$

where: the right and left operators U, W are now different, $p = m + n$, $q = m - n$ and $p \pm q$ are non-null *parameters*, with lifting of classical Hamiltonian mechanics [3]

$$A(t) = \left[e^{-t \times q \times (\partial H/\partial r^k) \times (\partial/\partial p_k)}\right] \times \times A(0) \times \left[e^{t \times p \times (\partial/\partial r^k) \times (\partial H/\partial p_k)}\right], \quad (6a)$$

$$\frac{dr^k}{dt} = p \times \frac{\partial H(t,r,p)}{\partial p_k}, \quad \frac{dp_k}{dt} = -q \times \frac{\partial H(t,r,p)}{\partial r^k} \quad (6b)$$

[2]Dissipative nuclear models and other theories representing dissipation via "imaginary potentials" in the Hamiltonian were initiated in the first half of the 20-th century and are still of widespread use nowadays. These theories are afflicted by problems of physical consistency so severe to prevent serious scientific applications. In this case the Hamiltonian is *nonhermitean*, $H = p^2/2m + i \times V \neq H^\dagger$; then, the Lie group (1a) becomes $A(t) = \{e^{i \times H \times t}\} \times A(0) \times \{e^{-i \times t \times H^\dagger}\}$; and Lie brackets (1b) are generalized into the triple system $idA/dt = [A, H, H^\dagger] = A \times H^\dagger - H \times A$. It then follows that dissipative models with "imaginary potentials" lose *all algebras* in the brackets of the time evolution, let alone all Lie algebras (because algebras, as currently understood in mathematics, require *bilinear* brackets verifying certain basic axioms). Under these conditions, words such as "neutrons" and "protons" are deprived of any physical meaning, because of: the impossibility of introducing the SU(2) symmetry necessary for the characterization of the spin; the violation of the Poincare symmetry necessary for the correct definition of the mass; and numerous additional flaws. Above all, quantities to be measured, such as the nonconserved energy H, are no longer observable (because non-Hermitean). On the contrary, the studies presented herein are based on the requirement that *all nonconserved quantities to be measured must be Hermitean as a necessary condition for observability* (see Sect. 1.7, Chapter 1, for additional inconsistencies of pre-existing nonconservative theories and Sect. 2.10, Chapter 1, for their resolution).

$$\frac{dA}{dt} = (A, H)_{\text{classical}} = p \times \frac{\partial A}{\partial r^k} \times \frac{\partial H}{\partial p_k} - q \times \frac{\partial H}{\partial r^k} \times \frac{\partial A}{\partial p_k}. \tag{6c}$$

Since the new brackets (A, B) *are not* antisymmetric by conception, they are indeed suitable to represent the *nonconservation (or time-rate-of-variation) of the energy*

$$i \times \frac{dH}{dt} = (H, H) = (p - q) \times H \times H \neq 0. \tag{7}$$

As one can see, the main idea of papers [1-3] is that *the Hamiltonian remains fully Hermitean, thus observable, while its nonconservation is characterized by the generalized mathematical structure of the theory.*

Evidently, the latter theory is no longer Lie. Nevertheless, brackets (A, B) remain bilinear, thus characterizing a consistent algebra which results in verifying the axioms of the covering *Lie-admissible algebras* first identified by the American mathematician A.A. Albert, although without specific realizations. A generic nonassociative algebra with brackets (A, B) is said to be Lie-admissible when the attached antisymmetric brackets are Lie,

$$[A\hat{;}B] = (A, B) - (B, A) = \text{Lie}, \tag{8}$$

(see Sect. 1.7 of Chapter 1 for details).

This means that, even though non-Lie, the new time evolutions preserve a well defined Lie content, as necessary for a *covering theory.* It turns out that the new theory also admits a well defined content of *Jordan algebras* because the attached symmetric brackets are Jordan

$$\{A\hat{;}B\} = (A, B) + (B, A) = \text{Jordan}. \tag{9}$$

The above structure realized, for the first time to my knowledge, Jordan's search for physical applications of his algebras. Regrettably, Refs. [1, 2, 3] on the first (p, q)-deformations of Lie's theory dating back to 1967 are generally ignored in the rather vast contemporary literature on the simpler q-deformations.

My Proposal at Harvard University to Construct the Hadronic Generalization of Quantum Mechanics under DOE Support

In 1967 I emigrated with my wife and daughter to the USA, where I soon discovered that, being still vastly unknown at that time in *mathematical*, let alone physical circles, Lie-admissible algebras were not conducive to my locating an academic position. I therefore dedicated myself to a decade of publishing a variety of papers on research lines fully conventional for the time.

In 1978, while at Harvard University, I resumed the research significant for this monograph, thanks to financial support from the U.S. Department of Energy, which I still remember with gratitude today. In fact, in 1978 I proposed [4] the construction of a generalization of quantum mechanics specifically built for a representation of hadrons as they are in the physical reality: extended, nonspherical, and deformable charge distributions possessing some of the highest densities measured in laboratory until now.

I suggested for the new mechanics the name of *hadronic mechanics* to indicate its primary application, the study of strongly interacting particles known as hadrons, with the understanding that the mechanics was expected to be equally applicable to other fields (in the same way as quantum mechanics for the atomic structure is also applicable to crystal and other fields).

Hadronic mechanics was subsequently studied by numerous authors among whom I indicate: S. Okubo, S. Adler, T. Gill, L. Lindesay, C. Wolf, and others in the USA; J. Ellis, M.E. Mavromatos, N. Nanopoulos at CERN, and others in Switzerland; J.V. Kadeisvili, Yu. Arestov and A.K. Aringazin in the USSR; R. Mignani, M. Gasperini, F. Cardone and others in Italy; A. Jannussis, F. Brodimas, C.N. Ktorides, and others in Greece; J. Lohmus, E. Paal, L. Sorgsepp and others in Estonia; D. Schuch, C.A.C. Dreismann and others in Germany; N. Ntibashirakandi, D.K. Callebaut, and others in Denmark; E. Trell and others in Sweden; D. Rapoport-Campodonico and others in Argentina; M. Nishioka and others in Japan; A.O.E. Animalu and others in Africa; M. Mijatovic, B. Veljanovski, and others in Yugoslavia; and others (see Vol. II of monographs [9] for comprehensive listings)[3].

The original proposal of 1978 was based on the following main generalizations:

I) General Lie-admissible Lifting, which can be expressed via the following generalization of Heisenberg's representation [4]

$$A(t) = U \times A(0) \times W^{\dagger} = e^{i \times H \times Q \times t} \times A(0) \times e^{-i \times t \times P \times H}, \qquad (10a)$$

$$i \, \frac{dA}{dt} = (A \hat{;} H)_{\text{operator}} = A < H - H > A = A \times P \times H - H \times Q \times A =$$

[3]The studies on hadronic mechanics have resulted to date in some 2,000 papers, about 20 monographs and some 40 volumes of conference proceedings for over 10,000 pages of research published in America, Europe, and Asia.

$$= (A \times T \times H - H \times T \times A) + (A \times W \times H + H \times W \times A) = \qquad (10b)$$
$$= [A \hat{,} H] + \{A \hat{,} H\},$$

$$H = H^\dagger, \quad P = Q^\dagger, \qquad (10c)$$

where $P = T + W$, $Q = T - W$ and $P \pm Q$ are nonsingular operators (or matrices), with complementary *lifting of the Schrödinger representation* [5]

$$H(r,p) > |\hat{\psi}\rangle = H(r,p) \times Q \times |\hat{\psi}\rangle = E \times |\hat{\psi}\rangle, \qquad (11a)$$

$$\langle\hat{\psi}| < H = \langle\hat{\psi}| \times Q \times H(r,p) = \langle\hat{\psi}| \times E', \quad E' \neq E, \qquad (11b)$$

and classical counterparts [4]

$$A(t) = \left[e^{-t\times(\partial H/\partial r^j)\times Q_i^j \times(\partial/\partial p_i)}\right] \times \\ \times A(0) \times \left[e^{t\times(\partial/\partial r^j)\times P_i^j \times(\partial H/\partial p_i)}\right], \qquad (12a)$$

$$\frac{dr^k}{dt} = P_i^k(t,r,p) \times \frac{\partial H(t,r,p)}{\partial p_i}, \\ \frac{dp_k}{dt} = -Q_k^i(t,r,p) \times \frac{\partial H(t,r,p)}{\partial r^i}, \qquad (12b)$$

$$\frac{dA}{dt} = (A \hat{,} H)_{\text{classical}} = \frac{\partial A}{\partial r^i} \times P_j^i \times \frac{\partial H}{\partial p_j} - \frac{\partial H}{\partial r^i} \times Q_j^i \times \frac{\partial A}{\partial p_j}, \qquad (12c)$$

under the condition that the antisymmetric brackets $[A\hat{,}B] = (A\hat{,}B) - (B\hat{,}A)$ are Lie and the symmetric operator brackets $\{A\hat{,}B\} = (A\hat{,}B) + (B\hat{,}A)$ are Jordan.

The needs to generalize the parameter into the operator form of Lie-admissible theories were numerous. First, Lie-admissible theory (5) clearly offered no possibility for a genuine lifting of Schrödinger's representation, while the broader theory (10) did admit the nontrivial liftings (11).

Also, the time evolution of the former, Eqs. (5a), is clearly *nonunitary*,

$$U = e^{i \times H \times p \times t}, \quad W = e^{i \times t \times q \times H}, \quad U \times W^\dagger \neq I. \qquad (13)$$

It is a useful exercise for the reader interested in studying the new chemistry to prove that the application of time evolution (5a) to itself yields precisely the operator time evolution (10). In turn, the application of the latter time evolution to itself preserves the Lie-admissible structure (although with different values of P and Q).

This is a confirmation of the fact that brackets $(A\hat{,}B)$ are the most general possible bilinear nonassociative brackets [because they are the most general possible combination of antisymmetric and symmetric brackets, as shown in Eqs. (10b)].

As a result, dynamical equations (10) are "directly universal," that is, inclusive of all infinitely possible theories with an algebra in the brackets of the time evolution directly in the coordinate frame of the experimenter.

The representation of time-rate-of-variations of nonconserved quantities is evidently of the type

$$i \times \frac{dH}{dt} = (H\hat{,}H) = H < H - H > H = H{\times}P{\times}H - H{\times}Q{\times}H \neq 0. \quad (14)$$

In particular, *Lie-admissible theories (10)-(14) are structurally irreversible, that is, irreversible for reversible Hamiltonians.*

II) General Lie-Isotopic lifting. While studying theory (10)-(14), I realized back in 1978 that the attached antisymmetric algebra *is not* characterized by the conventional Lie brackets $[A, B] = A \times B - B \times A$, but instead by the more general brackets

$$[A\hat{,}B] = (A, B) - (B, A) = A \times T \times B - B \times T \times A, \quad (15a)$$

$$T = P + Q = P^{\dagger} + Q^{\dagger} = Q + P = T^{\dagger}, \quad (15b)$$

which brackets verify the abstract Lie axioms although in a generalized form.

I therefore called the latter theory an *isotopy* of Lie's theory from the Greek meaning of being "axiom-preserving" [4]. I called the full Lie-admissible theory (10)-(13) a *genotopy* of Lie's theory [4] from the Greek meaning of being "axiom inducing."

The corresponding mechanics were called *isomechanics* and *genomechanics*, while the mathematics needed for the treatment of the corresponding theories were called *isomathematics* and *genomathematics*, respectively.

As a result of the latter property, in memoirs [4] of 1978 I proposed the following *Lie-isotopic lifting of Heisenberg's representation*

$$A(t) = U{\times}A(0){\times}U^{\dagger} = [e^{i\times H\times T\times t}]{\times}A(0){\times}[e^{-i\times t\times T\times H}], \quad T = T^{\dagger}, \quad (16a)$$

$$i\frac{dA}{dt} = [A\hat{,}H]_{\text{operator}} = A\hat{\times}H - H\hat{\times}A = A{\times}T{\times}H - H{\times}T{\times}A, \quad (16b)$$

where T is a nonsingular operator (or matrix), with corresponding lifting of Schrödinger's equation

$$H(r,p) \times T \times |\hat{\psi}\rangle = E \times |\hat{\psi}\rangle, \quad \langle\hat{\psi}| \times T \times H(r,p) = \langle\hat{\psi}| \times E, \quad (17)$$

and classical counterparts

$$A(t) = e^{-t\times(\partial H/\partial r^j)\times T_i^j\times(\partial/\partial p_i)} \times A(0) \times e^{t\times(\partial/\partial r^j)\times T_i^j\times(\partial H/\partial p_i)}, \quad (18a)$$

$$\frac{dr^k}{dt} = T_i^k(t,r,p) \times \frac{\partial H(t,r,p)}{\partial p_i}, \quad \frac{dp_k}{dt} = -T_k^i(t,r,p) \times \frac{\partial H(t,r,p)}{\partial r^i}, \quad (18b)$$

$$\frac{dA}{dt} = [A\hat{,}H]_{\text{classical}} = \frac{\partial A}{\partial r^i} \times T_j^i \times \frac{\partial H}{\partial p_j} - \frac{\partial H}{\partial r^i} \times T_j^i \times \frac{\partial A}{\partial p_j}. \quad (18c)$$

A dominant feature of the isomechanics is that of characterizing conventional total conservation laws in view of the antisymmetric character of the brackets, e.g.,

$$i \times \frac{dH}{dt} = H\hat{\times}H - H\hat{\times}H = H \times T \times H - H \times T \times H \equiv 0. \quad (19)$$

The theory is therefore suitable to represent *closed-isolated systems with conventional total conservation laws.*

Nevertheless, the theory is not purely Hamiltonian. In fact, it requires *two operators for the characterization of systems*, the conventional Hamiltonian $H = p^2/2m + V$ plus the new operator T. I therefore assumed that the conventional Hamiltonian represents all conventional potential interactions, while the operator T represents *new interactions not derivable from a potential.*

Note that *the above features imply a new notion of bound state with conventional total conservation laws, yet internal forces that are both conservative and nonconservative.* We merely have internal exchanges of energy, angular momentum and all other physical quantities, but always in such a way that the total quantities are conserved, the systems being isolated.

A classical example is given by the structure of Jupiter (considered isolated from the rest of the Solar system) which evidently verifies the conservation of the total angular momentum, yet one can directly observe via telescopes internal vortices with *varying angular momenta.* In my monograph on theoretical mechanics published by Springer-Verlag [14] I then used classical Hamiltonian mechanics for the study of planetary systems, and its isotopic covering for the study of the structure of individual planets.

Operator examples of closed-isolated non-Hamiltonian systems studied in my monograph [9] published by the Ukraine Academy of Sciences are given by hadrons, nuclei, or stars owing to their hyperdense structure which renders them conceptually equivalent to Jupiter. I used quantum mechanics for the study of the atomic structure and the covering hadronic mechanics for the structure of hadrons, nuclei, and stars. At any rate, the belief that the hadronic constituents can freely move within the hyperdense hadronic medium in the core of a star as the atomic constituents do, has no scientific credibility [12].

III) General, multi-valued Lie-admissible and Lie-isotopic liftings. More recently, by applying to my own studies the belief of the lack of final character of any theory, I realized that, despite their remarkable generality, theory (10)-(14) was still insufficient to represent complex systems such as biological structures, because recent studies have shown that they generally require irreversible and *multivalued* formulations.

I therefore proposed the still more general lifting of Lie-admissible theories into their multi-valued covering form [6]

$$A(t) = U \times A(0) \times U^\dagger = e^{i \times H \times \{Q\} \times t} \times A(0) \times e^{-i \times t \times \{P\} \times H}, \quad P = Q^\dagger, \tag{20a}$$

$$\begin{aligned} i\, dA/dt = (A \hat{,} H)_{\text{operator}} = A\{<\}H - H\{>\}A = \\ = A \times \{P\} \times H - H \times \{Q\} \times A = \\ = (A \times \{T\} \times H - H \times \{T\} \times A) + \\ +(A \times \{W\} \times H + H \times \{W\} \times A) = [A \hat{,} H] + \{A \hat{,} H\}, \end{aligned} \tag{20b}$$

$$\{P\} = \{P_1, P_2, P_3, \ldots\}, \quad \{Q\} = \{Q_1, Q_2, Q_3, \ldots\}, \tag{20c}$$

where $\{P\}$, $\{Q\}$ and $\{P \pm Q\}$ are now nonsingular (ordered) *sets* of operators (or matrices), with complementary lifting of Schrödinger's equations

$$H(r,p) \times \{P\} \times |\{\hat{\psi}\}\rangle = \{E\} \times |\{\hat{\psi}\}\rangle, \tag{21a}$$

$$\langle\{\hat{\psi}\}| \times \{Q\} \times H(r,p) = \langle\{\hat{\psi}\}| \times \{E\}, \tag{21b}$$

and classical counterparts [*loc. cit.*]

$$A(t) = \left[e^{-t \times (\partial H/\partial r^j) \times \{Q\}_i^j \times (\partial/\partial p_i)}\right] \times \tag{22a}$$

$$\times A(0) \times \left[e^{t \times (\partial/\partial r^j) \times \{P\}_i^j \times (\partial H/\partial p_i)}\right], \tag{22a}$$

$$\begin{aligned} \frac{dr^k}{dt} = \{P\}_i^k(t,r,p) \times \frac{\partial H(t,r,p)}{\partial p_i}, \\ \frac{dp_k}{dt} = -\{Q\}_k^i(t,r,p) \times \frac{\partial H(t,r,p)}{\partial r^i}, \end{aligned} \tag{22b}$$

$$\frac{dA}{dt} = (A \hat{,} H)_{\text{classical}} = \frac{\partial A}{\partial r^i} \times \{P\}_j^i \times \frac{\partial H}{\partial p_j} - \frac{\partial H}{\partial r^i} \times \{Q\}_j^i \times \frac{\partial A}{\partial p_j}. \tag{22c}$$

under the condition again that the attached antisymmetric (operator symmetric) brackets are of Lie (Jordan) type although in a broader multi-valued version.

The existence of the multi-valued Lie-isotopic particularization of the above theory is evident and will be implied hereon.

The Forgotten Historical Legacy of Lagrange, Hamilton, and Jacobi on the Origin of Irreversibility

There is little doubt that one of the most fundamental open problems of contemporary physics, directly relevant to chemistry, is an axiomatically consistent representation of the *irreversibility* of numerous physical and chemical systems.

The problem is created by the fact that quantum mechanics is structurally *reversible* in time, namely, the time reversal image of systems are as physical as the original systems. This feature is certainly verified for the systems for which quantum mechanics was built for, the atomic structure. However, the belief that the entire microscopic world is as irreversible as electron orbits is purely nonscientific.

Moreover, all action-at-a-distance, potential forces identified so far and, thus, all possible Hamiltonians, are also reversible. These features are in manifest disagreement with the evidence of the irreversibility of nature.

In view of the above occurrences, a virtually endless number of studies have been conducted during the 20-th century in the dream of reconciling the irreversibility of reality with the reversibility of quantum mechanics.

Even though I respected these studies, I never accepted them, and followed a basically different approach since my Ph.D. studies in physics.

In essence, I always accepted the exact validity of quantum mechanics for the conditions for which it was constructed, the description of *stable-reversible orbits of electrons in the structure of the hydrogen atom.*

However, the belief that the same discipline is equally exact for the dramatically different conditions of electrons in the core of a collapsing star, is so farfetched, to constitute the negation of science. In fact, the belief necessarily implies the local stability of electrons moving within hyperdense media with consequential local conservation of the angular momenta (from a pillar of the theory, the rotational symmetry), thus accepting in its totality the existence of the perpetual motion.

While my contemporaries attempted to adapt the irreversible physical reality to a beloved theory, I adopted the opposite approach of adapting the theory to physical reality. I therefore initiated the search of a generalization of quantum mechanics suitable to represent irreversibility.

At any rate, it "should" be evident, and therefore admitted, by all scholars in the field that a *classical, macroscopic, irreversible system cannot possibly be reduced to a finite collection of elementary particles all in reversible conditions (as requested by quantum mechanics), and, vice versa, a finite collection of elementary particles in reversible conditions cannot possibly yield an irreversible macroscopic system.* The need to generalize quantum mechanics for any serious study of irreversible systems is therefore mandatory.

During my Ph.D. studies in Turin, I had the opportunity of studying the original articles (in Italian) by Lagrange which were truly inspiring. The physics community of the 20-th century has used the "truncated Lagrange's equation," those without external terms, in a feverish dream of reducing the entire universe to actions-at-a-distance among isolated points.

By comparison, Lagrange had a dramatically broader and more realistic view, which incorporated the limited conceptions of the 20-th century as particular cases, but added an infinite class of unrestricted forces of contact type among extended particles which, as such, are not derivable from a potential, thus being *the origin of irreversibility.*

I also studied the original writings by Hamilton and discovered exactly the same dichotomy. Physics of the 20-th century has used the "truncated Hamilton's equations," namely, those without the historical external terms, thus adapting physical reality to an intentionally simplified theory.

By comparison, Hamilton had a much more adequate conception of nature. He argued that all potential forces should be represented with a function we call today the Hamiltonian, while all remaining forces and effects not derivable from a potential should be represented with external terms fully analogous to those introduced by Lagrange. Hamilton also assumed the external terms as the only possible representation of irreversibility[4].

By carefully *avoiding* textbooks in mechanics of the 20-th century (because dealing with the truncated equations *only*, as well known), I also studied the *original works* by Jacobi and other founders of analytic mechanics and found exactly the same majestic conception of nature (in fact, the original Jacobi theorem *is not* that treated in contempo-

[4]The adulteration of a Hamiltonian with the addition of irreversible terms $H = p^2/2m + V(r) \rightarrow H' = p^2/2m + V(r) + V'(t, \dots)$ would imply that "irreversibility has a potential energy", and other consequences outside science. This indicates the *necessity* for introducing additional entities besides the Hamiltonian, which is the original conception by Lagrange, Hamilton, and Jacobi and which, as we shall see, is the foundation of the our treatment of irreversible chemical processes studied in this monograph.

rary textbooks, but one primarily conceived for analytic equations with external terms, the case without external terms being trivial).

In this way I reached the view that *the physics of the 20-th century as failed to achieve a meaningful study on the origin of irreversibility because of the truncation of Lagrange's and Hamilton's analytic equations via the removal of their external terms.*

Subsequently, during my studies at Harvard University in the late 1970s I proved that the "true Hamilton's equations", those with external terms, are a particular case of Eqs. (12)[5],

$$\frac{d}{dt}\frac{\partial L(r,v)}{\partial r^k} - \frac{\partial L(r,v)}{\partial v^k} = F_k(t,r,v), \tag{23a}$$

$$\frac{dr^k}{dt} = \frac{\partial H(r,p)}{\partial p_i}, \quad \frac{dp_k}{dt} = -\frac{\partial H(r,p)}{\partial r^k} + F_k(t,r,p), \tag{23b}$$

$$P^i_j = \delta^i_j, \quad Q^j_i = \delta^j_i \left(1 + \frac{f_j}{\partial H/\partial r^j}\right). \tag{23c}$$

I also showed that the brackets of the true equations (23b) *do not* characterize a consistent algebra, and do not permit a consistent exponentiation to a group, precisely because of the external terms, thus preventing the construction of a covering analytic theory.

I therefore proposed the *identical* Lie-admissible reformulation of the true Hamilton equations as a *necessary* condition to restore a consistent algebra and group (see Sect. 2.10 of Chapter 1 for details). The *additive* external terms are identically reformulated via the *multiplicative* terms P and Q in the analytic equations, resulting in the consistent generalized algebras and groups of Eqs. (12).

The above occurrences had fundamental implications in all my subsequent studies. In fact, *the achievement of an algebraically and group theoretically consistent representation of the historical Hamilton equations with external terms set the foundations for an axiomatically and*

[5] I wrote two monographs [14] on the integrability conditions for the existence of a Lagrangian or a Hamiltonian, called *conditions of variational selfadjointness.* It turned out that Lagrange's and Hamilton's external terms in three dimensions are *essentially nonselfadjoint*, that is, they *do not* admit a Lagrangian or a Hamiltonian representation in the coordinates of the observer, thus confirming the need for external terms. The use of coordinate transformations $r \to r'(r,p)$, $p \to p'(r,p)$ to reduce the true analytic equations to the truncated ones is highly discouraged because the needed transformations are highly *nonlinear*, thus mapping the original inertial frames of the observers into highly noninertial ones, with consequential violation of Galilei's and Einstein's relativities. At any rate, such a reduction has a purely political-academic character because laboratory equipment and measuring apparata cannot evidently be put in transformed trajectories of the type $r' = A \times \exp[B \times r \times p]$, $p' = C \times p \times \log(D \times p^2)$; A, B, C, D constants.

physically consistent representation of irreversibility at all desired levels of study.

Subsequent studies reviewed in Sect. 2.10, Chapter 1, identified unique and unambiguous operator map of Eqs. (12) into (10), thus permitting the identification of irreversibility in the most elementary layer of nature, such as that of electrons in the core of stars.

I hope that chemists reading these introductory lines do not dismiss them as esoteric *physical* issues of no relevance to chemistry. As we shall see later on in this Preface as well as in the monograph, chemical reactions and numerous other events in organic and inorganic chemistry are strictly irreversible. In these introductory lines I am therefore addressing the issue of a *theory permitting an axiomatically consistent representation of irreversible chemical processes.*

Catastrophic Inconsistencies of Generalized Theories Treated with Conventional Mathematics

After achieving the above chain of generalized theories, I initiated their severe scrutiny for mathematical rigor and physical consistency. I discovered in this way that *all the above operator and classical generalized theories have mathematical and physical inconsistencies so grave to prevent any meaningful application, whenever the theories are treated with the conventional mathematics of quantum or Hamiltonian mechanics.* These inconsistencies are reviewed in Sect. 1.7 of Chapter 1, for which I suggested the incisive title of "Catastrophic Inconsistencies of Generalized Theories Formulated via Conventional Mathematics".

The origin of the inconsistencies is the *nonunitary structure* of the time evolutions which implies the *lack of invariance of units, eigenvalues, Hermiticity, probability, causality and other fundamental features.*

Since the isotopic, genotopic, and hyperstructural branches of hadronic mechanics are "directly universal" in their fields, the above catastrophic inconsistencies also apply to a rather large number of generalized theories attempted during the 20-th century.

Moreover, all attempts at achieving consistent liftings of the fundamental equations for the linear momentum, Eq. (2b), had failed for decades and, in their absence, hadronic mechanics was basically incomplete as well as insufficiently developed for applications.

The Mandatory Construction of New Mathematics

At that point I had two alternatives: abandon the dream of achieving a consistent covering of quantum mechanics, or search for a resolution of said inconsistencies. I opted for the second line of inquiries, and initiated its systematic study. It was easy to see that *the primary reason for the catastrophic inconsistencies was the use of "conventional" mathematics for the treatment of "generalized" theories.*

For instance, it was easy to prove that the use of the conventional Hilbert space defined over a conventional field of complex number implies the lack of conservation of the original Hermiticity under all nonunitary time evolutions, thus causing the loss of observables (see Sect. 1.7 of Chapter 1 for details). On the contrary, a suitable lifting of the Hilbert space and numbers does indeed permit the preservation of the original Hermiticity at all times, thus permitting a consistent representation of observables. At that point I had no other alternative but trying to build new mathematics for the resolution of said inconsistencies and the achievement of an axiomatic consistency comparable to that of conventional quantum mechanics.

I was assisted in this task by numerous mathematicians among whom I mention: H.C. Myung, D.B. Lin, M. Tomber, and others in the USA; Gr. Tsagas, D.S. Sourlas, T. Vougiouklis, and others in Greece; R. Miron, S. Vacaru, J.V. Kadeisvili and others in the USSR; R. Aslaner, S. Keles, and others in Turkey; C.-X. Jiang and others in China; N. Kamiya and others in Japan; and other mathematicians (see Vol. I of monograph [9]).

In this way, we constructed the following new mathematics [6]:

A) **Isomathematics**, for the consistent formulation of single-valued reversible Lie-isotopic theories;

B) **Genomathematics**, for the consistent formulation of single-valued irreversible Lie-admissible theories;

C) **Hypermathematics**, for the consistent formulation of multi-valued irreversible Lie-admissible theories;

plus the **isodual mathematics** [8] for the consistent formulation of the anti-isomorphic image of the above theories to represent antimatter beginning at the classical level.

The reader should be aware that the above new mathematics include, for necessary condition of consistency, *new numbers and fields, new vector and Hilbert spaces, new algebras and groups, new geometries and topologies,etc.*, all this for *each* of the above iso-, geno-, and hyper-mathematics, plus all their isoduals.

The main idea of the new mathematics [4] is the lifting of the conventional, trivial unit I of current use in mathematics, physics, and chemistry into a generalized unit which is the *inverse* of the isotopic, genotopic, or hyperstructural element (see Chapter 1 for technical details),

$$\hat{I} = 1/T = \hat{I}^{\dagger}, \tag{24a}$$

$$\hat{I}^{>} = 1/P, \quad {}^{<}\hat{I} = 1/Q, \quad \hat{I}^{>} = ({}^{<}\hat{I})^{\dagger}, \tag{24b}$$

$$\{\hat{I}^{>}\} = \{1\}/\{P\}, \quad \{{}^{<}\hat{I}\} = \{1\}/\{Q\}, \quad \{\hat{I}^{>}\} = (\{{}^{<}\hat{I}\})^{\dagger}. \tag{24c}$$

Then, the totality of the conventional mathematics of quantum mechanics has to be lifted into such a form to admit the new units as the correct left and right units at all levels. These liftings then yielded *iso-, geno- and hyper-numbers, iso-, geno- and hyper- spaces, iso-, geno-, and hyper-geometries,etc.*

The achievement of an invariance fully comparable to that of quantum mechanics can be seen from these introductory lines. In fact, whether conventional or generalized, the basic unit is the most fundamental invariant of all theories. The representation of nonpotential forces via generalized units then guarantees their invariance.

Stated in different terms, nonpotential interactions due to contact among extended bodies (such as a spaceship during re-entry in atmosphere or the deep overlapping of the wavepackets of particles) should be represented with anything, *except the Hamiltonian.* In fact, such a representation voids the physical meaning of the Hamiltonian as representing the total energy $H = p^2/2m + V$ in favor of abstract mathematical forms. Also, a Hamiltonian representation of systems with nonpotential forces exists in the fixed frame of the observer only in very special and limited cases (generally in one dimension only [14]).

Once the Hamiltonian is excluded for the representation of contact nonpotential interactions, the unit is evidently the best known solution not only because it guarantees invariance, but also because it achieves "direct universality," this time intended as the capability to represent all possible nonpotential forces directly in the frame of the experimenter.

Additional studies proved the capability of generalized units to represent all possible nonpotential forces which additionally are *nonlinear* (in the wavefunction) and *nonlocal* (e.g., of integral type over a surface or a volume). Recall that quantum mechanics is strictly linear, local, and potential. The generalization of the unit then offers the possibility of reaching a covering theory with all infinitely possible nonlinear, nonlocal, and nonpotential interactions.

After having initiated the studies in 1967 [1], the resolution of the catastrophic inconsistencies indicated above was achieved only in memoir [6] of 1996. A reason for such a long delay is that, after having constructed the needed new numbers and fields, new vector and Hilbert spaces, *etc.*, the generalized theories were still afflicted by un-acceptable inconsistencies, such as the lack of invariance of the generalized time evolutions under their own action.

A reason for the delay was that the inconsistencies existed where I would expect them the least, in the use of *conventional differentials and derivatives* for generalized theories. It was only after I reached in memoir [6] generalized formulations of the differential calculus for each of the broader theories, that I finally resolved all known mathematical and physical inconsistencies.

For instance, the invariant and axiomatically correct formulation of isomechanics required the following basic *isodifferential calculus* [6]

$$\hat{d}r^k = \hat{I}^k_i(t,r,p,\psi,\dots)\times dr^i, \quad \frac{\hat{\partial}}{\hat{\partial}r^k} = \hat{T}^i_k(t,r,p,\psi,\dots)\times\frac{\partial}{\partial r^i}, \tag{25}$$

with corresponding formulations of the Lie-isotopic dynamical equations.

The above isodifferential calculus then permitted the achievement, for the first time in memoir [6] of 1996, of an axiomatically consistent lifting of the fundamental equations for the linear momentum,

$$p_k\times T\times|\hat{\psi}\rangle = -i\times\hbar\times\hat{\partial}_k|\hat{\psi}\rangle = -i\times\hbar\times T^i_k(t,r,p,\psi,\dots)\times\partial_i|\hat{\psi}\rangle, \tag{26}$$

with corresponding *genodifferential and hyperdifferential calculi* for the remaining formulations.

The achievement of axiomatically consistent lifting of quantum mechanics was then a mere consequence, as first reported in memoir [7] of 1997.*After three decades from the initiation of the studies, and thanks to the contributions by mathematicians, theoreticians, and experimentalists too numerous to mention here, hadronic mechanics finally reached operational maturity in 1997 !*

The achievement of the above maturity included the proof that *the isotopic branch of hadronic mechanics verifies in their entirely all conventional quantum axioms and physical laws (Heisenberg's uncertainty, Pauli's exclusion principle, probability and causality laws, etc.)* In fact, I have stressed in the literature that *hadronic mechanics is not a "new theory" because it merely provides "new realizations" of the conventional abstract axioms of quantum mechanics.*

The proof of the above features is elementary for the isomechanics because its basic structure remains Lie. The same proof for the broader Lie-admissible theories was not simple because of the apparent lack of antisymmetric character of the brackets of the time evolution law. However, at a deeper analysis it emerged that, when properly formulated with their appropriate genomathematics, the Lie- admissible brackets turn out to be fully Lie, thus permitting the abstract identity of quantum mechanics with all generalized theories herein considered.

The reader is warned that *all papers in generalized theories with a nonunitary structure prior to 1996 are afflicted by the indicated catastrophic inconsistencies.*

Classification of Hadronic Mechanics

As a result of all the above studies, hadronic mechanics has nowadays achieved a maturity of formulation sufficient for practical applications and experimental verifications, and comprises the following branches:

1) Quantum mechanics (characterized by the Hamiltonian $H = p^2/2m + V$ and the trivial value I of the unit) representing isolated and reversible systems of point-like particles at large mutual distances with only action-at-a-distance, potential interaction, such as crystals;

2) The covering **isomechanics** *(characterized by two quantities, the Hamiltonian H and the isotopic element T) characterizing isolated and reversible systems of extended particles with action-at-a-distance, potential interactions represented by H plus contact, internal, nonpotential interactions represented by T, as expected in the structure of a hadron;*

3) The broader **genomechanics** *(characterized by three quantities, the Hamiltonian H plus the two operator P and Q) characterizing open, nonconservative and irreversible systems of extended particles with potential interactions represented by H and unrestricted contact-nonpotential-irreversible interactions represented by P in one direction of time and by Q in the reversed direction of time;*

4) the still broader multi-valued **hypermechanics** *(characterized in its most general possible form by three sets of quantities, a set of Hamiltonians $\{H\}$ plus two sets of operators $\{P\}$ and $\{Q\}$) representing complex multi-valued irreversible systems in open-nonconservative conditions, such as the growth in time of a cell or of a sea shell;*

*5) I also introduced an additional group of theories anti-isomorphic to all the preceding ones, called "***isodual***" (8), to represent antimatter beginning at the classical level, for the intent of resolving a further insufficiency of the physics of the 20-th century, the representation of matter at all possible levels (from Newtonian mechanics to quantum field theory), while representing antimatter solely in second quantization.*

Simple Method for the Explicit Construction of Hadronic Mechanics

Operational maturity of hadronic mechanics was achieved with the following method for the explicit construction of concrete hadronic models [7], which construction is applied several times in this monograph according to the following main steps:

1) Explicit construction of specific models of hadronic mechanics via a method applicable by anybody, without the necessary knowledge of advanced new mathematics, which consists of *the systematic application of nonunitary transformations* $U \times U^\dagger \neq I$ *to the totality of mathematical and physical aspects of any given unitary model.* For the isotopic case, the method yields the isounit

$$I \rightarrow \hat{I} = U \times I \times U^\dagger = 1/\hat{T},$$

the isoassociative product

$$\begin{aligned} A \times B \rightarrow \hat{A} \hat{\times} \hat{B} = U \times (A \times B) \times U^\dagger = \\ = (U \times A \times U^\dagger) \times (U \times U^\dagger)^{-1} \times (U \times B \times U^\dagger) = \hat{A} \times \hat{T} \times \hat{B}, \end{aligned}$$

the isonumbers

$$n \rightarrow \hat{n} = U \times n \times U^\dagger = n \times (U \times U^\dagger) = n \times \hat{I},$$

with isoproduct $\hat{n} \hat{\times} \hat{m}$, isoschrödinger equation

$$U \times H \times |\psi\rangle = (U \times H \times U^\dagger) \times (U \times U^\dagger)^{-1} \times (U \times |\psi\rangle) = \hat{H} \times \hat{T} \times |\hat{\psi}\rangle,$$

isohilbert product

$$\begin{aligned} \langle\psi| \times |\psi\rangle \times I \rightarrow \langle\hat{\psi}|\hat{\times}|\hat{\psi}\rangle = U \times (\langle\psi| \times |\psi\rangle \times I) \times U^\dagger = \\ = (\langle\psi| \times U^\dagger \times (U \times U^\dagger)^{-1} \times (U \times |\psi\rangle) \times (U \times I \times U^\dagger), \end{aligned}$$

etc. A *dual* nonunitary transform permits the explicit construction in all necessary details of the broader genotheory, with the lifting for motion forward in time $I \rightarrow \hat{I}^> = U \times I \times W^\dagger$, $U \times U^\dagger \neq I$, $W \times W^\dagger \neq I$, and the conjugate *different*, thus irreversible lifting for motion backward in time $I \rightarrow {}^<\hat{I} = W \times U^\dagger = (\hat{I}^>)^\dagger \neq \hat{I}^>$. Multi-valued, irreversible, hyperstructural theories are equally constructed in a very simple way via ordered sets of nonunitary transforms.

2) The proof that the models resulting from the above liftings are indeed invariant, that is, they admit the same numerical predictions at all times under the same conditions. The proof is also expressible in very elementary terms, under the conditions that the invariance *is not* studied with the mathematics of quantum mechanics, but rather with the appropriate broader mathematics. For instance, it is easy to see that the application of conventional nonunitary transforms *does not* leave invariant the above generalized structures, because, for $Z \times Z^\dagger \neq I$, $\hat{I} \rightarrow \hat{I}' = Z \times \hat{I} \times Z^\dagger \neq \hat{I}$. However, nonunitary transforms belong to the old theory. Their use in the covering theories is as inconsistent as trying to elaborate *quantum* mechanics via the *isomathematics*. For

consistency, quantum mechanics must be elaborated with conventional mathematics and isomechanics must be elaborated with isomathematics. This demands the reformulation of nonunitary transforms into the identical *isounitary law* $Z = \hat{Z} \times \hat{T}^{-1}$, $Z \times Z^\dagger \equiv \hat{Z}\hat{\times}\hat{Z} = \hat{Z}^\dagger\hat{\times}\hat{Z} = \hat{I}$, under which the invariance of the isounit is evident, $\hat{I} \rightarrow \hat{I}' = \hat{Z}\hat{\times}\hat{I}\hat{\times}\hat{Z}^\dagger \equiv \hat{I}$. The invariance of the isounit then implies the invariance of the entire isotheory, as we shall see in details.

3) The *projection* of the above *generalized* theories formulated on generalized spaces over generalized fields into a formulation on *conventional* spaces over conventional fields. This projection is useful for practical calculations, with the understanding that the sole correct formulation is the generalized one. In different terms, unlike quantum mechanics, hadronic mechanics admits *two* explicit treatments: the first treatment is that via isomathematics which implies the referral of all quantities to the *generalized* unit $\hat{I}$; the second treatment is that via its projection into conventional spaces, i.e., by referring all quantities to the *conventional* unit I. The latter formulation is indeed useful for numerical computations, as we shall see, but it is also a source of numerous misinterpretations and controversies, since scholars tend to assume as true conclusions derived in the latter setting, while the same conclusions are generally impossible in the correct formulations of generalized theories. A case fundamental for this monograph is the study of valence bonds via the isotopic branch of hadronic mechanics. Conclusions reached on this isotopic model via a conventional potential well on a conventional Hilbert space are as inconsistent and inconceivable for isomechanics as conclusions on the *atomic* structure reached via *isohilbert* spaces defined over the generalized isofields. Serious conclusions can be reached for conventional theories when treated via conventional mathematics, and for generalized theories when treated via their own generalized mathematics. Any mixing of different theories and/or their mathematics cannot yield serious scientific conclusions.

The Birth of Hadronic Superconductivity

Another event that had a major impact on this monograph is the birth of *hadronic superconductivity*, which was initiated by A.O.E. Animalu [10] in 1994, was then developed by Animalu and me [11], and also consisted of iso-, geno-, and hyper-formulations.

In essence, Animalu (who is one of the founders of hadronic mechanics) focused his attention on a main result of memoirs [4] dealing with the structure model of the π^0 meson as a bound state of one electron and one positron at short distances in singlet state with internal nonpotential interactions. Animalu confirmed the fact that the emerging Hulten force is so attractive to overcome all possible repulsive Coulomb forces, thus holding also for pairs of electrons represented via isomechanics, and not necessarily for electron-positron pairs.

As a result, Animalu [*loc. cit.*] formulated a basically novel model of the Cooper pair in superconductivity in which, for the first time to my knowledge, we have an *explicitly "attractive" force between the two identical electrons.* As well known, quantum mechanical models are strictly statistical in character, that is, representing a large number of Cooper pairs. The main point here is that the same quantum models cannot possibly represent the structure of *one* Cooper pair, as done by hadronic superconductivity.

Subsequently, Animalu and Santilli [11]: re-examined the main structural equations of the new isosuperconductivity; verified their compliance with the axioms of isomechanics; confirmed the nonlinear, nonlocal, and nonpotential character of the contact interactions due to deep wave-overlapping of the electrons; and pointed out the remarkable verification of the theory by experimental data in superconductivity.

Application of Hadronic Mechanics to Pauli's Exclusion Principle

A truly fundamental topic left completely open by the vast studies on quantum mechanics of the 20-th century is a scientific interpretation of Pauli's exclusion principle, that is, the capability of identical Fermions in general, and of electrons in particular, in the same energy level to "exclude" each other from possessing all identical quantum numbers. As well known, Pauli's principle is merely *imposed* on all models, although without any explanation of the mechanism and interactions causing said exclusions. As we shall see, the latter knowledge has truly major *industrial*, let alone scientific implications.

As clearly stated in the first proposal of 1978 [4], hadronic mechanics was built to achieve the missing scientific representation, that is, the identification of the interactions responsible for the exclusions, and their invariant mathematical representation in a form predicting numbers verifiable with experiments.

The main hypothesis is that *Pauli's exclusion principle is due to nonlinear, nonlocal, and nonpotential interactions originating from the overlapping of the wavepackets of Fermions.* These interactions are of *contact, zero-range type*, thus admitting no potential, and consequently being kilometrically beyond the descriptive capacities of quantum mechanics.

Vice-versa, any attempt to represent Pauli's exclusion principle via quantum mechanics is certainly faced with catastrophic inconsistencies.

This is due to the fact that the only possible quantum representations are those with a Hamiltonian, that is, via a potential. In turn, the granting of a potential energy to the interactions, say, between the two electrons of the helium can be easily proved to imply corresponding large deviations from the experimental data on spectral lines.

In reality, the proved complete inability by quantum mechanics to represent Pauli's exclusion principle is a direct confirmation of the lack of Hamiltonian character of the interactions responsible for its existence.

The assumption of the nonpotential-nonhamiltonian character of the interactions underlying Pauli's principle, formulated since the original proposal [4], eliminated this fundamental inconsistency *ab initio*. The subsequent representation of the interactions via the isounit then provides its axiomatically consistent and invariant version.

It was also identified in the original proposal [4] that *the nonpotential interactions due to overlapping of the wavepackets of Fermions are stable only in singlet coupling (antiparallel spins), because triplet coupling (antiparallel spins) cause repulsive forces.* This feature was derived from the property that wavepackets of two identical Fermions in the same energy level spinning one inside the other under antiparallel spins are "in phase" with each other, while wavepackets spinning one inside the other with parallel spin evidently imply drag forces due to the motion of one wavepacket against that of the other, thus resulting in repulsion. This feature is evidently in full agreement with physical reality, e.g., because the two electrons of the ground state of the helium can only have antiparallel spins. Hadronic mechanics merely confirmed that the parallel spin case is not possible because it is highly unstable.

The above features were rudimentary illustrated in the original proposal [4] with the "gear model," that is, with the coupling of two actual mechanical gears which, as well known, can only spin with antiparallel rotations, because parallel rotations would imply the snapping of the gear teeth.

The understanding that antiparallel couplings imply repulsion brought to the hypothesis that *the nonpotential interactions due to the overlapping of the wavepackets of identical Fermions in singlet coupling are "strongly" attractive (that is, attractive in large value, rather than in the conventional sense of strong interactions).*

In summary, the original proposal to build hadronic mechanics reached the following results for Pauli's exclusion principle: 1) The inapplicability of quantum mechanics for a scientific representation of the principle, due to catastrophic inconsistencies for the representation of the principle via a potential or a Hamiltonian; 2) The nature of the interactions responsible for the principle as being of contact, zero-range, nonpotential-nonhamiltonian type due to the wave-overlapping of the wavepackets of the identical Fermions; 3) The discovery that said nonpotential interactions are strongly repulsive for singlet couplings of identical Fermions in the same energy level, the singlet coupling being the only possible state; 4) The discovery that said nonpotential forces are strongly attractive for singlet couplings, as numerically verified in the structure model $\pi^0 = (\hat{e}^+, \hat{e}^-)_{\mathrm{HM}}$; and 5) The expectation that, in due time, the covering hadronic mechanics can indeed provide the desired scientific representation of the principle, with particular reference to a *representation of the deviations from quantum mechanical predictions on spectral lines which are well known to exist for all atoms other than the hydrogen, beginning with the helium.*

A truly fundamental point the reader is suggested to keep in mind during the entire analysis presented in this monograph, is that *the nonlinear, nonlocal, and nonpotential interactions of electrons originating from their wavepackets occur under the full assumption that their charge is indeed point-like.* In different terms, one of the most anti-scientific beliefs of the physics of the 20-th century (because manifestly preventing advances on basic issues) is that the point-like charge of electrons solely permits local theories with potential interactions. Such a religious belief, evidently intended to protect old theories, is possible solely under the additional assumption that the electron has a "point-like wavepacket," which is an evident nonscientific posture.

Sixteen years passed since the original proposal [4] without any additional study on Pauli's principle due to known academic opposition against the research here considered. In 1994 A.O.E. Animalu contacted me asking support for his application of hadronic mechanics to the Cooper pair in superconductivity, as reported in the preceding section. The studies were then continued by Animalu and Santilli [11].

The hadronic treatment of superconductivity is essentially given by *Cooper pairs conceived as a straight-line version of Pauli's principle for the two electrons of the helium.* In fact, the nonpotential interactions of wave-overlappings in singlet couplings are so strong to overcome the repulsive Coulomb force, and form a correlation-bond.

The above conception is confirmed by calculations conducted by Animalu [10] according to which *the Cooper pair has a dimension precisely of the order of Bohr's radius* (about 10^{-8} cm). The conception is also confirmed by rather solid experimental verifications presented in Refs. [10].

The conception of the Cooper pair via the isomechanical treatment of the historical Pauli's principle is not a pure academic curiosity, because it implies rather serious industrial possibilities, such as *the prediction of a new type of electric current that based on the creation and prop-*

agation of Cooper pairs, rather than individual electrons. To begin an understanding of the practical implications of the studies presented in this monograph, the reader is suggested to meditate a moment on the dramatic advantages of an electric current composed of Cooper pairs as compared to the conventional current used today. The Cooper pair has a magnetic moment which, in first approximation, can be assumed as being ignorable (due to the antiparallel coupling). By comparison, electrons have a very large magnetic moment (for particle standards). Since, during their motion through conductors, electric current experience electromagnetic interactions with the charge structure of atoms, it is evident that *a current of electron pairs in singlet coupling with null magnetic moments (Cooper pairs) experiences a dramatically smaller resistance than the same current composed by isolated electrons with large magnetic moments, which is the very principle of hadronic superconductivity.*

The possible practical creation and propagation of Cooper pairs is under separate industrial study, and will be reported at some future time.

In conclusion, by the mid 1990's hadronic mechanics had permitted significant advances toward a scientific representation of Pauli's exclusion principle for the atomic structure, as well as for the application of the principle to the Cooper pair in superconductivity. However, no application to the coupling of valance electron in chemistry existed at that time.

Insufficiencies of Quantum Chemistry

At the time of my memoirs [7] of 1997, I had not been exposed to chemistry since my undergraduate studies in physics of the late 1950s. It was at that time that I received an unexpected inquiry by the U. S. chemist D. D. Shillady requesting comments on my paper with A.O.E. Animalu [11]. I was impressed by the depth of his chemical knowledge. I therefore dedicated top priority to answer Shillady's numerous and penetrating questions on paper [10b] in the best way I could. In this way, after about 40 years, I was again exposed to chemistry.

My initial inability to understand basic theoretical assumptions of quantum chemistry were soon replaced by in-depth scrutinies, which then resulted in the identification of clear insufficiencies of the theory as presented in the first section of Chapter 1.

I was not surprised to see that, in the same way as I have been oblivious to chemistry for 40 years, chemists have also been oblivious to physics for essentially the same period of time. As such, most chemists were not aware of a number of basic developments in physics, which are of direct relevance to chemistry, among which I mention:

i) The historical process in nuclear physics that mandated the creation of a "new strong force" due to the excessive weakness of the exchange, van der Waals and other nuclear forces. In essence, chemistry had adopted for the molecular structure the exchange, van der Waals and other forces of purely nuclear type, but had ignored their weakness, thus missing the equivalent in molecular bonds of the strong forces in nuclear bonds. This comparative perspective created my first conviction that quantum chemistry missed the most important and dominant force in molecular structures.

ii) The dichotomy between orbital chemical theories, where the correlation is admitted for an arbitrary number of electrons, as compared to the experimental evidence that correlations solely occur between electron pairs. It appears that chemists remained un- informed of various studies on correlations in particle physics and superconductivity. This perspective motivated my second conviction that quantum chemistry misses means for the restriction of correlations to electron *pairs*, with rather serious consequences, such as the easily proved admission in the hydrogen and water molecules of an *arbitrary* number of hydrogen atoms.

iii) The lack in quantum chemistry of the contact interactions expected in deep overlappings of the extended wavepackets of valence electrons with point-like charge, as necessary in molecular structures but not in their atomic counterparts. This occurrence motivated my third conviction that quantum chemistry was constructed by blindingly using quantum mechanics in view of its known successes for *atomic* structures, but without a serious scientific appraisal of its limitations for the different conditions of *molecular* structures. In fact, the electron *wavepacket* can be well approximated as a point in the atomic structure, due to relatively large mutual distances, thus implying the applicability of quantum mechanics. In the transition to molecular bonds, such an approximation is no longer valid, because, by its very definition, valence bonds imply deep superposition of the wavepackets of the electrons, with consequential emergence of nonlinear, nonlocal and nonpotential interactions, which are kilometrically beyond any dream of scientific treatment via quantum mechanics.

iv) The adulterations of basic axioms of quantum mechanics and chemistry recently adopted in chemistry to achieve more accurate representations of molecular data, such as Gaussian screenings of the Coulomb law, variational methods, and other approaches, which are outside said axioms for numerous reasons. The Coulomb law is a well known fundamental *invariant* of quantum mechanics and chemistry, that is, it is not possible to change its structure under all infinitely possible transformations admitted by the theory, the *unitary* transformation $U \times U^\dagger = U^\dagger \times U = I$. Any modification of the Coulomb law can, therefore, be only reached via *nonunitary* transforms $U \times U^\dagger \neq I$. In turn, nonunitary maps imply the exiting of the axioms of quantum theories, whether physical or chemical. Thus, all Gaussian and other screenings-modifications of the basic Coulomb law are outside the basic axioms of quantum mechanics. In any case, these adulterations imply the inapplicability of the very notion of "quantum of energy," since the latter requires stable energy levels which are prohibited under screened Coulomb laws. Similarly, in order to claim on true scientific-nonpolitical grounds that a variational theory really belongs to quantum mechanics, one must prove that the resulting wavefunctions are indeed exact solutions of purely quantum mechanical Coulomb equations without *ad hoc* adulterations. It was easy to see that this is not the case. As an illustration, for the H_2^+ ion we have the following occurrences: 1) the known Coulomb quantum equation is unable to represent the binding energy correctly; 2) the variational method does indeed achieve a correct representation of said binding energy; thus 3) the wavefunction of the variational method cannot possibly be an exact solution of the Coulomb equation. These occurrences motivated my additional conviction that quantum chemistry is in the rather incredible state whereby the basic axioms of the theory have been long abandoned as a necessary condition to achieve meaningful representations of experimental data; yet the chemical theory was still dubbed "quantum" for evident political reasons. Stated in different terms, under my critical scrutiny, quantum chemistry of the 20-th century lacked the identification and treatment of the *nonunitary departures* from the strictly unitary quantum mechanics, whose need had been incontrovertibly established by *ad hoc* adulterations of its basic laws and principles to reach meaningful representations.

v) The rather serious inconsistencies implied by the lack of restrictions of correlation to electron pairs, such as the prediction via quantum electrodynamics that all molecules under an external magnetic field acquire a net magnetic polarity, thus being ferromagnetic, in dramatic disagreement with experimental evidence (see Sect. 1 of Chapter 1 of this monograph and Ref. [11] for details).

vi) The impossibility for quantum chemistry to provide "any" consistent theoretical representation of irreversible processes, such as chemical reactions. This is due to the evident reversible character of the very structure of the theory, as compared to the need for a structurally irreversible theory to describe irreversible processes. This occurrence is the result of the lack of propagation in chemistry of recent physical advances on the origin of irreversibility along the historical teaching by Lagrange and Hamilton indicated earlier.

vii) The impossibility for quantum chemistry to provide serious representations of biological systems. Quantum mechanics notoriously holds only for perfectly rigid, perfectly conservative, and perfectly reversible, thus eternal systems, such as a hydrogen atom or of a crystal. Scholars who believe that such a theory can effectively represent biological systems exit the boundary of science. In fact, the strict implementation of quantum mechanics implies that our body should be perfectly rigid, perfectly conserved, and perfectly reversible, thus reaching the status of an eternal God. There is a limit in the widespread desire to preserve old theories at whatever cost, beyond which limit one reaches shear scientific corruption. This is the case for the belief of the exact validity of quantum mechanics and chemistry in biology.

Construction of Hadronic Chemistry

After having reached the conviction that quantum chemistry was in need of a number of basic revisions, as a result of some two decades of research on hadronic mechanics, it was easy for me to suggest to Shillady the lifting of quantum chemistry into a nonunitary formulation with progressive levels of complexity. By following the basic features of hadronic mechanics, I proposed the construction of *hadronic chemistry* with the following main branches:

I) Isochemistry (characterized by the isomathematics in which the generalized unit is Hermitean) for a novel, axiomatically consistent representation of closed-isolated systems with linear and nonlinear, local and nonlocal, and potential as well as nonpotential internal forces, such as molecular structures. In this case conventional action-at-a-distance, potential interactions can be represented with the Hamiltonian $H = p^2/2m + V$, while contact nonpotential interactions due to deep wave-overlapping of valence electrons are represented by the isotopic element T (or, equivalently, the isounit $\hat{I} = 1/T$). Compatibility between the conventional, unitary, *atomic* structure, and the broader nonunitary *valence* structure (which is crucial to preserve the notion of quantum of energy where applicable) is easily reached by imposing that the generalized unit rapidly recovers the conventional quantum unit for distances bigger than 1 fm. The basic isoschrödinger equations can be obtained via a nonunitary transform of the conventional equations (2), which is assumed to characterize precisely the generalized unit

$$U \times U^\dagger = \hat{I} \neq I, \tag{27a}$$

$$U \times H \times |\psi\rangle = (U \times H \times U^\dagger) \times (U \times U^\dagger)^{-1} \times (U \times |\psi\rangle) = \\ = \hat{H} \times \hat{T} \times |\hat{\psi}\rangle = U \times E \times |\psi\rangle = E \times |\hat{\psi}\rangle, \tag{27b}$$

$$\hat{H} = U \times H \times U^\dagger, \quad |\hat{\psi}\rangle = U \times |\psi\rangle, \quad \hat{T} = (U \times U^\dagger)^{-1}. \tag{27c}$$

The isoschrödinger equations can then be written for $\hbar = 1$ (see Chapter 1 for details)

$$\hat{i}\hat{\times}\hat{\partial}_{\hat{t}}|\hat{\psi}\rangle = \hat{H}\hat{\times}|\hat{\psi}\rangle = \hat{H}(r,p) \times \hat{T}(t,r,p,\hat{\psi},\dots) \times |\hat{\psi}\rangle = \\ = \hat{E}\hat{\times}|\hat{\psi}\rangle = E \times |\hat{\psi}\rangle, \tag{28a}$$

$$\hat{p}_k\hat{\times}|\hat{\psi}\rangle = \hat{p}_k \times \hat{T} \times |\hat{\psi}\rangle = -i \times \hat{\partial}_k|\hat{\psi}\rangle = -i\hat{T}_k^i \times \partial_i|\hat{\psi}\rangle, \tag{28b}$$

with equivalent conjugate equations here ignored. Note the capability of the above isoequations to represent all infinitely possible potential interactions, as well as nonlinear, nonlocal and nonpotential forces. Note also that in this case the basic product is $A\hat{\times}B = A \times \hat{T} \times B$ with isounit $\hat{I}$, $\hat{I}\hat{\times}A = A\hat{\times}\hat{I} = A$, for which the traditional product "2×2" is no longer equal to 4, but rather to a generally integrodifferential quantity depending form the selected isounit $\hat{I}$. This illustrates the departure of isomathematics from conventional mathematics.

II) Genochemistry (characterized by the broader genomathematics with nonhermitean genounits), for a novel, axiomatically consistent description of irreversible chemical processes with unrestricted interactions, as expected in chemical reactions. In this case all conventional, reversible, potential interactions are represented with the Hamiltonian, while irreversibility originates from the nonhermitean character of Lagrange's and Hamilton's "external terms" represented with the real-valued and nonsymmetric matrix Q for motion forward in time, and the conjugate matrix $P = Q^\dagger$ for motion backward in time. In this case the basic equations in both directions of time can be obtained via *two* different nonunitary transforms, one per each ordered product to the right and to the left

$$U \times U^\dagger \neq I, \quad W \times W^\dagger \neq I, \quad U \times W^\dagger = \hat{I}^>, \quad W \times U^\dagger = {}^<I, \tag{29a}$$

$$U \times H \times \ |\psi\rangle = (U \times H \times W^\dagger) \times (U \times W^\dagger)^{-1} \times (U \times |\psi\rangle) = \\ = \hat{H}^> \times Q \times |\hat{\psi}^>\rangle, \tag{29b}$$

$$\langle\psi| \times H \times W = \langle\psi| \times W \times (W \times U^\dagger)^{-1} \times (U^\dagger \times H \times W) = \\ = \langle{}^<\hat{\psi}| \times P \times {}^<\hat{H}, \tag{29c}$$

$$\hat{H}^> = U \times H \times W^\dagger, \quad Q = (U \times W^\dagger)^{-1}, \quad |\hat{\psi}^>\rangle = U \times |\psi\rangle, \tag{29d}$$

and can be written for the forward direction of time

$$\hat{H}^> > |\hat{\psi}^>\rangle = \hat{H}^> \times \hat{Q} \times |\hat{\psi}^>\rangle = \hat{E}^> > |\hat{\psi}^>\rangle, \tag{30a}$$

$$\hat{p}_k^> > |\hat{\psi}^>\rangle = \hat{p}_k^> \times \hat{Q} \times |\hat{\psi}^>\rangle = -i \times \partial_k^> |\hat{\psi}\rangle = -i\hat{Q}_k^i \times \partial_i |\hat{\psi}\rangle, \tag{30b}$$

with conjugate expression for the backward direction of time. Note: the structural irreversibility of the theory, that is, irreversibility for all reversible Hamiltonians; the need in this case of *two* products, one ordered to the right and the other to the left, $A > B = A \times \hat{Q} \times B$ and $A < B = A \times \hat{P} \times B \neq A > B$, with corresponding genounits $\hat{I}^> = 1/\hat{Q}$ and ${}^<\hat{I} = 1/\hat{P}$; and the fact that the traditional product "2×2" not only yields generalized values, but also different values for different ordering, since, in general, $2 < 2 \neq 2 > 2$.

III) Hyperchemistry (characterized by the still broader hypermathematics with hyperunits characterized by ordered sets of nonhermitean operators), for a novel, axiomatically consistent representation of irreversible biological structures with multi-valued unrestricted internal processes. The basic equations can be obtained via a multi-valued extension of the two nonunitary transforms of genochemistry and can be written for motion forward in time

$$\begin{gathered}\{U\} \times \{U^\dagger\} \neq I, \quad \{W\} \times \{W^\dagger\} \neq I, \quad \{U\} \times \{W^\dagger\} = \{\hat{I}^>\}, \\ \{W\} \times \{U^\dagger\} = \{{}^<I\}, \quad \{\hat{H}^>\} > |\hat{\psi}^>\rangle = \{E\} \times |\hat{\psi}^>\rangle,\end{gathered} \tag{31a}$$

$$\{\hat{p}_k^>\} > |\hat{\psi}^>\rangle = -i \times \{\partial_k^>\} |\hat{\psi}\rangle. \tag{31b}$$

In the latter case the product "2×2" yields *two different sets of values*, one set for motion forward in time, and the other for motion backward in time, with intriguing possibilities, e.g., initiating the understanding of a DNA code via a number theory whose basic unit can be a countable set of generalized values.

Main Features of Hadronic Chemistry

As it is the case for hadronic mechanics, *the isotopic branch of hadronic chemistry verifies, by conception and construction, all conventional laws and principles of quantum chemistry, and only realizes them in a broader way.*

Moreover, also by conception and construction, *in their explicit realizations, hadronic and quantum chemistry coincide everywhere, except for new effects due to the correlation of valence electrons at small distances.*

Finally, it should be indicated up-front that *quantum and hadronic chemistry coincide at the abstract, realization-free level, also by conception and construction, and hadronic chemistry merely provides "new realizations" of the abstract axioms of quantum chemistry. Therefore, criticisms on the axiomatic structure of hadronic chemistry "are" criticisms on the axiomatic structure of quantum chemistry.*

As an explicit illustration, the axiomatic structure of a quantum mechanical or chemical eigenvalue equation is that of *right, associative module* "$H \times |\psi\rangle$" on a Hilbert space of states $|\psi\rangle$. This essentially consists of the associative multiplication to the right of an operator H times the state $|\psi\rangle$, where the associative character is expressed by the property $H_1 \times (H_2 \times |\psi\rangle) = (H_1 \times H_2) \times |\psi\rangle = H_1 \times H_2 \times |\psi\rangle$.

Hadronic chemistry preserves the above axiomatic structure of the eigenvalue equation. In fact, we also have a right associative module $H \hat{\times} |\hat{\psi}\rangle$ $(= H \times \hat{T} \times |\hat{\psi}\rangle$, $\hat{T}$ fixed), in which associativity is also characterized by the properties $H_1 \hat{\times} (H_2 \hat{\times} |\hat{\psi}\rangle) = (H_1 \hat{\times} H_2) \hat{\times} |\hat{\psi}\rangle = H_1 \hat{\times} H_2 \hat{\times} |\hat{\psi}\rangle$. In particular, $\hat{T}$ is positive-definite, as we shall see. Therefore, at the abstract, realization free level, all distinctions are lost between "$H \times |\psi\rangle$" and "$H \hat{\times} |\hat{\psi}\rangle$," thus implying the abstract identity between quantum and hadronic chemistry.

Practical Construction of Iso-, Geno- and Hyper-Chemical Models

It is important for the reader interested in studying the new chemistry to know up-front that *the new mathematics are not needed in practical applications of hadronic chemistry.* In fact, they are needed only to prove the axiomatic consistency and invariance of the new chemistry, which verification, once conducted, does not need to be repeated for each application owing to its generality.

A virtually endless number of specific iso-, geno- and hyper-models can be constructed very easily by merely subjecting conventional quantum chemical models to nonunitary transforms along dynamical equations (28), (30a), and (31), with the subsidiary conditions that said transforms are nonunitary only in the region of valence bond, while they are unitary in all remaining regions (as a condition to admit quantum chemistry identically outside valence bonds).

A fundamental illustration of isochemistry is the nonunitary transform needed to represent the interactions of the deep overlapping of the extended wavepackets of two valence electrons with point charges when in singlet couplings, which transform can be written under the assumption that the valence bond is restricted to distances of the order of 1 fm (see later for better treatments),

$$U \times U^{\dagger} = \begin{cases} I & \text{for } r > 1 \text{ fm}, \\ e^{\hat{\psi}/\psi \int d^3 r \hat{\psi}_{\uparrow}(r) \times \hat{\psi}_{\downarrow}(r)} & \text{for } r \leq 1 \text{ fm}, \end{cases} \quad (32)$$

where $\hat{\psi}$ represents the hadronic wavefunction of valence electron pairs; ψ represents the conventional quantum wavefunction; generalized unit (32) realizes the condition of recovering quantum chemistry identically for $r > 1$ fm; nonlinear interactions are clearly expressed by the presence of wavefunctions in the exponent; and the all important nonlocal interactions are represented by the integral. As one can see, for sufficiently large mutual distances between the valence electrons, the integral in the exponent is identically zero, the transform is unitary everywhere, and hadronic chemistry recovers quantum chemistry identically at all distances.

Numerous additional applications and realizations of the generalized units exist in the literature of hadronic mechanics depending on the case at hand[6]. Geno- and hyper-models can be constructed accordingly and will be presented in future papers.

It is also important for the reader to know up-front how *isochemistry admits as simple particular cases all the modifications of the Coulomb law recently used in chemistry.* For instance, all infinitely possible screenings of Coulomb law are easily obtained via the above nonunitary rule,

[6]Another simple example is *the nonunitary transform permitting a representation of the actual, extended, nonspherical and deformable charge distributions of hadrons, e.g., of the spheroidal ellipsoidal type* $U \times U^{\dagger} = \hat{I} = \text{Diag.}(n_1^2, n_2^2, n_3^2, n_4^2)$, where the n_k's are *functions* of various local quantities evidently characterizing the semiaxes of the ellipsoids and n_4 represents the density of the hadron considered. Representations of more general shapes exist with nondiagonal isounits [9]. Note that the representation of hadrons as "extended" is notoriously impossible for the mathematical structure of quantum mechanics, let alone the representation of "nonspherical" and "deformable" shapes, which representations would imply the violation of the entire their beginning with its fundamental rotational symmetry. As one can see, these representations become elementary for the covering hadronic mechanics. One of the immediate and direct applications and experimental verifications is *the achievement of an "exact" representation of "all" total nuclear magnetic moments* which, as remarked in footnote 1, is impossible for quantum mechanics. In fact, we merely have an explicit representation of the historical hypothesis that protons and neutrons can experience a deformation of their shape when under the strong forces of a nuclear structure. In turn, such (generally small) deformations imply alterations of the intrinsic, magnetic moments (called "mutations" for reasons related to the underlying algebras [1]) which mutations, still in turn, permit exact representations of the magnetic moments of *all* nuclei. Note that we are referring to deformations of shape without any alteration of the intrinsic angular momentum (spin) of protons and neutrons. The representation of magnetic moments is achieved in full compatibility with conventional quantum mechanical laws, such as Pauli's exclusion principle (see Vol. III of Refs. [9] and literature quoted therein for all details).

as illustrated via the screening with one Gaussian

$$U \times U = Ae^{Br}, \quad U(1/r)U^{\dagger} = Ae^{Br}/r, \quad A, \ B = \text{constants}. \qquad (33)$$

Along the same lines, it is easy to prove that the wavefunctions of variational methods, which cannot possibly satisfy unadulterated quantum equations for the reasons indicated earlier, are indeed particular cases of isochemical equations.

In summary, we can say that the main elements of hadronic chemistry originating from nonunitary transforms of quantum structures, yield Gaussian screenings, variational methods, and several other approaches already existing in the technical literature.

Their true nonunitary structure has not been identified in the conventional chemical literature for various reasons. A primary function of hadronic chemistry is that of identifying such a nonunitary structure, constructing the most general class of models with such a nonunitary structure, and provide a formulation which is axiomatically consistent *per se*, while being compatible with conventional quantum laws.

In so doing, hadronic chemistry achieves "direct universality" for the representation of all possible nonunitary structures, while achieving invariance, that is, the prediction of the same numerical value for the same quantity under the same conditions but at different times.

The Main Assumption of Hadronic Chemistry: the Correlation-Bond of Valence Electron Pairs into the "Isoelectronium"

As recalled earlier, the molecular models produced by quantum chemistry in the 20-th century are afflicted by truly serious insufficiencies, or sheer inconsistencies, such as: lack of a force sufficiently "strong" to account for the actual strength of molecular bonds; admission of an arbitrary number of hydrogen atoms in the hydrogen and water molecules; insufficiently accurate representation of molecular binding energies and other molecular characteristics; prediction that all molecules are ferromagnetic; and other serious flaws.

A primary motivation of my interest in constructing hadronic chemistry has been to reach a resolution of the above insufficiencies in a form compatible with experimental data.

The primary means for the achievement of this objection is the main assumption of hadronic chemistry, according to which *pairs of valence electrons from different atoms correlate and bond themselves into a singlet quasiparticle state we have called isoelectronium.*

The name of isoelectronium was selected, first, as "electronium" intended as a complement of the "positronium," to indicate that the constituents are electrons, and, second to indicate that the state is a manifestation of isochemistry. The prefix "iso," therefore, indicates that the interactions responsible for the correlation- bound are the repulsive Coulomb force plus the attractive nonpotential forces due to wave-overlappings in single couplings. More technically, the prefix "iso" indicates that the constituents *are not* ordinary electrons e^-, but rather *isoelectrons* $\hat{e}^-$, that is, electrons with new renormalizations caused by the new interactions.

It is important to know up-front that the isoelectronium is another step in the study of Pauli's exclusion principle via hadronic mechanics. As recalled earlier, hadronic mechanics has permitted an initial yet meaningful quantitative representation of Pauli's principle for the two electrons of the ground state of the helium. Subsequently, the hadronic treatment of the Cooper pair turned out to be a straight-line realization of Pauli's principle of the same electron pair of the helium, only free to move as part of an electric current. In particular, the size of the Cooper pair resulted in being of the same order of magnitude as Bohr's radius.

The isoelectronium is another manifestation of Pauli's exclusion principle, this time, expressed for electron pairs of two different atoms. First, as we shall see, nonunitary transform (32) introduces a new contact *nonpotential* force between pairs of valence electrons, which force is so attractive to overcome the repulsive Coulomb force. This produces the first concrete and explicit *attractive* force between valence electrons known to me, since the average of all other forces admitted by quantum chemistry is null in semiclassical approximation, as we shall see. The new nonpotential force results in being the "strong molecular force" that is missing in conventional molecular models, as indicated earlier.

It is evident that the notion of isoelectronium restricts correlations solely to valence electron *pairs*, thus explaining *ab initio* the impossibility of more than two H-atoms in the hydrogen and water molecules.

The achievement by hadronic chemistry of an essentially exact representation of molecular characteristics should be evident already from these introductory lines. Recall that molecular models based on Gaussian screenings of the Coulomb laws have already achieved a significant improvement of the representation of molecular features. But these models are particular cases of the much broader class of nonunitary screenings permitted by hadronic chemistry. Therefore, the desired accuracy in the representation of experimental data by hadronic chemistry is merely reduced to the proper selection of the basic nonunitary transform.

The resolution of the other insufficiencies of current molecular models will be discussed in the monograph.

A few comments are in order with respect to the open problem of the *mainline of the isoelectronium*, on which there exist seemingly different views. It is important to understand that these differences are due to differences in the assumptions rather than in the structure.

A first group of colleagues assumes that the isoelectronium is a largely unstable quasiparticle with a meanlife of the order of 10^{-16} sec. I believe that this view is correct under its underlying main assumption, the restriction of the correlation-bond radius to about 10^{-13} cm.

However, the reader should consider the alternative view offered by Pauli's exclusion principle, in which case the mainline of the isoelectronium must be equal to the duration of applicability of the same principle. Under the assumption that the diatomic molecule here considered is unperturbed and in its ground state, it is logical to assume that the two valence electrons of the isoelectronium are in the same energy level. In this case, Pauli's exclusion principle must hold for the life of the unperturbed ground state of the molecule. Under these assumptions, *the isoelectronium has an infinite lifetime.*

Mutatis mutandae, the assumption that the two valence electrons are in the same unperturbed energy level, plus the assumption of a finite lifetime of the isoelectronium, would directly implies a violation of Pauli's exclusion principle, evidently because the two assumptions would admit electrons in the same energy level with identical features.

As one can see, the assumption that the isoelectronium for the hydrogen molecule is a manifestation of Pauli's exclusion principle with infinite lifetime at correlation-bonding length of the order of the orbital size is compatible with the assumption that the isoelectronium has a small meanlife when said correlation-bonding length is restricted to about one Fermi.

Experimental Verifications of Hadronic Chemistry

After achieving the main structural elements of hadronic chemistry, the subsequent steps were rather natural. I knew from the study of ball lightings during my high school years that electrons can bond to each other despite their Coulomb repulsion. By 1978, I had worked out the main elements for their theoretical representation, including the nonunitary character of the theory and the emerging strong binding force [4]. These ideas had seen their applications and experimental verification in various studies, including A.O.E. Animalu's construction of isosuperconductivity and a novel interpretation of the Cooper pair [10]. At the Sanibel Symposium of 1996, I presented an unpublished paper on the natural extension of these studies to a new model of molecular bonds based on the singlet coupling of pairs of valence electrons.

A difficult part was to prove that hadronic chemistry could represent experimental data in a way much more accurate than that of quantum chemistry. This task was achieved by Santilli and Shillady, as reviewed in Chapters 4 and 5.

By looking in retrospective, the confrontation of hadronic chemistry with experimental data resulted in being beyond my most optimistic expectations because:

- *Hadronic chemistry has resolved all insufficiencies and inconsistencies of quantum chemistry known to me;*

- *Hadronic chemistry has achieved representations from first un-adulterated principles of binding energies, electric and magnetic moments and other molecular data which are accurate to several digits;*

- *Hadronic chemistry has permitted new industrial applications, which are unthinkable for quantum chemistry (see Chapter 7).*

Above all, I believe that the above successes of hadronic chemistry constitute the most compelling evidence to date on the insufficiencies of quantum mechanics and the need to seek a broader theory.

Proposed Experiments

Hadronic chemistry is at its infancy and so much remains to be done. In this monograph, I have made a genuine effort to present plausible alternatives on various issues yet to be finalized. In particular, I have made an effort in presenting different views without assuming a final position one way or another.

Among a variety of intriguing problems to be resolved in due time, I mention here *the open problem of the meanlife of the isoelectronium* discussed above. My personal view is that the final resolution of this problem *cannot* be achieved on *theoretical* grounds alone, and requires the conduction of comprehensive *experiments* on the photoproduction of electrons from the hydrogen gas, as discussed in the concluding remarks of Chapters 4 and 5. The measurements of the percentage of emitted electron pairs compared to that of individual electrons would then provide a numerical value of the percentage of stability of the isoelectronium, of course, when the constituents are at short distance only.

I assume the reader is familiar with similar photoproduction experiments which have indicated the emission of electron pairs, although for the helium and not for the hydrogen. These experiments confirm that the deep correlation of electron pairs in singlet coupling, by no means, originates in the molecular structure, because it exists already in the two electrons of helium; the pairing then persists in the Cooper pair in superconductivity; thereafter, the pairing occurs in the coupling of valence electrons in molecular structures; and, finally, the bonding of electrons in ball lighting as well as in various other conditions.

I should stress that, to be really valid, *the proposed experiment must use a variety of photon frequencies, beginning first with the frequency of separation of the hydrogen molecule, then passing to the characteristic frequency of the isoelectronium, and of course, including the frequency of the disintegration of the isoelectronium.*

To achieve a truly final experimental resolution on the meanlife of the isoelectronium, similar experiments should be additionally conducted on other molecules. In fact, it is conceivable that the considered meanlife may vary with the type and size of molecules.

Application of Hadronic Chemistry to New Clean Energies and Fuels

The most important motivation I had for the construction of hadronic chemistry is the same as that at the foundations of all the studies here considered: the need to identify new clean energies and fuels in face of increasingly alarming environmental problems caused by the disproportionate combustion of about 74 millions barrels of crude oil per day. The outcome of my research in these industrial applications of hadronic mechanics and chemistry is summarized in Chapter 7.

The reader will be able to see in this way the real insufficiencies of quantum mechanics and chemistry in face of major societal problems, and, in particular, the impossibility for the contemporary model of molecular structure to predict really *new* methods for the production of environmentally acceptable new energies and fuels at a price competitive over that of fossil fuels.

The reader will also be able to see what appears to be the most salient relevance of the new isochemical model of molecular structures: the prediction and quantitative treatment of a basically new method for a highly efficient separation of the water and other liquids or, more generally, the utilization of energy in liquid molecules, while jointly producing a fuel which is cleaner and cheaper than gasoline.

The culmination of research is given by the insufficiency of the isochemical molecular model itself, and the birth of a new *chemical species*, the first I know since the discovery of the valence in the 1800s, which new species I have tentatively called *magnecules*. It essentially consists of new stable clusters of molecules, and/or dimers, and/or individual atoms under a new strongly attractive internal bond originating from the polarization of the orbits of peripheral atomic electrons. Magnecules are identifiable via clear peaks in chromatographic equipment which peaks remain unidentified following the largest possible search among known molecules, while admitting no infrared signature for gases, or no ultraviolet signature for liquids other than those of the individual constituents.

Predictably, the discovery of a new chemical species creates a host of intriguing open problems, including new means of storing energy in gaseous and liquid molecules, new combustion processes, new analytic methods for the study of magnecules, *etc.*

In conclusion, in the traditional meaning of scientific advances, the resolution by hadronic chemistry of basic open problems of quantum chemistry, rather than reaching a final stage, has created a variety of new intriguing problems for further future advances.

RUGGERO MARIA SANTILLI
THE INSTITUTE FOR BASIC RESEARCH
PALM HARBOR, FLORIDA, USA
NOVEMBER 10, 2000

References

[1] Santilli, R.M.: Nuovo Cimento **51**, 570 (1967).

[2] Santilli, R.M.: Suppl. Nuovo Cimento **6**, 1225 (1968).

[3] Santilli, R.M.: Meccanica **1**, 3 (1968).

[4] Santilli, R.M.: Hadronic J. **1**, 224, 574 and 1267 (1978).

[5] Myung, H.C. and Santilli, R.M.: Hadronic J. **5**, 1277 and 1367 (1983).

[6] Santilli, R.M.: Rendiconti Circolo Matematico Palermo, Suppl. **42**, 7 (1996).

[7] Santilli, R.M.: Found. Phys. **27**, 625 and 1159 (1997).

[8] Santilli, R.M.: Hyperfine Interactions **109**, 63 (1997).

[9] Santilli, R.M.: *Elements of Hadronic Mechanics*, Vols. **I** and **II** (1995), and **III** (in preparation), Ukraine Academy of Sciences, Kiev.

[10] Animalu, A.O.E.: Hadronic J. **17**, 379 (1994).

[11] Animalu, A.O.E. and Santilli, R.M.: Int. J. Quantum Chemistry **29**, 175 (1995).

[12] Santilli, R.M.: Hadronic J. **21**, 789 (1998).

[13] Santilli, R.M.: J. New Energy **4**, issue 1 (1999).

[14] Santilli, R.M.: Foundations of Theoretical Mechanics, Vol. **I** (1978) and **II** (1978), Springer-Verlag, New York, Heidelberg.

[15] Aringazin, A.K.: Hadronic J. **23**, 57 (2000).

[16] Santilli, R.M. and Shillady, D.D.: Int. J. Hydrogen Energy, **24**, 943 (1999). Santilli, R.M. and Shillady, D.D.: Int. J. Hydrogen Energy, **24**, 173 (2000).

Acknowledgments

The author wishes to thank Prof. T. Nejat Veziroglu of the University of Miami in Coral Gables, Florida, Editor in Chief of *International Journal of Hydrogen Energy* (Elsevier Science, Oxford, England), his Editorial Board, and his reviewers for an outstanding editorial processing and review of the first two papers on hadronic chemistry, which represent the birth of the new discipline (see Ref. [16] of the Preface).

Additional special thanks are due to Prof. M. van der Merwe, Editor in Chief of *Foundations of Physics* and *Foundations of Physics Letters*, his Editorial Board and his reviewers for the publication of a final series of paper on the background hadronic mechanics without which this monograph would not have seen the light of the day (see Ref. [7] of the Preface).

Further thanks are due to Prof. D.V. Ahluwalia, Editor of *International Journal of Modern Physics A* and *B* and *Modern Physics Letters*, for outstanding editorial processing of additional papers providing crucial foundations of the new discipline (see the references of Chapter 2).

Additional thanks are due to Professors D.D. Shillady, A.K. Aringazin and M.G. Kucherenko for basic contributions in the applications and verifications of hadronic chemistry, without which this monograph could not have been written.

Special thanks are also due to Mr. Leon Toups, President of *USMagnegas, Inc.*, Mr. John Stanton, President of *EarthFirst Technologies, Inc.*, of Largo, Florida, and all members of their companies, for invaluable financial and personal support in the conduction of innovative experiments needed for the verification of the new chemical species of gas magnecules predicted by hadronic chemistry.

Additional special thanks are due to Mr. J. Herskee, Senior Vice President, Dr. Thomas McGee, Senior Vice President, and Mr. Kenneth L. Purzycki, Director Fragrances, *Givaudan-Roure Corporation*,

Teaneck, New Jersey, for providing samples of their fragrance oils that proved to be crucial for the construction and detection of liquid magnecules. Additional thanks are due to Dr. Konrad Lerch, Director of the *Givaudan-Roure Research Laboratory*, Dubendorf, Switzerland, for providing the first photographic evidence of magnecules in liquids (see Figure 8.19).

Additional special thanks are due to numerous analytic laboratories, such as: *National Technical Systems*, McClellan Air Force Base, Sacramento, California; *Pinellas County Forensic Laboratory*, Largo, Florida; *Givaudan-Roure Research Laboratory*, Dubendorf, Switzerland; *Marine Science Laboratory*, University of South Florida in St. Petersburg; Chemistry Laboratory of Florida *International University*, Miami; *SGS U.S. Testing Company*, Fairfield, New Jersey; *Spectra Laboratory*, Largo, Florida; and other laboratories.

Special thanks are due to Dr. David Hamilton of the *U.S. Department of Energy*, A.O.E. Animalu and J.V. Kadeisvili of the *Institute for Basic Research*, Palm Harbor, Florida, for critical reviews of the manuscript, and numerous invaluable suggestions for improvements and clarifications.

Thanks are also due to Mrs. D. Zuckerman for long hours spent in the linguistic control of the manuscript, and to the staff of the division of the *Institute for Basic Research* at the *Karaganda State University*, Kazakstan, under the Directorship of Prof. A.K. Aringazin, for providing the LaTeX version of the manuscript as well as a final technical review.

Thanks are finally due to Dr. Sabine Freisem and her associates at the Editorial Office of *Kluwer Academic Publishers*, Dordrecht, The Netherlands, for an outstanding editorial review and finalization of the manuscript.

Needless to say, the author is solely responsible for the content of this monograph due to numerous additions and revisions implemented in the final version.

Chapter 1

INTRODUCTION

1. Axiomatic Consistency of Quantum Chemistry

One of the most salient characteristics of *quantum chemistry* (see, e.g., representative books [1]) is its majestic axiomatic consistency when expressed via the axioms of the underlying *quantum mechanics.*

Such a consistency is due to the fact that the fundamental Heisenberg's time evolution in its finite form characterizes a *Lie group of unitary transforms* on a Hilbert space $\mathcal{H}$ with states $|\psi\rangle$ and inner product $\langle\phi| \times |\psi\rangle$ over a field $\mathbb{C}(c, +, \times)$ of complex numbers c with conventional commutative addition "+" and associative multiplication "×", additive unit 0, and multiplicative unit 1,

$$A(t) = U(t) \times A(0) \times U^\dagger(t) = e^{i\times H\times t} \times A(0) \times e^{-i\times t\times H}, \qquad (1.1a)$$

$$|\psi(t)\rangle = U \times |\psi(t_0)\rangle = e^{iH\times t} \times |\psi(t_0)\rangle, \qquad (1.1b)$$

$$U \times U^\dagger = U^\dagger \times U = I. \qquad (1.1c)$$

The corresponding infinitesimal versions is given by the celebrated Heisenberg and Schrödinger representations

$$i\frac{dA}{dt} = [A, H] = A \times H - H \times A, \qquad (1.2a)$$

$$[r^i, p_j] \times |\psi\rangle = \delta^i_j \times |\psi\rangle,$$

$$H \times |\psi\rangle = E \times |\psi\rangle, \quad H = H^\dagger, \quad E \in \mathbb{R}(n, +, \times), \qquad (1.2b)$$

$$p_k \times |\psi\rangle = -i\hbar\partial_k|\psi\rangle, \qquad (1.2c)$$

where: H is the Hermitean, thus observable, Hamiltonian representing the total energy; $\mathbb{R}(n, +, \times)$ is the field of real numbers n; and the brackets $[A, B] = A \times B - B \times A$ characterize the famous Lie product.

The axiomatic consistency of quantum chemistry is then guaranteed by Lie's theory on Hilbert spaces over the field of complex numbers, which theory is one of the most rigorous scientific edifices of the 20-th century. In particular, Lie's theory guarantees that quantum chemistry, when formulated along the above lines, admits the same numerical predictions at all times under the same conditions.

The latter feature, which is evidently necessary for the practical value of any theory, is expressed by the following *invariance of the time evolution under unitary transformations*

$$U \times H \times |\psi\rangle = U \times H \times U^{\dagger} \times U \times |\psi\rangle = \\ = H' \times |\psi'\rangle = E' \times |\psi\rangle, \quad E' \equiv E, \tag{1.3a}$$

$$U \times \{e^{i \times H \times t}\} \times U^{\dagger} = U \times \{I + i \times H \times t/1! + \\ +(i \times H \times t)^2/2! + \ldots\} \times U^{\dagger} = e^{i \times H' \times t}, \tag{1.3b}$$

$$U \times (A \times B) \times U^{\dagger} = A' + B', \quad K' = U \times K \times U^{\dagger}, \quad K = A, B, \tag{1.3c}$$

$$U \times [A, B] \times U^{\dagger} = A' \times B' - B' \times A', \tag{1.3d}$$

$$U \times [r^i, p_j] \times |\psi\rangle = [r'^i, p'_j] \times |\psi'\rangle = \delta^i_j \times |\psi'\rangle. \tag{1.3e}$$

Therefore, quantum chemistry possesses invariant functions and transforms used in the data elaborations of the theory. It then follows that the theory admits: *invariant units of space, time, energy, etc.*, thus being applicable to measurements; *invariant Hermiticity*, thus admitting the same observable at all times; and *unique and invariant numerical predictions*.

In summary, when based on the true axioms of quantum mechanics without *ad hoc* adulterations, quantum chemistry has a majestic axiomatic consistency, and provide invariant results without any internal inconsistency.

Evidently, we are not in a position to review in details the above axiomatic aspects. A technical knowledge of quantum mechanics and Lie's theory is essential for a technical understanding of this monograph.

2. Scope of These Studies

Despite the above majestic axiomatic consistency and achievements of historical proportions throughout the 20-th century, any belief that quantum chemistry can describe exactly all chemical conditions existing in the universe implies the exiting of the boundaries of science.

After one century of attempts, quantum chemistry still remains afflicted by a number of basically unresolved limitations, insufficiencies, or sheer inconsistencies, which are well known to experts in the field, but are generally ignored in graduate courses *and* in the specialized technical literature.

During the last part of the 20-th century, quantum chemical research has been based on *ad hoc* adulterations of the majestic axiomatic structure of the theory outlined above, with particular reference to departures from its most fundamental features, such as the Coulomb law and its unitary invariance.

As an example, various chemical research has been based on the so-called *screening of the Coulomb law* via Gaussians or other functions $f(r)$, $V(r) = \pm e^2/r \rightarrow f(r) \times e^2/r$, for the intent of improving the insufficient representation of molecular experimental data of the conventional theory.

Such screenings and other departures constitute *ad hoc* adulterations of the axiomatic structure of quantum chemistry, since they can only be reached via *nonunitary transforms*, thus exiting the boundary of "quantum chemistry," as technically known.

A fundamental property of quantum chemistry is precisely *the invariance of the Coulomb law under unitary transforms*,

$$V(r) = \pm\frac{e^2}{r} \rightarrow V'(r') = U(r'(r)) \times V(r) \times U^\dagger(r'(r)) = V(r') = \pm\frac{e^2}{r'},$$
$$U \times U^\dagger = U^\dagger \times U = I.$$

Therefore, the screening $V(r) = \pm e^2/r \rightarrow f(r) \times e^2/r$ can *only* be reached via the nonunitary transform $U \times U^\dagger = f(r)$. The exiting from the axiomatic structure of quantum chemistry is then incontrovertible. The important issue here is that nonunitary transforms have "catastrophic inconsistencies" studied in details in Sect. 1.7, which inconsistencies have to be studied in details prior to claiming scientific results.

Moreover, the notion of "quantum" of energy requires, for its very existence, *stable quantized orbits*, which are only admitted by the Coulomb law. *Ad hoc* modified laws of the type $f(r) \times e^2/r$ *do not* admit stable orbits, e.g., the Bohr's orbits are impossible under such adulteration. Therefore, Gaussians and other screenings of the Coulomb law prevent the very notion of "quantum" of energy, thus rendering equivocal the preservation of the very name of "quantum chemistry."

Additional insufficiencies originate from the comparison between the axiomatic structure of quantum chemistry and reality. The theory is notoriously *linear* (in the wavefunctions), *local-differential* (from its very topology, e.g., that of Mackay's imprimitivity theorem), and *potential-Hamiltonian-unitary* (namely, representing systems solely admitting a potential in a unitary time evolution). As a consequence, chemical structures are abstracted into a finite set of isolated points with only action-at-a-distance, potential interactions representable with a conventional Hamiltonian.

As we shall see, the above limited representational capabilities imply serious insufficiencies in the contemporary conception of molecular bonds. In fact, valence electrons do have a *point-like structure* but only for the *charge*, while possessing well-defined *extended wavepackets*. During the correlation of the valence electrons at short distance we therefore have the necessary conditions of deep mutual overlapping of their wavepackets, which are known to be: *nonlinear* (in the wavefunction); *nonlocal-integral* (in the sense that the interaction exists over a finite volume which is not reducible to a finite set of isolated points); and *nonpotential-nonhamiltonian-nonunitary* (in the sense that the interaction is of contact/zero-range type for which the notions of potential or Hamiltonian have no mathematical or physical meaning, thus resulting in a nonunitary structure, as we shall see in detail).

Other differences between the axiomatic structure of quantum chemistry and reality are so large and visible to be properly called dramatic. We are referring here to:

- The *structural reversibility in time of quantum chemical laws and axioms*, i.e., the causal admissibility of their time reversal image, as compared to

- The *manifest irreversibility of various chemical processes*, i.e., the violation of causality by their time reversal images, as evident for chemical reactions, biological processes, *etc.*

Under the above serious structural insufficiencies, as well as the additional insufficiencies identified in the next section, it is evident that any attempt at reconciling the above structural problems with quantum chemistry is political-nonscientific. Rather than adapting reality to a preferred theory, science can only be served by modifying the theory to achieve a better representation of reality.

A main objective of this monograph is the submission of a generalization-broadening of quantum chemistry which:

1) coincides with quantum chemistry everywhere except in the region of valence electron bonds;

2) extends the applicability of the theory to include an invariant representation of the most general possible class of nonunitary transforms representing short-range nonlinear, nonlocal, and nonpotential interactions due to the overlapping of the wavepackets of valence electrons; and

3) preserves the validity for closed-isolated reversible systems of all established physical laws at all distances and conditions, including those at short distances, such as Heisenberg's uncertainty principle, Pauli's exclusion principle, probability and causality laws, *etc.*

In this monograph we shall then introduce a new model of molecular bonds for the hydrogen molecule (Chapter 4), the water and other molecules (Chapter 5); show that the new model permits a representation of experimental data on binding energies, electric and magnetic moments much more accurate than that permitted by quantum chemistry; and indicate the resolution of at least some of the insufficiencies of current chemical descriptions.

Our main line of research is that of assuming the axiomatic consistency of quantum chemistry as the foundations of our generalized theories, and construct covering theories via *broader realizations* of the same axioms solely applicable for broader conditions.

Above all, the most fundamental requirement of our studies is that of *preserving the abstract Lie axioms, and merely submit broader realizations valid for linear and nonlinear, local and nonlocal, potential and nonpotential, Hamiltonian and nonhamiltonian effects.* As we shall see, this condition alone guarantees that the generalized chemical theory submitted in this monograph has exactly the same axiomatic consistency as that of the conventional theory.

3. Insufficiencies of Quantum Chemistry for Molecular Structures

Let us now identify the insufficiencies of quantum chemistry for the simplest possible class of systems, those that are *isolated from the rest of the universe*, thus verifying conventional *conservation laws* of the total energy, total linear momentum, *etc.*, and are *reversible* (namely, their time reversal image is as physical as the original system).

The most representative systems of the above class are given by *molecules*, here generically defined as aggregates of atoms under a valence bond [1]. Despite undeniable achievements, *quantum chemical models of molecular structures* have the following fundamental insufficiencies:

1.3.1) Lack of a sufficiently strong binding force. As it is well known, the average of all Coulomb forces among the atoms constituting a molecule is identically null at the semiclassical level, thus resulting in the absence of any attractive force at all. As an example, the currently used Schrödinger equation for the H_2 molecule is given by the familiar

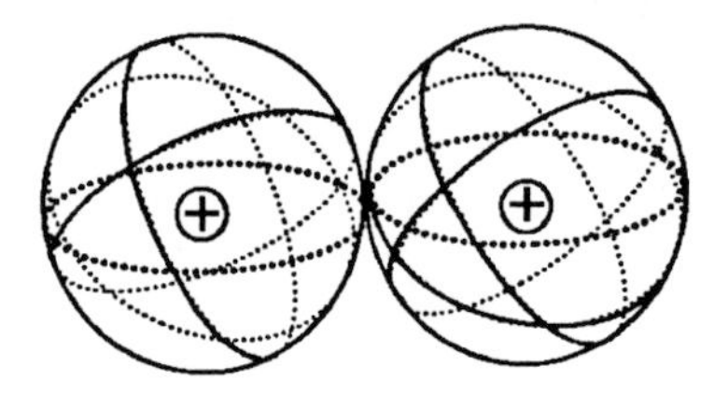

Figure 1.1. A schematic view of the contemporary *physical* (and *not* chemical) conception of the simplest possible molecule, the hydrogen molecule at absolute zero degree temperature, here conceptually represented as two H-atom next to each other in the absence of any rotational or vibrational motion. Therefore, the spheres represent the spherical distribution of the *orbits* (and *not* of the orbitals) of the individual electrons. Conceptual chemical visualizations, e.g., that in terms of orbitals (which are probability distributions, rather than physical orbits), have no implications for the content of this section.

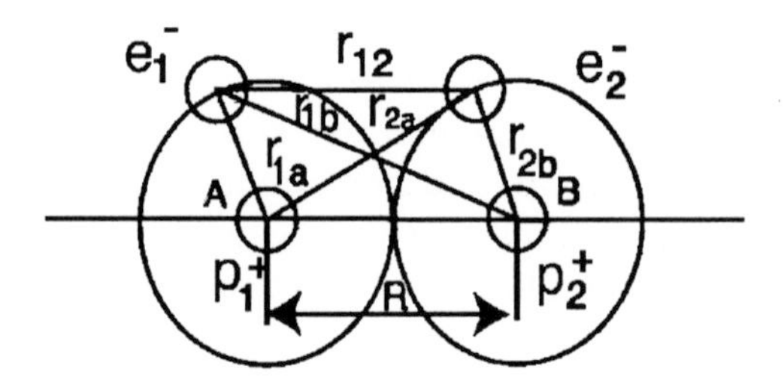

Figure 1.2. A schematic view of the fact that the total Coulomb force among two atoms of a molecular structure is *identically null* in semiclassical approximation. As a consequence, conventional Coulomb interactions cannot be credible grounds for molecular bonds.

expression [1],

$$\left(-\frac{\hbar^2}{2\mu_1}\nabla_1^2 - \frac{\hbar^2}{2\mu_2}\nabla_2^2 - \frac{e^2}{r_{1a}} - \frac{e^2}{r_{2a}} - \frac{e^2}{r_{1b}} - \frac{e^2}{r_{2b}} + \frac{e^2}{R} + \frac{e^2}{r_{12}}\right)|\psi\rangle = E|\psi\rangle, \tag{1.4}$$

which contains the Coulomb attraction of each electron by its own nucleus, the Coulomb attraction of each electron from the nucleus of the other atom, the Coulomb repulsion of the two electrons, and the Coulomb repulsion of the two protons. It is easy to see that in semiclassical average, the two attractive forces of each electron from the nucleus of the other atom are compensated by the average of the two repulsive forces between the electrons themselves and those between the protons, under which Eq. (1.1) reduces to two independent *neutral* hydrogen atoms *without* attractive interaction, as depicted in Fig. 1.1,

$$\left[\left(-\frac{\hbar^2}{2\mu_1}\nabla_1^2 - \frac{e^2}{r_{1a}}\right) + \left(-\frac{\hbar^2}{2\mu_2}\nabla_2^2 - \frac{e^2}{r_{2a}}\right)\right] |\psi\rangle = E|\psi\rangle. \tag{1.5}$$

In view of the above occurrence, quantum chemistry tries to represent molecular bonds via *exchange, van der Waals and other forces* [1]. However, the latter forces were historically introduced for *nuclear* structures in which they are known to be *very weak*, thus being grossly insufficient to provide a true representation of molecular bonds. In fact, it is now part of history that, due to the insufficiencies of exchange, van der Waals and other forces, nuclear physicists were compelled to introduce the *strong force*. As an illustration, calculations show that, under the currently assumed molecular bonds, the molecules of a tree leaf should be decomposed into individual atomic constituents by a weak wind of the order of 10 miles per hour. To put it in a nutshell, after a century of research, *quantum chemistry still misses in molecular structures the equivalent of the strong force in nuclear structures.*

1.3.2) Quantum chemistry admits an arbitrary number of atoms in the hydrogen, water and other molecules. This additional insufficiency is proved beyond scientific doubt by the fact that the exchange, van der Waals, and other bonding forces used in current molecular bonds were constructed in nuclear physics for the specific purpose of admitting an arbitrary number of constituents. When the same forces are used for molecular structures, they also admit an arbitrary number of constituents. As specific examples, when applied to the structure of the hydrogen or water molecules, any graduate student in chemistry can prove that, under exchange, van der Waals and other forces of nuclear type, the hydrogen, water and other molecules admit an *arbitrary* number of hydrogen atoms, that is, rather than explaining the reason why nature has selected the molecules H_2 and H_2O as the sole possible, current molecular models equally admit the molecules H_5, H_{23}, H_7O, HO_{21}, $H_{12}O_{15}$, *etc.* in dramatic disagreement with experimental evidence. The same inconsistencies are independently reached by the fact that correlations in current chemical models are not restricted to *pairs of valence electrons*, but hold for an arbitrary number of electrons.

1.3.3) Quantum chemistry has been unable to achieve an exact representation of molecular binding energies and other molecular characteristics via un-adulterated axiomatic principles. After about one century of attempts, quantum chemistry still misses a historical 2% in the representation of the binding energy of hy-

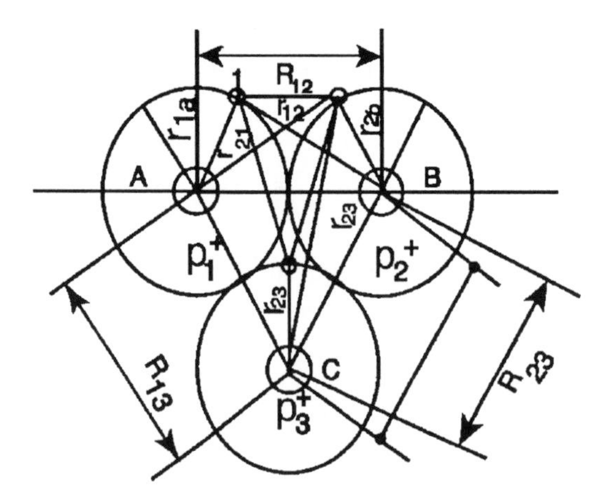

Figure 1.3. A schematic view of the fact that the current conception of the structure of the hydrogen molecule (Figs. 1.1 and 1.2) admits a third hydrogen atom with molecular structure H_3 and, consequently, an arbitrary number of H-atoms, thus admitting molecules of the generic type H_5, H_{27}, *etc.* ... As we shall see in Chapter 8, a structure with the weight of three H atoms has indeed been detected, although misinterpreted due to various reasons discussed in Chapter 8. Here we want merely to indicate that the structure H_3, if conceived as a molecule, should also admit H_4, H_5, *etc.*, which have never been detected. This feature is in dramatic disagreement with experimental evidence because the sole possible stable structure of the hydrogen molecule is H_2. This inconsistency is due to the fact that quantum chemical models of molecular structures use bonding forces such as the exchange, van der Waals and other forces, which were conceived in *nuclear* physics for an *arbitrary* number of constituents. Other reasons are due to the lack of restriction of correlation to only pairs of valence electrons, as established in reality. As a consequence, quantum chemistry lacks the equivalent in molecular structures of the strong force in nuclear structures. Quantum chemistry also lacks any direct or indirect means to limit the number of constituents of a molecular structure, thus confirming the inconsistency of Fig. 1.2.

drogen, water and other basic molecules, despite the use of some of the best methods, such as self-consistent methods [1].

1.3.4) More accurate representations of binding energies violate basic quantum axioms and physical laws. As recalled in the preceding section, a number of attempts have been conducted, which do indeed achieve a more accurate representation of binding energies, although such a representation is reached via a number of mathematical schemes, such as the so-called *Gaussian screening of the Coulomb law* [1]. As we shall study in detail in Sect. 1.7, these schemes imply a *necessary violation of quantum axioms and physical laws.* In fact, the Coulomb law is a fundamental invariant of quantum mechanics. It then follows that no screening at all is possible within the rigid axiomatic structure of the theory and its underlying *unitary transforms* outlined in Sect. 1.1. The only possibility of screening the Coulomb law is then via the use of *nonunitary transforms* indicated earlier, i.e.,

$$F = \pm\frac{e^2}{r} \rightarrow U \times (\pm\frac{e^2}{r}) \times U^\dagger = \pm\frac{e^2}{r}, \quad U \times U^\dagger = I, \tag{1.6a}$$

$$F = \pm\tfrac{e^2}{r} \rightarrow W \times (\pm\tfrac{e^2}{r}) \times W^\dagger = \pm e^{A \times r} \times \tfrac{e^2}{r}, \quad W \times W^\dagger = e^{A \times r} \neq I. \tag{1.6b}$$

It then follows that *Gaussian screenings of the Coulomb law imply the abandonment of the entire axiomatic structure of quantum chemistry.*

1.3.5) Quantum chemistry cannot provide a meaningful representation of thermodynamical properties. The missing 2% in the representation of binding energies is misleadingly small, because it corresponds to about 1,000 Kcal/mole while an ordinary thermodynamical reaction implies an average of 20 Kcal/mole. No scientific calculation can be conducted when the error is a factor of the order of 50.

1.3.6) Computer usages in quantum chemical calculations require excessively long periods of time. This is notoriously due to the slow convergence of conventional quantum series, a feature that persists to this day despite the availability of powerful computers.

1.3.7) Quantum chemistry has been unable to explain the correlation of valence electrons into pairs. The correlation of valence electrons solely into pairs originates from Pauli's exclusion principle (which is *assumed* by quantum chemistry without any explanation), and then acquires a deep role in molecular structures. The inability by quantum chemistry to restrict correlations solely to valence pairs is then passed to orbital theories, which work well at semi-empirical levels, but which remain afflicted by yet unresolved problems, such as admitting the correlation among *many electrons* while experimental evidence establishes that the correlation solely occurs among *electron pairs*. The insufficiency then cause other inconsistencies, such as the admission of an arbitrary number of atoms in all molecules, as indicated earlier.

1.3.8) Quantum chemistry has been unable to reach an exact representation of the electric and magnetic dipole and multipole moments of the hydrogen, water and other molecules. These representations lack a few percentages for the hydrogen molecule and larger percentages for bigger molecules. The departures of the prediction of quantum chemistry from reality are such that said representations have at times even the *wrong sign* [1].

1.3.9) Quantum chemistry predicts that all molecules are ferromagnetic. This inconsistency is a consequence of the most rigorous

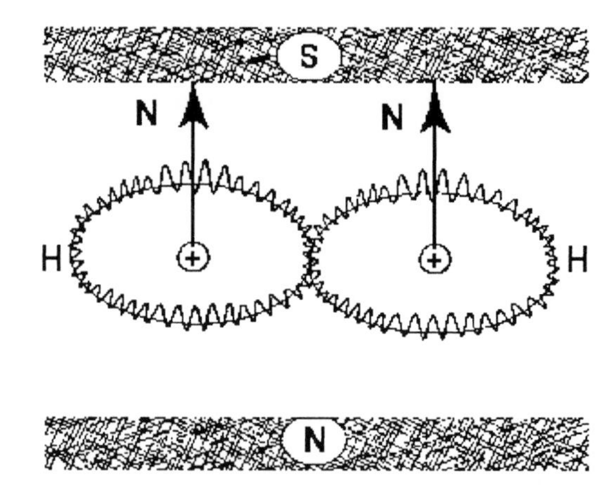

Figure 1.4. A schematic view of the prediction by quantum chemistry that all molecules are ferromagnetic, here expressed for the case of the simplest possible hydrogen molecule H-H at absolute zero degrees temperature. The prediction is an unavoidable consequence of the current conception of molecular structure (Fig. 1.1), in which each atom preserves its individuality while the only acting forces are exchange and other forces of nuclear type. The most rigorous discipline of this century, quantum electrodynamics, then establishes that, under an external magnetic field, the orbits of all valence electrons acquire the same polarization, resulting in a total net magnetic polarity $H_\uparrow$-$H_\uparrow$, which is in dramatic disagreement with experimental evidence.

discipline of this century, quantum electrodynamics, which establishes that, under an external magnetic field, the orbits of valence electrons must be polarized in such a way to offer a magnetic polarity opposite to that of the external field (a polarization that generally occurs via the transition from a space to a toroidal distribution of the orbitals). As it is well known, the individual atoms of a molecule preserve their individuality in the current model of chemical bonds. As a result, quantum electrodynamics predicts that the valence electrons of the individual atoms of a molecular bond must acquire polarized orbits under an external magnetic field. As a result, quantum chemistry predicts that the application of an external magnetic field South-North, to hydrogen H-H, water H-O-H and other molecules implies their acquisition of the net total, opposite polarity North-South, $H_\uparrow$-$H_\uparrow$, $H_\uparrow$-$O_\uparrow$-$H_\uparrow$, *etc.*, which is in dramatic disagreement with experimental evidence.

No serious advance in chemistry can occur without, first, the admission, and, then, a detailed scientific study of the above insufficiencies.

4. Insufficiency of Quantum Chemistry for Chemical Reactions

Insufficiencies far greater than those outlined so far occur whenever attempting to study via quantum mechanics and chemistry structures and/or processes that are *irreversible* (namely, whose time reversal image is not physical, e. g., because it would violate causality or the principle of conservation of the energy). Irreversibility is typically the case of *chemical reactions* at large, such as

$$H_2 + \frac{1}{2}O_2 \rightarrow H_2O. \tag{1.7}$$

As indicated earlier, the insufficiency is due to the fact that *quantum mechanics and chemistry are structurally reversible*, that is, all their mathematical axioms and physical laws are fully reversible in time. This prevents any scientifically rigorous representation of irreversible systems and processes.

At the same time, all known potential interactions are fully reversible in time, with no exception known to the author. This additional occurrence prevents a scientific-quantitative representation of irreversible processes via a time-reversal-violating Hamiltonian.

At any rate, one of the most important teaching in the history of science is that by Lagrange, Hamilton, and Jacobi who pointed out that *irreversibility cannot be represented with a potential*, and, for this reason, they formulated analytic mechanics with *external terms representing contact-nonpotential interactions which are precisely the origin of irreversibility.*

In the planetary and atomic structures, evidently, there is no need for external terms in the analytic equations, since all acting forces are of potential type. In fact, these systems admit an excellent approximation as being made-up of *massive points moving in vacuum without collisions.* In these fields, the historical analytic equations were "truncated" with the removal of the external terms.

In view of the successes of the planetary and atomic models, the entire scientific development of the 20-th century was restricted to the "truncated analytic equations," without any visible awareness that *they do not* represent the scientific conception by the originators of analytic mechanics.

The above historical, conceptual and technical scenario prevents any attempt at exact representing irreversible chemical problems via quantum chemistry, evidently because that theory is strictly potential-Hamiltonian.

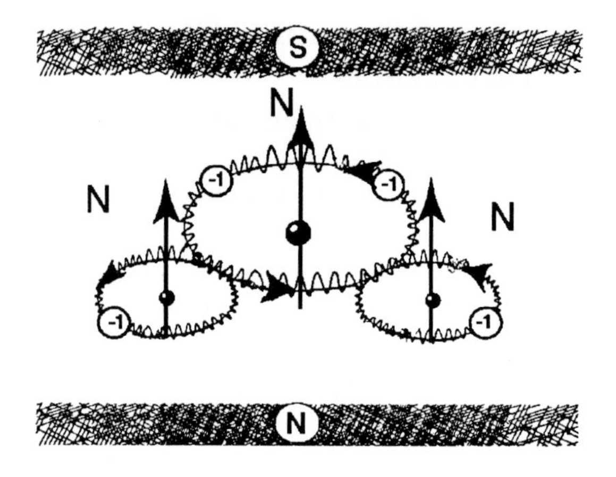

Figure 1.5. A schematic view of the prediction by quantum chemistry that water is ferromagnetic, in dramatic disagreement with experimental evidence. In fact, quantum chemistry does not restrict the correlation of valence electrons to pairs, as a result, the individual valence electrons of the water molecule remain essentially independent. Quantum electrodynamics then demands the polarization of all valence electrons under an external magnetic field, resulting in the net magnetic polarity of this figure, and consequential ferromagnetic character. All forces currently admitted by quantum chemistry, such as exchange forces, van der Waals forces, *etc.*, cannot evidently remove the inconsistency, e.g., because of their lack of restriction of correlation to valence pairs, as well known.

Therefore, no serious scientific advance on irreversible processes in general, and irreversible chemical reactions in particular, can be achieved without first *generalizing quantum chemistry into a structurally irreversible theory*, that is, a theory which is irreversible for all possible, reversible Hamiltonians.

5. Insufficiencies of Quantum Chemistry for Biological Structures

Any belief of the validity of quantum mechanics and chemistry in biology would simply be beyond boundary of science. This is due to the fact that the discrepancies between the predictions of quantum mechanics and chemistry and the evidence in biology are so large to prevent quantitative scientific descriptions.

Besides being reversible, quantum mechanics and chemistry can only represent *perfectly rigid systems*, as well known from the fundamental rotational symmetry. As a consequence, the representation of biological systems via quantum mechanics and chemistry would imply that our body should be perfectly rigid, without any possibility of introducing deformable-elastic structures, because the latter would imply catastrophic inconsistencies in the basic axioms.

Moreover, another pillar of quantum mechanics and chemistry is the verification of total conservation laws, for which Heisenberg's equation of motion became historically established. In fact, the time evolution of an arbitrary quantity A is given by

$$i\frac{dA}{dt} = [A, H] = A \times H - H \times A, \tag{1.8}$$

under which expression we have the conservation law of the total energy, $i\ dH/dt = H \times H - H \times H \equiv 0$.

A basic need for a scientific representation of biological structures is instead the representation of *the time-rate-of-variations of physical characteristics*, such as size, weight, density, *etc.* Unlike atomic-molecular systems, for which quantum mechanics and chemistry were built, biological systems *grow* in time, and then *age*, thus changing in time.

The irreconcilable incompatibility between the *variations in time* of biological structures and the *conservation in time* of quantum mechanical and chemical systems as per Eq. (1.8) establishes the structural inapplicability of quantum mechanics and chemistry for a scientific description of biological structures.

When passing to deeper studies, the insufficiencies of quantum mechanics and chemistry emerge even more forcefully. As an example, quantum mechanics and chemistry can well represent the *shape* of sea shells, but not their *growth in time*. In fact, computer simulations [2] have shown that, when the geometric axioms of quantum mechanics and chemistry (those of the Euclidean geometry) are imposed as *exactly* valid, sea shells first grow in a deformed way, and then crack during their growth.

The latter results should not be surprising to readers with an inquisitive mind, again, because the growth of sea shells is an *irreversible* and nonconservative process while the geometric axioms of quantum mechanics and chemistry are perfectly *reversible* and *conservative*, as indicated earlier, thus resulting in an irreconcilable structural incompatibility, this time at the geometric level without any conceivable possibility of reconciliation.

Additional studies have established that the insufficiencies of quantum mechanics and chemistry in biology are much deeper than the above, and invest the *mathematics* underlying these disciplines. In fact, Illert [2] has established that the minimally correct representation of the growth in time of sea shells requires the *doubling of the Euclidean axes*. However, sea shells are perceived by the human mind (via our three Eustachian tubes) as growing in our *three-dimensional* Euclidean space. Santilli [2] has shown that the only known resolution of such a dichotomy is that via *multi-valued mathematics*, that is, mathematics in which operations such as product, addition, *etc.* produce a *set of values*, rather than one single value as in quantum mechanics and chemistry (see next Chapter for details).

At any rate, the belief that the simplistic mathematics underlying quantum mechanics and chemistry can explain the complexity of the DNA, has no scientific credibility, the only serious scientific issue being the search for suitable generalizations.

In conclusion, science will never admit "final theories." No matter how valid any given theory may appear at any point in time, its structural broadening for the description of more complex conditions is only a matter of time. This is the fate also of quantum mechanics and chemistry, which, despite outstanding achievements, cannot possibly be considered as "final theories" for all infinitely possible conditions existing in the universe.

After all, following only a few centuries of evolution, rather than having reached a "final stage," science is only at its infancy.

6. The Central Topic of Study of This Monograph

In view of the insufficiencies outlined above, in this monograph we introduce three generalizations-broadening of quantum chemistry of progressively increasing complexity, in the hope of resolving at least some of the insufficiencies outlined above for molecules, chemical reactions, and biological structures, respectively. The results will then be applied in Chapters 7 and 8 to new clean energies and fuels.

Among a number of possibilities, *we follow the historical teaching by Einstein, Podolsky, and Rosen [3a] (E-P-R) on the "lack of completion" of quantum mechanics and, therefore, of quantum chemistry.*

Among a variety of possibilities, we study a possible "completion" with *the representation of deep wave-overlappings of the wavepackets of valence electrons in chemical bonds.* In different terms, our main assumption is that quantum chemistry is "incomplete" because it treats valence electrons as points (which is necessary for consistency from the very topological structure of the theory), or, equivalently, because it restricts the study solely to the *point-like charge, and ignores altogether the extended wavepackets.*

As is well known, all massive particles (beginning with the electrons) have a wavepacket of the order of 1 fm (10^{-13} cm). We therefore assume quantum mechanics and chemistry as being *exactly* valid at distances sufficiently bigger than 1 fm, while we seek possible, generally small

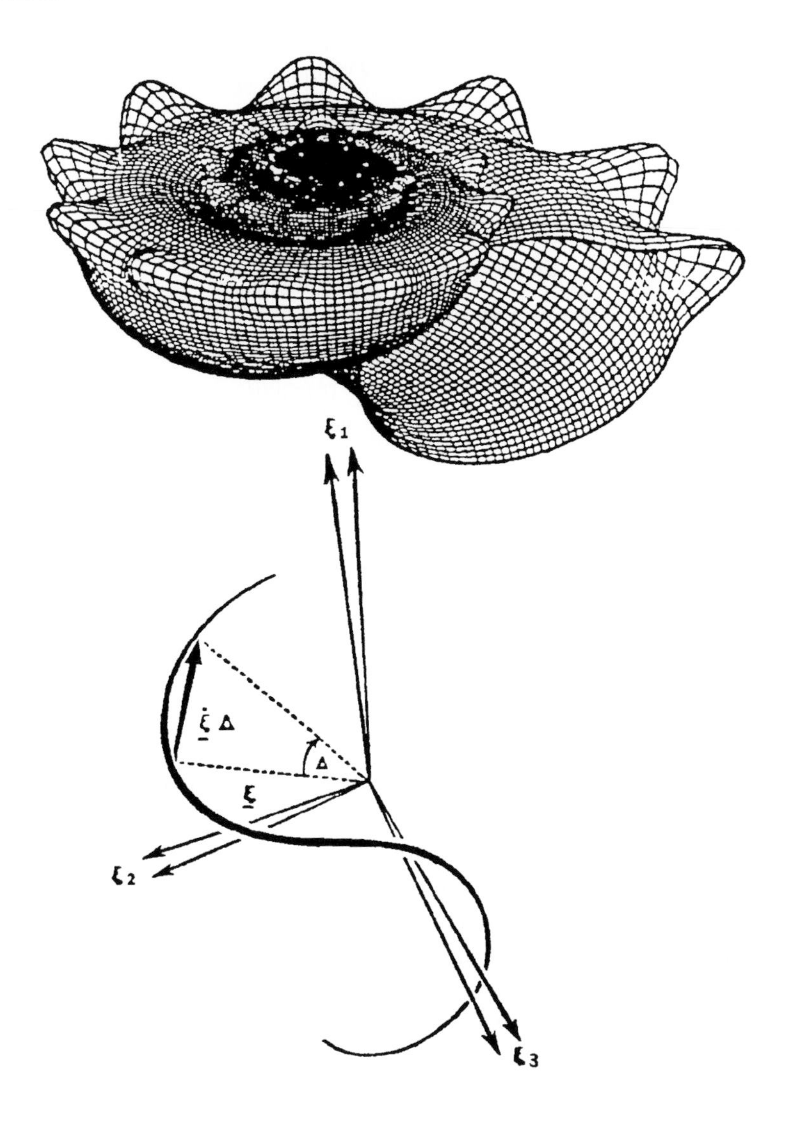

Figure 1.6. A computer visualization of the shape of a sea shell, the *Angaria Delphinium* from Ref. [2] based on the *exact validity* of the geometric axioms of quantum mechanics and chemistry, those of the Euclidean geometry. As one can see, the representation of the *shape* is rather accurate. Nevertheless, the same visualization becomes grossly in contradiction with evidence when extended to the *growth in time* of the same sea shell. Illert [2] has proved that the correct representation of the growth requires a 3×3-*dimensional space*, i.e., the doubling of our conventional Cartesian coordinates. But this mathematical representation has to be compatible with our 3-dimensional perception of the sea shell. Santilli [2] has shown that the above dichotomy establishes the *multi-valued* character of biological structure at large, namely, their need for a mathematics dramatically more general than that currently used in quantum mechanics and chemistry (see Sect. 2.4 for details).

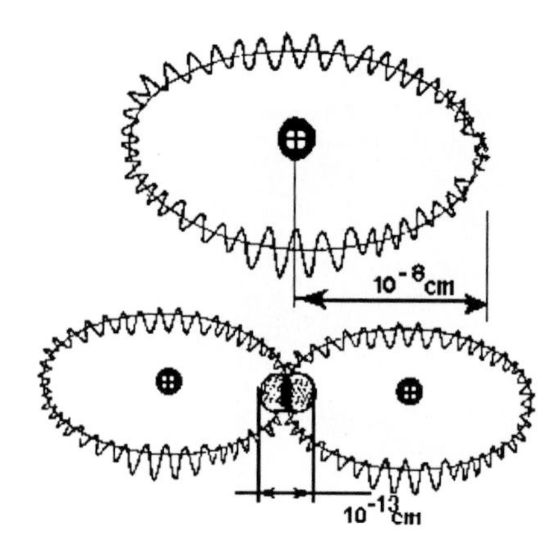

Figure 1.7. A schematic view of the main conceptual difference between the structure of a hydrogen *atom* and that of a hydrogen *molecule*. In the former case, we have large mutual distances as compared to the size of the wavepackets, thus implying the exact validity of quantum mechanics. In the latter case, we have instead the appearance of conditions of deep overlapping of the wavepackets of the electrons that are absent in the atomic structure, while being beyond any hope of scientific treatment via quantum mechanics and chemistry.

corrections at mutual distances of the order of 1 fm or less due to the deep wave-overlappings of valence electrons.

At a deeper level, we assume quantum mechanics as being exact for the structure of the individual hydrogen atoms. In this case the size of the wavepacket of the electron can be effectively ignored due to the large distance of the Coulomb interaction with the nucleus. This implies the *exact* validity of the axiomatic structure of Sect. 1.1, and permits the effective applicability of a linear, local and potential theory, resulting in an accurate description of the hydrogen atom, as historically established.

In the transition from the structure of the hydrogen atom to the structure of the hydrogen molecule, there is the additional presence of the deep overlapping of the wavepackets of the valence electrons at short distance which overlapping is completely absent for the hydrogen structure. Therefore, we argue that, while remaining exactly valid for the *atomic* structure, quantum mechanics itself, let alone quantum chemistry, needs a suitable "completion" for the description of the hydrogen *molecule* due to the appearance of conditions and interactions beyond the descriptive capacity of its basic axioms (Sect. 1.1).

The central topic of study of this monograph is, therefore, a quantitative invariant representation of the deep overlappings of the wavepackets of valence electrons.

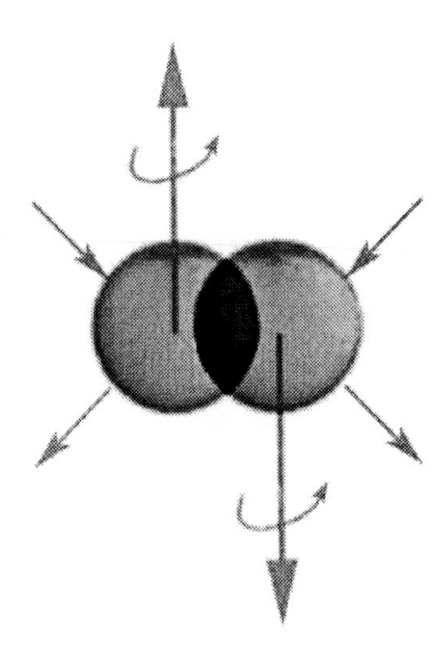

Figure 1.8. A schematic view of the fundamental conditions studied in this memoir, *the deep overlapping of the extended wavepackets of valence electrons with point-like charge in singlet coupling,* as required by Pauli's principle. Recall that, for quantum mechanics, the electrons are points and, thus, the conditions of this figure have no meaning at all. However, this point character generally refers to the *charge* structure of the electron. For the covering hadronic mechanics, the point-like charge structure of the electrons persists, with the additional study of the *overlapping* of the *extended wavepackets.* Also, quantum mechanics can only describe electrons via dynamical equations which are *linear, local, and potential.* On the contrary, hadronic mechanics is capable of representing in an invariant way the main characteristics of wave-overlapping as being *nonlinear, nonlocal, and nonpotential.* As we shall see in Chapters 3 and 4, the latter conditions imply the birth of a basically novel strongly attractive force between the valence electrons with consequential new conception of valence bonds and molecular structures capable of essentially exact representations of experimental data.

These physical conditions are known to be (see Fig. 1.8):

1) *nonlinear* (i.e., dependent on powers of the wavefunctions greater than one);

2) *nonlocal-integral* (i.e., dependent on integrals over the volume of wave-overlapping which, as such, cannot be reduced to a finite set of isolated points);

3) *nonpotential* (i.e., consisting of "contact" interactions, that is, contact caused by the actual physical contact, with consequential "zero range" for which the notion of potential energy has no mathematical or physical sense);

4) *nonhamiltonian* (i.e., not representable via a Hamiltonian, thus requiring additional terms); and, consequently,

5) *nonunitary* (i.e., the time evolution *violates* unitarity conditions (1.1c)).

Note that the nonlocality is expected from the extended character of the *wavepackets* of the valence electrons under a fully point-like charge structure.

In short, *we study a "completion" of quantum mechanics and chemistry via the addition of effects at distances of the order of 1 fm (only), which are assumed to be nonlinear, nonlocal, nonpotential, nonhamiltonian, and nonunitary.*

Note also that the condition of *nonunitarity is necessary* for novelty, because it is necessary to exit the class of equivalence of quantum chemistry. In different terms, the assumption of a unitary representation of wave-overlappings implies the *prohibition* of any novelty, because the theory would be fully within the axioms of existing doctrines.

As we shall see in Chapters 2 and 3, the above representation of deep overlapping of valence electrons in singlet coupling permits:

1′) The introduction of a basically new, strongly attractive, non-Coulomb force among pairs of valence electrons which can be written in first approximation for the case of the hydrogen molecule via a Hulten term in the equation

$$\left[\left(-\frac{\hbar^2}{2\mu_1}\nabla_1^2 - \frac{e^2}{r_{1a}}\right)+\left(-\frac{\hbar^2}{2\mu_2}\nabla_2^2 - \frac{e^2}{r_{2a}}\right)-V_0\frac{e^{-r_{12}/R}}{1-e^{-r_{12}/R}}\right]|\psi\rangle = E\times|\psi\rangle. \tag{1.9}$$

As one can see, the above force is completely absent in current models of molecular bonds, and can indeed provide the equivalent of the strong nuclear force currently missing in molecular structures.

2′) The explanation why the hydrogen molecule admits only two atoms. Once coupled into the singlet state of Fig. 1.8, valence electrons will repel any additional electrons, thus restricting the number of correlated electrons to two atoms only.

3′) The new bonding force between valence electrons permits a representation of binding energies accurate to the seventh digit (sic), as we shall see in Chapters 4 and 5.

4′) The representation here considered will be derivable from first axiomatic principles without *ad hoc* manipulations.

5′) Representations of binding energies accurate to the seventh digit will permit accurate thermochemical calculations.

6′) Computer calculations based on the new binding force (1.9) converge at least 1,000 faster than conventional calculations.

7′) The correlation of valence electrons is solely restricted to pairs at all levels.

8′) The model permits representations of electric and magnetic moments correct in sign as well as accurate to the seventh digit (sic).

9′) The emerging new model of molecular structure eliminates the inconsistent prediction that all molecules are ferromagnetic.

In short, we shall show in this monograph that the study of the nonlinear, nonlocal and nonpotential conditions of deep overlappings of the wavepackets of valence electrons offers realistic possibilities of resolving all insufficiencies 1.3.1)–1.3.9) of quantum chemistry outlined in Sect. 1.3.

7. Catastrophic Inconsistencies of Generalized Nonunitary Theories on Conventional Mathematics

The next aspect of fundamental importance for these studies is that the treatment of nonlinear, nonlocal and nonunitary effects via the conventional mathematics of quantum mechanics and chemistry admits physical and mathematical inconsistencies so serious to be simply catastrophic, as studied in detail in Refs. [4].

The author has dedicated his research life to the construction of axiomatically consistent and invariant generalizations of quantum mechanics for the treatment of nonlinear, nonlocal, and nonpotential effects (see representative papers [5] and monographs [6], independent physics papers [7], mathematical papers [8], applications and experimental verifications [9, 10] and independent review monographs [11]). The results of these studies remain, however, generally unknown in physical circles, let alone in the specialized chemical literature, thus warranting a review in this section.

To begin, let us recall that a theory is said to be *equivalent to quantum mechanics* when it can be derived from the latter via any possible unitary transform (verifying certain conditions of topological smoothness hereon ignored for simplicity).

As a consequence, a *necessary and sufficient condition for a theory to be inequivalent to quantum mechanics is that it must be outside the class of unitary equivalence of the theory*, that is, the new theory must be connected to quantum mechanics via a *nonunitary transform* on a Hilbert space $\mathcal{H}$ over the field of complex numbers $\mathbb{C} = \mathbb{C}(c, +, \times)$,

$$U \times U^{\dagger} \neq I. \tag{1.10}$$

As such, true generalized theories must have a *nonunitary structure*, e.g., their time evolution must verify law (1.10), rather than laws (1.1c).

During his undergraduate studies in physics at the University of Torino, Italy, and as part of his Ph. D. thesis, Santilli [5a, 5b] published in 1967 the following structural generalization of the basic equations (1.1), the first in scientific records (which became known two decades later as "q-deformations," although without generally quoting of their origination in Ref. [5a]),

$$A(t) = e^{i \times H \times q \times t} \times A(0) \times e^{-i \times t \times p \times H}, \tag{1.11a}$$

$$\begin{aligned} i\, dA/dt = (A, B)_{\text{operator}} &= p \times A \times H - q \times H \times A = \\ &= m \times (A \times B - B \times A) + n \times (A \times B + B \times A) = \\ &= m \times [A, B] + n \times \{A, B\}, \end{aligned} \tag{1.11b}$$

where $p = m + n, \quad q = m - n$ and $p + q$ are non-null parameters (and "$\times$" represents again the conventional associative product of numbers or matrices, $AB \equiv A \times B, \quad A \times (B \times C) \equiv (A \times B) \times C$).

After an extensive research in European mathematics libraries, the brackets $(A, B) = p \times A \times B - q \times B \times A$ resulted to be *Lie-admissible* according to Albert, that is, the brackets are such that the attached antisymmetric product $(A, B) - (B, A) = (p + q) \times [A, B]$ is Lie. At that time (1967), only three articles on this subject had appeared in the field in the sole mathematical literature.

Subsequently, while at Harvard University under DOE support, Santilli proposed the following most general possible Lie-admissible theory [Ref. [5e], Eqs. (4.15.34) and (4.18.11)],

$$A(t) = U \times A(0) \times U^{\dagger} = e^{i \times H \times Q \times t} \times A(0) \times e^{-i \times t \times P \times H}, \quad P = Q^{\dagger}, \tag{1.12a}$$

$$\begin{aligned} i\, dA/dt = (A \hat{,} B)_{\text{operator}} &= A < B - B > A = \\ &= A \times P \times H - H \times Q \times A = \\ &= (A \times T \times B - B \times T \times A) + (A \times W \times B + B \times W \times A) = \\ &= [A \hat{,} B] + \{A \hat{,} B\}. \end{aligned} \tag{1.12b}$$

It is an instructive exercise for the reader interested in learning the formalism of these studies to prove that:

1) time evolutions (1.11) and (1.12) are nonunitary, $U \times U^\dagger \neq I$;

2) the application of another nonunitary transform $R \times R^\dagger \neq I$ to structure (1.11) yields precisely the broader structure (1.12) by essentially transforming the parameters p and q into the operators $P = p \times (R \times R^\dagger)^{-1}$ and $Q = q \times (R \times R^\dagger)^{-1}$; and

3) the application of any additional nonunitary transforms $S \times S^\dagger \neq I$ to structure (1.12) preserves its Lie-admissible character, although with different versions of the P and Q operators. This essentially proves the following:

Theorem 1.1: *General Lie-admissible laws (1.12) are "directly universal" in the sense of containing as particular cases all infinitely possible nonunitary generalizations of quantum mechanical equations ("universality") directly in the frame of the observer ("direct universality"), while admitting a consistent algebra in their infinitesimal form (i.e., a product verifying the distributive and scalar laws).*

The above theorem can be equally proved by noting that the product $(A\hat{,}B)$ is the most general possible "product" of an "algebras" as commonly understood in mathematics (namely, a bilinear composition law verifying the right and left distributive and scalars laws). In fact, the product $(A\hat{,}B)$ consists of the most general possible combination of Lie and Jordan products, thus admitting as particular cases *all* known algebras, and related physical-chemical models, including: associative algebras, Lie algebras, Jordan algebras, alternative algebras, supersymmetric algebras, Kac-Moody algebras, *etc.*

Despite their unquestionable mathematical beauty, theories (1.11) and (1.12) possess the following catastrophic physical inconsistencies [4]:

Theorem 1.2: *All theories possessing a nonunitary time evolution formulated on conventional Hilbert spaces $\mathcal{H}$ over conventional fields of complex numbers $\mathbb{C}(c, +, \times)$ do not admit consistent physical applications because:*

1) They do not possess invariant units of time, space, energy, etc., thus lacking physically meaningful application to experimental measurements;

2) They do not conserve Hermiticity in time, thus lacking physically meaningful observables;

3) They do not possess unique and invariant numerical predictions;

4) They generally violate probability and causality laws; and

5) They violate the basic axioms of Einstein's special relativity.

A comprehensive treatment is presented in report [4g]. We here recall only the following main points.

The basic geometric units of physical and chemical theories are not abstract mathematical notions, because they embody the most fundamental quantities, the basic units. For instance, the basic unit of the three-dimensional Euclidean space $I = \text{Diag.}(1, 1, 1)$ in actuality represents in a dimensionless form the basic units of length used per each direction of space, $I = \text{Diag.}(1 \text{ cm}, 1 \text{ cm}, 1 \text{ cm})$. But nonunitary transforms do not preserve, by conception, the geometric unit, $U \times U^\dagger \neq I$, e.g., for the Euclidean unit we may have

$$\begin{aligned} I = \text{Diag.}(1 \text{ cm}, 1 \text{ cm}, 1 \text{ cm}) \rightarrow \\ \rightarrow I' = U \times I \times U^\dagger = \text{Diag.}(7 \text{ cm}, 9.2 \text{ cm}, -2 \text{ cm}). \end{aligned} \tag{1.13}$$

The lack of invariance of the basic units for all nonunitary theories, and their consequential lack of applicability to measurements, then follows.

Similarly, it is easy to prove that the condition of Hermiticity at the initial time, $(\langle\phi| \times H^\dagger) \times |\psi\rangle \equiv \langle\phi| \times (H \times |\psi\rangle)$ is violated at subsequent times for all theories with nonunitary structures when formulated on $\mathcal{H}$ over $\mathbb{C}$. This additional catastrophic inconsistency (called *Lopez's lemma* [4b]), can be expressed by

$$\begin{aligned} &[\langle\psi| \times U^\dagger \times (U \times U^\dagger)^{-1} \times U \times H \times U^\dagger] \times U|\psi\rangle = \\ &= \langle\psi| \times U^\dagger \times [(U \times H \times U^\dagger) \times (U \times U^\dagger)^{-1} \times U|\psi\rangle] = \\ &= (\langle\hat{\psi} \times T \times \hat{H}^\dagger) \times |\hat{\psi}\rangle = \langle\hat{\psi}| \times (\hat{H} \times T \times |\hat{\psi}\rangle), \\ &|\hat{\psi}\rangle = U \times |\psi\rangle, \quad \hat{T} = (U \times U^\dagger)^{-1} = T^\dagger, \\ &\hat{H}^\dagger = T^{-1} \times \hat{H} \times T \neq H. \end{aligned} \tag{1.14}$$

As a result, all nonunitary theories do not admit physically meaningful observables.

Assuming that the preceding inconsistencies can be by-passed with some adulteration or manipulation, nonunitary theories still remain with additional catastrophic inconsistencies. An additional one is the lack of invariance of numerical predictions.

Suppose that the considered nonunitary theory is such that $U \times U^\dagger_{[t=0]} = 1$ and $U \times U^\dagger_{[t=15min]} = 15$, and that the theory originally predicts at time $t = 0$, say, an eigenvalue of 2 eV. It is then easy to see that the same theory predicts under the same conditions the *different* eigenvalue 30 eV at $t = 15$ minutes, thus having no physical value of any type. In fact, we have

$$U \times U^\dagger[t = 0] = I, \quad U \times U^\dagger[t = 15 \text{ min}] = 15, \tag{1.15a}$$

$$U \times H \times |\psi\rangle = (U \times H \times U^\dagger) \times (U \times U^\dagger)^{-1} \times (U \times |\psi\rangle) =$$
$$= H' \times T \times |\hat{\psi}\rangle = U \times E \times |\psi\rangle = E \times (U \times |\psi\rangle) = E \times |\hat{\psi}\rangle, \tag{1.15b}$$
$$H' = U \times H \times U^\dagger, \quad T = (U \times U^\dagger)^{-1},$$

$$H'|\hat{\psi}\rangle\,|_{t=0} = 2 \text{ eV} \times |\hat{\psi}\rangle\,|_{t=0}, \quad T = 1\,|_{t=0}, \tag{1.15c}$$

$$H' \times |\hat{\psi}\rangle\,|_{t=15\,\text{m}} = 2 \text{ eV} \times (U \times U^\dagger) \times |\hat{\psi}\rangle\,|_{t=15\,\text{m}} = \\ = 30 \text{ eV} \times |\hat{\psi}\rangle\,|_{t=15\,\text{m}}\,. \tag{1.15d}$$

Probability and causality laws are notoriously based on the unitary character of time evolution and the invariant decomposition of the unit. Their violation for all nonunitary theories is then evident. It is an instructive exercise for the reader interested in learning hadronic chemistry to identify a specific example of nonunitary transforms for which the effect *precedes* the cause.

The violation of the basic axioms of Einstein's special relativity by all nonunitary theories is so evident as to require no comment.

The most fundamental drawback of the theories considered is their *lack of invariance.* This can be best illustrated with the lack of invariance of the general Lie-admissible laws (1.12), which admit as a particular case all other nonunitary theories. In fact, under nonunitary transforms $U \times U^\dagger \neq I$ we have, e.g., the lack of invariance of the Lie-admissible brackets,

$$U \times (A \hat{,} B) \times U^\dagger = U \times (A < B - B > A) \times U^\dagger = (U \times A \times U^\dagger) \times \\ \times [(U \times U^{-1}) \times (U \times P \times U^\dagger) \times (U \times U^\dagger)^{-1}] \times (U \times B \times U^\dagger) - \\ -(U \times B \times U^\dagger) \times [(U \times U^{-1}) \times (U \times Q \times U^\dagger) \times (U \times U^\dagger)^{-1}] \times \\ \times (U \times A \times U^\dagger) = A' \times P' \times B' - B' \times Q' \times A' = \\ = A' <' B' - B' >' A'. \tag{1.16}$$

The above rules confirm the preservation of a Lie-admissible structure under the most general possible transforms, thus confirming the direct universality of laws (1.12) as per Theorem 1.1. The point is that *the formulations are not invariant* because

$$P' = (U \times U^{-1}) \times (U \times Q \times U^\dagger) \times (U \times U^\dagger)^{-1} \neq P, \tag{1.17a}$$

$$Q' = (U \times U^{-1}) \times (U \times Q \times U^\dagger) \times (U \times U^\dagger)^{-1} \neq Q. \tag{1.17b}$$

The invariance of quantum mechanics follows from the fact that the associative product "$\times$" is not changed in Eqs. (1.3c). By comparison, the generalized products "$<$" and "$>$" of broader theories (1.12) are changed by the transformations, including the time evolution of the theory itself. The same results also holds for other nonunitary theories, as the reader is encouraged to verify.

It is important to know that corresponding catastrophic inconsistencies exist for all *classical* counterparts of nonunitary theories, namely, *classical noncanonical formulations* [4g]. The lack of canonical character of the time evolution then implies the lack of conservation of the basic units, the lack of unique and invariant numerical predictions, *etc.*

The mathematical inconsistencies of nonunitary and noncanonical theories are equally catastrophic. All these theories are formulated *over a given field of numbers.* Whenever the theory is either noncanonical at the classical level or nonunitary at the operator level, the first noninvariance is that of the basic unit, that is, *the basic unit of the underlying field.*

The lack of conservation of the unit then implies the loss of the basic field of numbers in which the theory is constructed. It then follows that the entire axiomatic structure of the theory as formulated at the initial time, is no longer applicable at subsequent times.

For instance, the formulation of a nonunitary theory on a conventional Hilbert space has no mathematical sense because that space is defined over the field of complex numbers. The loss of the latter under nonunitary transforms then implies the loss of the former.

Similarly, the formulation of a classical noncanonical theory over the Euclidean space is afflicted by catastrophic mathematical inconsistencies, because, for evident reason of axiomatic consistencies, the Euclidean space requires an invariant field of real numbers.

In short, the lack of invariance of the fundamental unit under noncanonical or nonunitary time evolutions implies the collapse of the entire mathematical formulation, without known exception.

Theorem 1.3: *All classical theories with a noncanonical time evolutions formulated on conventional metric spaces over conventional fields do not possess a consistent mathematical structure because of the lack of invariance in time of the unit, fields, vector and metric spaces, etc.*

The best illustration is given by the Lie-admissible reformulation of the historical analytic equations, *Lagrange's and Hamilton's equations with external terms* [14] (see monograph [6c, 6d]), which are "directly universal" for all conceivable classical equations of motion. Yet, they do not possess an invariant structure, and, as formulated above, have no known physical application.

The same fate occurs for the *Birkhoffian generalization of Hamiltonian mechanics* presented in monograph [6b]. In fact, such a mechanics is a *noncanonical* image of the conventional Hamiltonian mechanics of the contemporary literature (that with the truncated external terms). As such, they are noninvariant under their own action with consequential activation of the inconsistencies of Theorem 1.3.

The reader should be aware that the above physical inconsistencies apply not only for Eqs. (1.12) but also for a large number of generalized theories, as expected from the direct universality of the former. It is of the essence to identify in the following at least the most important representative cases of physically inconsistent theories, to prevent their possible application in chemistry (see Ref. [4g] for detail and literature):

1) Dissipative nuclear theories [12a] represented via an imaginary potential in nonhermitean Hamiltonians, $H = H_0 = iV \neq H^\dagger$, and consequential violation of structures (1.1). These theories *lose all algebras in the brackets of their time evolution (requiring a bilinear product) in favor of a triple system,* $i\, dA/dt = A \times H - H^\dagger \times A = [A, H, H^\dagger]$. This implies the complete loss of notions such as "protons and neutrons" as conventionally understood, trivially, because their definition (*e.g.*, for the spin) mandates the presence of a consistent algebra in the brackets of the time evolution.

2) Statistical theories with an external collision term C [12b] and equations of motion of the density $i\, d\rho/dt = \rho \odot H = [\rho, H] + C$, $H = H^\dagger$, which violate the conditions for the product $\rho \odot H$ to characterize any algebra, as well as the existence of exponentiation law (1.1a), let alone the conditions of unitarity.

3) The so-called "q-deformations" of the Lie product $A \times B - q \times B \times A$, where q is a non-null scalar [12c], which deformations are a trivial particular case of Santilli's (p, q)-deformations (1.11).

4) The so-called "k-deformations" [12d] which are a relativistic version of the q-deformations, thus also being a particular case of general structures (1.11).

5) The so-called "star deformations" [12e] with generalized associative product $A \star B = A \times T \times B$, T fixed, and generalized Lie product $A \star B - B \star A$, which violate structures (1.1) and *coincide* with Santilli's Lie-isotopic laws.

6) Deformed creation-annihilation operators theories [12f, 12g].

7) Nonunitary statistical theories [12h].

8) Irreversible black holes dynamics with Santilli's Lie-admissible structure (1.16) [12i].

9) Noncanonical time theories [12j].

10) The so-called supersymmetric theories [12k] with product $(A, B) = [A, B] + \{A, B\} = (A \times B - B \times A) + (A \times B + B \times A)$ which are an evident particular case of Santilli's Lie-admissible product (1.12b) with $T = 1$ and $W = I$.

11) The Kac-Moody superalgebras [12l] which are also a particular case of Santilli's Lie-admissible product (1.12b) with $T = I$ and W a phase factor.

12) String theories [12m] which have a supersymmetric structure with consequential time evolution, which is non-canonical at the classical level and nonunitary at the operator level.

13) Grand unified theories on a curved manifold [12m] whose time evolution is also noncanonical at the classical level and nonunitary at the operator level.

All the above theories have a nonunitary structure formulated via conventional mathematics and therefore have the catastrophic physical inconsistencies of Theorem 1.2.

Additional generalized theories were attempted via *the relaxation of the linear character of quantum mechanics* [4]. These theories are essentially based on eigenvalues equations with the structure $H(t, r, p, |\psi\rangle) \times |\psi\rangle = E \times |\psi\rangle$ (i.e., H depends on the wavefunction).

Even though mathematically impeccable and possessing a seemingly unitary time evolution, these theories also possess rather serious physical drawbacks, such as: they violate the superposition principle of quantum mechanics, thus being strictly inapplicable to composite systems such as a molecule; they violate the fundamental Mackay imprimitivity theorem which is necessary for the applicability of Galilei's or Einstein's relativities; and possess other drawbacks [4] so serious as to prevent any application.

Yet another type of broader theory is *Weinberg's theory with a nonassociative envelope* [12n]. This is a theory in which the associative enveloping product $A \times B$ of the Lie product $[A, B] = A \times B - B \times A$ is replaced with a certain *nonassociative* product $A \odot B$, $\quad A \odot (B \odot C) \neq (A \odot B) \odot C$.

The latter theory violates Okubo's No-Quantization Theorem [4a], which prohibits the use of nonassociative envelopes because of catastrophic physical consequences, such as the loss of equivalence between the Schrödinger and Heisenberg representations (the former remains associative, while the latter becomes nonassociative, thus resulting in inequivalence).

Weinberg's theory also has additional serious inconsistencies, studied in detail in Ref. [4d], such as: the complete absence of any unit at all, let alone its lack of invariance (because of the lack of a nontrivial quantity

E such that $E \odot A \equiv A \odot E = A, \quad \forall A$) with consequential impossible applications to measurements; the violation of Galilei's and Einstein's relativities (because of the loss of exponentiation); and others.

Several authors also attempted *the relaxation of the local-differential character of quantum mechanics* [4]. The attempts were done via the addition of "integral potentials" $V = \int d\tau \Gamma(\tau, \dots)$ in a Hamiltonian.

These theories are structurally flawed on both mathematical and physical grounds. In fact, the nonlocal extension is elaborated via the conventional mathematics of quantum mechanics which, beginning with its topology, is strictly local-differential, thus implying fundamental *mathematical* inconsistencies. Nonlocal interactions are in general of contact type, for which the notion of a potential has no physical meaning, thus resulting in rather serious *physical* inconsistencies.

In conclusion, by the early 1980's Santilli had identified classical and operator generalized theories [5, 6] which are directly universal in their fields, with a plethora of simpler versions by various other authors. However, all these theories subsequently resulted in having no physical meaning because they are noninvariant when elaborated with conventional mathematics.

8. Hadronic Mechanics

While at Harvard University under DOE support, Santilli proposed in memoirs [5c, 5d, 5e] of 1978 the construction of the *hadronic generalization of quantum mechanics*, or *hadronic mechanics* for short, via the following branches:

I) Isotopic branch of hadronic mechanics, which is characterized by the following generalization of the basic equations of Sect. 1.1 on $\mathcal{H}$ over $\mathbb{C}$, first proposed in Ref. [5e], Eqs. (4.15.49),

$$A(t) = U \times A(0) \times U^\dagger = e^{i \times H \times T \times t} \times A(0) \times e^{-i \times t \times T \times H}, \tag{1.18a}$$

$$i \frac{dA}{dt} = [A \hat{,} H]_{\text{operator}} = A \hat{\times} H - H \hat{\times} A = A \times T \times H - H \times T \times A, \tag{1.18b}$$

$$T = T^\dagger > 0, \quad U \times U^\dagger \neq I, \tag{1.18c}$$

where T is a nonsingular operator (or matrix), with classical counterparts [5d]

$$A(t) = e^{t \times T_i^j \times (\partial H / \partial r^i) \times (\partial / \partial p_j)} \times A(0) \times e^{-t \times T_i^j \times (\partial / \partial r^i) \times (\partial H / \partial r_j)}, \tag{1.19a}$$

$$\frac{dr^k}{dt} = T_i^k(t, r, p) \times \frac{\partial H(t, r, p)}{\partial p_i}, \quad \frac{dp_k}{dt} = -T_k^i(t, r, p) \times \frac{\partial H(t, r, p)}{\partial r^i}, \tag{1.19b}$$

$$\frac{dA}{dt} = [a\hat{,}H]_{\text{classical}} = \frac{\partial A}{\partial r^i} \times T^i_j \times \frac{\partial H}{\partial p_j} - \frac{\partial H}{\partial r^i} \times T^i_j \times \frac{\partial A}{\partial p_j}, \qquad (1.19c)$$

where the term "isotopic" was suggested [5d] in its Greek meaning to indicate the "axiom-preserving character" of the new theory. In fact, the new brackets $[A\hat{,}B] = A\hat{\times}B - B\hat{\times}A = A \times T \times B - B \times T \times A$ do indeed verify the Lie algebras axioms, although in a generalized form.

A most dominant feature of theory (1.18) is that of being *nonunitary*, Eq. (1.18c), due to the general lack of commutativity between H and T. Therefore, the theory is outside the class of equivalence of quantum mechanics.

Equivalently, theory (1.18) admits nonpotential-nonhamiltonian forces represented by the isotopic element T, which is indeed nonhamiltonian in nature. In particular, besides the condition of Hermiticity and positive-definiteness, $T = T^\dagger > 0$, Eqs. (1.18) leave unrestricted the functional dependence of the isotopic operator, which can be of the form $T = T(r, v, \psi, \partial\psi, \dots)$. Therefore, the nonpotential interactions represented by T can be nonlinear in the wavefunctions, as well as nonlocal-integral, as desired.

Another important feature of theory (1.18) is that of being *structurally reversible, that is, reversible for reversible Hamiltonian*, as it is the case for quantum mechanics. This is due to conjugation in time via the operation of Hermiticity, which voids possible attempts at representing irreversibility via time-dependent isotopic elements $T = T(t, \dots) \neq T(-t, \dots)$.

Still another important feature of the above theory is that of characterizing conventional total conservation laws, evidently in view of the totally antisymmetric character of the fundamental brackets, e.g.,

$$i \times \frac{dH}{dt} = H\hat{\times}H - H\hat{\times}H = H \times T \times H - H \times T \times H \equiv 0. \qquad (1.20)$$

Therefore, the isotopic branch of hadronic mechanics is particularly suited to represent *closed-isolated reversible systems with linear and nonlinear, local and nonlocal, potential and nonpotential internal forces.*

It is evident that theories (1.18) and (1.19) are promising for a deeper understanding of molecular structures because, in addition to all conventional quantum chemical features represented with the Hamiltonian H, the theories admit a quantitative representation of the conditions of deep mutual overlappings of the wavepackets of valence electrons via isotopic elements we shall study later.

II) Genotopic branch of hadronic mechanics, which is based on the following further generalization of the dynamical equations of

Sect. 1.1, first proposed in Ref. [5e], Eqs. (4.15.34) and (4.18.11):

$$A(t) = U \times A(0) \times U^\dagger = e^{i \times H \times Q \times t} \times A(0) \times e^{-i \times t \times P \times H}, \quad P = Q^\dagger, \quad (1.21a)$$

$$\begin{aligned} i\tfrac{dA}{dt} &= (A \hat{,} B)_{\text{operator}} = A < B - B > A = \\ &= A \times P \times H - H \times Q \times A = \\ &= (A \times T \times B - B \times T \times A) + (A \times W \times B + B \times W \times A) = \\ &= [A \hat{,} B] + \{A \hat{,} B\}, \end{aligned} \quad (1.21b)$$

$$P \neq Q, \quad P = Q^\dagger, \quad U \times U^\dagger \neq I, \quad (1.21c)$$

where $P = T + W$, $Q = T - W$ and $P + Q$ are nonsingular nonhermitean operators (or matrices), with classical counterparts [5d, 5f],

$$A(t) = e^{t \times (\partial H/\partial p_j) \times P^i_j \times (\partial/\partial r^i)} \times A(0) \times e^{-t \times (\partial/\partial p_j) \times Q^i_j \times (\partial H/\partial r^i)}, \quad (1.22a)$$

$$\frac{dr^k}{dt} = P^k_i(t, r, p) \times \frac{\partial H(t, r, p)}{\partial p_i}, \quad \frac{dp_k}{dt} = Q^i_k(t, r, p) \times \frac{\partial H(t, r, p)}{\partial r^i}, \quad (1.22b)$$

$$\frac{dA}{dt} = (A \hat{,} B)_{\text{classical}} = \frac{\partial A}{\partial r^i} \times P^i_j \times \frac{\partial H}{\partial p_j} - \frac{\partial H}{\partial r^i} \times Q^i_j \times \frac{\partial A}{\partial p_j}, \quad (1.22c)$$

under the condition that the antisymmetric brackets attached to the generalized brackets, $[A \hat{,} B] = (A \hat{,} B) - (B \hat{,} A)$, are Lie.

The term "genotopic" was suggested [5d] from its Greek meaning to indicate the "axiom-inducing character" of the theory. In fact, the original Lie character is now lost in favor of the covering axioms of Lie-admissibility.

It is evident that theory (1.21) is nonunitary as the simpler version (1.18), although the theory is now *structurally irreversibility*, that is, it is irreversible for all reversible Hamiltonians, precisely as desired. Moreover, the irreversibility is represented by the nonhermiticity of the genotopic elements P and Q, rather than by the Hamiltonian, also as desired.

Another dominant characteristic of the above broader theory is that of representing *time-rate-of-variations* of physical characteristics (rather than the usual conservation laws), evidently in view of the fact that the dynamical brackets are no longer antisymmetric by conception. We then have the time-rate-of-variation of the total energy

$$i \times \frac{dH}{dt} = (H \hat{,} H) = H < H - H > H = H \times P \times H - H \times Q \times H \neq 0. \quad (1.23)$$

It is evident that theories (1.21) and (1.22) are promising for a deeper understanding of irreversible processes and chemical reactions because, in addition to including all conventional quantum chemical treatment via the usual Hamiltonian H, they admit two additional genotopic elements P and Q suitable to represent irreversibility.

III) Hyperstructural branch of hadronic mechanics, which is based on the following multi-valued equations first proposed by Santilli in the more recent memoir [5k]

$$A(t) = U \times A(0) \times U^\dagger = e^{i \times H \times \{Q\} \times t} \times A(0) \times e^{-i \times t \times \{P\} \times H}, \quad P = Q^\dagger, \tag{1.24a}$$

$$\begin{aligned} i\frac{dA}{dt} &= (A\hat{,}H)_{\text{operator}} = A < H - H > A = \\ &= A \times \{P\} \times H - H \times \{Q\} \times A = \\ &= (A \times \{T\} \times H - H \times \{T\} \times A) + \\ &+ (A \times \{W\} \times H + H \times \{W\} \times A) = \\ &= [A\hat{,}H] + \{A\hat{,}H\}, \end{aligned} \tag{1.24b}$$

$$\{P\} = \{P_1, P_2, P_3, \dots\} \neq \{Q\} = \{Q_1, Q_2, Q_3, \dots\}, \tag{1.24c}$$

where $\{P\} = \{T\} + \{W\}$, $\{Q\} = \{T\} - \{W\}$ and $\{P \pm Q\}$ are now non-singular sets of operators (or matrices), with classical counterparts [5k]

$$A(t) = e^{t \times (\partial H/\partial p_j) \times \{P\}^i_j \times (\partial/\partial r^i)} \times \tag{1.25a}$$

$$\times A(0) \times e^{-t \times (\partial/\partial p_j) \times \{Q\}^i_j \times (\partial H/\partial r^i)},$$

$$\frac{dr^k}{dt} = \{P\}^k_i(t, r, p) \times \frac{\partial H(t, r, p)}{\partial p_i}, \quad \frac{dp_k}{dt} = -\{Q\}^i_k(t, r, p) \times \frac{\partial H(t, r, p)}{\partial r^i}, \tag{1.25b}$$

$$\frac{dA}{dt} = (A\hat{,}H)_{\text{classical}} = \frac{\partial A}{\partial r^i} \times \{P\}^i_j \times \frac{\partial H}{\partial p_j} - \frac{\partial H}{\partial r^i} \times \{Q\}^i_j \times \frac{\partial A}{\partial p_j}. \tag{1.25c}$$

under the condition again that the attached antisymmetric brackets are of Lie type although in a multi-valued version.

The term "hyperstructural" was submitted [5k] in the contemporary mathematical meaning of being "multi-valued." The above third broader branch of hadronic mechanics evidently preserves the nonunitary irreversible character of the genotopic theory, but adds the vast degrees of freedom related to multi-valuedness, thus permitting quantitative studies of biological structures, as we shall see.

The achievement of an invariant formulation of nonlinear, nonlocal, and nonpotential interactions required the construction of *new mathematics* of progressively increasing complexity submitted by Santilli [5d, 5k] under the names of *iso-, geno-, and hyper-mathematics.* These new mathematics will be reviewed in the next chapter.

The main motivation for the new mathematics is essentially the following. The basic unit I of quantum mechanics and chemistry is the unit of the universal enveloping operator algebra with conventional product $A \times B(= AB), I \times A = A \times I = A$, for all possible elements A.

When the associative product is lifted in the form $A \hat{\times} B = A \times T \times B$, with T a fixed operator (or matrix), it is evident that I is no longer the unit of the theory, $I \hat{\times} A = T \times A \neq A$, thus resulting in the catastrophic physical and mathematical inconsistencies of the preceding section whenever conventional mathematics is used.

The achievement of invariance requires, first, the identification of the new generalized unit of the theory, and, then, the reconstruction of the entire mathematics with respect to that unit, thus implying new numbers and fields, new vector and Hilbert spaces, new algebras and geometries, *etc.*

By the mid 1990's the above new mathematics had been sufficiently developed. However, the various branches of hadronic mechanics still lacked the crucial invariance. After laborious studies, the problem resulted in being where one would expect it the least, in the *ordinary differential* calculus, which resulted in depending on the assumed basic unit, evidently, when the latter is no longer a trivial constant.

Mathematical maturity was finally reached only in report [5k] of 1996, which achieved the necessary liftings of the ordinary differential calculus. The invariant formulation of hadronic mechanics was then reached in the subsequent memoirs [5l, 5n].The reader should therefore be aware that *all publications on hadronic mechanics prior to Refs. [5k, 5l, 5n] of 1996-1997 are physically inconsistent because they lack the crucial invariance necessary for consistent applications.*

In summary, hadronic mechanics is an image of quantum mechanics formulated via the novel iso-, geno- and hyper-mathematics, and their isoduals, with corresponding *iso-, geno-, and hyper-mechanics* for the representation of *single-valued reversible, single-valued irreversible, and multi-valued irreversible systems*, respectively.

It should be indicated that *hadronic mechanics coincides with quantum mechanics at the abstract, realization-free level in all its iso-, geno-, and hyper-branches.* In fact, hadronic mechanics merely provides broader realizations of the conventional quantum axioms of Sect. 1.1.

In particular, *hadronic mechanics preserves all conventional laws and principles of quantum mechanics,* such as Pauli's exclusion principle, Heisenberg's uncertainties, causality, probability laws, *etc.*

Also, *hadronic mechanics constitutes a 'completion' of quantum mechanics much along the celebrated argument by Einstein-Podolsky-Rosen [3a], and provides a concrete and explicit realization of the theory of "hidden variables" [3c]*, which hidden variables are realized precisely by the isotopic element $\lambda = T$. The latter aspect is studied in detail in Ref. [3e] and the results are omitted here for brevity.

We should finally mention that hadronic mechanics admits additional novel mathematics called *isodual mathematics* [5k] for the invariant formulation of *antimatter* at all levels [5o]. It then follows that *hadronic mechanics admits eight classical branches (classical Hamiltonian, iso-, geno- and hypermechanics for the description of matter plus their isoduals for the description of antimatter), as well as eight corresponding operator branches with eight, corresponding interconnecting maps [called conventional, iso-, geno- and hyper-quantization and their isoduals].*

Since in this monograph we shall solely consider *matter*, no isodual mathematics and related theory will be considered for brevity. The reader should however be aware that isodual mathematics is emerging as being particularly useful for quantitative representations of biological events, such as bifurcations. The study of isodual theories is therefore recommended to researcher in biology.

Hadronic mechanics nowadays possesses applications and experimental verifications in virtually all branches of sciences, including particle physics, nuclear physics, astrophysics, superconductivity, chemistry, gravitation, and cosmology. Memoir [6k] provides a comprehensive review. A few representative applications are indicated in the following Figs. 1.9–1.20 without treatment.

In inspecting these results the reader should keep in mind that a primary objective of hadronic mechanics is that of achieving new structure models of hadrons, nuclei and molecules as a necessary condition for the prediction and development of new clean energies and fuels, since the identification of all possible energies with existing models has been exhausted long ago (see Sect. 7.7). The new structure model of molecules is the primary scope of this monograph. The new structure models of hadrons and nuclei have been identified elsewhere (see the recent memoir [6k]), and they cannot be reviewed here in technical details to prevent a prohibitive length of this monograph.

In order to minimize misrepresentations in the inspection of the figures below, it is important to know that hadronic mechanics assumes as exact the SU(3)-color models of hadronic classification, while it assumes that quarks are composite, thus admitting other *elementary* constituents.

In fact, on one side, the experimental verification of the Mendeleev-type SU(3)-color classification of hadrons into families is beyond scientific doubt. By contrast, the belief that quarks are the elementary constituents of hadrons is afflicted by a plethora of inconsistencies and unsolved problems. In the final analysis, the reader should remember that, by their very definition, quarks are purely mathematical representations of a purely mathematical unitary symmetry defined in the purely mathematical unitary space without any possibility of rigorous definition in our spacetime.

Moreover, the sole scientific possibility of defining physical masses is via the second Casimir invariant of the Poincaré symmetry. But the latter symmetry cannot admit quarks. Therefore, on rigorous scientific grounds, quark masses are purely mathematical parameters defined in the purely mathematical unitary space without any possibility of being defined in our spacetime.

Similarly, a truly scientific (exact) confinement of the un-observable hypothetical quarks has been shown to be impossible, contrary to the experimental evidence that quarks do not exist in an isolated, directly detectable form.

Under these premises, any firm belief that quarks exist in our spacetime is purely nonscientific.

On the contrary, the assumption that quarks are mathematical objects with a composite structure, as studied by hadronic mechanics, resolves all known insufficiencies of hadron physics, by achieving a harmonious symbiosis between the SU(3)-color classification of hadrons and the structure of each individual hadron of a given family with physical particles fully definable in our spacetime, generally identifiable in the spontaneous decay with the lowest mode.

Note that the above study is strictly prohibited by quantum mechanics for well known historical reasons. For instance, the smallest known hadron, the π^0, admits the spontaneous decay with the lowest model $\pi^0 \rightarrow e^+ + e^-$. Therefore, it is natural to assume that the π^0 is a bound state of one electron and one positron. Such an assumption is strictly prohibited by quantum mechanics because it would require "positive" binding energies, and imply other inconsistencies. However, under the assumption that, when members of a hadronic structure, electrons and positrons obey the broader physical laws and symmetries of hadronic mechanics, all inconsistencies are resolved [6k].

A similar case occurs for the neutron which decays spontaneously into a proton, an electron and an antineutrino. Therefore, it is natural to assume that the neutron is a bound state of one proton and one electron, as originally conceived by Rutherford (a "compressed hydrogen atom in the core of a star"). Again, such an assumption is strictly prohibited by quantum mechanics for numerous reasons known since Pauli's times. In fact, Rutherford's "hypothesis" of the neutron was accepted, although his "conception" of the neutron was rejected.However, all historical objections against the latter are resolved when the constituents obey the generalized laws and symmetries of hadronic mechanics [6k].

Rather than being a purely esoteric academic aspect, the issue as to whether the constituents of the neutron are the hypothetical quarks or the physical proton and electron has direct societal relevance because the neutron is the largest reservoir of clean energy available to mankind, as reviewed in Sect. 7.7. The conjecture that quarks are the elementary constituents of the neutron prohibits the practical utilization of such energy for various technical reasons [6k]. On the contrary, the assumption that the proton and electrons are physical constituents of the neutron (again, in a generalized form) directly permits the prediction and industrial utilization of said inextinguishable clean energy. The same quark conjecture prevents numerous additional new clean energies of nuclear type.

In summary, the conjecture that quarks are the "elementary" constituents of hadrons, unless subjected to a severe scientific scrutiny due to its plethora of inconsistencies, constitutes one of the biggest threats to contemporary society because it prohibits, or otherwise jeopardizes, the prediction and industrial development of new clean energies so much needed by mankind, the understanding being that the assumption of quarks as "composite" mathematical objects is fully acceptable on scientific and societal grounds.

9. Hadronic Superconductivity

An understanding of the isospecial relativity requires the knowledge that the new discipline and its underlying new mathematics are applicable in fields beyond nuclear physics, particle physics and astrophysics. Another field of applicability is superconductivity.

The Cooper (or electron) pairing in superconductivity is a physical system requiring *attractive* interactions among two *identical* electrons via the intermediate action of Cuprate ions, as confirmed by recent evidence of their tunnelling together and other data.

There is little doubt that the current quantum explanation via interactions mediated by "phonons" is afflicted by a number of open issues. In fact, phonons represent elementary heat excitations-oscillations in a crystal. In this case it is difficult to understand how photons can be propagated in vacuum from atom to atom in the fixed lattice of a crystal without collisions, and, even assuming that this is possible, it is difficult to understand how phonons can create an *attraction* between pairs of identical electrons.

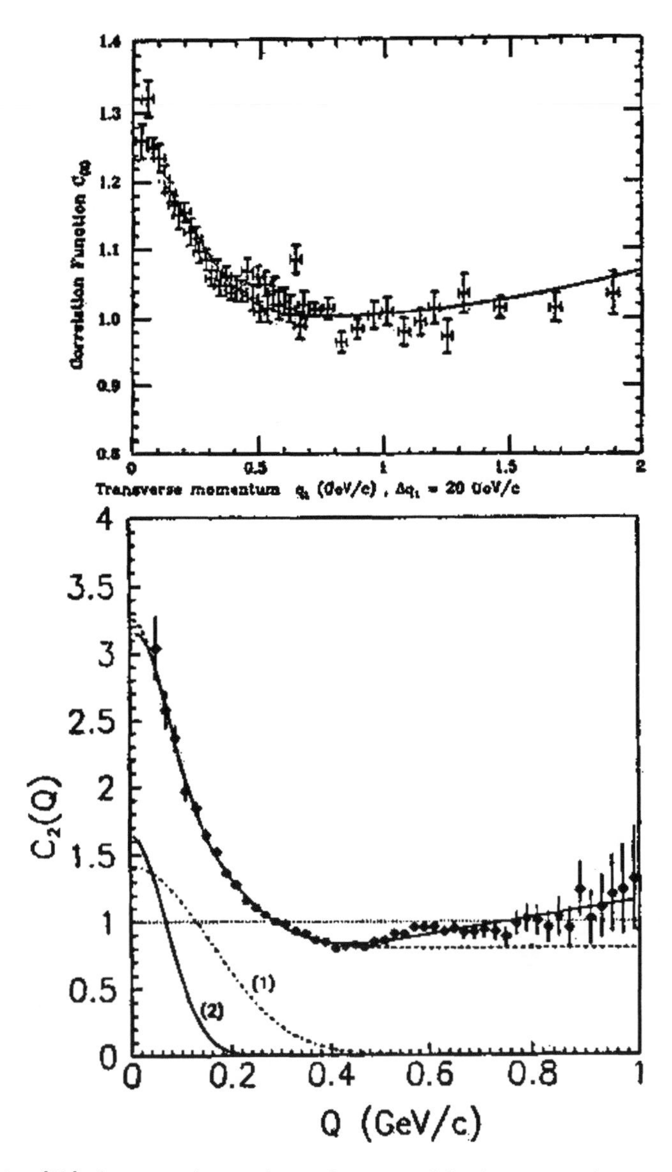

Figure 1.9. [6k] An experimental verification of hadronic mechanics in particle physics particularly important for this monograph, *the exact fits of the experiment data on the Bose-Einstein Correlation in proton-antiproton annihilation at high energy (upper fit) and at low energy (lower fit).* Quantum mechanics can only represent such experimental data via the introduction of *several arbitrary parameters* of unknown physical origin (called "chaoticity parameters"). The two-point correlation function itself is outside the capability of quantum mechanics, because of the need of off-diagonal elements in vacuum expectation values which admit none. The latter elements are the very origin of the correlation, and are fully admitted by hadronic expectation values for nondiagonal isotopic elements $\hat{T}$.

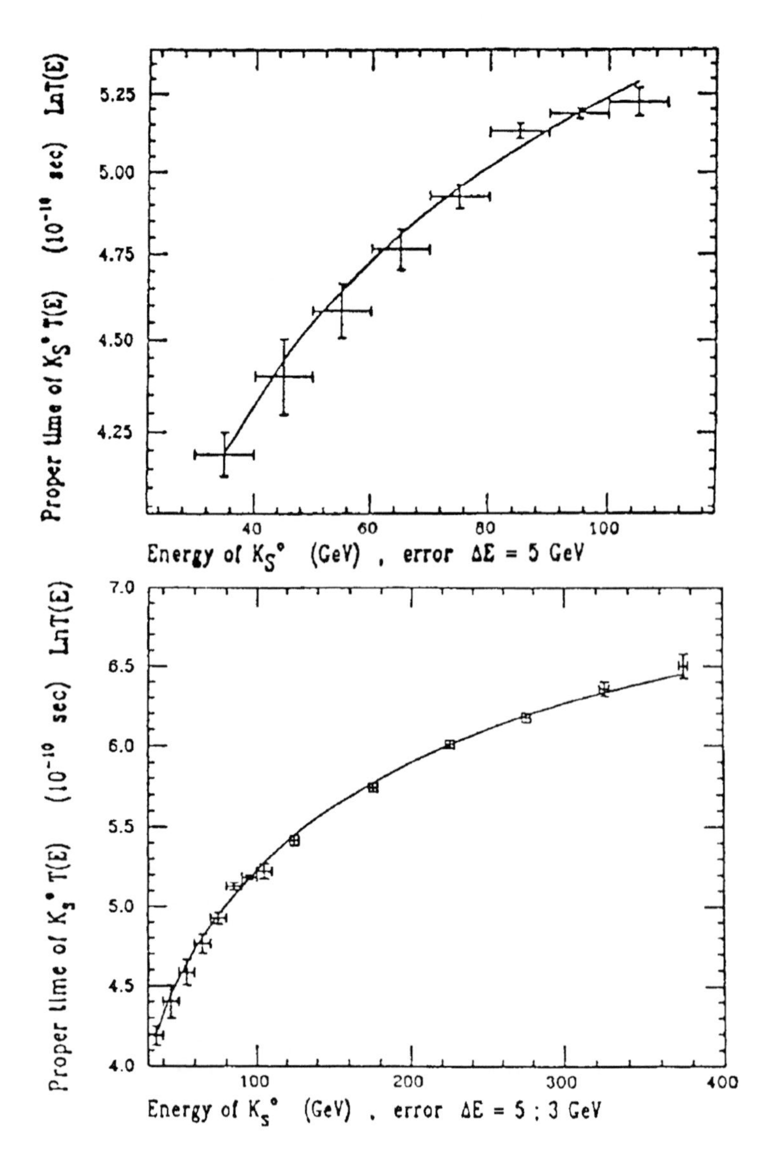

Figure 1.10. [6k] Another representative experimental verification of hadronic mechanics in particle physics, *the exact fit of the experimental data on the anomalous behavior of the meanlife of the (unstable) kaon with energy.* The upper figure represents the exact fit of the experimental data from 0 to 100 GeV, and the lower figure represents the exact fit from 0 to 400 GeV. This experimental verification is important for this monograph because it establishes the existence of nonlinear, nonlocal, and nonpotential effects in the interior of kaons, similar to those in the Bose-Einstein Correlation, which effects are at the foundations of the new isochemical model of valence bonds presented in this monograph.

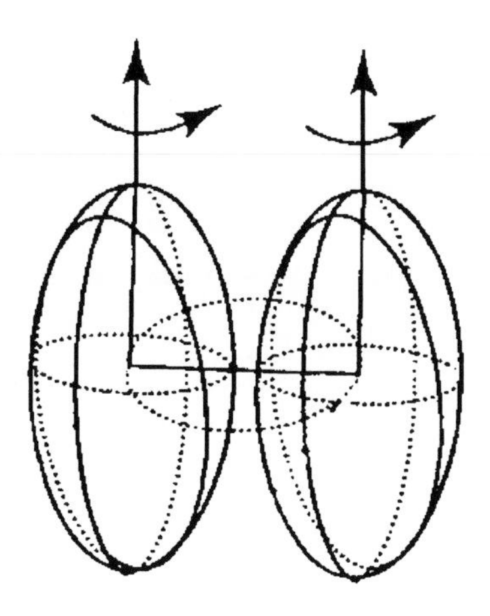

Figure 1.11. [6k] A schematic view of an experimental verification of hadronic mechanics in nuclear physics, *the first exact-numerical representation of all nuclear magnetic moments.* Quantum mechanics must represent protons and neutrons as perfectly rigid and perfectly spherical charge distributions, as a necessary condition not to violate a pillar of the theory, the rotational symmetry O(3). In so doing, however, quantum mechanics has been unable to reach an exact representation of nuclear magnetic moments despite large (and expensive) efforts conducted throughout the past century. In fact, about 1% is still missing in the representation of the magnetic moment of the deuteron after all possible corrections, while increasingly bigger (and somewhat embarrassing) deviations exist for bigger nuclei. Hadronic mechanics represents protons and neutrons as they are in the physical reality, extended, nonspherical, and deformable, spheroidal ellipsoids. Such a representation is achieved via the isounit $\hat{I} = \text{Diag.}(n_1^2, n_2^2, n_3^2, n_4^2)$, where n_1^2, n_2^2, n_3^2 represent the spheroidal semiaxes (normalized to 1 for the sphere), and n_4^2 permits the first representation of the density of nucleons (also normalized to 1 for the vacuum). The deformation of the charge distribution of nucleons then implies a corresponding alteration of their intrinsic magnetic moments when in a nuclear structure. Still in turn, the latter alteration permits an exact-numerical representation of all total nuclear magnetic moments (that of the deuteron is represented in this figure with about 1% prolate deformation). This verification of hadronic mechanics also implies a significant industrial-ecological application currently under study, the possible recycling of radioactive nuclear wastes via their stimulated decay.

At any rate, particle physics of the 20-th century has identified all possible particles. Yet, this branch of physics has no evidence of phonons, as well as of the interactions electron-phonon.

The Cooper pair (CP) is an excellent physical system to test the effectiveness of the isotopic methods at large. Comprehensive studies along these lines have been conducted by Animalu [10a] who has introduced a nonlinear, nonlocal, and nonhamiltonian theory called *Animalu's iso-*

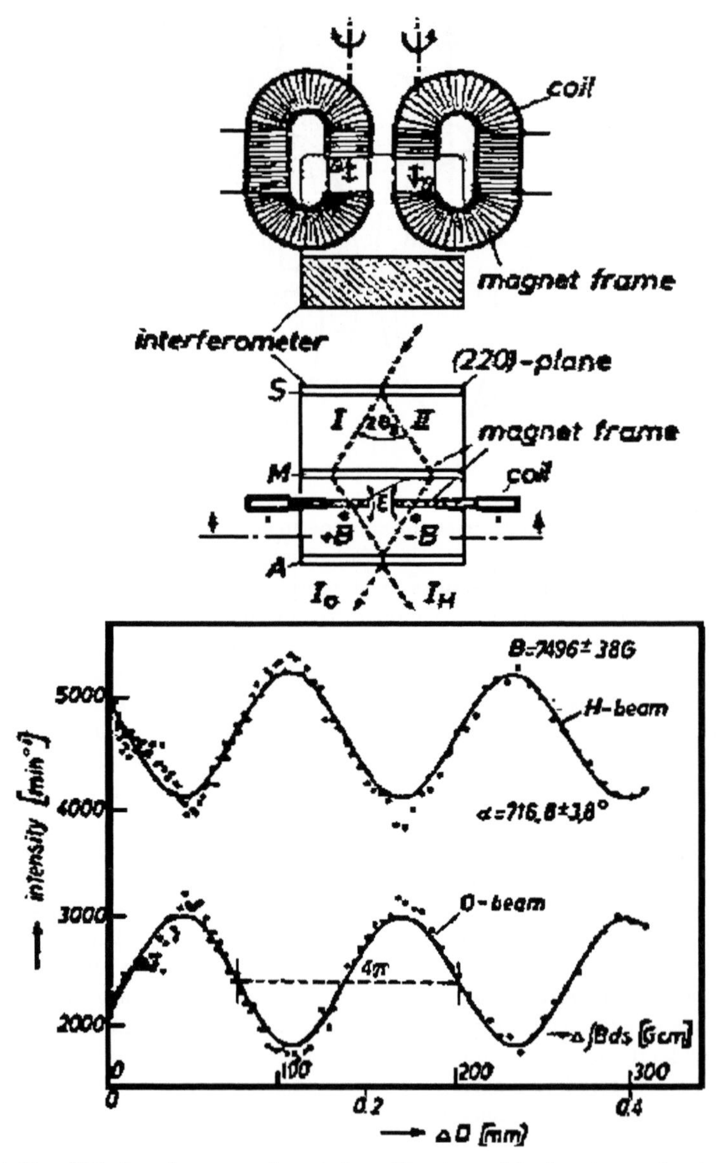

Figure 1.12. [6k] Another experimental verification of hadronic mechanics in nuclear physics, *the exact numerical representation of the experimental data on the 4π-spinorial symmetry of thermal neutrons when passing through intense external electromagnetic fields, such as those in the vicinity of nuclei.* This event is also related to the lack of perfect rigidity of neutrons. When exposed to intense external fields, neutrons experience a deformation of their shape, that alters their intrinsic magnetic moments. Consequently, thermal neutrons do not perform two complete spin flips as expected for conventional features. Again, hadronic mechanics permits an exact representation of the experimental data via the isounit of the preceding figure, while reconstructing as exact the spinorial symmetry SU(2).

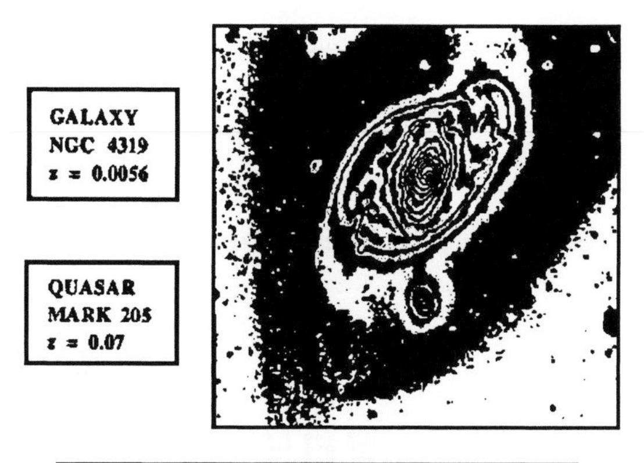

GAL.	ω'_1	QUASAR	B	$\hat{\omega}_2$
NGC	0.018	UB1	31.91	0.91
		BSO1	20.25	1.46
NGC 470	0.008	68	87.98	1.58
		68D	67.21	1.53
NGC 1073	0.004	BSO1	198.94	1.94
		BSO2	109.98	0.60
		RSO	176.73	1.40
NGC 3842	0.020	QSO1	14.51	0.34
		QSO2	29.75	0.95
		QSO3	41.85	2.20
NGC 4319	0.0056	MARK205	12.14	0.07
NGC 3067	0.0049	3C232	82.17	0.53

Figure 1.13. [6k] An example of experimental verification of hadronic mechanics in astrophysics, *the exact-numerical representation of the large differences in cosmological redshifts between certain quasars (such as Mark 205) and their associated galaxy (MGC 4319), when they are physically connected according to clear experimental evidence via gamma spectroscopy, in which case they should have the same redshift.* According to special relativity the speed of light is a universal constant. According to the covering isospecial relativity, the speed of light is a local variable depending on the characteristics of the medium in which it propagates. The astrophysical evidence here considered is, therefore, evidence that light exits the quasars already redshifted due to its decrease in their huge chromospheres. As a consequence, redshift is not necessarily a measure of the expansion of the universe, as currently believed.

superconductivity which is in remarkable agreement with experimental data, and possesses intriguing and novel predictive capacities.

We summarize below the recent presentation of the Cooper Pair (CP) by Animalu and Santilli [10b] at the 1995 Sanibel Symposium held in Florida represented with the symbol

$$\mathrm{CP} = (\hat{e}^-_\uparrow, \hat{e}^-_\downarrow)_{\mathrm{HM}}, \tag{1.26}$$

where $\hat{e}^-$ represents the isoelectron, that is, the ordinary electron when described via the isomechanics. For brevity we outline only the non-

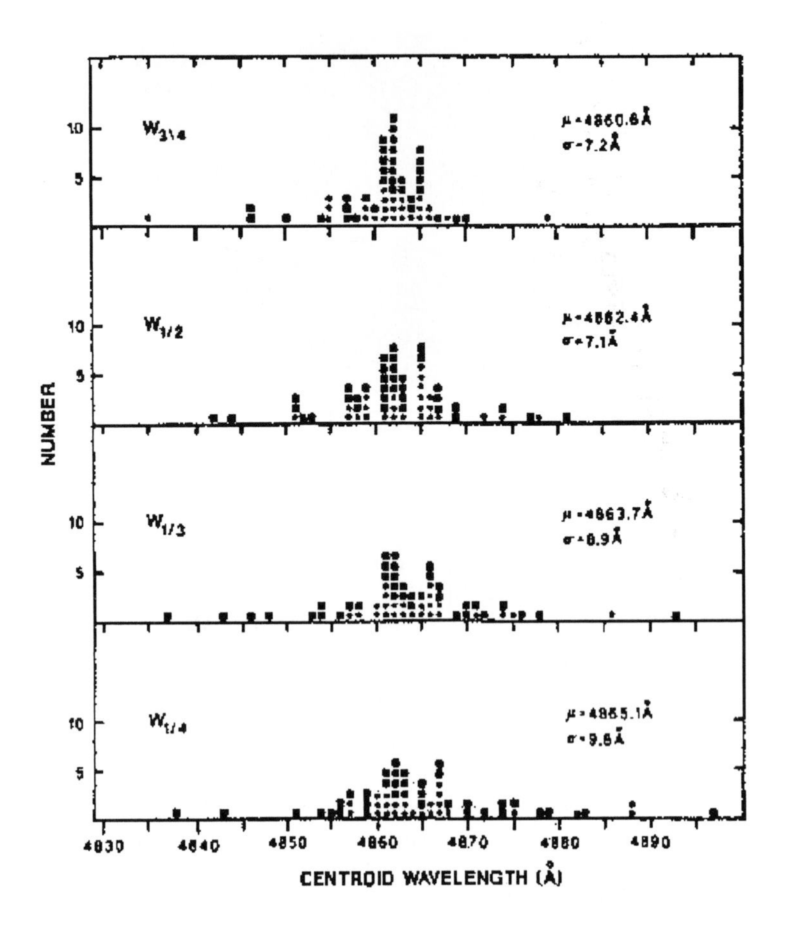

Figure 1.14. [6k] Another experimental verification in astrophysics of hadronic mechanics and its isospecial relativity: *the exact-numerical representation of quasars' internal redshifts and blueshifts.* In essence, for each given quasar, the cosmological redshift is not constant, but depends on the frequency of the light, with an internal redshift (further increase of the cosmological redshift) for the infrared spectrum, and an internal blueshift (decrease of the cosmological redshift) for the ultraviolet spectrum. These internal redshifts and blueshift are evidently incompatible with special relativity since they imply different speeds of light in the interior of the quasars chromospheres for different frequencies. By comparison, isospecial relativity predicts these events and represents them exactly. Other cosmological implications of hadronic mechanics and its underlying isominkowskian geometry are: the first cosmological model with an exact *symmetry*, the direct product of the Poincarè-Santilli isosymmetry for matter and its isodual for antimatter; the first inclusion of biological systems in cosmology; total, identically null characteristics of the universe (null total mass, null total time, *etc.*) for equal amounts of matter and antimatter (as a consequence of the isodual mathematics), thus without discontinuity at creation; the lack of need of the "missing mass"; and other intriguing implications.

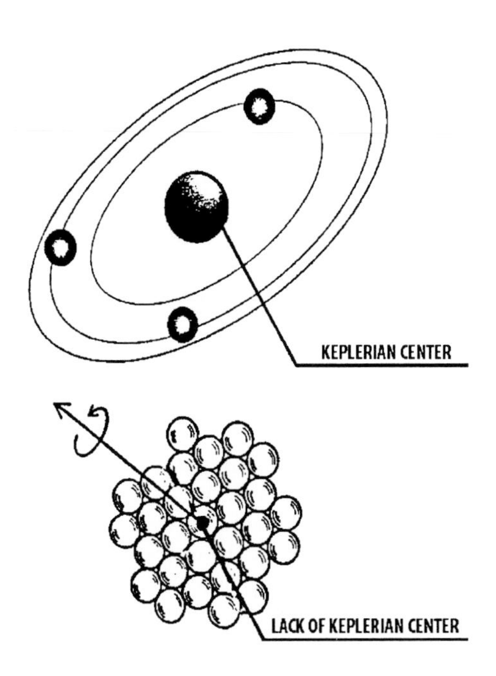

Figure 1.15. [6k] An illustration of another experimental verification of hadronic mechanics with industrial applications indicated in Chapiter 7, a *representation of the dramatic structural differences between the atomic structure, which is Keplerian with exact validity of the Poincaré symmetry and special relativity, versus the structures of hadrons, nuclei and molecules which have no Keplerian center, thus implying a necessary breaking of the Poincaré symmetry and the inapplicability of special relativity.* Hadronic mechanics has achieved an exact representation of hadronic and nuclear structures without Keplerian center via the isotopies of the Poincarè symmetry and special relativity. A non-Keplerian representation of molecular structures is presented in this monograph. In turn, such a representation is at the foundation of the new clean energies and fuels outlined in Chapter 7.

relativistic profile, and refer to the quoted literature for the relativistic extension.

Model (1.26) has a number of intriguing implications and possibilities. In fact, the model is essentially that for the hadronic bound state of the π^0 meson submitted by Santilli in the original proposal to build hadronic mechanics (Ref. [5e], Sect. 5)

$$\pi^0 = (\hat{e}^+_\uparrow, \hat{e}^-_\downarrow)_{\mathrm{HM}}, \tag{1.27}$$

because the emerging nonlocal-nonpotential interactions result in being attractive irrespective of whether the Coulomb interaction is attractive or repulsive. The existence of a consistent model for the structure CP

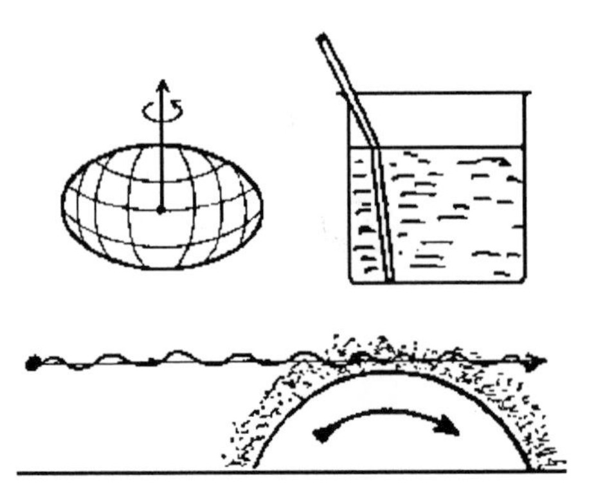

Figure 1.16. [6k] A schematic view of the capability by Santilli's isospecial relativity to represent the interior behavior of light as it occurs in the physical reality, with a *locally varying speed*. The representation is reached via a noncanonical transform at the classical level or a nonunitary transform at the operator level of *all* aspects of special relativity, beginning with the basic four-dimensional unit and then passing to numbers, spaces, *etc.* For the case of the Minkowski metric, the simplest possible isotopic lifting implies the map of the conventional metric representing the constant speed of light in vacuum, $m = \text{Diag.}(1,1,1,-c^2)$ into the isometric $\hat{m} = \text{Diag.}(1,1,1,-c^2/n^2)$, where n is the usual index of refraction. Therefore, isospecial relativity achieves a direct representation (that is, a representation via the geometry itself) of locally varying speeds of light, while admitting special relativity as a particular case for $n = 1$. More general liftings of the Minkowskian geometry permit the representation of bound systems without Keplerian center because generalized units represent contact nonpotential interactions, thus forcing the constituents to be in contact among each other, exactly as it occurs in the structure of hadrons, nuclei and molecules. Isospecial relativity is, therefore, the most fundamental theory underlying all studies presented in this monograph.

$= (\hat{e}^-_\uparrow, \hat{e}^-_\downarrow)_{\text{HM}}$, where the Coulomb interaction is repulsive, then provides additional support for the hadronic model $\pi^0 = (\hat{e}^+_\uparrow, \hat{e}^-_\downarrow)_{\text{HM}}$, where the Coulomb interaction is attractive.

Consider one electron with charge $-e$, spin up $\uparrow$ and wavefunction $\psi_\uparrow$ in the field of another electron with the same charge, spin down $\downarrow$ and wavefunction $\psi_\downarrow$ considered as *external*. Its Schrödinger equation is given by the familiar expression

$$H_{\text{Coul.}} \times \psi(t,r) = \left(\frac{1}{2m} p_k p^k + \frac{e^2}{r}\right) \times \psi_\uparrow(t,r) = E_0 \times \psi_\uparrow(t,r), \quad (1.28a)$$

$$p_k \times \psi_\uparrow(t,r) = -i \times \partial_k \psi_\uparrow(t,r), \quad (1.28b)$$

where m is the electron rest mass. The above equation and related wavefunction $\psi_\uparrow(t,r)$ represent *repulsion*, as well known. We are interested

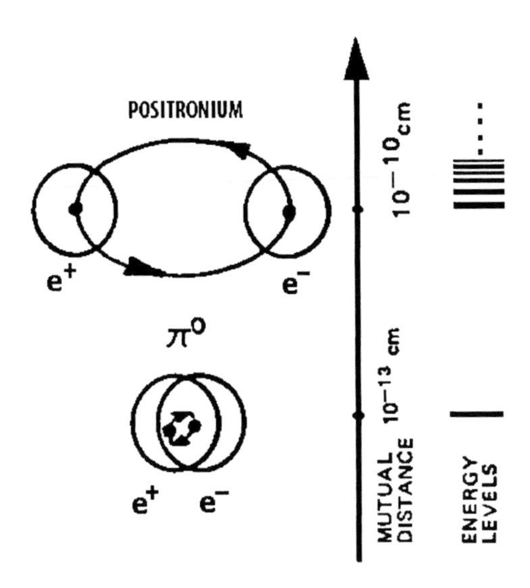

Figure 1.17. [6k] An illustration of a structure model which is possible for hadronic mechanics, yet strictly prohibited by quantum mechanics, *the π^0 meson as a bound state at 1 fm of one electron and one positron.* This model is impossible for quantum mechanics for numerous reasons. For instance, quantum bound states have a negative binding energy, thus implying a total mass generally smaller than the sum of the masses of the constituents, which requirement, for the case of the π^0, would imply a rest energy less than 1 MeV, while the π^0 has 138 MeV rest energy. Hadronic mechanics resolves this and all other limitations of quantum mechanics, by achieving a structure model which is capable of an *exact* representation of the *totality* of the characteristics of the π^0, including rest energy, charge radius, meanlife, charge, spin and their parity, magnetic moment, as well as its decay with the lowest mode $\pi^0 \to e^+ + e^-$, which results in being a tunnel effect of the constituents. It should be stressed that, when members of the π^0 structure, electrons *are not* the conventional particles as detected in laboratory until now and as characterized by the Poincaré symmetry, but are instead *isoelectrons*, i.e., particles characterized by the Poincaré-Santilli isosymmetry. The above model of the smallest possible hadron has been extended to all mesons, which have resulted in being hadronic bound states of electrons and positrons. The new structure model of hadrons is compatible with current quark theories, under the assumption that quarks are composites, and actually resolves their vexing inconsistencies, e.g., the lack of a rigorous confinement of quarks (which is easily reached by hadronic mechanics via the incoherence between external and internal Hilbert spaces).

in the physical reality in which there is *attraction* represented by a new wavefunction here denoted $\hat{\psi}_\uparrow(t, r)$.

By recalling that quantum mechanical Coulomb interactions are invariant under unitary transforms, the map $\psi_\uparrow \to \hat{\psi}_\uparrow$ is representable by a transform $\hat{\psi} = U\psi$ which is *nonunitary*, $U \times U^\dagger = U^\dagger U = \hat{I} \neq I$,

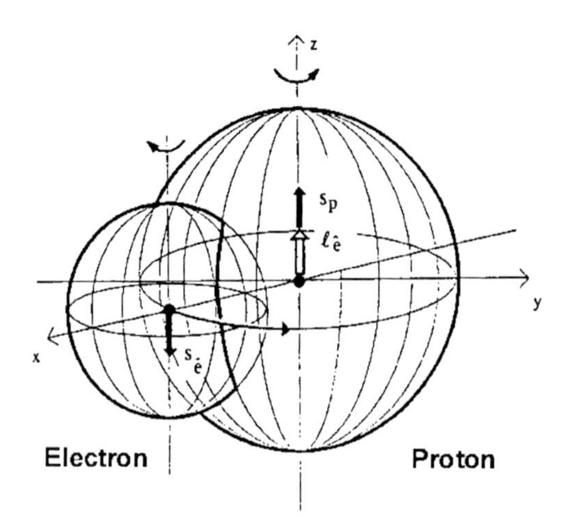

Figure 1.18. [6k] An illustration of another structure model which is strictly impossible for quantum mechanics, yet readily achieved by hadronic mechanics, *the neutron as originally conceived by Rutherford, namely, a hydrogen atom "compressed to the size of the proton" in the core of a star* (a bound state of a proton and an electron at 1 fm distance). Quantum mechanics is unable to achieve this representation because of a host of inconsistencies, such as: need for a "positive" binding energy (to reach the rest energy of the neutron from the lighter proton and electron); impossibility to represent the spin 1/2 of the neutron from a bound state of two particles each having spin 1/2; inability to represent the magnetic moment, charge radius and meanlife of the neutron; *etc.* All these inconsistencies have been resolved by hadronic mechanics. The reduction of the structure of the neutron to physical particles then permitted the construction of a similar model for unstable baryons, which model resulted in being compatible with quark theories under the assumption that quarks are composite. It should be noted that the inability by quantum mechanics to represent Rutherford's conception of the neutron had negative scientific implications of historical proportions, because it signaled the abandonment of hadronic constituents as physical particles, and their assumption as hypothetical undetectable particles, such as quarks. The occurrence also had ecological implications of large proportions, because it prohibited the study of the neutron as the biggest reservoir of clean energy available to mankind. The occurrence finally obstructed the study of new forms of clean nuclear energy, which are now under study via hadronic mechanics (see Chapter 7).

where $\hat{I}$ has to be determined (see below). This activates *ab initio* the applicability of hadronic mechanics as per Sect. 1.8. The first step of the proposed model is, therefore, that of transforming system (1.28) in

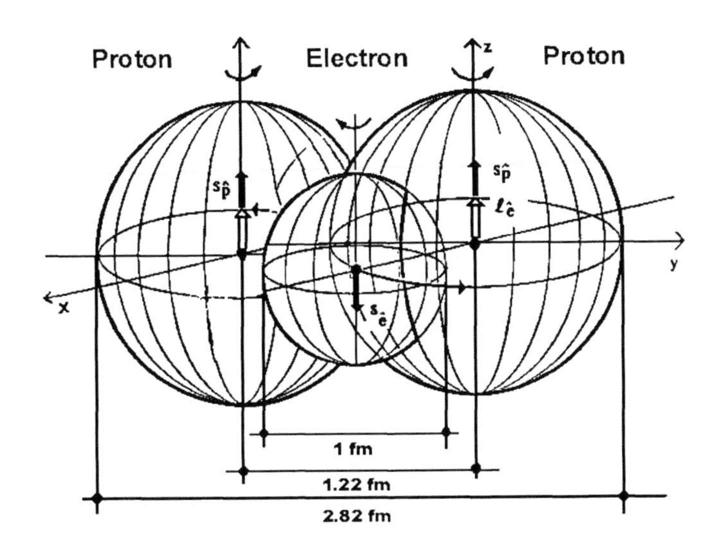

Figure 1.19. [6k] Another fundamental problem which was left unresolved by quantum mechanics throughout the entire studies of the 20-th century is the spin one of the deuteron. The basic axioms of quantum mechanics requires that the grand state of a two-body system of spin 1/2 particles must have spin zero. This additional problem was resolved by the covering hadronic mechanics. In fact, from the reduction of the neutron to a hadronic bound state of a proton and an electron of the preceding figure, it follows that *the deuteron is a three-body system composed by two isoprotons and one isoelectron verifying Santilli's isospecial relativity and the Poincaré-Santilli isosymmetry*. The latter model resulted in being capable of representing the *totality* of the characteristics of the deuteron, including spin, magnetic moment and other features not represented by quantum mechanics. The above deuteron model was then extended to all nuclei, resulting in a new hadronic structure model of nuclei which are reduced to isoprotons and isoelectrons while recovering the conventional model of protons and neutrons *as a first approximation*. It should be stressed that, far from being an academic esoteric topic, the reduction of nuclei to isoprotons and isoelectrons has truly fundamental implications of direct societal relevance, because it implies new forms of clean energies which are impossible for the old proton-neutron model. By keeping in mind our alarming environmental problems, the new hadronic structure model of nuclei cannot be dismissed on ground of personal opinions without raising serious problems of scientific accountability.

$\psi_\uparrow$ into a new system in $\hat{\psi}_\uparrow = U \times \psi_\uparrow$ where U is nonunitary,

$$\begin{aligned} U \times H_{\text{Coulomb}} \times U^\dagger \times (U \times U^\dagger)^{-1} \times U \times \psi_\uparrow(t, r) = \\ = \hat{H}_{\text{Coulomb}} \times T \times \hat{\psi}_\uparrow(t, r) = \\ = \left(\tfrac{1}{2m}\hat{p}_k \times T \times \hat{p}^k + \tfrac{e^2}{r}\hat{I}\right) \times T \times \hat{\psi}_\uparrow(t, r) = E \times \hat{\psi}_\uparrow(t, r), \end{aligned} \tag{1.29a}$$

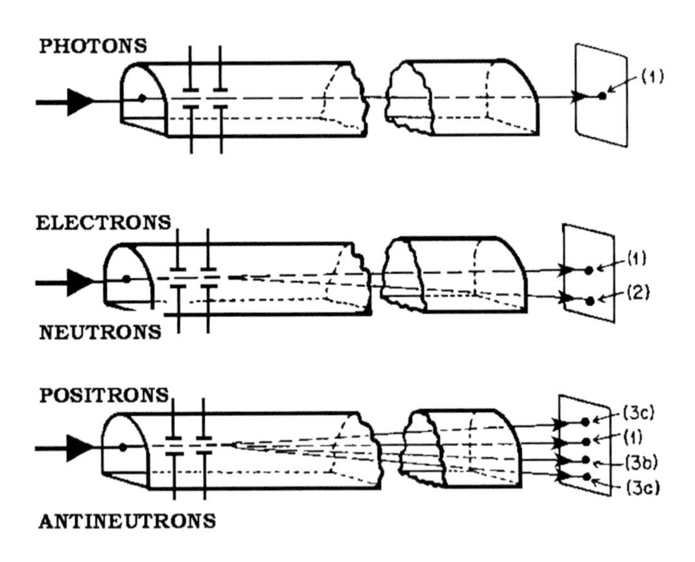

Figure 1.20. [6k] A schematic view of *the prediction by hadronic mechanics that antimatter in the gravitational field of matter, or vice versa, experiences repulsion (antigravity).* The prediction is permitted by the resolution of all known arguments against antigravity thanks to the use of the isodual mathematics. The prediction has been suggested by Santilli to be tested via a *horizontal* vacuum tube 50 m long and 1 m in diameter in which: 1) collimated photons are first passed through the tunnel to identify the point of no gravity on a scintillator screen at the end of the tunnel; 2) a beam of very low energy electrons is then passed through the same collimation to identify the displacement due to gravity on the same end scintillator plate, which displacement is visible with the naked eye for milli-electron Volts energy of the electrons; and 3) a beam of very low energy positrons is then passed through the same collimation. The upward displacement of the positrons due to gravity would then establish the existence of antigravity in a way visible to the naked eye. Irrespective of whether positive or negative, the test would have far reaching implications for all of science.

$$\hat{p}_k \times T \times \hat{\psi}_\uparrow(t,r) = -i \times T^i_k \times \partial_i \hat{\psi}_\uparrow(t,r). \tag{1.29b}$$

System (1.29) is incomplete because it misses the interaction with the Cu^{z+} ion represented by the familiar term $-ze^2/r$ [10]. The latter are not transformed (i.e., they are conventionally quantum mechanical) and, therefore, they should be merely added to the transformed equations (1.29). The formal equations of the proposed model CP = $(e^-_\uparrow, e^-_\downarrow)_{HM}$ are therefore given by

$$\left(\frac{1}{2m}\hat{p}_k \times T \times \hat{p}^k + \frac{e^2}{r} \times \hat{I} - z\frac{e^2}{r}\right) \times T \times \hat{\psi}_\uparrow(t,r) =$$

$$= \frac{1}{2m} \hat{p}_k \times T \times \hat{p}^k \times T \times \hat{\psi}_\uparrow + \frac{e^2}{r} \hat{\psi}_\uparrow - z \frac{e^2}{r} \times T \times \hat{\psi}_\uparrow(t, r) = \tag{1.30}$$
$$= E \times \hat{\psi}_\uparrow(t, r), \quad \hat{p}_k \times T \times \hat{\psi}_\uparrow(t, r) = -i \times T_k^i \times \partial_i \hat{\psi}_\uparrow(t, r).$$

In order to achieve a form of the model confrontable with experimental data, we need an explicit expression of the isounit $\hat{I}$. Among various possibilities, Animalu [10] selected the simplest possible isounit for the problem at hand, today called *Animalu's isounit*, which we write in form

$$\hat{I} = e^{-\langle \hat{\psi}_\uparrow | \hat{\psi}_\downarrow \rangle \psi_\uparrow / \hat{\psi}_\uparrow} \approx 1 - \langle \hat{\psi}_\uparrow | \hat{\psi}_\downarrow \rangle \psi_\uparrow / \hat{\psi}_\uparrow + \ldots,$$
$$\hat{T} = e^{+\langle \hat{\psi}_\uparrow | \hat{\psi}_\downarrow \rangle \psi_\uparrow / \hat{\psi}_\uparrow} \approx 1 + \langle \hat{\psi}_\uparrow | \hat{\psi}_\downarrow \rangle \psi_\uparrow / \hat{\psi}_\uparrow + \ldots, \tag{1.31}$$

under which Eqs. (1.30) can be written

$$\frac{1}{2m} \hat{p}_k T \hat{p}^k T \hat{\psi}_\uparrow - (z-1) \frac{e^2}{r} \hat{\psi}_\uparrow - \tag{1.32}$$
$$- z \frac{e^2}{r} \langle \hat{\psi}_\uparrow | \hat{\psi}_\downarrow \rangle (\psi_\uparrow / \hat{\psi}_\uparrow) \hat{\psi}_\uparrow(t, r) = E \hat{\psi}_\uparrow.$$

Now, it is well known from quantum mechanics that the radial part of $\psi_\uparrow$ in the ground state ($L = 0$) behaves as

$$\psi_\uparrow(r) \approx A e^{-r/R}, \tag{1.33}$$

where A is (approximately) constant and R is the coherence length of the pair. The radial solution for $\hat{\psi}_\uparrow$ also in the ground state is known from Eqs. (5.1.21), p. 837, Ref. [5e] to behave as

$$\hat{\psi}_\uparrow \approx B \frac{1 - e^{-r/R}}{r}, \tag{1.34}$$

where B is also approximately a constant. The last term in the l.h.s. of Eq. (1.34) behaves like a *Hulten potential*

$$V_0 \times \frac{e^{-r/R}}{1 - e^{-r/R}}, \quad V_0 = e^2 \langle \hat{\psi}_\uparrow | \hat{\psi}_\downarrow \rangle. \tag{1.35}$$

After substituting the expression for the isomomentum, the radial isoschrödinger equation can be written

$$\left(-\frac{\hat{I}}{2 \times \hat{m}} r^2 \frac{d}{dr} r^2 \frac{d}{dr} - (z-1) \frac{e^2}{r} - V_0 \frac{e^{-r/R}}{1 - e^{-r/R}} \right) \times \hat{\psi}_\uparrow(r) = E \times \hat{\psi}_\uparrow(r), \tag{1.36}$$

where $\hat{m}$ is the isorenormalized mass of the isoelectron.

The solution of the above equation is known from Ref. [5e], Sect. 5.1. The Hulten potential behaves at small distances like the Coulomb potential,

$$V_{\mathrm{Hulten}} = V_0 \times \frac{e^{-r/R}}{1 - e^{-r/R}} \approx V_0 \times \frac{R}{r}. \tag{1.37}$$

At distances smaller than the coherent length of the pair, Eq. (1.36) can therefore be effectively reduced to the form

$$\left(-\frac{1}{2 \times \hat{m}} r^2 \frac{d}{dr} r^2 \frac{d}{dr} - V \frac{e^{-r/R}}{1 - e^{-r/R}} \right) \times \hat{\psi}_{\uparrow}(r) = E \times \hat{\psi}_{\uparrow}(r), \tag{1.38a}$$

$$V = V_0 \times R + (z - 1) \times e^2, \tag{1.38b}$$

with general solution, boundary condition and related spectrum (see Ref. [5c], pp. 837-838)

$$\hat{\psi}_{\uparrow}(r) = {}_2F_1(2\times\alpha+1+n, 1-\alpha, 2\times\alpha+1, e^{-r/R}) e^{-\alpha \times r/R} \frac{1 - e^{-r/R}}{r}, \tag{1.39a}$$

$$\alpha = (\beta^2 - n^2)/2n > 0, \quad \beta^2 = \hat{m} \times V \times R^2/\hbar^2 > n^2, \tag{1.39b}$$

$$E = -\frac{\hbar^2}{4 \times \hat{m} \times R^2} \left(\frac{\hat{m} \times V \times R^2}{\hbar^2} \frac{1}{n} - n \right)^2, \quad n = 1, 2, 3, \ldots \tag{1.39c}$$

where we have reinstated $\hbar$ for clarity.

Santilli [5e] identified the numerical solution of Eqs. (1.39) for the hadronic model $\pi^0 = (\hat{e}^+_{\uparrow}, \hat{e}^-_{\downarrow})_{\mathrm{HM}}$ (in which there is evidently no contribution from the Cuprate ions to the constant V), by introducing the parameters

$$k_1 = \hbar/2 \times \hat{m} \times R \times c_0, \quad k_2 = \hat{m} \times V \times R^2/\hbar, \tag{1.40}$$

where c_0 is the speed of light in vacuum. Then,

$$V = 2 \times k_1 \times k_2^2 \times \hbar \times c_0/R, \tag{1.41}$$

and the total energy of the state $\pi^0 = (e^+_{\uparrow}, e^-_{\downarrow})_{\mathrm{HM}}$ becomes in the ground state (which occurs for $n = 1$ for the Hulten potential)

$$\begin{aligned} E_{\mathrm{tot},\pi^0} &= 2 \times k_1 \times [1 - (k_2 - 1)^2/4] \times \hbar \times c_0/R = \\ &= 2 \times k_1(1 - \varepsilon^2) \times \hbar \times c_0/R. \end{aligned} \tag{1.42}$$

The use of the total energy of the π^0 (135 MeV), its charge radius ($R \approx 10^{-13}$ cm) and its meanlife ($\tau \approx 10^{-16}$ sec), then yields the values (Eqs. (5.1.33), p. 840, Ref. [5e])

$$k_1 = 0.34, \quad \varepsilon = 4.27 \times 10^{-2}, \tag{1.43a}$$

$$k_2 = 1 + 8.54 \times 10^{-2} > 1. \qquad (1.43b)$$

Animalu [10a] identified the solution of Eqs. (1.39) for the Cooper pair by introducing the parameters

$$k_1 = \varepsilon \times F \times R/\hbar \times c_0, \quad k_2 = KR/\varepsilon_F, \qquad (1.44)$$

where ε_F is the iso-Fermi energy of the isoelectron (that for hadronic mechanics).

The total energy of the Cooper pair in the ground state is then given by

$$E_{\text{Tot, Cooper pair}} = 2 \times k_1 \times [1 - (k_2 - 1)^2/4] \times \hbar \times c_0/R \approx k_2 \times T_c/\theta_D, \quad (1.45)$$

where θ_D is the Debye temperature.

Several numerical examples were considered in Refs. [10]. The use of experimental data for aluminum,

$$\theta_D = 428^0 K, \quad \varepsilon_F = 11.6 \text{ eV}, \quad T_c = 1.18^0 K, \qquad (1.46)$$

yields the values

$$k_1 = 94, \quad k_2 = 1.6 \times 10^{-3} < 1. \qquad (1.47)$$

For the case of $YBa_2Cu_3O_{6-\chi}$ the model yields [*loc. cit.*]

$$k_1 = 1.3z^{-1/2} \times 10^{-4}, \quad k_2 = 1.0 \times z^{1/2}, \qquad (1.48)$$

where the effective valence $z = 2(7 - \chi)/3$ varies from a minimum of $z = 4.66$ for $YBa_2Cu_3O_{6.96}$ ($T_c = 91^0 K$) to a maximum of $z = 4.33$ for $YBa_2Cu_3O_{6.5}$ ($T_c = 20^0 K$). The general expression predicted by hadronic mechanics for $YBa_2Cu_3O_{6-\chi}$ is given by (Eq. (5.15), p. 373, Ref. [10a])

$$T_c = 367.3 \times z \times e^{-13.6/z}, \qquad (1.49)$$

and it is in remarkable agreement with experimental data (see Figs. 1.21–1.23).

A few comments are now in order. The above Animalu-Santilli model of the Cooper pair is indeed nonlinear, nonlocal and nonpotential. In fact, the nonlinearity in $\hat{\psi}_\uparrow$ is expressed by the presence of such a quantity in Eqs. (1.31). The nonlocality is expressed by the term $\langle \hat{\psi}_\uparrow | \hat{\psi}_\downarrow \rangle$ representing the overlapping of the wavepackets of the electrons, and the nonpotentiality is expressed by the presence of interactions, those characterized by the isounit, which are outside the representational capabilities of the Hamiltonian H. This illustrates the necessity of using hadronic mechanics or other similar nonhamiltonian theories (provided

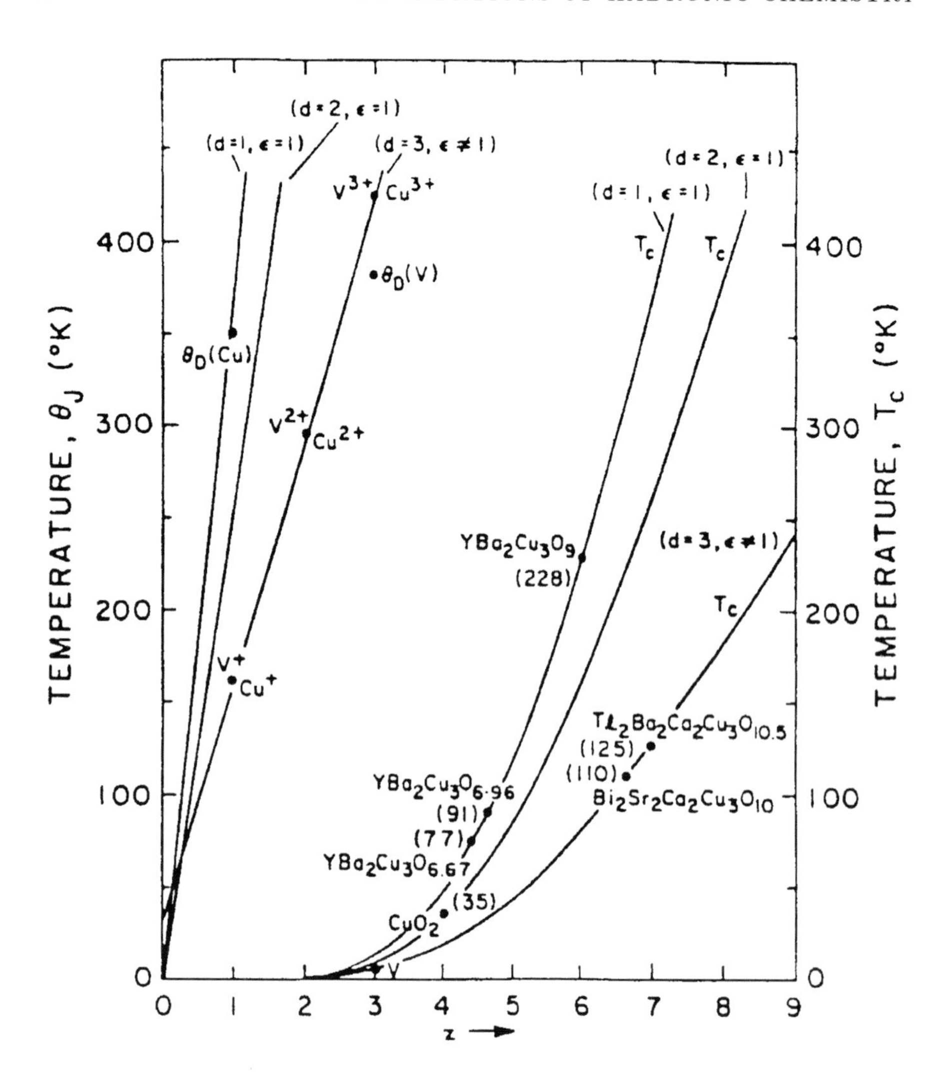

Figure 1.21. A reproduction of Fig. 10 of Ref. [10a] illustrating the remarkable agreement between the predicted dependence of T_c from the effective valence z of ions (continuous curve) and the experimental values on the "jellium temperature" for various compounds (solid dots).

that they are physically consistent), because of the strictly linear-local-potential character of quantum mechanics.

Note that, whenever the wave-overlapping is no longer appreciable, i.e., for $\langle\hat{\psi}_\uparrow|\hat{\psi}_\downarrow\rangle = 0$, $\hat{I} \equiv I$, quantum mechanics is recovered identically as a particular case, although without attraction.

The mechanism of the creation of the *attraction* among the *identical* electrons of the pair via the intermediate action of Cuprate ions is a general law of hadronic mechanics according to which *nonlinear-nonlocal-nonhamiltonian interactions due to wave-overlappings at short distances are always attractive in singlet couplings and such to absorb Coulomb interactions, resulting in total attractive interactions irrespective of whether the Coulomb contribution is attractive or repulsive.* As noted earlier, the Hulten potential is known to behave as the Coulomb one at small distances and, therefore, the former absorbs the latter.

Alternatively, we can say that within the coherent length of the Cooper pair, the Hulten interaction is stronger than the Coulomb force. This results in the overall attraction. Thus, the similarities between the model for the π^0 and that for the Cooper pair are remarkable. The applicability of the same model for other aspects should then be expected, such as for a deeper understanding of the valence, and will be studied in the next chapters.

Another main feature of the model is characterized also by a general law of hadronic mechanics, that *bound state of particles due to wave-overlappings at short distances in singlet states suppress the atomic spectrum of energy down to only one possible level.* The Hulten potential is known to admit a *finite* number of energy levels. Santilli's [5e] solution for the π^0 shows the suppression of the energy spectrum of the positronium down to only one energy level, 135 MeV of the π^0 for $k_2 > 1$. Similarly, the solutions for the Cooper pair [10] also reduce the same finite spectrum down to only one admissible level.

Excited states are indeed admitted, but they imply large distances R for which nonlinear-nonlocal-nonhamiltonian interactions are ignorable. This implies that all excited states are conventionally quantum mechanical, that is, they *do not* represent the π^0 or the Cooper pair. Said excited states represent instead the discrete spectrum of the ordinary positronium, or the continuous spectrum of repulsive Coulomb interactions among the two identical electrons.

Alternatively, we can say that, in addition to the conventional, quantum mechanical, Coulomb interactions among two electrons, there is *only one additional system* of hadronic type with *only one energy level* per each couple of particles, one for $\pi^0 = (e^+_\uparrow, e^-_\downarrow)_{\mathrm{HM}}$ and the other for the Cooper pair, $\mathrm{CP} = (e^-_\uparrow, e^-_\downarrow)_{\mathrm{HM}}$.

The case of possible triplet couplings also follows a general law of hadronic mechanics. While singlets and triplets are equally admitted in quantum mechanics (read, coupling of particles at large mutual distances under their point-like approximation), this is no longer the case for hadronic mechanics (read, couplings of particles when repre-

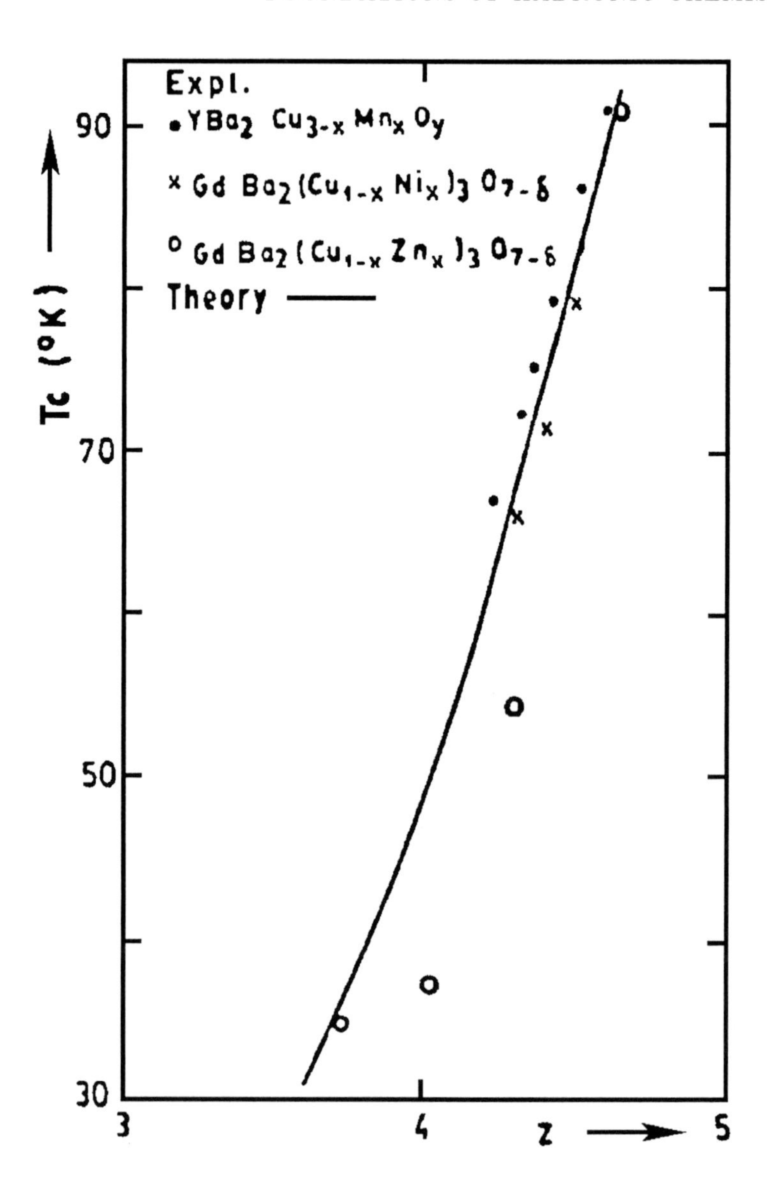

Figure 1.22. A reproduction of Fig. 5, p. 380 of Ref. [10a] showing the agreement between the prediction of isosuperconductivity for the doped 1:2:3 Cuprates and the experimental data.

sented as being extended and at mutual distances smaller than their wavepackets/wavelengths). In fact, *all triplet couplings of particles under nonlinear-nonlocal-nonhamiltonian interactions are highly unstable, the only stable states being the singlets.*

Table 1. $YBa_2Cu_{3-x}Mn_xO_y$
(After N.L. Saini *et al.*, Int. J. Mod. Phys. B6, 3515 (1992)

x	y	z	T_c (theory)	T_c (expt.)
0.00	6.92	4.613	88.9	91
0.03	6.88	4.541	83.5	86.6
0.09	6.87	4.447	76.7	79.0
0.15	6.91	4.387	72.6	75.0
0.21	6.92	4.312	67.6	72.0
0.30	6.95	4.212	61.3	67.0

Note: T_c (theory) $= 367.3z\exp(-13.6/z)$, where the effect of replacing Cu_3 by $Cu_{3-x}Mn_x$ is obtained by replacing 3 by $(3-x)+2x=3+x$, which lowers the effective valence (z) on Cu^{z+} ions to $z=2y/(3+x)$.

Table 2. $GdBa_2(Cu_{1-x}Ni_x)_3O_{7-\delta}$
(After, Chin Lin *et al.*, Phys. Rev. B42, 2554 (1990))

x	$y=7-\delta$	z	T_c (theory)	T_c (expt.)
0.000	6.96	4.640	91.0	91
0.025	6.96	4.527	82.4	79
0.050	6.96	4.419	74.8	71
0.075	6.96	4.316	67.9	65

Note: T_c(theory) $= 367.3z\exp(-13.6/z)$, $z=2y/3(1+x)$ as discussed in Table I.

Table 3. $GdBa_2(Cu_{1-x}Zn_x)_3O_{7-\delta}$
(After, Chin Lin *et al.*, Phys. Rev. B42, 2554 (1990))

x	$y=7-\delta$	z	T_c (theory)	T_c (expt.)
0.000	6.96	4.640	91.0	91
0.025	6.96	4.309	67.4	54
0.050	6.96	4.009	49.0	37
0.075	6.96	3.737	36.1	35

Figure 1.23. A reproduction of the tables of p. 379, Ref. [10a] illustrating the agreement between the predictions of the model with experimental data from other profiles.

This law was first derived in Ref. [5e] via the "gear model", i.e., the illustration via ordinary mechanical gears which experience a highly repulsive force in triplet couplings, while they can be coupled in a stable way only in singlets. The possibility of applying the model to a deeper understanding of Pauli's exclusion principle is then consequential, and will be studied in Chapters 4 and 5.

The connection between the proposed model and the conventional theory of the Cooper pair is intriguing. The constant in the Hulten potential can be written

$$V_0 = \hbar\omega, \tag{1.50}$$

where ω is precisely the (average) *phonon frequency*. The total energy can then be rewritten

$$E_{Tot} = 2\times\varepsilon_F - E \approx 2\times k_1\times k_2\times\hbar\times c_0/R(e^{1/N\times V}-1), \tag{1.51}$$

where $N \times V$ is the (dimensionless) *electron-phonon coupling constant.*

In summary, a main result of studies [10] is that *the conventional representation of the Cooper pair via a mysterious "phonon" can be reformulated without any need of such a particle, resulting in a real, sufficiently strong attraction between the identical electron, which is absent in the phonon theory.*

The above model of the Cooper pair see its true formulation at the relativistic level because it provides a *geometrization* of the Cooper pair, better possibilities for novel predictions and the best possible experiments tests. These profiles [10] will not be reviewed for brevity.

References

[1] Boyer, D.J.: *Bonding Theory*, McGraw Hill, New York (1968) [1a]. Hanna, M.W.: *Quantum Mechanical Chemistry*, Benjamin, New York (1965) [1b]. Pople, J.A. and Beveridge, D.L.: *Approximate Molecular Orbits*, McGraw Hill, New York (1970) [1c]. Linnet, J.W.: *The Electronic Structure of Molecules, A New Approach*, Methuen Press, London and Wiley, New York, (1964) [1d].

[2] Illert, C. and Santilli, R.M.: *Foundations of Conchology*, Hadronic Press (1996).

[3] Einstein, A. Podolsky, B. and Rosen, N.: Phys. Rev. **47**, 777 (1935) [3a]. Von Neumann, J.: *The Mathematical Foundations of Quantum Mechanics*, Princeton Univ. Press, Princeton, N. J., 1955 [3b]. Bell, J.S.: Physics **1**, 195 (1965) [3c]. Bohm, D.: *Quantum Theory*, Dover Publ., New York (1979) [3d]. Santilli, R.M.: Acta Appl. Math. **50**, 177 (1998) [3e].

[4] Okubo, S.: Hadronic J. **5**, 1667 (1982) [4a]. Lopez, D.F.: in *Symmetry Methods in Physics* (Memorial Volume dedicated to Ya.S. Smorodinsky), A.N. Sissakian, G.S. Pogosyan, and S.I. Vinitsky (ed.), J.I.N.R., Dubna, Russia, p. 300 (1994); and Hadronic J. **16**, 429 (1993) [4b]. Jannussis, A. and Skaltzas, D.: Ann. Fond. L. de Broglie **18**, 1137 (1993) [4c]. Jannussis, A., Mignani, R. and Santilli, R.M.: Ann. Fond. L. de Broglie **18**, 371 (1993) [4d]. Schuch, D.: Phys. Rev. **A55**, 955 (1997) [4e]. Santilli, R.M.: Modern Phys. Letters **13**, 327 (1998) [4f]. Santilli, R.M.: Intern. J. Modern Phys. **A14**, 3157 (1999) [4g].

[5] Santilli, R.M.: Nuovo Cimento **51**, 570 (1967) [5a]. Supl. Nuovo Cimento **6**, 1225 (1968) [5b]. Mechanica **1**, 3 (1968) [5c]. Hadronic J. **1**, 224 [5d], 574 [5e] and 1267 (1978) [5f]. Hadronic J. **3**, 440 (1979) [5g]. Lett. Nuovo Cimento **37**, 509, (1983) [5h]. Hadronic J. **8**, 25 and 36 (1985) [5i]. Algebras, Groups and Geometries **10**, 273 (1993) [5j]. Rendiconti Circolo Matematico Palermo, Suppl. **42**, 7 (1996) [5k]. Found. Phys. **27** 625 (1997) [5l]. Found. Phys. Letters **4**, 307 (1997) [5m]. Found. Phys. **27**, 1159 (1997) [5n]. Hyperfine Interactions **109**, 63 (1997) [5o]. Intern. J. Modern Phys. **D7**, 351 (1998) [5p]. in *Modern Modified Theories of Gravitation and Cosmology*, E.I. Guendelman (ed.); Hadronic J. **2**, 113 (1998) [5q]. in *Proceedings of the VIII M. Grossmann Meeting on General Relativity*, Jerusalem, June 1997, World Scientific, in press (1999) [5r]. Algebras, Groups and Geometries **15**, 473 (1998) [5s]. IBR Web Site http://www.i-b-r.org, pages 18 [5t] and 19 [5u].

[6] Santilli, R.M. *Foundations of Theoretical Mechanics*, Vol. **I** (1978) [6a] and **II** (1983) [6b], Springer-Verlag, Heidelberg-New York. *Lie-Admissible Approach to the Hadronic Structure*, Vol. **I** (1978) [6c] and **II** Hadronic Press (1982) [6d]. *Isotopic Generalization of Galilei's and Einstein's Relativities*, Vols. **I** [6e] and **II** [6f], Hadronic Press (1991). *Elements of Hadronic Mechanics*, Vol. **I** [6g] and **II** (1995) [6h] and **III**, in preparation [6i], Ukraine Academy of Sciences, Kiev. *Isotopic, Genotopic and Hyperstructural Method in Theoretical Biology*, Ukraine Academy of Sciences, Kiev (1996) [6j]; Journal of New Energy **4**, issue 1 (1998) [6k].

[7] Adler, S.L.: Phys. Rev. **17**, 3212 (1978). George, Cl., Henin, F., Mayne, F. and Prigogine, I.: Hadronic J. **1**, 520 (1978). Okubo, S.: Hadronic J. **3**, 1 (1979). Fronteau, J., Tellez Arenas, A. and Santilli, R.M.: Hadronic J. **3**, 130 (1978). Ktorides, C.N., Myung, H.C. and Santilli, R.M.: Phys. Rev. **D22**, 892 (1982). Kalnay, A.J.: Hadronic J. **6**, 1 (1983). Gasperini, M.: Hadronic J. **6**, 935 and 1462 (1993), **7**, 650 and 951 (1984). Mignani, R.: Lett. Nuovo Cimento **39**, 413 (1984). Constantoupoulos, J.D. and Ktorides, C.N.: J. Phys. **A17**, L29 (1984). Nishioka, M.: Nuovo Cimento **82A**, 351 (1984). Rapoport-Campodonico, D.: Algebras, Groups and Geometries **8**, 1 (1991). Jannussis, A., Brodimas, G. and Mignani, R.: J. Phys. **A24**, L775 (1991). Jannussis, A., Mignani, R. and Skaltsas, D.: Physica **A187**, 575 (1992). Gill, T., Lindesay, J. and Zachary, W.W.: Hadronic J. **17**, 449 (1994). Jannussis, A., Mijatovic, M. and Veljanowski, B.: Physics Essays **4**, 202 (1991). Aringazin, A.K.: Hadronic J. **12**, 71 (1989). Aringazin, A.K., *et al.*: in *Frontiers of Fundamental Physics*, M. Barone and F. Selleri (ed.), Plenum, New York, p. 153 (1995). Ellis, J.: in *Proceedings of the Seventh Marcel Grossmann Meeting on General Relativity*, R.T. Jantsen, G. MacKaiser and R. Ruffini (ed.), World Scientific, p. 755 (1996). Ellis, J., Mavromatos, N.E. and Nanopoulos, D.V.: in *Proceedings of the Erice Summer School, 31st Course: From Superstrings to the Origin of Space-Time*, World Scientific (1996).

[8] Tsagas, Gr.T. and Sourlas, D.S.: Algebras, Groups and Geometries **12**, 1 and 67 (1995), and Algebras, Groups and Geometries **13**, 129 (1996). Rapoport-Campodonico, D.: in *New Frontiers in Physics*, T.L. Gill (ed.), Hadronic Press (1996). Aslaner, R. and Keles, S.: Algebras, Groups and Geometries **14**, 211 (1997). Vacaru, S.: Algebras, Groups and Geometries **14**, 225 (1997). Lin, E.B.: Hadronic J. **11**, 81 (1988). Jiang, C.X.: Algebras, Groups and Geometries **15**, 351 (1998). Trell, E.: Algebras, Groups and Geometries **15**, 299 (1998). Kamiya, N. *et al.*: Hadronic J. **16**, 68 (1993). Aslander R. and Keles, S.: Algebras, Groups and Geometries **14** (1997), in press. Kadeisvili, J.V.: Algebras, Groups and Geometries **9**, 283 and 319 (1992). In *Symmetry Methods in Physics* (Dedicated to Smorodinsky), Pogosyan, G. *et al.* (ed.), JINR, Dubna (1994). Indian J. Appl. of Math. **25**, 7 (1995). Math. Methods in Applied Sciences **19**, 362 (1996). Algebras, Groups and Geometries **13**, 171 (1996). Acta Appl. Math. **50**, 131 (1998). IBR Web Site http://www.i-b-r.org, page 18.

[9] Arestov Yu. *et al.*: Found. Phys. Letters **11**, 483 (1998) [9a]. Santilli, R.M.: in *Collection of Papers on the Lorentz Symmetry*, V.V. Dvoeglazov (ed.), Nova Publisher, in press [9b]. Cardone, F. *et al.*: Phys. **C18**, L61 and L141 (1992) [9c]. Santilli, R.M.: Hadronic J. **15**, 1 (1992) [9d]. Mignani, R.: Lett. Nuovo Cimento **39**, 413 (1982) [9e]. Kalnay, A.J.: Hadronic J. **6**, 1 (1983) [9f]. Santilli, R.M.: Comm. Theor. Phys. **4**, 123 (1995) [9g]. Santilli, R.M.: Intern. J. Phys. , in

press [9h]. Santilli, R.M.: Hadronic J. **17**, 311 (1994) [9j]. Mignani, R.: Physics Essays **5**, 531 (1992) [9k]. Santilli, R.M.: in *Frontiers of Fundamental Physics*, M. Barone and F. Selleri (ed.), Plenum Press (1994) [9l]. USMagnegas Web Site http://www.usmagnegas.com [9m].

[10] Animalu, A.O.E.: Hadronic J. **17**, 379 (1994) [10a]. Animalu, A.O.E. and Santilli, R.M.: Int. J. Quantum Chemistry **29**, 175 (1995) [10b].

[11] Tsagas Gr. and Sourlas, D.: *Mathematical Foundations of the Lie-Santilli Theory,* Ukraine Academy of Sciences, Kiev (1992) [11a]; Kadeisvili, J.V.: *Santilli's Isotopies of Contemporary Algebras, Geometries and Relativities*, Ukraine Academy of Sciences, Kiev, 2-nd edition (1997) [11b]; Aringazin, A.K., Jannussis, A., Lopez, D.F., Nishioka, M. and Veljanosky, B.: *Santilli's Lie-Isotopic Generalization of Galilei's and Einstein's Relativities*, Kostarakis Publisher, Athens (1990) [11c]; Lôhmus, J., Paal, E. and Sorgsepp, L.: *Nonassociative Algebras in Physics*, Hadronic Press, Palm Harbor, FL, USA (1994) [11d]. Vacaru, S.: *Interactions, Strings and Isotopies in Higher Order Anisotropic Superspaces*, Hadronic Press (1999) [11e]. Jiang, C.X.: *An Introduction to Santilli's Isonumber Theory*, Hadronic Press, in press [11f].

[12] Blatt, J.M. and Weiskopf, V.F.: *Theoretical Nuclear Physics*, Wiley, New York (1964) [12a]. Band, W.: *An Introduction to Quantum Statistics*, van Nostrand, Princeton (1955) [12b]. Biedernharn, L.C.: J. Phys. **A22**, L873 (1989); Macfarlane, A. J.: Phys. **A22**, L4581 (1989) [12c]. Lukierski, J., Viowiski, A. and Ruegg, H.: Phys. Lett. **B293**, 344 (1992); Lukierski, J., Ruegg, H. and Ruhl, W.: Phys. Lett. **B13**, 357 (1993) [12d]. Bayer, F., Flato, M., Fronsdal, C., Lichnerowicz, A. and Sternheimer, D.: Ann. Phys. **111**, 61 and 111 (1978); Fronsdal, C.: Rep. Math. Phys. **15**, 111 (1978) [12e]. Jannussis, A. *et al.*: Nuovo Cimento **B103**, 17 and 537 (1989) [12f]. S. Sebawe Abdallah *et al.*: Physica **A163**, 822 (1990) [12g]. George, Cl., Henin, F., Mayne, F. and Prigogine, I.: Hadronic J. **1**, 520 (1978) [12h]. Ellis, J., Mavromatos, N.E. and Nanopoulos, D.V.: in *Proceedings of the Erice Summer School, 31st Course: From Superstrings to the Origin of Space-Time*, World Scientific (1996) [12i]. Caldirola, P.: Nuovo Cimento **3**, 297 (1956); Jannussis, A. *et al.*: Lett. Nuovo Cimento **29**, 427 (1980); Lee, D.: Phys. Rev. Lett. **122**, 217 (1983) [12j]. Mahopatra, R.N. *Unification and Supersymmetries: The Frontiers of Quark-Lepton Physics*, 2-nd ed., Springer-Verlag, (1992) [12k]. Kac, V.G.: *Infinite Dimensional Lie Algebras*, Birkhauser, New York (1983) [12l]. Sanchez, N.S. (ed.): *String Theory in Curved Spacetime*, World Scientific, Singapore (1998) [12m]. Weinberg, S.: Ann. Phys. **194**, 336 (1989). Jordan, T.F.: Ann. Phys. **225**, 83 (1993) [12m].

[13] Shektman, V.: *The Real Structure of High T_c Superconductivity,* Springer-Verlag, New York (1993) [6a]. Zhang, M.-L. and Zhao, Z.-X.: Phys. Lett. **A207**, 362 (1995).

[14] Lagrange, L.: *Mechanique Analytique* (1788), reprinted by Gauthier-Villars, Paris (1888) [14a]. Hamilton, W.R.: *On a General Method in Dynamics* (1834), reprinted in: *Hamilton's Collected Works*, Cambridge Univ. Press (1940) [14b].

[15] Sinanoglu, O.: Proc. Nat. Acad. Sciences **47**, 1217 (1961) [15a]. Lewis, G.N.: J. Am. Chem. Soc. **38**, 762 (1916) [15b]. Langmuir, I.: J. Am. Chem. Soc. **41**, 868

(1919) [15c]. Boyd, R.J. and Yee, M.C.: J. Chem. Phys. **77**, 3578 (1982) [15d]. Cann, N.M. and Boyd, R.J.: J. Chem. Phys. **98**, 7132 (1992) [15e].

Chapter 2

ELEMENTS OF ISO-, GENO-, AND HYPER-MATHEMATICS AND THEIR ISODUALS

1. Introduction

In the preceding chapter we have shown that no genuine broadening of quantum mechanics, superconductivity, and chemistry is possible without exiting the classes of equivalence of these theories, which equivalence classes are given by all possible *canonical transformations* for classical formulations, and *unitary* transformations for operator theories, hereon referred in the unified notation $U \times U^\dagger = U^\dagger \times U = I$.

Therefore, genuinely broader classical theories must have *noncanonical structures*, while truly advanced operator theories require *nonunitary structures*, hereon referred with the unified notation $U \times U^\dagger \neq I$.

In the preceding chapter we pointed out that noncanonical theories are necessary for the representation of classical, contact, zero-range, resistive interactions, such as those experienced by a space-ship during re-entry in Earth's atmosphere. In fact, the equations of motion of these systems are necessarily nonhamiltonian in the reference frame of the experimenter, as studied in details in monographs [2a] via the integrability conditions for the existence of a Hamiltonian, the so-called *conditions of variational selfadjointness.*

We then pointed out that, along similar lines, nonunitary theories are necessary to represent the resistive interactions experienced by an extended proton when moving in the core of a star. Equivalently, nonunitary theories are necessary to represent contact interactions due to deep wave-overlappings of the wavepackets of particles. In fact, the latter interactions too are of contact type, thus admitting no potential. As a consequence, these interactions cannot be represented with a Hamiltonian $H = p^2/2 \times m + V(r,p) = H^\dagger$, thus exiting the class of unitary equivalence with time evolution $U = \exp[i \times H \times t]$, $U \times U^\dagger = U^\dagger \times U = I$.

In the preceding chapter we have also provided evidence of the existence of these contact nonhamiltonian and nonunitary effects in particle physics, nuclear physics, astrophysics, gravitation, cosmology, and superconductivity. This evidence has been presented in preparation of the main topic of study of this monograph: *the interaction of valence electrons in molecular bonds*, which is expected to be precisely of nonhamiltonian-nonunitary type due to deep wave-overlappings of the wavepackets of valence electrons which are absent in the atomic structure.

In the preceding chapter we have furthermore indicated that all noncanonical-nonunitary theories are afflicted by catastrophic physical and mathematical inconsistencies when formulated with conventional mathematics, that is, with conventional numbers, fields, metric spaces, Hilbert spaces, *etc.* In particular, theories with noncanonical-nonunitary time evolutions predict different numerical values for the same measurement under the same conditions at different times, e.g., the length of a given rigid rod under resistive forces changes in time under a conventional noncanonical representation. As a result, *noncanonical-nonunitary theories formulated with conventional mathematics have no known physical or chemical value.*

The above scenario leaves no other alternative than that of constructing *new mathematics* specifically conceived to treat noncanonical-nonunitary theories in an invariant way.

The latter task was initiated by Santilli in 1978 [1a] with the suggestion to construct the novel *iso-* and *geno-mathematics*, with the yet broader *hyper-mathematics* and their *isoduals* added later (see the recent account [1e]). These new mathematics were then developed by various mathematicians and theoreticians, and have now reached operational maturity. Hadronic chemistry is the image of conventional quantum chemistry under the novel iso-, geno-, and hyper-mathematics and their isoduals.

In this chapter we outline the elements of these broader iso-, geno-, and hyper-mathematics with the understanding that scholars interested in acquiring an in depth technical knowledge of hadronic chemistry should also study the original literature in the field. A knowledge of conventional numbers and fields, metric and Hilbert spaces, algebras and groups, geometries and topologies, *etc.* is assumed.

For less mathematically inclined readers, we present in the following chapter very simple methods based on noncanonical-nonunitary transforms which permit the explicit construction of noncanonical-nonunitary chemical theories sufficient for all practical applications.

2. Elements of Isomathematics

2.1. Isounits and Isoproducts

As we shall see in the next chapter, the achievement of invariance for theories with nonunitary time evolution $U \times U^\dagger \neq I$ requires the assumption of $\hat{I} = U \times U^\dagger$ as the new generalized unit. *Santilli isomathematics* is then given by the reconstruction of the entire conventional mathematics in such a way as to admit $\hat{I}$, rather than I, as the correct left and right new unit at all levels.

As we shall see in Chapter 3, isomathematics characterizes the *isotopic branch of hadronic chemistry*, or *isochemistry* for short. This is the branch of hadronic chemistry suitable for the representation of closed-isolated systems which are reversible in time, and admit internal interactions of conventional, long range, linear, local, and potential type, as well as broader short range interactions of nonlinear, nonlocal, and non-potential type due to the deep wave-overlappings of the wavepackets of valence electrons.

The most fundamental notion of hadronic chemistry is, therefore, the generalization of the trivial n-dimensional unit $I = (I^i_j) = \text{Diag}(1, 1, 1, \ldots)$, $i, j = 1, 2, \ldots, n$ of current use in chemistry (such as the Euclidean unit $I = \text{Diag.}(1, 1, 1)$), which, for the case of isochemistry, is lifted into a nowhere singular, positive-definite, $n \times n$-dimensional matrix $\hat{I}$ whose elements have an arbitrary, nonlinear and integral dependence on all needed variables, such as time t, space coordinates r, momenta p, wave-functions ψ, their derivatives $\partial\psi$, and any other needed quantity [1a],

$$I = \text{Diag.}(1, 1, 1) > 0 \rightarrow \hat{I} = (\hat{I}^j_i) = \hat{I}(t, r, p, \psi, \partial\psi, \ldots) = 1/\hat{T} > 0. \quad (2.1)$$

Recall that conventional fields of numbers contain *two units*, the *additive unit* 0, which is such that for any number n we have $n + 0 = 0 + n = n$, and the *multiplicative unit* I, which is such that for any number n we have $n \times I = I \times n = n$. Therefore, lifting (2.1) implies the *generalization of the multiplicative unit*.

Consider then the conventional associative product currently used in chemistry, $A \times B = AB$, where A, B are generic quantities (e.g., numbers, vector-fields, operators, *etc.*). It is easy to see that the generalized unit $\hat{I}$ cannot remain a unit under product $A \times B$ because $A \times \hat{I} \neq A$. Therefore, for consistency, the conventional associative product $A \times B$ must be lifted in an amount which is the *inverse* as that of the unit [1a],

$$A \times B \rightarrow A \hat{\times} B = A \times \hat{T} \times B, \quad \hat{T} \text{ fixed for all products}, \quad (2.2)$$

under which $\hat{I}$ is indeed the correct left and right new unit,

$$I \times A = A \times I = A \rightarrow \hat{I} \hat{\times} A = A \hat{\times} \hat{I} = A, \tag{2.3}$$

for all elements A of the considered set. In this case (only) $\hat{I}$ is called the *isounit* and $\hat{T}$ is called the *isotopic element*.

Liftings (2.1), (2.2), and (2.3) were called *isotopic* [1a] in the Greek sense of preserving all original axioms. In fact, $\hat{I}$ preserves all topological features of I, such as nonsingularity, Hermiticity, and positive-definiteness, while the new product $A \hat{\times} B$ preserves the original associative axiom, $A \hat{\times} (B \hat{\times} C) = (A \hat{\times} B) \hat{\times} C$, for which reason the product $A \hat{\times} B$ is called *isoassociative product* or *isoproduct* for short.

Note that other generalizations of the associative product, such as $\hat{T} \times A \times B$ or $A \times B \times \hat{T}$ would violate the general axioms of an algebra (right and left distributive and scalar laws [*loc. cit.*]).

Examples of isounits were given in Chapter 1. Additional examples will be provided in subsequent sections and chapters. At this mathematical stage it is sufficient to think of the isounits $\hat{I}$ as arbitrary, nonsingular, positive-definite, $n \times n$-dimensional matrices. As such, isounits can always be diagonalized into the form $\hat{I} = \text{Diag.}(A, B, C, \ldots)$, where A, B, C, ..., are positive-definite constants or functions which can be subjected to a variety of interpretations.

The simplest possible, yet instructive interpretation is that in which $\hat{I}$ is three-dimensional and represents the *extended, nonspherical, and deformable shape* of the particle considered, in which case the functions $A = n_1^2$, $B = n_2^2$, and $C = n_3^2$ represent the semiaxes n_1, n_2, n_3 of deformable spheroidal ellipsoids (see Sect. 2.3 below). Nondiagonal isounits $\hat{I}$ can then represent more complex shapes.

The simplest possible interpretation when $\hat{I}$ is four-dimensional, $\hat{I} = \text{Diag.}(A, B, C, D)$, is that in which A, B, C represent the shape and D represents the *density* of the particle considered.

A more general class of isounits has the structure

$$\hat{I} = \text{Diag.}(A, B, C, D) \times \Gamma(t, r, \psi, \ldots),$$

where A, B, C represent the shape, D represents the density and $\Gamma = \exp F(t, r, \psi, \ldots)$ represent nonlinear, nonlocal and nonpotential interactions (see next chapter).

Recall that *quantum mechanics can only represent perfectly spherical and perfectly rigid bodies* as a necessary condition not to violate its pillar, the rotational symmetry. In particular, the perfect spheridicity is represented via the *basic unit* of the Euclidean geometry, $I = \text{Diag.}(1, 1, 1)$, where $1, 1, 1$ are the semiaxes of the perfect sphere. The lifting $I = \text{Diag.}(1,1,1) \rightarrow \hat{I} = \text{Diag.}(A,B,C)$ is then useful to represent *nonspherical and deformable shapes*, besides providing their sole possible *invariant description.*

Recall also that *quantum mechanics provides no geometric or other description of the density of the particle considered, besides the density of the vacuum*, the latter one being normalized to the value 1 in the 4-th component of the *basic unit* $I = \text{Diag.}(1,1,1,1)$ of the Minkowskian space with metric $\eta = \text{Diag.}(1,1,1,-1)$. The lifting $I = \text{Diag.}(1,1,1,1) \rightarrow \hat{I} = \text{Diag.}(A,B,C,D)$ is then useful to represent the density D of any physical medium other than the vacuum. The normalization of D to one for the vacuum then implies the well known realization $D = n_4^2$ where n_4 is the familiar *index of refraction of light* with local speed c_0/n_4, where for $n_4 = 1$ one recovers the speed of light in vacuum c_0. As one can see, *the isotopies permit a direct geometrization of locally varying speeds of light depending on the medium in which it propagates* (see [1d] for a comprehensive study).

Finally, recall that *quantum mechanics has no possibility to represent contact nonpotential interactions.* This is due to the fact that the Hamiltonian has the structure $H = p^2/2 \times m + V(r,p)$. As such, the Hamiltonian can only represent action-at-a-distance interactions with a potential V. Any change in the above structure in attempting to represent nonpotential interactions in a way compatible with quantum mechanics implies the necessary loss of H as the total energy, with major structural implications.

By comparison, *the only known invariant representation of nonlinear, nonlocal, and nonpotential interactions is that via the lifting of the basic unit* $I = 1 \rightarrow \hat{I} = \Gamma(t,r,p,\psi,\ldots)$. In fact, the representation via the isounit is necessarily of nonpotential type, while being the only known invariant representation, trivially, because the unit is the basic invariant of any theory.

This illustrates the importance of the isounits for the central topic of this monograph: *the invariant representation of the nonlinear, nonlocal, and nonpotential interactions expected in the deep overlapping of the wavepackets of electrons in valence bonds, which representation is completely missing in quantum chemistry.*

In particular, this monograph is devoted to the task of showing that the above representation implies a *basically new model of molecular structures which has permitted the representation of molecular characteristics exact to the seventh digit* (*sic*), while quantum chemical model generally miss 2% (Sect. 1.2).

It is hoped that the above comments illustrate the need of generalizing the basic unit of quantum chemistry to includes features, such as nonspherical-deformable shapes, densities, and nonpotential interactions, which are beyond any possibility of scientific treatment by quantum chemistry.

Once the necessity of generalizing the basic unit is understood, the necessity of generalizing the entire mathematical structure of quantum chemistry becomes an evident condition for consistency, thus giving rise to the birth of isomathematics.

In the following section we shall identify the most important mathematical notions which are necessary under the liftings $I \to \hat{I} = 1/\hat{T} > 0$ and $A \times B \to A \hat{\times} B = A \times \hat{T} \times B$.

2.2. Isonumbers and Isofields

Let $\mathbb{F} = \mathbb{F}(a, +, \times)$ be a field of ordinary real numbers $a = n$, $\mathbb{F} = \mathbb{R}(n, +, \times)$, complex numbers $a = c$, $\mathbb{F} = \mathbb{C}(c, +, \times)$, or quaternions $a = q$, $\mathbb{F} = \mathbb{Q}(q, +, \times)$. The *isofields* $\hat{\mathbb{F}} = \hat{\mathbb{F}}(\hat{a}, \hat{+}, \hat{\times})$, first introduced by Santilli [1b] (see also the independent study [4a]), are sets with elements $\hat{a} = a \times \hat{I}$, called *isonumbers*, where $a \in \mathbb{F}$ and $\hat{I}$ is generally outside $\mathbb{F}$, equipped with the *isosum* $\hat{a} \hat{+} \hat{b} = (a + b) \times \hat{I}$ and the *isoproduct* $\hat{a} \hat{\times} \hat{b} = \hat{a} \times \hat{T} \times \hat{b} = (a \times b) \times \hat{I}$. It is easy to see that $\hat{\mathbb{F}}$ verifies all axioms of a field, including most importantly closure under the associative and distributive laws. Therefore, the lifting $\mathbb{F} \to \hat{\mathbb{F}}$ is an isotopy.

All operations of $\hat{\mathbb{F}}$ are lifted in a simple way, yielding the *isopower*, *isoquotient*, *isosquare root*, *isomodulus*, *etc.* [1b]

$$\hat{a}^{\hat{n}} = \hat{a} \hat{\times} \hat{a} \hat{\times} \ldots \hat{\times} \hat{a} = (a^n) \times \hat{I}, \tag{2.4a}$$

$$\hat{a} \hat{/} \hat{b} = (\hat{a}/b) \times \hat{I} = (a/b) \times \hat{I}, \tag{2.4b}$$

$$\hat{a}^{\hat{\frac{1}{2}}} = a^{\frac{1}{2}} \times \hat{I}, \quad \hat{a}^{\hat{\frac{1}{2}}} \hat{\times} \hat{a}^{\hat{\frac{1}{2}}} = a \times \hat{I}, \tag{2.4c}$$

$$\hat{|} \hat{a} \hat{|} = |a| \times \hat{I}. \tag{2.4c}$$

The most important point in the theory of isonumbers is that *the abstract axioms of a field of numbers do not necessarily require that the unit must assume the trivial value +1 dating back to biblical times, because the same axioms also admit an arbitrary value of the unit, provided that it is positive-definite.*

In fact, the set of isoreal numbers identified in the original proposal [1b] $\hat{n} = n \times \hat{I}$, where n is an arbitrary real number and $\hat{I}$ can be a positive-definite number, matrix, operator, *etc.*, when equipped with the isosum $\hat{n} \hat{+} \hat{m} = (n + m) \times \hat{I}$ and the isoproduct $\hat{n} \hat{\times} \hat{m} = (n \times m) \times \hat{I}$ verify exactly the same axioms as those for real numbers n.

Despite the similarities between numbers and isonumbers, the latter theory has important implications studied in detail in Jiang's monograph

[4a]. For the limited scope of this presentation it is sufficient to note that *prime numbers depend on the assumed unit.* For instance, the number 4 becomes prime when the isounit has the value $\hat{I} = 3$.

Similarly, traditional statements such as "$2 \times 2 = 4$" have sense only for conventional mathematics and quantum chemistry, but they have no mathematical meaning for isomathematics and related isochemistry without first identifying the assumed unit and the assumed product. In fact, for $\hat{I} = 3, \quad 2 \hat{\times} 2 = 12$.

In summary, conventional quantum chemistry is based on conventional numbers. By comparison, the broader hadronic chemistry is based on *new numbers*, those with a positive-definite but otherwise arbitrary unit $\hat{I}$. This is not a mere mathematical curiosity, but the *only* known way for the *invariant* representation of nonspherical-deformable shapes, densities, and nonlinear, nonlocal and nonpotential interactions due to the deep overlapping of the wavepackets of valence electrons.

2.3. Isospaces and Isogeometries

As well known, metric (or pseudo-metric) spaces $S = S(r, m, \mathbb{R})$ with local coordinates r and metric m are defined over the field of real numbers $\mathbb{R} = \mathbb{R}(n, +, \times)$. This is the case for the basic carrier space of chemistry, the three-dimensional Euclidean space $E = E(r, \delta, \mathbb{R})$ with coordinates $r = \{r^k\} = \{x, y, z\}, \ k = 1, 2, 3$, metric $\delta = \{\delta_{ij}\} = \mathrm{Diag.}(1, 1, 1)$ and basic unit $I = (I^i_j) = \mathrm{Diag.}(1, 1, 1)$. The lifting of numbers $\mathbb{R} \to \hat{\mathbb{R}}$ then requires a necessary compatible lifting of the all metric (or pseudo-metric) spaces.

The *isoeuclidean spaces* $\hat{E} = \hat{E}(\hat{r}, \hat{\delta}, \hat{\mathbb{R}})$, first introduced by Santilli in Ref. [1c] (see also the recent account [1d]), are metric spaces with *isocoordinates* $\hat{r} = r \times \hat{I} = \{r^k\} \times \hat{I}$ and *isometric* $\hat{\Delta} = \hat{\delta} \times \hat{I}, \ \hat{\delta} = \{\hat{\delta}_{ij}\} = \{\hat{T}^k_i \times \delta_{kj}\}$ with interval defined over $\hat{\mathbb{R}}$

$$
\begin{aligned}
\hat{r}^{\hat{2}} = \hat{r}^k \hat{\times} \hat{r}_k = \hat{r}^i \hat{\times} \hat{\Delta}_{ij} \hat{\times} \hat{r}^j = \\
= (r^i \times \hat{I}) \times \hat{T} \times (\hat{\delta}_{ij} \times \hat{I}) \times \hat{T} \times (r^j \times \hat{I}) = \\
= (r^i \times \hat{\delta}_{ij} \times r^j) \times \hat{I} \in \hat{\mathbb{R}}.
\end{aligned}
\tag{2.5}
$$

It has been proved [*loc. cit.*] than $\hat{E}$ preserves all axioms of the conventional space E, and therefore, the lifting $E \to \hat{E}$ is an isotopy. In essence, the axioms of the Euclidean geometry are realized in such a way as to admit a metric with an unrestricted functional dependence, $\hat{\delta} = \hat{\delta}(t, r, p, \psi, \partial\psi, \ldots)$. Note that the isometric contains as particular cases all possible metrics with the same signature, including the Riemannian metric, yet the axioms are Euclidean. Note also that the isounit is the same for both isospaces and their underlying isofield.

The *isoeuclidean geometry* [1c] is the geometry of the isoeuclidean spaces elaborated via the entire isomathematics. The reader should be aware that, while Christoffel's symbols, covariant derivatives, and the other tools of the Riemannian geometry are inapplicable to the conventional Euclidean geometry because the latter is flat, the same tools apply in full; to the covering isoeuclidean geometry [1d, 1e], evidently due to the explicit dependence of the isometric on the local coordinates r. As a result, the lifting from the Euclidean geometry to its isotopic covering is, by far, nontrivial.

In reality, as proved in Refs. [1d, 1e], the isoeuclidean geometry permits a symbiotic unification of the conventional Euclidean and Riemannian geometry in space coordinates, which are now differentiated by the selection of the *unit*, rather than the metric, as conventionally done.

A similar isotopic unification holds for the Minkowskian and Riemannian geometry for spacetime, which are geometrically unified into the *isominkowskian geometry* [1d, 1e], by therefore, permitting a symbiotic unification of special and general relativities, with a resolution of at least some of the vexing problems that have afflicted general relativity during the 20-th century.

The reader should also be aware of the availability of other isogeometries, such as the *isosymplectic geometry* [1e] which is important for the isoquantization to be studied in the next chapter. Regrettably, we are not in a position to review these geometries to prevent a prohibitive length of this monograph.

Hereon we shall adopt the notation that all quantities with a "hat," such as $\hat{r}$, $\hat{H}$, *etc.*, are defined on isospace over isofields, while all quantities without a "hat," such as r, H, *etc.*, are defined on conventional spaces over conventional fields. Note that the isoproduct of $\hat{r}$ by an arbitrary quantity A is conventional, $\hat{r}\hat{\times}A = (r \times \hat{I}) \times \hat{T} \times A \equiv r \times A$. We shall therefore tacitly assume that $\hat{r}$ is the conventional coordinate, only written in isospace over isofield, an assumption which is confirmed by the invariant (2.5) in which the use of r is sufficient.

We shall also assume that contractions such as $\hat{r}^k\hat{\times}\hat{r}_k$ are in isospace (i.e., requiring the isometric $\hat{\delta}$) while contractions such as $\delta_{ij} \times \hat{r}_i \times \hat{r}_j$ or $\hat{T}^i_k \times \hat{r}^k$ are in ordinary space (i.e., requiring the conventional metric δ).

A fundamental notion of the isoeuclidean geometry is that of *generalized space units*. In fact, the Euclidean unit $I = \text{Diag.}(1, 1, 1)$ represents in a dimensionless form the units used for each of the three Cartesian axes, e.g., $I = \text{Diag.}(1 \text{ cm}, 1 \text{ cm}, 1 \text{ cm})$. Since the isounit is positive-definite, it can always be diagonalized into the form

$$\hat{I} = \text{Diag.}(n_1^2, n_2^2, n_3^2), \quad \hat{T} = \text{Diag.}(n_1^{-2}, n_2^{-2}, n_3^{-2}), \quad n_k > 0. \tag{2.6}$$

This indicates that isospaces imply not only the assumption of new units $n_k^2 \times \text{cm}$, but also the fact that such units are generally different for different axes.

When the carrier space $\hat{E}$ is extended to include time via the Kronecker product $\hat{E}(\hat{t}, \hat{\mathbb{R}}_t) \times E(\hat{r}, \hat{\delta}, \hat{\mathbb{R}})$, this implies the additional assumption of a new, positive-definite, one-dimensional unit of time, resulting in the total spacetime unit

$$\hat{I}_{\text{Tot}} = \hat{I}_{\text{space}} \times \hat{I}_{\text{time}} = \{\text{Diag.}(n_1^2, n_2^2, n_3^2)\} \times \{n_4^2\}. \tag{2.7}$$

In summary, the fundamental notion of hadronic chemistry is a new conception of spacetime, called *isospacetime* [1d] with *isocoordinates* $\hat{r} = \{r^k\} \times \hat{I}$ and *isotime* $\hat{t} = t \times \hat{I}_t$ defined on corresponding isofields.

The broadening of the basic units evidently requires a corresponding broadening of the notion of distance into the *isodistance*, which, under realization (2.6), can be written [1c, 1d]

$$\hat{D} = \left[\frac{(r_{1x} - r_{2x})^2}{n_1^2} + \frac{(r_{1y} - r_{2y})^2}{n_2^2} + \frac{(r_{1z} - r_{2z})^2}{n_3^2} \right]^{1/2} \times \hat{I}. \tag{2.8}$$

As we shall see, hadronic chemistry can be introduced via "hidden" degrees of freedom of conventional chemistry. This point can be first illustrated by indicating that isospaces originate from the following degrees of freedom of the Euclidean metric and of related unit [*loc. cit.*]

$$\delta \to \hat{\delta} = n^{-2} \times \delta, \quad I \to \hat{I} = n^2 \times I, \tag{2.9}$$

where n is a non-null real parameter, under which the Euclidean interval remains invariant [*loc. cit.*]

$$\begin{aligned} r^2 = (r^i \times \delta_{ij} \times r^j) \times I \equiv [r^i \times (n^{-2} \times \delta_{ij}) \times r^j] \times (n^2 \times I) = \\ = (r^i \times \hat{\delta}_{ij} \times r^j) \times \hat{I} = \hat{r}^{\hat{2}}. \end{aligned} \tag{2.10}$$

The general form of isospaces emerges when extending the parameter n to a 3×3 positive-definite matrix $\hat{I}$. The classical formulation of isochemistry is based on the above "hidden" degree of freedom of the Euclidean geometry.

Therefore, the fundamental geometric invariant of isochemistry is

$$[\text{Length}]^2 \times [\text{Unit}]^2 = \text{invariant}, \tag{2.11}$$

where the identities $I \times I = I$ and $\hat{I} \hat{\times} \hat{I} = \hat{I}$ are assumed. We are at freedom to change lengths and units under the above invariance law, provided that the changes are *inverse* of each other.

Note that, while the unit is lifted in the form $I = \text{Diag.}(1,1,1) \rightarrow \hat{I} = \text{Diag.}(n_1^2, n_2^2, n_3^2)$, the metric is lifted in the inverse form $\delta = \text{Diag.}(1,1,1) \rightarrow \hat{\delta} = \text{Diag.}(n_1^{-2}, n_2^{-2}, n_3^{-2})$. In this case, the perfect sphere is lifted precisely to the nonspherical and deformable shapes indicated in Sect. 2.2.2, $r^2 = r^i \times \delta_{ij} \times r^2 = (x^2 + y^2 + z^2) \times I \rightarrow \hat{r^2} = r^i \times \hat{\delta}_{ij} \times r^j \times \hat{I} = (x^2/n_1^2 + y^2/n_2^2 + z^2/n_3^2) \times \hat{I}$.

An understanding of isomathematics requires the knowledge that the quantity $\hat{r^2} = r^i \times \hat{\delta}_{ij} \times r^j \times \hat{I} = (x^2/n_1^2 + y^2/n_2^2 + z^2/n_3^2) \times \hat{I}$ is the perfect sphere when computed in isospace, called *isosphere*, while the ellipsoidical shape occurs only in its *projection* on conventional fields. This is due to the fact that each semiaxis is lifted in the amount $1_k \rightarrow n_k^2$ while the corresponding unit is lifted in the *inverse* amount $1_k \rightarrow 1/n_k^2$. In view of the basic invariant (2.11), *the numerical values of the semiaxes of the isosphere remains* 1.

The relativistic extension of the above Euclidean isotopies is based on the lifting of the 4-dimensional unit of the Minkowskian geometry into the 4-dimensional isounit $I = \text{Diag.}(1,1,1,1) \rightarrow \hat{I} = \text{Diag.}(n_1^2, n_2^2, n_3^2, n_4^2)$, $n_\mu > 0$, the lifting of the Minkowski metric into the isometric $\eta = \text{Diag.}(1,1,1,-c_0^2) \rightarrow \hat{\eta} = \text{Diag.}(1/n_1^2, 1/n_2^2, 1/n_3^2, -c_0^2/n_4^2)$, and the lifting of the light cone into the *light isocone*, $x^2 + y^2 + z^2 - c_0^2 \times t^2 = 0 \rightarrow x^2/n_1^2 + y^2/n_2^2 + z^2/n_3^2 - c_0^2 \times t^2/n_4^2 = 0$. The representation by the isotopies of the local character of the speed of light within physical media, $c = c_0/n_4$, is then evident (see Ref. [1d] for a detailed study).

Despite their simplicity, the physical and chemical implications of the above results are far reaching. Recall that the notorious "universal constancy of the speed of light" is a philosophical abstraction, because light has a speed c_0/n_4 depending on the features of the medium in which it propagates, by having one speed in our atmosphere, a different speed in water, a yet different speed in glass, *etc.* The construction of a relativistic model based on the "universal constancy of the speed of light" cannot, therefore, be exact when propagation of impulses or events occurs within physical media, such as that due to wave-overlappings of valence electrons. Even though numerically small, corrections due to the real value of the speed of light implies the possible birth of new effects and related new technologies, as we shall see in the final part of this monograph.

Note also that the above invariance renders isogeometries compatible with our sensory perception, thus permitting intriguing advances in biology, e.g., the use of the isoeuclidean geometry to represent the *growth* of sea shells in time which *is not* quantitatively possible via the conventional Euclidean geometry, as indicated in Chapter 1 [2e]. An important aspect is that the internal generalized geometry, if of isotopic type, is compatible with our Euclidean perception. By comparison, the use of other (e.g., curved) geometries would be detected by our sensory perception and, as such, be inapplicable because contrary to visual observation.

The above occurrence illustrates the expected role of isomathematics in chemistry, because permitting a structurally broader formalism while being axiom-preserving, thus including the compatibility with our sensory perception of the Euclidean axioms.

The reader should be aware that the implications of the new notion of isospacetime are far reaching. For instance, any given biological structure seen from a conventional observer (say, a sea shell) has different dimensions as well as a different shape for an internal isotopic observer, because of invariance (2.11); locomotion can occur via a local change of the geometry without any applied external force, because of the inherent change of distance (2.8); and even greater peculiarities occur when one adds the isotopies of time, which permit quantitative interpretations of bifurcations via the four possible motions in time, forward and backward to future and past times (see [2e] for detailed studies and, later on, Fig. .2.1).

It should be stressed that the isoeuclidean geometry is here presented for the sole intent of providing a *mathematical* elaboration of the isotopies of quantum chemistry, because our physical reality remains in the conventional Euclidean space. In fact, upon completing the desired mathematical elaboration in isospace over isofields, physically measurable quantities are obtained in their *projection* on conventional spaces over conventional fields.

2.4. Isodifferential Calculus

One of the reasons for the delay in the construction of hadronic mechanics since its proposal in 1978 was due to difficulties in identifying the origin of the non-invariance of the initial formulation, that is, the lack of prediction of the same numerical values for the same quantities under the same conditions, but at different times. Eventually such non-invariance resulted in being where expected the least, in the ordinary differential calculus.

In memoir [1e] of 1996 Santilli resolved the problem of invariance by introducing for the first time the *isodifferential calculus* with *space isodifferentials*

$$\hat{d}\hat{r}^k = \hat{I}^k_i \times d\hat{r}^i, \qquad (2.12a)$$

$$\hat{d}\hat{r}_k = \hat{T}_k^i \times d\hat{r}_i, \tag{2.12b}$$

and *space isoderivatives*

$$\begin{aligned}\hat{\partial}_k = \hat{\partial}\hat{/}\hat{\partial}\hat{r}^k = \partial\hat{/}(\hat{I}_i^k \times \partial\hat{r}^i) = \hat{T}_k^i \times \partial\hat{/}\partial\hat{r}^i = \\ = (\hat{T}_k^i \times \partial/\partial\hat{r}^i) \times \hat{I} = \hat{T}_k^i \times \partial/\partial r^i,\end{aligned} \tag{2.13a}$$

$$\begin{aligned}\hat{\partial}^k = \hat{\partial}\hat{/}\hat{\partial}\hat{r}_k = \partial\hat{/}(\hat{T}_k^i \times \partial\hat{r}_i) = \hat{I}_i^k \times \partial\hat{/}\partial\hat{r}_i = \\ = (\hat{I}_i^k \times \partial/\partial\hat{r}_i) \times \hat{I} = \hat{T}_i^k \times \partial/\partial r_i,\end{aligned} \tag{2.13b}$$

with basic properties

$$\partial\hat{r}^i\hat{/}\partial\hat{r}^j = \hat{\delta}_j^i = \delta_j^i \times \hat{I}, \tag{2.14}$$

and, by using (2.13b), the *isolaplacian*

$$\hat{\partial}_k\hat{\partial}^k = \hat{\partial}_i \times \hat{\delta}^{ij} \times \hat{\partial}_j = \partial_i \times \delta^{ij} \times \hat{T}_j^k \times \partial_k. \tag{2.15}$$

Note that isoderivatives commute when formulated on isospace over isofields, i.e., $\hat{\partial}_i\hat{\partial}_j = \hat{\partial}_j\hat{\partial}_i$, as necessary for an isotopy. However, their projections in conventional spaces E over conventional field $\mathbb{R}$ do not necessarily commute because of the possible r-dependence of $\hat{T}$, $(\hat{T}_i^m \times \partial_m) \times (\hat{T}_j^n \times \partial_n) \neq (\hat{T}_j^m \times \partial_m) \times (\hat{T}_i^n \times \partial_n)$. This dichotomy is a general rule for all isotopic theories because they admit *two* generally different interpretations, that on isospaces over isofields (i.e., computed with respect to generalized units), and that of its projection on conventional spaces over conventional fields (i.e., computed with respect to conventional units).

The *time isodifferentials* and *isoderivatives* are defined accordingly via the use of the (scalar) isounit of time $\hat{I}_t = 1/\hat{T}_t$

$$\hat{d}\hat{t} = \hat{I}_t \times d\hat{t}, \quad \hat{\partial}_t = \hat{T}_t \times \partial_t = \hat{T}_t \times \partial/\partial t. \tag{2.16}$$

The (indefinite) *isointegrals* are trivially defined by $\hat{\int} = \int \hat{T}$, so that

$$\hat{\int} \hat{d}\hat{r} = \int \hat{T} \times \hat{I}\, d\hat{r} = \hat{r}. \tag{2.17}$$

The underlying broader notion of continuity is indicated in the forthcoming Sect. 2.7.

2.5. Isohilbert Spaces

As it is well known, the Hilbert space $\mathcal{H}$ used in conventional chemical bonds is expressed in terms of the states $|\psi\rangle$, $|\phi\rangle, \ldots$, with normalization

$\langle\psi| \times |\psi\rangle = 1$ and inner product $\langle\phi| \times |\psi\rangle = \int dr^3 \phi^\dagger(r) \times \psi(r)$ defined over the field of complex numbers $\mathbb{C} = \mathbb{C}(c, +, \times)$.

The lifting of the latter, $\mathbb{C}(c, +, \times) \rightarrow \hat{\mathbb{C}}(\hat{c}, \hat{+}, \hat{\times})$, requires a compatible lifting of $\mathcal{H}$ into the *isohilbert space* $\hat{\mathcal{H}}$ with *isostates* $|\hat{\psi}\rangle, |\hat{\phi}\rangle, \ldots$, *isoinner product* and *isonormalization*

$$\langle\hat{\psi}|\hat{\times}|\psi\rangle \times \hat{I} = \left[\hat{\int} d\hat{r}^3 \; \hat{\psi}^\dagger(\hat{r}) \times \hat{T} \times \hat{\psi}(\hat{r})\right] \times \hat{I} \in \hat{\mathbb{C}}, \qquad \langle\hat{\psi}|\hat{\times}|\hat{\psi}\rangle = 1, \tag{2.18}$$

first introduced by Myung and Santilli in 1982 [3a] (see also monographs [2c, 2d] for a comprehensive study).

It is easy to see that the isoinner product is still inner (because $\hat{T} > 0$). Thus, $\hat{\mathcal{H}}$ is still Hilbert and the lifting $\mathcal{H} \rightarrow \hat{\mathcal{H}}$ is an isotopy. Also, it is possible to prove that *isohermiticity coincides with conventional Hermiticity*. As a result, all quantities which are observable for conventional quantum chemistry remain so for our hadronic chemistry, and *vice versa*.

For consistency, the conventional eigenvalue equation $H \times |\psi\rangle = E \times |\psi\rangle$ must also be lifted into the *isoeigenvalue form* [3a]

$$\hat{H}\hat{\times}|\hat{\psi}\rangle = \hat{H} \times \hat{T} \times |\hat{\psi}\rangle = \hat{E}\hat{\times}|\hat{\psi}\rangle = = (E \times \hat{I}) \times \hat{T} \times |\hat{\psi}\rangle = E \times |\hat{\psi}\rangle, \tag{2.19}$$

where, as one can see, the final results are ordinary numbers. Note the *necessity* of the isotopic action $\hat{H}\hat{\times}|\hat{\psi}\rangle$, rather than $\hat{H} \times |\hat{\psi}\rangle$. In fact, only the former admits $\hat{I}$ as the correct unit, $\hat{I}\hat{\times}|\hat{\psi}\rangle = \hat{T}^{-1} \times \hat{T} \times |\hat{\psi}\rangle \equiv |\hat{\psi}\rangle$.

It is possible to prove that *the isoeigenvalues of isohermitean operators are isoreal*, i.e., they have the structure $\hat{E} = E \times \hat{I}, \quad E \in \mathbb{R}(n, +, \times)$. As a result all real eigenvalues of conventional quantum chemical bonds remain real for hadronic chemistry.

We also recall the notion of *isounitary operators* as the isooperators $\hat{U}$ on $\hat{\mathcal{H}}$ over $\hat{\mathbb{C}}$ satisfying the isolaws

$$\hat{U}\hat{\times}\hat{U}^\dagger = \hat{U}^\dagger\hat{\times}\hat{U} = \hat{I}. \tag{2.20}$$

where we have used the identity $\hat{U}^{\hat{\dagger}} \equiv \hat{U}^\dagger$. We finally indicate the notion of *isoexpectation value* of an isooperator $\hat{H}$ on $\hat{\mathcal{H}}$ over $\hat{\mathbb{C}}$

$$\langle\hat{H}\rangle = \frac{\langle\hat{\psi}|\hat{\times}\hat{H}\hat{\times}|\hat{\psi}\rangle}{\langle\hat{\psi}|\hat{\times}|\hat{\psi}\rangle}. \tag{2.21}$$

It is easy to see that *the isoexpectation values of isohermitean operators coincide with the isoeigenvalues*, as in the conventional case.

Note also that *the isoexpectation value of the isounit is the isounit*, $\langle \hat{I} \rangle = \hat{I}$, provided, of course, that one uses the isoquotient (otherwise $\langle \hat{I} \rangle = I$).

The isotopies of quantum chemistry studied in the next chapter are based on the following novel invariance property of the conventional Hilbert space [1f], here expressed in term of a non-null scalar n independent from the integration variables,

$$\langle \hat{\phi} | \times | \hat{\psi} \rangle \times I \equiv \langle \hat{\phi} | \times n^{-2} \times | \hat{\psi} \rangle \times (n^2 \times I) = \langle \phi | \hat{\times} | \psi \rangle \times \hat{I}. \tag{2.22}$$

It then follows that the isotopies of quantum chemistry are based on a concrete realization of the theory of "hidden variables", that is, via the explicit realization of the hidden *variable* in term of the isotopic *operator*, $\lambda = \hat{T}$, and isoeigenvalue law (2.19). Note that the isoeigenvalue E evidently depends on the assumed value of $\lambda = \hat{T}$ per each $\hat{H}$. The "hidden" character of the realization is best expressed by the identity of the basic abstract axioms of quantum and hadronic mechanics and chemistry.

The latter topic has fundamental relevance because the known argument by Einstein, Podolski, and Rosen on the "lack of completion of quantum mechanics" dioscussed in Chapter 1 is evidently extendable in its entirety to the "lack of completion of quantum chemistry." Hadronic chemistry and its new model of molecular structures presented in Chapters 4 and 5 are, therefore, a "completion" of quantum chemistry and conventional molecular models precisely along the historical teaching by Einstein, Podolsky and Rosen. The "completion" is achieved via the invariant treatment of nonlinear, nonlocal, and nonpotential interactions which are completely absent in quantum chemistry. For these aspects and related references, one may consult Ref. [1g] or monograph [2d].

Note that new invariances (2.10) and (2.22) have remained undetected throughout the 20-th century because they required the prior discovery of *new numbers*, those with arbitrary units.

2.6. Isoperturbation Theory

We are now sufficiently equipped to illustrate the computational advantages in the use of isotopies. Under sufficient continuity conditions, *all perturbative and other series which are conventionally divergent ones (weakly convergent) can be turned into convergent (strongly convergent) forms via the use of isotopies with sufficiently small isotopic element (sufficiently large isounit)*,

$$|\hat{T}| \ll 1, \quad |\hat{I}| \gg 1. \tag{2.23}$$

The emerging perturbation theory was first studied by Jannussis and Mignani [3b], and then studied in more detail in monograph [2d] under the name of *isoperturbation theory.*

Consider a Hermitean operator on $\mathcal{H}$ over $\mathbb{C}$ of the type

$$H(k) = H_0 + k \times V, \quad H_0 \times |\psi\rangle = E_0 \times |\psi\rangle, \tag{2.24a}$$

$$H(k) \times |\psi(k)\rangle = E(k) \times |\psi(k)\rangle, \quad k \gg 1. \tag{2.24b}$$

Assume that H_0 has a nondegenerate discrete spectrum. Then, conventional perturbative series are *divergent*, as well known. In fact, the eigenvalue $E(k)$ of $H(k)$ up to second order is given by

$$\begin{gathered} E(k) = E_0 + k \times E_1 + k^2 \times E_2 = \\ = E_0 + k \times \langle\psi| \times V \times |\psi\rangle + k^2 \times \sum_{p \neq n} \frac{|\langle\psi_p| \times V \times |\psi_n\rangle|^2}{E_{0n} - E_{0p}}, \end{gathered} \tag{2.25}$$

But under isotopies we have

$$H(k) = H_0 + k \times V, \quad H_0 \times \hat{T} \times |\tilde{\psi}\rangle = \tilde{E}_0 \times |\tilde{\psi}\rangle, \quad \tilde{E}_0 \neq E_0, \tag{2.26a}$$

$$H(k) \times \hat{T} \times |\hat{\psi}(k)\rangle = \tilde{E}(k) \times |\hat{\psi}(k)\rangle, \quad \tilde{E} \neq E, \quad k > 1. \tag{2.26b}$$

A simple lifting of the conventional perturbation expansion then yields

$$\begin{gathered} \tilde{E}(k) = \tilde{E}_0 + k \times \tilde{E}_1 + k^2 \times \tilde{E}_2 + \hat{O}(k^2) = \\ = \tilde{E}_0 + k \times \langle\tilde{\psi}| \times \hat{T} \times V \times \hat{T} \times |\tilde{\psi}\rangle + \\ + k^2 \times \sum_{p \neq n} \frac{|\langle\hat{\psi}_p| \times \hat{T} \times V \times \hat{T} \times |\hat{\psi}_n\rangle|^{\hat{2}}}{\tilde{E}_{0n} - \tilde{E}_{0p}}, \end{gathered} \tag{2.27}$$

whose convergence can be evidently reached via a suitable selection of the isotopic element, e.g., such that $|\hat{T}| \ll k$.

As an example, for a positive-definite constant $\hat{T} \ll k^{-1}$, expression (2.27) becomes

$$\begin{gathered} \tilde{E}(k) = \tilde{E}_0 + k \times \hat{T}^2 \times \langle\hat{\psi}| \times V \times |\psi\rangle + k^2 \times T^5 \times \\ \times \sum_{p \neq n} \frac{|\langle\psi_p| \times V \times |\psi_n\rangle|^2}{\tilde{E}_{0n} - \tilde{E}_{0p}}. \end{gathered} \tag{2.28}$$

This shows that the original divergent coefficients $1, k, k^2, \ldots$ are now turned into the manifestly convergent coefficients $1, k \times T^2, k^2 \times T^5, \ldots,$ with $k > 1$ and $\hat{T} \ll 1/k$, thus ensuring isoconvergence for a suitable selection of $\hat{T}$ for each given k and V.

A more effective reconstruction of convergence can be seen in the algebraic approach following the outline of forthcoming Sect. 2.9. At this introductory stage, we consider a divergent canonical series,

$$\begin{aligned} A(k) = A(0) + k \times [A, H]/1! + \\ + k^2 \times [[A, H], H]/2! + \ldots \to \infty, \quad k > 1, \end{aligned} \tag{2.29}$$

where $[A, H] = A \times H - H \times A$ is the familiar Lie product, and the operators A and H are Hermitean and sufficiently bounded. Then, under the isotopic lifting the preceding series becomes

$$\hat{A}(k) = \hat{A}(0) + k \times [A \hat{,} H]/1! + k^2 \times [[A \hat{,} H] \hat{,} H]/2! + \cdots \leq |N| < \infty, \tag{2.30a}$$

$$[A \hat{,} H] = A \times \hat{T} \times H - H \times \hat{T} \times A, \tag{2.30b}$$

which holds, e.g., for the case $T = \varepsilon \times k^{-1}$, where ε is a sufficiently small positive-definite constant.

In summary, the studies on the construction of hadronic mechanics have indicated that the apparent origin of divergences (or slow convergence) in quantum mechanics and chemistry is their lack of representation of nonlinear, nonlocal, and nonpotential effects because when the latter are represented via the isounit, full convergence (much faster convergence) can be obtained.

As we shall see in Chapters 4 and 5, the proposed new model of chemical bond does indeed verify the crucial condition (2.23), by permitting convergence of chemical computations at least one thousand times faster than those of quantum chemistry, with evident advantages, e.g., a drastic reduction of computer time.

2.7. Isofunctional Analysis

A necessary condition for the consistency of isochemistry is that it is elaborated with the novel *isofunctional analysis* first studied by Kadeisvili [3c] who introduced the notion of *isocontinuity*.

Isofunctional analysis consists of conventional and special functions and transforms re-constructed in such a way to admit the quantity $\hat{I}$ as the correct left and right unit.

As an indication, the use in hadronic chemistry of conventional trigonometric functions, Legendre polynomials, Fourier and Laplace transforms, *etc.*, would lead to a host of inconsistencies that often remain undetected by nonexperts in the field. The inconsistencies also occur in the use of elementary notions, such as exponentiation, determinant, trace, *etc.*

We cannot possibly review here the isofunctional analysis and must, for brevity, refer to monograph [2c]. We merely mention the *isotrigonometric functions*, first studied by Santilli [2c], which are defined in a two-dimensional isoeuclidean space with diagonal isounit

$$\hat{E}(\hat{r}, \hat{\delta}, \hat{\mathbb{R}}), \quad \hat{r} = \{x, y\} \times \hat{I}, \quad \hat{I} = \text{Diag.}(n_1^2, n_2^2), \quad n_k > 0, \tag{2.31a}$$

$$\hat{r}^{\hat{2}} = (x^2/n_1^2 + y^2/n_2^2) \times \hat{I} \in \hat{\mathbb{R}}. \tag{2.31b}$$

Under isopolar coordinates, the above isospace is turned into the *isogauss plane* with isounit $\hat{I}_{\hat{\theta}} = n_1 \times n_2$. Conventional angles $\theta \in \mathbb{R}$ are then turned into *isoangles* $\hat{\theta} = \theta \times \hat{I}_{\hat{\theta}} = \theta \times (n_1 \times n_2) \in \hat{\mathbb{R}}_{\hat{\theta}}$. Conventional trigonometric functions then acquire the form [2c]

$$\text{isocos}\,\hat{\theta} = [n_1 \times \cos(\theta \times n_1 \times n_2)] \times \hat{I}, \tag{2.32a}$$

$$\text{isosin}\,\hat{\theta} = [n_2 \times \sin(\theta \times n_1 \times n_2)] \times \hat{I}, \tag{2.32b}$$

and properties

$$\begin{gathered} \text{isocos}^{\hat{2}}\hat{\theta}\hat{/}n_1^2 + \text{isosin}^{\hat{2}}\hat{\theta}\hat{/}n_2^2 = \\ = (\cos^2(\theta \times n_1 \times n_2) + \sin^2(\theta \times n_1 \times n_2) \times \hat{I} = \hat{I}. \end{gathered} \tag{2.33}$$

Expression (2.33) characterizes the *isocircle*, i.e., the perfect circle in isospace with *isopolar coordinates* [*loc. cit.*]

$$\hat{x} = \hat{r}\hat{\times}\text{isocos}\,\hat{\theta}, \quad \hat{y} = \hat{r}\hat{\times}\text{isosin}\,\hat{\theta}. \tag{2.34}$$

In fact, in the transition $E \to \hat{E}$ we have the lifting of the semiaxes of the circle $1_k \to n_k^{-2}$, thus resulting in ellipses. But the related units are lifted by the *inverse* amount, $1_k \to n_k^2$, thus preserving the original perfect circle in accordance with invariance law (2.11).

The *isospherical coordinates* are then given by [2c]

$$\begin{gathered} \hat{x} = \hat{r}\hat{\times}\text{isosin}\,\hat{\phi}\hat{\times}\text{isocos}\,\hat{\theta}, \\ \hat{y} = \hat{r}\hat{\times}\text{isosin}\,\hat{\phi}\hat{\times}\text{isosin}\,\hat{\theta}, \\ \hat{z} = \hat{r}\hat{\times}\text{isocos}\,\hat{\phi}, \end{gathered} \tag{2.35}$$

where $\hat{\phi} = \phi \times n_3$.

For brevity, we regret to refer the interested reader to the above quoted literature for remaining aspects of the isofunctional analysis. Additional notions of the isofunctional analysis will be introduced when needed during the course of our analysis.

2.8. Isolinearity, Isolocality, Isocanonicity and Isounitarity

It is important to see that, despite their physical inequivalence, *the isotopies of Hilbert spaces preserve the conventional linearity, locality, and unitarity*. As we shall see, the achievement of invariance in the description of nonlinear, nonlocal, and noncanonical-nonunitary interactions is due precisely to the reconstruction on isospaces over isofields of the conventional structures.

To begin, the isotopies of Hilbert spaces are highly nonlinear in the wavefunctions (and their derivatives), as evident from isoeigenvalues equation $\hat{H}(\hat{r},\hat{p})\hat{\times}\hat{T}(\hat{r},\hat{p},\hat{\psi},\partial\hat{\psi},\ldots) \times |\hat{\psi}\rangle = E \times |\hat{\psi}\rangle$. Yet, the theory is i*isolinear*, i.e., it verifies the linearity conditions in isospace. In fact, for all possible $\hat{a}, \hat{b} \in \hat{\mathbb{C}}$ and $|\hat{\phi}\rangle, |\hat{\psi}\rangle \in \hat{\mathcal{H}}$, we have the identity

$$\hat{A}\hat{\times}(\hat{a}\hat{\times}|\hat{\psi}\rangle + \hat{b}\hat{\times}|\hat{\phi}\rangle) = \hat{a}\hat{\times}\hat{A}\hat{\times}|\hat{\psi}\rangle + \hat{b}\hat{\times}\hat{A}\hat{\times}|\hat{\phi}\rangle. \tag{2.36}$$

The implications of the above property are far reaching. Recall from Sect. 1.7 that conventionally formulated nonlinear theories cannot be used for the study of composite systems because of the violation of the superposition principle (which follows form the violation of linearity). This implies that *quantum chemistry cannot permit consistent nonlinear formulations of molecular bonds, thus implying the dramatic reduction of reality to linear theories*. Isolinearity resolves this inconsistency, by permitting for the first time rigorous formulation of *nonlinear molecular models*, precisely because of the reconstruction of the superposition principle on isospaces over isofields.

A similar situation occurs for *locality*. In fact, the isotopies of Hilbert spaces are *nonlocal-integral* because interactions of that type are admitted in the isounits, as we shall see. Nevertheless, isohilbert spaces are *isolocal*, i.e., they verify the condition of locality on isospaces over isofields. In particular, the isotopies of conventional theories are everywhere local-differential except at the isounit. On more technical grounds, the isotopies imply the birth of a new *integro-differential topology* first submitted by Tsagas and Sourlas [3d] (see also Santilli [1e]).

By recalling that quantum chemistry is strictly local-differential, the above new topology has fundamental relevance inasmuch as it permits mathematically rigorous quantitative studies of the nonlocal-integral component of valence bonds due to wave-overlappings.

The reconstruction of canonicity on isospace over isofields, called *isocanonicity*, is studied in the next section.

Next, recall that quantum chemistry is said to be unitary in the sense that the only allowed transformations are of the unitary type, $U \times U^{\dagger} = U^{\dagger} \times U = I$. By comparison, the isotopies of Hilbert spaces are *nonunitary* because they admit interactions not representable with a Hamiltonian, as we shall see. Nevertheless, the isotopies reconstructs unitarity on isospace over isofields, a property called *isounitarity*. In fact, nonunitary transforms can always be rewritten in the *identical* isotopic form [2c]

$$W \times W^{\dagger} = \hat{I} = 1/\hat{T} \neq I, \quad W = \hat{W} \times \hat{T}^{1/2}, \tag{2.37a}$$

$$W \times W^{\dagger} \equiv \hat{W} \hat{\times} \hat{W}^{\dagger} = \hat{W}^{\dagger} \hat{\times} \hat{W} = \hat{I}. \tag{2.37b}$$

The reconstruction of unitarity on isospace over isofields also has deep implications for the mathematical structure as well as practical applications of hadronic chemistry. In fact, said reconstruction permits for the first time the invariant treatment of *nonpotential* interactions and effects which are completely absent in quantum chemistry.

In summary, a serious understanding of hadronic chemistry requires the knowledge that its underlying novel mathematics permits rigorous, invariant representations of new features, effects, and interactions which are strictly prohibited by quantum chemistry, as it is typically the case for all covering new theories versus their original particularization.

2.9. Lie-Santilli Isotheory

The fundamental algebraic structure of contemporary chemistry, Lie's theory, is linear, local-differential, and canonical in its classical realization, or unitary in its operator form (see, e.g., monographs [6]). As such, Lie's theory is insufficient to characterize the desired nonlinear, nonlocal-integral, and noncanonical-nonunitary component of valence bonds.

In a series of studies dating back to 1978, Santilli [1a] (see also monographs [2b]) proposed the isotopies of the entire Lie theory (universal enveloping associative algebras, Lie algebras, Lie groups, transformation and representation theory, *etc.*). The emerging broader theory is today called the *Lie-Santilli isotheory*. For independent reviews one may consult monographs [4].

By conception and construction, the Lie-Santilli isotheory *is not* a new theory, but merely a *new realization* of the abstract axioms of Lie's theory. Also, recall that all simple Lie algebras (over a field of characteristic zero) are known from Cartan's classification. Therefore, the isotopies of Lie's theory cannot possibly produce new algebras, and have been constructed instead to produce *novel nonlinear, nonlocal, and noncanonical-nonunitary realizations* of known Lie algebras.

The preservation of the abstract Lie axioms under nonlinear, nonlocal, and noncanonical-nonunitary realizations is permitted by the reconstruction of linearity, locality, and canonicity-unitarity on isospaces over isofields, thus illustrating the methodological significance of the latter.

The main lines of the Lie-Santilli isotheory can be summarized as follows. Let $\xi(L)$ be the universal enveloping associative algebra of an n-dimensional Lie algebra L with generators $X = \{X_k\} = \{X_k^{\dagger}\}$, unit

$I = \text{Diag.}(1,1,1,\ldots)$, associative product $X_i \times X_j$, infinite-dimensional basis I, X_k ,$X_i \times X_j, i \leq j, X_i \times X_j \times X_k$, $i \leq j \leq k, \ldots$, (Poincaré-Birkhoff-Witt theorem), and related exponentiation $e^{iXw} = I + (i \times X \times w)/1! + (i \times X \times w) \times (i \times X \times w)/2! + \ldots$, $w \in \mathbb{R}$.

The *universal enveloping isoassociative algebra* $\hat{\xi}(L)$, first introduced by Santilli in Ref. [1a], is the isotopic image $\hat{\xi}(L)$ of $\xi(L)$ with isounit $\hat{I}$, the same generators $\hat{X}_k = X_k$ only computed on isospace, isoassociative product $\hat{X}_i \hat{\times} \hat{X}_j$, infinite dimensional isobasis (isotopic Poincaré-Birkhoff-Witt theorem [*loc. cit.*])

$$\hat{I}, \ \hat{X}_k, \ \hat{X}_i \hat{\times} \hat{X}_j, \ i \leq j, \ \ \hat{X}_i \hat{\times} \hat{X}_j \hat{\times} \hat{X}_k, \ i \leq j \leq k, \ldots, \tag{2.38}$$

and *isoexponentiation* (for $\hat{X} = X$)

$$\hat{e}^{i \times X \times w} \equiv \hat{e}^{\hat{i} \hat{\times} \hat{X} \hat{\times} \hat{w}} = \{e^{i \times X \times \hat{T} \times w}\} \times \hat{I}, \tag{2.39}$$

where $w = \{w_k\} \in \mathbb{R}$, and $\hat{w} = w \times \hat{I} \in \hat{\mathbb{R}}$ denotes the *isoparameters.*

Let L be the Lie algebra homomorphic to the antisymmetric algebra $[\xi(L)]^-$ of $\xi(L)$ over a field $\mathbb{F}(a, +, \times)$ of real, complex, or quaternionic numbers a with familiar Lie's Second Theorem $[X_i, X_j] = X_i \times X_j - X_j \times X_i = C_{ij}^k \times X_k$. The *Lie-Santilli isoalgebra*, first introduced in Ref. [1a], is the isospace $\hat{L}$ with elements $\hat{X}_k = X_k = X_k^\dagger$ on $\hat{\mathcal{H}}$ over $\hat{\mathbb{F}}$ with the *isocommutation rules*

$$[\hat{X}_i \hat{,} \hat{X}_j] = \hat{X}_i \hat{\times} \hat{X}_j - \hat{X}_j \hat{\times} \hat{X}_i = \hat{C}_{ij}^k \hat{\times} \hat{X}_k, \tag{2.40}$$

whose brackets satisfy *Lie's axioms in the isotopic form*

$$[\hat{A} \hat{,} B] = -[\hat{B} \hat{,} \hat{A}], \tag{2.41a}$$

$$[\hat{A} \hat{,} [\hat{B} \hat{,} \hat{C}]] + [\hat{B} \hat{,} [\hat{C} \hat{,} \hat{A}]] + [\hat{C} \hat{,} [\hat{A} \hat{,} \hat{B}]] = 0, \tag{2.41b}$$

and the *isodifferential rules*

$$[\hat{A} \hat{\times} \hat{B} \hat{,} \hat{C}] = \hat{A} \hat{\times} [\hat{B} \hat{,} \hat{C}] + [\hat{A} \hat{,} \hat{C}] \hat{\times} \hat{B}. \tag{2.42}$$

Let G be the (connected) Lie transformation group characterized by the "exponentiation" of L into the elements $U(w) = e^{i \times X \times w}$ with familiar laws $U(w) \times U(w') = U(w + w')$, $U(w) \times U(-w) = U(0) = I$. Then, the (connected) *Lie-Santilli isotransformation group* $\hat{G}$, first submitted in Ref. [1a], is the "isoexponentiation" of $\hat{L}$ according to Eqs. (2.39) with isotopic laws

$$\begin{aligned} \hat{x}' = \hat{U}(\hat{w}) \hat{\times} \hat{x} = \hat{e}^{\hat{i} \hat{\times} \hat{X} \hat{\times} \hat{w}} \hat{\times} \hat{x} = \\ = \{e^{i \times X \times \hat{T} \times w}\} \times \hat{I} \times \hat{T} \times \hat{x} = \{e^{i \times X \hat{\times} \hat{T} \times w}\} \times \hat{x}, \end{aligned} \tag{2.43a}$$

$$\hat{U}(\hat{w})\hat{\times}\hat{U}(\hat{w}') = \hat{U}(\hat{w}+\hat{w}'), \quad \hat{U}(\hat{w})\hat{\times}\hat{U}(-\hat{w}) = \hat{U}(\hat{0}) = \hat{I}. \qquad (2.43b)$$

We finally recall the *isotopic Baker-Campbell-Hausdorff theorem* [1a]:

$$\hat{U}_1\hat{\times}\hat{U}_2 = \{e_{\hat{\xi}}^{\hat{X}_1}\}\hat{\times}\{e_{\hat{\xi}}^{\hat{X}_2}\} = \hat{U}_3 = e_{\hat{\xi}}^{\hat{X}_3}, \qquad (2.44a)$$

$$\hat{X}_3 = \hat{X}_1 + \hat{X}_2 + [\hat{X}_1\hat{,}\hat{X}_2]/2 + [(\hat{X}_1 - \hat{X}_2)\hat{,}[\hat{X}_1\hat{,}\hat{X}_2]]/12 + \ldots, \qquad (2.44b)$$

which assures that the isoalgebra $\hat{L}$ is closed under isocommutation rules whenever the original algebra is closed (that is, $\hat{L}$ does not yield new generators) for all possible nonsingular isotopic elements $\hat{T}$.

The classical realization (see Chapter 3 for detail) of a one-dimensional connected Lie-Santilli isogroup is given by in the unified notation $b = \{b^\mu\} = \{r^k, p_k\}$, $\mu = 1, 2, \ldots, 6$, [1a, 2c],

$$\hat{b}(\hat{0}) = \hat{U}(\hat{t})\hat{\times}\hat{b} = \{\hat{e}^{\hat{i}\hat{\times}\hat{t}\hat{\times}\hat{\partial}\hat{H}\hat{/}\hat{\partial}\hat{b}^\mu\hat{\times}\hat{\omega}^{\mu\nu}\hat{\times}\hat{\partial}\hat{/}\hat{\partial}\hat{b}^\nu}\}\hat{\times}\hat{b}(0), \qquad (2.45)$$

where $\hat{\omega} = \omega \times \hat{I}$, and $\omega^{\mu\nu}$ is the conventional Lie tensor. In the neighborhood of the isounit we then have the equation

$$\hat{d}\hat{b}^\mu\hat{/}\hat{d}\hat{t} = \hat{\omega}^{\mu\nu}\hat{\times}\hat{\partial}\hat{H}\hat{/}\hat{\partial}\hat{b}^\nu. \qquad (2.46)$$

where one should note the use of both space and time isoderivatives.

The operator realization (see Chapter 3 for details) of a one-dimensional connected Lie-Santilli isogroup on $\hat{\mathcal{H}}$ over $\hat{\mathbb{C}}$ is given by [1a, 2c]

$$\begin{aligned}\hat{A}(\hat{t}) &= \{\hat{e}^{\hat{i}\hat{\times}\hat{H}\hat{\times}\hat{t}}\}\hat{\times}\hat{A}(\hat{0})\hat{\times}\{\hat{e}^{-\hat{i}\hat{\times}\hat{t}\hat{\times}\hat{H}}\} = \\ &= \{e^{i\times\hat{H}\times\hat{T}\times t}\} \times \hat{A}(0) \times \{e^{-i\times t\times\hat{T}\times\hat{H}}\},\end{aligned} \qquad (2.47)$$

which admits in the neighborhood of the isounit the equation

$$\hat{i} \times \hat{d}\hat{A}\hat{/}\hat{d}\hat{t} = [\hat{A}\hat{,}\hat{H}] = \hat{A}\hat{\times}\hat{H}\hat{-}\hat{H}\hat{\times}\hat{A} = \hat{A} \times \hat{T} \times \hat{H} - \hat{H} \times \hat{T} \times \hat{A}. \quad (2.48)$$

A few comments are now in order. The nontriviality of the Lie-Santilli isotheory over the conventional formulation is established by *the appearance of the isotopic element $\hat{T}$ in the exponent of the group structure*, Eqs. (2.44). This guarantees that the isotheory has the most general possible nonlinear, nonlocal-integral, and nonhamiltonian-nonunitary structure although reformulated in an identical isolinear, isolocal, and isounitary form.

Properties (2.44) ensure that, under sufficient regularity and continuity conditions, the isotopies $\hat{L}$ or $\hat{G}$ of a given Lie algebra L or group G preserve the original connectivity and dimension. Moreover, from the positive-definiteness of the isounit $\hat{I} > 0$, it is easy to see that the isoalgebras and isogroups are locally isomorphic to the original algebras and groups, $\hat{L} \approx L, \hat{G} \approx G$.

The latter isomorphism is of paramount importance for hadronic chemistry. In fact, they assure the preservation of all conventional physical laws and relativities. As an example, the preservation of Pauli's exclusion principle can be anticipated already at this algebraic level from the preservation of the *exact* SU(2)-spin symmetry under nonlinear, nonlocal, and nonunitary interactions, provided that they are treated via the isomathematics (see Chapter 4 for detail). The same occurs for all space-time symmetries, such as the Galilei and Poincaré symmetries, or for internal symmetries, such as the isospin symmetry (see Refs. [2c, 2d] for detail).

As a matter of fact, a general property of the Lie-Santilli isotheory for which it was also constructed is the *reconstruction of exact spacetime and internal symmetries when believed to be conventionally broken.*

As an example, the rotational symmetry SO(2) is generally believed to be broken for ellipses (2.33) while, in actuality, the symmetry is exact when formulated on isospace $\hat{E}$ over the isofield $\hat{\mathbb{R}}$. In fact, expression (2.33) represents an ellipse only on ordinary space E over the ordinary field $\mathbb{R}$, while it becomes the perfect circle on $\hat{E}$ over $\hat{\mathbb{R}}$, thus confirming the exact character of $S\hat{O}(2)$.

Note that the use of quantum deformations implies the transition to algebras which *are not* isomorphic to the original algebras, thus implying the assumption of new, yet unknown physical laws and relativities.

Despite the *mathematical similarities* between the conventional and isotopic theories, the Lie and Lie-Santilli theories are *physically inequivalent.* In fact, the classical realization of the isotopic theory is characterized by a *noncanonical* transformation of the conventional realization (see Chapter 3) while the operator realization is characterized by a *nonunitary* transform of the conventional operator realization (see also Chapter 3). As such, the Lie-Santilli isotheory is outside the class of equivalence of the conventional Lie theory.

Note that the conventional Lie theory admits only one interpretation, that on conventional spaces over conventional fields. On the contrary, the Lie-Santilli isotheory admits *two* interpretations, that on isospaces over isofields, and its projection on conventional spaces over conventional fields.

As a final comment, we should indicate that the most rigorous way for deriving the new model of chemical bonds proposed in this monograph is that via the isotopies of the Galilean symmetry for the nonrelativistic case and the Poincaré symmetry for the relativistic counterpart. These isotopies are studied in detail in monograph [2f] (see also memoir [3f] and monograph [4c] for independent reviews).

We regret to be unable to study the latter approach in this monograph (and ignore so many other important topics, such as Clifford algebras and their isotopies) because their technical level would be beyond the scope of this presentation.

3. Elements of Genomathematics

3.1. Introduction

Some of the most fundamental problems left unresolved by the sciences of the 20-th century, including chemistry, *are the origin and invariant treatment of irreversible processes*, such as ordinary chemical reactions (which are generally irreversible in the sense that their image under the reversion of time cannot naturally occur).

Irreversibility remained essentially unresolved because, on one side, irreversible processes such as chemical reactions are generally *nonconservative*, while, on the other side, the theories used for their study, such as quantum mechanics and chemistry, can only represent strictly *conservative* systems. In fact, the *only* time evolution admitted by quantum mechanics and chemistry is that of Heisenberg's equations, which imply the *necessary* conservation of the energy, $i \times dH/dt = [H, H] = H \times H - H \times H \equiv 0$.

A lifetime of studies dedicated by Santilli on irreversibility has established that the insufficiencies here considered are due to the underlying *mathematics*. In fact, all axioms of quantum mechanics, such as those of the Lie theory (which represents the time evolution), are intrinsically reversible in time. Therefore, all attempt for quantitative representations of irreversible processes via a mathematics which is structurally reversible cannot be qualified as truly scientific.

It is easy to see that the same faith occurs for the broader isomathematics. In fact, its basic axioms, such as those of the Lie-Santilli isotheory, are also fully reversible in time, as a necessary condition to have an isotopy. Therefore, no really scientific attempt can be claimed by representing irreversible processes via isomathematics.

The above occurrences mandated the construction of a yet broader mathematics, today known as *Santilli genomathematics*, which was specifically conceived and constructed for quantitative representations of irreversible processes. In particular, by conception and construction, genomathematics possesses *structurally irreversible axioms*, that is, axioms which are irreversible even under the use of reversible interactions, potentials and Hamiltonians.

Santilli proposed the algebraic foundations of genomathematics during his Ph. D. studies of 1967 [5a, 5b, 5c]. Advances in the field were then done: in 1978, with the genotopies of Lie's theory [1a]; in 1979, with the initiation of the studies on the structure of the theory [5d]; in 1983, with the first formulation of genospaces [1c]; and in 1993 with the first known formulation of genonumbers [1b]. Mathematical maturity of genomathematics was reached in the memoir [1e] of 1996, while physical maturity, including the resolution of the catastrophic inconsistencies of Sect. 1.7, was reached in Ref. [5e] of 1997. As a result of all these efforts, genomathematics has reached today maturity for applications.

Rather oddly in view of the manifestly important mathematical, physical, and chemical implications, Santilli remains to this day the initiator and the most active author of works on genomathematics, the sole other original contributions being of indirect algebraic nature, as we shall see.

In this section we identify the most important elements of genomathematics, with the understanding that an in depth knowledge can be solely acquired via the study of the original contributions.

Genomathematics will then be assumed in the next chapter as the foundations of the *genotopic branch of hadronic chemistry*, or *genochemistry* for short.

For the less mathematically inclined reader, we shall present in the next chapter very simple methods based on nonunitary transforms which permits the explicit construction of genochemistry in all details needed for applications.

We finally note in these introductory lines that the term *genotopies* was introduced by Santilli in 1978 in the Greek meaning of being "axiom inducing". Therefore, while the isotopies are maps preserving the original axioms, the genotopies are broader maps violating the original axioms for broader formulations.

3.2. Main Structural Lines of Genomathematics

It may be helpful for the noninitiated reader to begin our presentation with the main structural lines of genomathematics and its irreversible structure, and then pass to a study of its specific notions.

A guiding principle is the construction of genomathematics as a covering of isomathematics, that is, as a broader mathematics admitting isomathematics as a particular case, yet possessing an irreversible structure in its general formulation.

The main idea of genomathematics is that its structure is inherent in the *conventional* Lie theory. A one-parameter connected Lie transformation group realized via a Hermitean generator $X = X^\dagger$ on a Hilbert space $\hat{\mathcal{H}}$ over a field $\mathbb{F}$ has in reality the structure of a *bi-module* (also called *spit-null extension* [1d]).

In nontechnical terms, a Lie bimodule is essentially characterized by an *action from the left* $U^>$ and an *action from the right* $^<U$ with explicit realization and interconnecting conjugation

$$A(w) = U^> > A(0) < {}^<U =$$

$$= \{e^{i\times X^> \times w}\} > A(0) < \{e^{-i\times w\times {}^<X}\} = \tag{2.49a}$$

$$= (I^> + i \times X^> \times w + \ldots) > A(0) < ({}^<I - i \times w \times {}^<X + \ldots),$$

$$U^> = ({}^<U)^\dagger = U, \quad X^> = ({}^<X)^\dagger = X, \quad I^> = {}^<I = I, \tag{2.49b}$$

where w is a Lie parameter and the multiplications $>$ and $<$ represent conventional associative products ordered to the right and to the left, respectively.

The infinitesimal version in the neighborhood of the unit then acquires the familiar form

$$\begin{aligned} i \times dA/dw = i \times [A(dw) - A(0)]/dw = \\ = A < X - X > A = A \times X - X \times A, \end{aligned} \tag{2.50}$$

which clarifies the important bimodular notion that, in $A \times X = A < X$ ($X \times A = X > A$), the operator X acts from the right (from the left).

The bimodular Lie structure can therefore be written as the bi-group $\{{}^<U, U^>\}$ acting on the a Hilbert bimodule $\{{}^<\mathcal{H}, \mathcal{H}^>\}$ over the bi-fields $\{{}^<\mathbb{F}, \mathbb{F}^>\}$, where ${}^<U = (U^>)^\dagger = U$, ${}^<\mathcal{H} = \mathcal{H}^> = \mathcal{H}$, ${}^<\mathbb{F} = \mathbb{F}^>$.

For the conventional formulation of Lie's theory, the above bimodular structure is generally ignored, even in the most advanced treatises in the field (see, e.g., the theoretical presentation [6a] or the mathematical monograph [6b]), because it is redundant and inessential since the modular action to the right and to the left are interconnected by the rules

$$X^> \times \psi^> = X \times \psi = -{}^<\psi < {}^<X = -\psi \times X, \tag{2.51}$$

where $\psi^> \in \mathcal{H}^>$, ${}^<\psi \in {}^<\mathcal{H}$, $X^>$ is an element of the universal enveloping associative algebra $\xi^>(L)$ of the considered Lie algebra $L \approx [\xi^>(L)]^-$ for the action to the right and ${}^<X \in {}^<\xi(L)$ for the action to the left.

Since $\mathcal{H}^> = {}^<\mathcal{H} = \mathcal{H}$, and $\xi^>(L) = {}^<\xi(L) = \xi(L)$, the *birepsentations* of the bimodular structure $\{{}^<\xi(L), \xi^>(L)\}$ over $\{{}^<\mathcal{H}, \mathcal{H}^>\}$ can be effectively reduced to the *one-sided representations*, or just *representations* for short, of $\xi(L)$ over $\mathcal{H}$, as well known [6]. However, as we shall see shortly, the original bimodular structure of Lie's theory is no longer trivial for broader realizations of the axioms.

Iso- and geno-mathematics were proposed [1a] on the basis of the mere observation that *the abstract axioms of Lie's bimodular structure (2.51) do not necessarily require that the multiplications $>$ and $<$ are conventional because these multiplications can also be generalized, provided that they remain associative.* In other word, the abstract axiomatic structure of the action from the right, $U^> > A(0)$ is that of a *right modular associative action*, with no restriction on the realization of the product $>$ other than the associativity law, and the same occurs for the action from the left $A(0) < {}^<U$.

The simplest possible broadening of the conventional formulation of Lie's theory is given by the *isotopies of Lie's theory* studied in the preceding sections. These isotopies are essentially characterized by the lifting of the conventional right modular associative product into the form of the preceding section, $U^> A(0) = U^> \times T \times A(0)$ with conjugate version from the left $A(0) < {}^<U = A(0) \times T \times {}^<U$ where $T = T^\dagger$ is a fixed, well behaved, nowhere singular and Hermitean matrix or operator of the same dimension of the considered representation. Its inverse $\hat{I} = T^{-1}$ is then a fully acceptable, generalized, left and right unit, $A^> > \hat{I} = \hat{I} < {}^<A = A$ for all possible elements A of the considered algebra.

The isotopic realization of Lie's axioms (2.49) is then given by the expression studied in the preceding section,

$$\begin{gathered} A(w) = U^> > A(0) < {}^<U = \\ = \{e^{i \times X^> \times w}\} > A(0) < \{e^{-i \times w \times {}^<X}\} = \\ = \{e^{i \times X^> \times \hat{T} \times w}\} \hat{\times} A(0) \hat{\times} \{e^{-i \times w \times \hat{T} \times {}^<X}\}, \\ X = X^\dagger, \quad \hat{T} = \hat{T}^\dagger, \end{gathered} \tag{2.52}$$

with infinitesimal form

$$i \times dA/dw = i \times [A(dw) - A(0)]/dw = A \times \hat{T} \times X - X \times \hat{T} \times A. \tag{2.53}$$

The isotopies then require, for mathematical and physical consistency, the reconstruction of the *entire* Lie theory with respect to the new unit $\hat{I}$ and isoproduct $A > B = A < B = A \hat{\times} B = A \times T \times B$, including: numbers and fields; metric and Hilbert spaces; Lie algebras, groups and symmetries; transformation and representation theories; *etc.*

The isotopies of Lie's theory also possess a trivial bimodular structure, in the sense that its two-sided representations can always be effectively reduced to the one-sided isorepresentations via a rule equivalent to (2.51).

The *genotopies of Lie's theory*, first proposed in Ref. [1a], and then studied in the various references quoted above, originates from the broader *realization* of the abstract Lie bimodule (2.49) under the name of *genoassociative multiplication and unit* (or *genoproduct* and *genounit* for short),

$$A > B = A \times P \times B, \quad A < B = A \times Q \times B, \tag{2.54a}$$

$$\hat{I}^{>} = P^{-1}, \quad I^{>} > A = A > \hat{I}^{>} = A, \tag{2.54b}$$

$$^{<}I = Q^{-1}, \quad ^{<}I < A = A < {}^{<}I = A, \tag{2.54c}$$

$$\hat{I}^{>} = P^{-1} = ({}^{<}\hat{I})^{\dagger} = (Q^{-1})^{\dagger}, \tag{2.54d}$$

where $P \neq Q$ are well behaved, everywhere invertible, nonhermitean matrices or operators generally realized via real-valued nonsymmetric matrices of the same dimension of the considered Lie representation. Moreover, it is requested that $P + Q$ and $P - Q$ are nonsingular to preserve a well defined Lie and Jordan content, respectively.

Genoproducts and genounits (2.54) characterize the following more general realization of the abstract axioms (2.49) first proposed in Refs. [5a, 1a] and today called *Lie-Santilli genogroups*

$$\begin{aligned} A(t) &= U^{>} > A(0) < {}^{<}A = \\ &= \{e_{>}^{i\times X\times w}\} > A(0) < \{{}_{<}e^{-i\times w\times X}\} = \\ &= \{e^{i\times X\times P\times w}\} \times A(0) \times \{e^{-i\times w\times Q\times X}\} = \\ &= (I^{>} + i\times X\times w + \ldots) > A(0) < ({}^{<}I - i\times w\times {}^{<}X + \ldots), \end{aligned} \tag{2.55a}$$

$$\begin{aligned} U^{>} &= ({}^{<}U)^{\dagger}, \\ X^{>} &= ({}^{<}X)^{\dagger} = {}^{<}X = X, \\ P^{>} &= P = ({}^{<}Q)^{\dagger} = Q^{\dagger}, \end{aligned} \tag{2.55b}$$

$$\hat{I}^{>} = P^{-1} = ({}^{<}I)^{\dagger} = (Q^{-1})^{\dagger}, \tag{2.55c}$$

with infinitesimal version in the neighborhood of the genounits characterized by the *Lie-Santilli genoalgebras* [*loc. cit.*]

$$\begin{aligned} i \times [A(dw) - A(0)]/dw &= (A, B) = \\ = A < X - X > A &= A \times P \times X - X \times Q \times A = \\ &= (A \times \hat{T} \times B - B \times \hat{T} \times A) + \\ &+ (A \times \hat{W} \times B + B \times \hat{W} \times A) = \\ &= [A\hat{,}B] + \{A\hat{,}B\}, \end{aligned} \tag{2.56}$$

where $P = \hat{T} + W$, $Q = W - \hat{T}$, and we have used the *genoexponentiation to the right and to the left*

$$\begin{aligned} e_{>}^{i\times X\times w} &= I^{>} + i \times X \times w/1! + (i \times X \times w) > (i \times X \times w)/2! + \cdots = \\ &= \{e^{i\times X\times P\times w}\} \times I^{>}, \end{aligned} \tag{2.57a}$$

$$e_<^{i\times w\times X} = {}^<I + i\times X\times w/1! + (i\times X\times w) > (i\times X\times w)/2! + \cdots =$$
$$= {}^<I \times \{e^{i\times w\times Q\times X}\}. \qquad (2.57b)$$

It is at this point where the essential bimodular character of axioms (2.49) acquire their full light because no longer effectively reducible to a one-sided form. In fact, we can still represent structure (2.55) with the bimodule $\{{}^<U, U^>\}$ acting on the bi-isospaces $\{{}^<\mathcal{H}, \mathcal{H}^>\}$ over certain bi-isofields $\{{}^<\mathbb{F}, \mathbb{F}^>\}$ (to be identified below). However, ${}^<\mathcal{H}$ is now nontrivially different than $\mathcal{H}^>$. As a result, *the representation theory of the Lie-Santilli genogroups has an essential bimodular structure which cannot be reduced to one-sided isorepresentations.*

It should be indicated that the (nonassociative) algebra with product $(A, B) = A\times P\times B - B\times Q\times A$ of time evolution (2.56) is a *Lie-admissible algebra* as defined by the American mathematician A.A. Albert [7] (see also Refs. [5]). A generally nonassociative algebra A with abstract elements a, b, c, ... (e.g., nonsingular matrices) and abstract product "ab" is said to be Lie-admissible when the attached antisymmetric algebra U^-, which has the same elements as U equipped with the new product $[a, b]_U = ab - ba$, is a Lie algebra.

The simplest possible Lie-admissible algebra is the associative algebra ξ of $n\times n$-dimensional matrices A, B, C, ... and conventional associative product $A\times B$, $A\times(B\times C) = (A\times B)\times C$. In fact, Lie's algebras L are constructed precisely as the attached antisymmetric algebras ξ^- of ξ with the familiar Lie product $[A, B]_A = A\times B - B\times A$.

The nonassociative algebra A with the product of Eq. (2.56) are Lie-admissible in the more general sense that the attached antisymmetric product is *Lie-isotopic*, $[A, B]_A = (A, B) - (B, A) = [A, B]_A = A\times\hat{T}\times B - B\times\hat{T}\times A$.

Note that *the Lie-admissible algebras with product (A, B), first introduced in Ref. [1a] of 1978, are the most general known algebras as currently understood in mathematics (characterized by a bilinear product which verifies the left and right distributive and scalar laws).* In fact, the algebras with product (A, B) include as particular case, depending on the assumed values of P and Q, all known algebras such as: associative algebras, Lie and Jordan algebras, Lie-isotopic and Jordan isotopic algebras, supersymmetric algebras, Kac-Moody algebras, *etc.*

Note the crucial role of the conjugation $U^> = ({}^<U)^\dagger$ (or $\hat{I}^> = ({}^<\hat{I})^\dagger$) at all levels, including the conventional, isotopic, and genotopic levels. In fact, it is easy to see that *the violation of said conjugation implies the violation of Lie's axioms*, resulting in the catastrophic inconsistencies indicated in Sect. 1.7.

To understand the reversible or irreversible character of the above structures, let us assume hereon in Eqs. (2.55) that: *the product* $>$ represent motion forward in time while the product $<$ represents motion backward in time; $H = p^2/2m + V(r)$ represents the total energy; and $w = t$ represent time. Recall that *all known potentials are invariant under time reversal*, including electric, magnetic, gravitational, and other potentials. Consequentially, *all known Hamiltonian are also time reversal invariant*, $H(t) \equiv H(-t)$. Recall finally, at this simple operator level, that time reversal invariance requires not only the change of time direction, but also the operation of Hermitean conjugation.

It is then evident that the conventional realization of Lie's axioms (2.49) is indeed time reversal invariant. As an illustration, suppose that A represents the linear momentum, $p = m \times v = m \times dr/dt$, which evidently changes sign under time reversal, $p(-t) = -p$. It is then evident that Heisenberg's equations (2.50) is time reversal invariant, i.e.,

$$\begin{aligned} i^\dagger \times dp(-t)/d(-t) = p(-t) \times H - H \times p(-t) = \\ = -i \times dp/dt = -(p \times H - H \times p). \end{aligned} \tag{2.58}$$

It is also evident that when $\hat{T}$ is independent of time, the isotopic realization of Lie's axioms is equally reversible under realization $\hat{w} = \hat{t} = \hat{t} \times \hat{I}_t$, again because all equations remain invariant under time reversal $t \to -t$, as illustrated by the case of the linear momentum

$$\begin{aligned} i^\dagger \times dp^\dagger(-t)/d(-t) = \\ = p^\dagger(-t) \times \hat{T}^\dagger \times H^\dagger - H^\dagger \times \hat{T}^\dagger \times p^\dagger(-t) = \\ = -i \times dp/dt = -(p \times \hat{T} \times H - H \times \hat{T} \times p). \end{aligned} \tag{2.59}$$

It should be noted that the assumption of an isotopic element $\hat{T}$ explicitly dependent on time and not invariant under time reflection, $\hat{T} = \hat{T}(t, \ldots) \neq \hat{T}(-t, \ldots)$, does yield an irreversible theory, but only for the closed-conservative case because the Hamiltonian is conserved irrespective of these characteristics of $\hat{T}$, i.e., $i \times dH/dt = H \times \hat{T} \times H - H \times \hat{T} \times H \equiv 0$. This realization of the Lie-Santilli isotheory is, therefore, recommendable for all closed-isolated systems with internal entropy, e.g., for the description of the structure of stars when isolated from the rest of the universe [2b].

It is finally evident that all the preceding time reversal invariances are lost at the broader level of genotopies. In fact, for the same case of the linear momentum, we have from Eqs. (2.56)

$$i^\dagger \times dp^\dagger(-t)/d(-t) =$$
$$= p^\dagger(-t) \times P^\dagger \times H^\dagger - H^\dagger \times Q^\dagger \times p^\dagger(-t) = \tag{2.60}$$
$$= -i \times dp/dt = -(p \times Q \times H - H \times P \times p),$$

namely, *the genotopies of Lie's theories are irreversible because the operation of time reversal interchanges the different operators* P *and* Q, *thus resulting in equations for motion forward in time which are different than those for motion backward in time.*

Note that, the above irreversibility holds under the *lack* of conservation of the energy, $i \times dH/dt = H \times P \times H - H \times Q \times H \neq 0$, as necessary for all open-nonconservative irreversible systems, such as chemical reactions.

The above result signals the achievement of the main objective of this section: the only known mathematics which is structurally irreversible, that is, irreversible for all infinitely possible reversible potentials and Hamiltonians. We reach in this way a mathematics offering realistic possibilities for quantitative scientific studies of irreversible chemical processes, to be studied later on.

It is evident that, to achieve consistency, the *entire* Lie's theory must be lifted into a dual genotopic form, with no known exception. In the following we provide a rudimentary review of the resulting genomathematics.

3.3. Genounits and Genoproducts

The fundamental elements of genomathematics are:

1) The assumptions of *two different, generalized, nonsingular, and nonhermitean units, the genounit* $\hat{I}^> = 1/\hat{Q}$ *for motion forward in time, and* $^<\hat{I}$ *for motion backward in time*, as in Eqs. (2.54);

2) The assumption of *two corresponding, ordered, generalized products, the genoproduct* $A > B = A \times Q \times B$, Q *fixed, which must be solely used for all motions forward in time, and the genoproduct* $A < B = A > P > B$, *which must be solely used for all motions backward in time*, under which $\hat{I}^>$ and $^<\hat{I}$ are the left and right units, respectively, also according to Eqs. (2.54); and

3) The assumption of the interconnected property $\hat{I}^> = P^{-1} = (^<\hat{I})^\dagger = (Q^{-1})^\dagger$.

Note that the latter assumption is crucial to avoid the catastrophic inconsistencies of Sect. 1.7.

A few comments are now in order on the genounits. Recall from Sect. 2.2 that the isounits are real-valued, positive-definite *diagonal* matrices of the type $\hat{I} = \text{Diag.}(n_1^2, n_2^2, n_3^2, n_4^2)$, where n_1^2, n_2^2, n_3^2 characterizes the *shape* of particles as spheroidal ellipsoids, and n_4^2 characterizes the *density* of the particle considered.

The realization of genounits $\hat{I}^>$ and ${}^<\hat{I}$ used in application is given by real-valued, yet *nondiagonal and nonsymmetric matrices*, that is, having the same terms $n_1^2, n_2^2, n_3^2, n_4^2$ of $\hat{I}$, *plus* at least one additional function $F(t, r, p, \psi, \ldots)$ in an off-diagonal place. As we shall see in the next chapters, the latter off-diagonal function is ideally suited to represent external nonpotential forces in the analytic equations. The representation of irreversibility is then *guaranteed* by the nonsymmetric nature of the genounits, while the invariant nature of the representation is ensured by its implementation via the unit of the theory.

A few comments are also due on the ordering of the product to the left or to the right. As well known, products of two quantities are generally done in quantum mechanics and chemistry without any consideration of their ordering. This is due to the fact that the operation "$2 > 3$" (namely, 2 multiplying 3 from the left) yields exactly the same result of the operation "$2 < 3$" (namely, 3 multiplying 2 from the right).

No ordering of the multiplications is also needed for isomathematics and its application, again, because the isoproduct of two quantities from the left and from the right yields the same result.

In the transition to genomathematics, the situation is different because now the two different multiplications of the two quantities, such as "$2 > 3$" and "$2 < 3$", yield different results. As a consequence, the consistent treatment of genotheories requires the assumption of either the ordering of the multiplication to the right or to the left, that is, *the restriction of the totality of the multiplication to only one of the two*, a restriction hereon assumed.

As we shall see, the ordering of the multiplication results in being truly fundamental for the representation of irreversibility, because, as indicated earlier, the ordered product $>$ is assumed to represent motion forward in time, and its conjugate $<$ is assumed to represent motion backward in time.

We can therefore say that the inability by quantum mechanics and chemistry to represent irreversibility can be ultimately traced to the identity of their products to the left and to the right.

3.4. Genonumbers and Genofields

Let $\mathbb{F} = \mathbb{F}(a, +, \times)$ be a conventional field of real $\mathbb{R}$, complex $\mathbb{C}$ or quaternionic $\mathbb{Q}$ numbers a with additive unit 0, multiplicative unit $I = 1$, sum $a + b$ and product $a \times b$. The *genofields to the right* $\mathbb{F}^> =$

$\mathbb{F}^>(a^>, +^>, \times^>)$, first introduced by Santilli [1b] in 1993, are rings with elements $a^> = a \times I^>$ called *genonumbers*, where a is an element of $\mathbb{F}$, and $I^> = P^{-1}$ is a well behaved, everywhere invertible and nonhermitean quantity generally outside $\mathbb{F}$, equipped with all operations ordered to the right, i.e., the *ordered genosum to the right, ordered genoproduct to the right, etc.*,

$$a^> +^> b^> = (a+b) \times I^>, \tag{2.61a}$$

$$a^> \times^> b^> = a^> > b^> = a^> \times \hat{Q} \times b^> = (a \times b) \times I^>, \tag{2.61b}$$

genoadditive unit to the right $0^> = 0$ and *multiplicative genounit to the right* $I^>$.

The *genofields to the left* ${}^<\mathbb{F} = {}^<\mathbb{F}({}^<a, {}^<+, {}^<\times)$ [1b] are rings with genonumbers ${}^<a = {}^<I \times a$, all operations ordered to the left, including genosum ${}^<a^< + {}^<b = {}^<I \times (a+b)$, genoproduct ${}^<a < {}^<b = {}^<a \times Q \times {}^<b = {}^<I \times (a \times b)$, *etc.*, with *additive genounit to the left* ${}^<0 = 0$ and *multiplicative genounit to the left* ${}^<I = Q^{-1}$.

A *bigenofield* is the structure $\{{}^<\mathbb{F}, \mathbb{F}^>\}$ with corresponding bielements, biunits, bioperations, *etc.*, acting to the left and right under the condition $\hat{I}^> = ({}^<I)^\dagger$.

It is easy to prove that each individual genofield to the right $\mathbb{F}^>$ or to the left ${}^<\mathbb{F}$ is a field isomorphic to the original field $\mathbb{F}$. Thus, the liftings $\mathbb{F} \rightarrow \mathbb{F}^>$, $\mathbb{F} \rightarrow {}^<\mathbb{F}$ and $\{\mathbb{F}, \mathbb{F}\} \rightarrow \{{}^<\mathbb{F}, \mathbb{F}^>\}$ are axiom-preserving.

Recall that in the definition of fields (and isofields) there is no need for the ordering of the multiplication because multiplications from the left or from the right coincide, i.e., $a > b = a < b$ (even for non-commutative isofields such as the quaternions and isoquaternions).

For the case of genofields, the separation of ordered multiplications to the right and to the left is necessary because, even though genomultiplications remain commutative, their numerical values are different, i.e.,

$$a > b = b > a \quad \text{and} \quad a < b = b < a, \tag{2.62a}$$

$$a > b = a \times P \times b \neq a < b = a \times Q \times b. \tag{2.62b}$$

An important discovery of Ref. [1b] is that *the axioms of a field are still verified if all operations are ordered to the left, and, separately, to the right.* The rigorous and invariant representation of irreversibility via genomathematics is permitted precisely in view of the latter property, which permits the initiation of irreversibility in the most basic quantities, such as units and products. All subsequent mathematical structures built on numbers will automatically preserve the same axiomatization of irreversibility.

3.5. Genospaces and Genogeometries

Let $S = S(r, g, \mathbb{R})$ be a conventional n-dimensional metric or pseudo-metric space with local chart $r = \{r^k\}, k = 1, 2, \ldots, n$, nowhere singular, real-valued and symmetric metric $g = g(r, \ldots)$ and invariant $r^2 = (r^t \times g \times r) \times I$ (where t denotes transposed) defined over the conventional field of real numbers $\mathbb{R} = \mathbb{R}(a, +, \times)$.

The n-dimensional *genospaces to the right* $S^> = S^>(r^>, G^>, \mathbb{R}^>)$, first introduced by Santilli in Refs. [1c, 2c, 1e] are vector spaces with local *genocoordinates to the right* $r^> = r \times I^>$, *genometric* $G^> = Q \times g \times I^> = (g^>) \times I^>, g^> = Q \times g$, and *genoinvariant to the right*

$$(r^>)^{2>} = (r^>)^t > (G^>) > r^> = [r^t \times (g^>) \times r] \times I^> \in \mathbb{R}^>, \qquad (2.63)$$

which, for consistency, must be a genoscalar to the right with structure $n \times I^>$ and be an element of the genofield $\mathbb{R}^>$. Note also that genospaces and their underlying genofields must have, for consistency, the same genounit to the right $I >= Q^{-1}$ where Q is given by an everywhere invertible, real-valued, non-symmetric $n \times n$ matrix.

The n-dimensional *genospaces to the left* ${}^<S = {}^<S({}^<r, {}^<+, {}^<\mathbb{F})$ [*loc. cit.*] are genospaces over genofields with all operations ordered to the left and a common $n \times n$-dimensional genounit to the left ${}^<I = P^{-1}$ which is generally different than that to the right, but verifying the interconnecting condition $P = Q^\dagger$.

The *bigenospaces* are the structures $\{{}^<S, S^>\}$ with bigenocoordinates, *etc.*, defined over the bigenofield $\{{}^<\mathbb{R}, \mathbb{R}^>\}$ under the condition $I^> = ({}^<I)^\dagger$.

It is easy to prove that genospaces to the right $S^>$ and, independently, those to the left ${}^<S$ (thus bigenospaces $\{{}^<S, S^>\}$) are locally isomorphic to the original spaces $S(\{S, S\})$.

In fact, the original metric g is lifted in the form $g^> \to Q \times g$, but the unit is lifted by the *inverse* amount $I \to I^> = P^{-1}$ thus preserving the original axioms (because the invariant is $(\text{length})^2 \times (\text{unit})^2$), and the same occurs for the other cases.

The best way to see the local isomorphism between conventional and genospaces is by noting that the latter are the result of the following novel degree of freedom of conventional spaces (here expressed for the case of a scalar complex function Q)

$$\begin{aligned} r^t \times g \times r \times I &\equiv r^t \times g \times r \times Q \times Q^{-1} \equiv (r^t \times g^> \times r) \times I^> \equiv \\ &\equiv P^{-1} \times P \times (r^t \times g \times r \times I) \equiv {}^<I \times (r \times {}^<g \times r^t). \end{aligned} \qquad (2.64)$$

This confirms the validity of the basic invariant (2.11) for genospaces too.

Genogeometries [1d, 1e] are the geometries of genospaces. We indicated in Sect. 2.2 that the isotopies of conventional geometries are nontrivial because they permit the unification of different geometries with the same signature, thus unifying Euclidean and Riemannian geometries.

The genotopies of conventional geometries have even deeper implications. In fact, the *genogeometries admit nonsymmetric metrics in a way compatible with the original axioms.* In turn, the admission of nonsymmetric metrics permits the first known, consistent, geometric treatment of irreversibility.

Consider, as an illustration, the Euclidean geometry. Its well known inability to characterize irreversibility originates from the complete reversible character of its basic axioms, the fully symmetric nature of its *diagonal* metric $\delta = \text{Diag.}(1, 1, \ldots, 1)$, and the consequential applicability of the same metric for both motions forward and backward in time.

Consider now the *genoeuclidean geometry.* By conception and construction, it admits *two* different metrics, the genometric $\hat{\delta}^{>} = Q \times \delta$ for motion forward in time, and the conjugate genometric ${}^{<}\hat{\delta} = \delta \times P$ for motions backward in time. Since $\hat{\delta}^{>} \neq {}^{<}\hat{\delta}$, the capability by the genoeuclidean geometry to provide a geometrically consistent characterization of irreversibility is evident.

As an intriguing comment for geometers, the genotopies have disproved the belief kept throughout the 20-th century that the axioms of the Riemannian geometry solely admit symmetric metrics. In fact, the *genoriemannian geometry* [1e] admits exactly the same axioms as the conventional geometry, yet the metric is not symmetric. The above erroneous belief is due to the un-necessary assumption that the unit of the geometry must be trivial symmetric unit $I = \text{Diag.}(1,1,\ldots,1)$. If the same axioms are instead realized via a generalized unit which is not symmetric, the metric admitted by the axioms must also be nonsymmetric.

The reader should be aware that, again, the achievement of nonsymmetric metrics under the Riemannian axioms is of truly fundamental character for a *consistent gravitational treatment of irreversibility.*

3.6. Genodifferential Calculus

The *genodifferential calculus to the right* on a genospace $S^{>}(r^{>}, \mathbb{R}^{>})$ over $\mathbb{R}^{>}$, first introduced by Santilli in Ref. [1e], is the image of the conventional differential calculus characterized by the expressions (where we have ignored for notational simplicity the multiplication to the right by $I^{>}$),

$$dr^k \rightarrow d^{>}r^k = (I^{>})^k_i \times dr^i, \tag{2.65a}$$

$$dr_k \rightarrow d^{>}r_k = P^i_k \times dr_i, \tag{2.65b}$$

$$\partial/\partial r^k \to \partial^>/\partial^> r^k = P^i_k \times \partial/\partial r^i, \tag{2.65c}$$

$$\partial/\partial r_k \to \partial^>/\partial^> r_k = (I^>)^k_i \times \partial/\partial r_i, \tag{2.65d}$$

where all operations are ordered to the right, and main properties

$$\partial^> r^i \hat{/}^> \partial^> r^j = \delta^i_j \times \hat{I}^>, \quad \partial^> r_i \hat{/}^> \partial^> r_j = \delta^j_i \times \hat{I}^>, \quad etc. \tag{2.66}$$

The *genodifferential calculus to the left* is the conjugate of the preceding one for the genounit to the left ${}^<I \neq I^>$. The *bigenodifferential calculus* is that acting on $\{{}^<S, S^>\}$ over $\{{}^<\mathbb{R}, \mathbb{R}^>\}$ under the subsidiary condition $I^> = ({}^<I)^\dagger$.

It is easy to see that the genocalculus to the right and, independently, that to the left preserve all original properties, such as commutativity of the second-order derivative, *etc.*

3.7. Genohilbert Spaces

Let $\mathcal{H}$ be a conventional Hilbert space with states $|\psi\rangle, |\phi\rangle, \ldots$, inner product $\langle\phi| \times |\psi\rangle$ over the field $\mathbb{C} = \mathbb{C}(c, +, \times)$ of complex numbers, and normalization $\langle\psi| \times |\psi\rangle = 1$.

A *genohilbert space to the right* $\mathcal{H}^>$ [1e] is a right genolinear space with genostates $|\psi^>\rangle, |\phi^>\rangle, \ldots$, *genoinner product* and *genonormalization to the right*

$$\langle\phi^>| > |\psi^>\rangle = \langle\phi^>| \times Q \times |\psi^>\rangle \times I^> \in \mathbb{C}^>(c^>, +^>, \times^>), \tag{2.67a}$$

$$\langle\phi^>| > |\psi^>\rangle = \hat{I}^>, \tag{2.67b}$$

defined over a genocomplex field to the right $\mathbb{C}^>(c^>, +^>, \times^>)$ with a common genounit $\hat{I}^> = Q^{-1}$.

A *genohilbert space to the left* ${}^<\mathcal{H}$ [*loc. cit.*] is the left conjugate of $\mathcal{H}^>$ with left genounit ${}^<I = P^{-1}$ generally different than $I^>$. A *bigenohilbert space* is the bistructure $\{{}^<\mathcal{H}, \mathcal{H}^>\}$ over the bigenofield $\{{}^<\mathbb{C}, \mathbb{C}^>\}$ under the conjugation $I^> = ({}^<I)^\dagger$.

Again, the right-, left-, and bi-genohilbert spaces are locally isomorphic to the original space $\mathcal{H}$. In fact, the original inner product is lifted by the amount $\langle| \times |\rangle \to \langle| \times P \times |\rangle$, but the underlying unit is lifted by the *inverse* amount, $1 \to P^{-1}$, thus leaving the original axiomatic structure unchanged.

The *genooperator theory* [2c] is a consequence of the following degree of freedom of conventional Hilbert spaces (where P and Q are assumed to be independent from the integration variable for simplicity)

$$\begin{gathered}\langle\phi| \times |\psi\rangle \equiv \langle\psi| \times |\psi\rangle \times P \times P^{-1} \equiv \\ \equiv \langle\phi| \times P \times |\psi\rangle \times P^{-1} = \langle\phi| < |\psi\rangle \times {}^<I \equiv \\ \equiv \langle\phi| \times |\psi\rangle \times Q \times Q^{-1} \equiv \langle\phi| > |\psi\rangle \times I^>.\end{gathered} \tag{2.68}$$

Evidently, the above property is the Hilbert space counterpart of the novel invariance (2.64). It should be noted that new invariances (2.64) and (2.68) have remained undetected since Riemannian's and Hilbert's times, respectively, because they required the prior discovery of *new numbers*, those with an arbitrary, generally nonhermitean unit.

Note that genomathematics permits a *second type of concrete realization of hidden variables and completion* of quantum mechanics and chemistry according to the historical teaching by Einstein, Podolsky, and Rosen, besides the completion permitted by the isotopies. Further realizations of hidden variables and completions are permitted by the hypermathematics and isodual mathematics of the remaining sections [1g].

3.8. Genolinearity, Genolocality, and Genounitarity

Genolinear operators to the right are operators A, B, ..., of an genoenveloping algebra to the right $\xi^>$ verifying the condition of genolinearity, i.e., linearity on $\mathcal{H}^>$ over $\mathbb{C}^>$, and a similar occurrence holds for the left case.

A similar case also holds for *genolocality*, which, as it was the case for isolocality, represents the topological property of locality everywhere except at the genounits.

Similarly, we have the notions of *genounitary operators to the right and to the left*

$$U^> > U^{>\dagger} = U^{>\dagger} > U = I^>, \tag{2.69a}$$

$$^<U < {}^<U^\dagger = {}^<U^\dagger < {}^<U = {}^<I. \tag{2.69b}$$

An important property is that all operators X which are originally Hermitean on $\mathcal{H}$ over $\mathbb{C}$ remains Hermitean on $\mathcal{H}^>$ over $\mathbb{C}^>$, or on $^<\mathcal{H}$ over $^<\mathbb{C}$. Thus, *genotopies preserve the original observables.* This property is due to the fact that the condition of genohermiticity on $\mathcal{H}^>$ reads $X^{\dagger>} = Q \times Q^{-1} \times X^\dagger \times Q \times Q^{-1} = X^\dagger$, as the reader can verify.

The above mathematical property has the fundamental meaning in chemistry of *permitting, apparently for the first time, the Hermiticity-observability of physical quantities when nonconserved.* In fact, all other methods for nonconservative systems (such as dissipative nuclear models) are based on the addition of an imaginary potential to the Hamiltonian which, as such, loses its Hermiticity-observability.

Another important property is that, under sufficient topological conditions, any conventionally nonunitary operator U on $\mathcal{H}$ over $\mathbb{C}$, $U \times U^\dagger \neq I$, can *always* be *identically* written in a genounitary form to the right or to the left, via the simple rule [11f]

$$U = (U^>) \times Q^{1/2} \quad \text{or} \quad P^{1/2} \times (^<U). \tag{2.70}$$

The reader should be aware that the entire theory of linear operators on a Hilbert space must be lifted into a genotopic form for consistency. For instance, conventional operations, such as TrX, Det$X$, *etc.* can be easily proved to be inapplicable for genomathematics, and must be replaced with the corresponding genoforms. The same happens for *all* conventional and special functions and transforms. Regrettably, we have to refer the reader to monograph [2c] for details.

3.9. Lie-Santilli Genotheory

We are now equipped to present the central mathematical structure of genochemistry, the *genotopies of Lie's theory*, first introduced by Santilli in Ref. [1a] (see also the subsequent works [1e, 2], today known as *Lie-Santilli genotheory* [4].

Consider the conventional Lie theory with ordered N-dimensional basis of Hermitean operators $X = \{X_k\}$, parameters $w = \{w_k\}$, universal enveloping associative algebra $\xi = \xi(L)$, Lie algebra $L \approx [\xi(L)]^-$, corresponding, (connected) Lie transformation group G on metric space $S(r, \delta, \mathbb{F})$ with local coordinates $r = \{r^k\}$ over a field $\mathbb{F}$.

The *Lie-Santilli genotheory* is here defined as a step-by-step bimodular lifting of the conventional Lie theory on bigenospaces over bigenofields, and includes:

1) The *universal genoenveloping algebra to the right* $\xi^>(L)$ of an N-dimensional Lie algebra L with ordered basis $X^> \equiv X = \{X_k\}, k = 1, 2, \ldots, N$, genounit $\hat{I}^> = Q^{-1}$, genoassociative product $X_i > X_j = X_i \times Q \times X_j$ and infinite-dimensional genobasis characterized by the *genotopic Poincaré-Birkhoff-Witt theorem to the right*,

$$\begin{aligned} I^> = Q^{-1}, \quad X_k, \quad X_i > X_j \quad (i \leq j), \\ X_i > X_j > X_k \quad (i \leq j \leq k), \ldots, \end{aligned} \tag{2.71}$$

and genoexponentiation (2.57a); the *universal genoassociative algebra to the left* $^<\xi(L)$ with genounit $^<I = P^{-1}$ and genoproduct $X_i < X_j = X_i \times P \times X_j$, with infinite-dimensional genobasis characterized by the *genotopic Poincaré-Birkhoff-Witt theorem to the left*,

$$\begin{aligned} ^<\hat{I} = P^{-1}, \quad X_k, \quad X_i < X_j \quad (i \leq j), \\ X_i < X_j < X_k \quad (i \leq j \leq k), \ldots, \end{aligned} \tag{2.72}$$

and genoexponentiation to the left (2.57b); the *bigenoenvelope* is the bistructure $\{^<\xi, \xi^>\}$ defined on corresponding bigenospaces and bigenofields under the condition $\hat{I}^> = (^<\hat{I})^\dagger$.

2) A *Lie-Santilli genoalgebra* is a bigenolinear bigenoalgebra defined on $\{{}^<\xi, \xi^>\}$ over $\{{}^<\mathbb{F}, \mathbb{F}^>\}$ with Lie-admissible product

$$\begin{aligned}(X_i, X_j) &= X_i < X_j - X_j > X_i = \\ &= X_i \times P \times X_j - X_j \times Q \times X_i.\end{aligned} \tag{2.73}$$

3) A (connected) *Lie-Santilli genotransformation group* is the bi-set $\{{}^<G, G^>\}$ of bigenotransforms on $\{{}^<S, S^>\}$ over $\{{}^<\mathbb{F}, \mathbb{F}^>\}$ with genounits ${}^<\hat{I} = (\hat{I}^>)^\dagger$

$$\begin{aligned}r^{>\prime} = (U^>) > r^> = (U^>) \times Q > r \times I^> = V \times r \times \hat{I}^>, \\ U^> = V \times \hat{I}^>,\end{aligned} \tag{2.74a}$$

$$\begin{aligned}{}^<r' = {}^<r < ({}^<U) = {}^<I \times r \times P \times ({}^<U) = {}^<I \hat{\times} r \times W, \\ {}^<U = {}^<\hat{I} \times W,\end{aligned} \tag{2.74b}$$

verifying the following conditions: genodifferentiability of the maps $G^> > S^> \to S^>$ and ${}^<S \leftarrow {}^<S < {}^<G$, invariance of the genounits and genolinearity, with realizations $U^> = \exp_>(i \times w \times X)$ and ${}^<U = \exp_<(-i \times w \times X)$, genolaws

$$\begin{aligned}U^>(w^>) > U^>(w^{>\prime}) = U^>(w^> + w^{>\prime}), \\ U^>(w^>) > U^>(-w^>) = U^>(0^>) = I^>,\end{aligned} \tag{2.75}$$

and Lie-admissible algebra (2.56) in the neighborhood of the genounits $\{{}^<\hat{I}, \hat{I}^>\}$.

Recall from Sect. 2.2 that, at the abstract realization-free level, the conventional Lie product $[A, B] = A \times B - B \times A$ and the Lie-Santilli product $[A\hat{,}B] = A\hat{\times}B - B\hat{\times}A = A\times\hat{T}\times B - B\times\hat{T}\times A$ coincide, evidently in view of the fact that both products are totally antisymmetric and verify the Jacobi law. It should be noted that, unexpectedly, *the Lie-admissible product* $(A, B) = A < B - B > A = A \times P \times B - B \times Q \times A$ *verifies the Lie axioms when defined on* $\{{}^<\xi, \xi^>\}$ *over* $\{{}^<\mathbb{F}, \mathbb{F}^>\}$, *despite its apparent lack of antisymmetry.*

In fact, the genoenvelopes to the left ${}^<\xi$ and to the right $\xi^>$ are isomorphic to the original envelope ξ, thus implying ${}^<I(A, B) = (A, B)I^>$ i.e., the value of the genoproduct $A < B = A \times P \times B$, when computed with respect to the genounit ${}^<I = P^{-1}$, is equal to that of the genoproduct $A > B = A \times Q \times B$ measured with respect to the genounit $I^> = Q^{-1}$.

One of the most important mathematical properties of this section, which is an evident consequence of the preceding analysis, is that *Lie-Santilli genogroups coincide at the abstract level with the original Lie-transformation groups.*

Note that the generators of the original Lie algebra are not lifted under genotopies, evidently because they represent conventional physical quantities, such as energy, linear momentum, angular momentum, *etc.* Only the *operations* defined on them are lifted. Note also that, when the conjugation $P = Q^\dagger$ is violated, the Lie axioms are lost. Note also that the genotheory is highly nonlinear, because the elements P and Q have un unrestricted functional dependence, thus including that on wavefunctions. Nevertheless, genomathematics reconstructs linearity in genospaces over genofields. The same happens for nonlocality, noncanonicity, nonunitary, and irreversibility [1f]. In fact, on genospaces over genofields, genotheories are fully linear, local, canonical or unitary and reversible. Departures from these axiomatic properties occur only in their *projection* over conventional spaces and fields.

Needless to say, we have been able to present in this section only the rudiments of the needed genomathematics, with the understanding that its detailed study is rather vast indeed.

4. Hypermathematics

As indicated in Sect. 2.2, *isomathematics* is only a particular case of the broader *genomathematics* of the preceding section. In turn, genomathematics is a particular case of the still more general multivalued *hypermathematics* [1e, 2e]. As we shall see, isomathematics is sufficient for the treatment of *reversible systems* with potential and nonpotential internal effects, such as molecular structures; genomathematics is necessary for the representation of *irreversible processes* caused by potential and nonpotential forces, such as chemical reactions; and hypermathematics has been worked out to represent *biological systems.*

The fundamental notions of hypermathematics are those of *hyperunit to the right and to the left*, which are given by a finite and ordered *set* of nowhere singular nonhermitean genounits,

$$\{\hat{I}^{>}\} = \{\hat{I}_1^{>}, \hat{I}_2^{>}, \ldots\} = \{1/Q_1, 1/Q_2, \ldots\}, \tag{2.76a}$$

$$\{^{<}\hat{I}\} = \{^{<}\hat{I}_1, {^{<}\hat{I}_2}, \ldots\} = \{1/P_1, 1/P_2, \ldots\}, \tag{2.76b}$$

with corresponding ordered *hypermultiplications to the right and to the left*,

$$A\{>\}B = \{A \times Q_1 \times B, A \times Q_2 \times B, \ldots\}, \tag{2.77a}$$

$$A\{<\}B = \{A \times P_1 \times B, A \times P_2 \times B, \ldots\}, \tag{2.77b}$$

under which the hyperunits $\{\hat{I}^{>}\}$ and $\{^{<}\hat{I}\}$ are the correct left and right units.

The above hyperunits and hyperproducts then imply the existence of a corresponding *hypermathematics*, which is characterized by *hypernumbers and hyperfields, hyperspaces and hypergeometries, hyperalgebras and hypergroups, etc.*

As an illustration, recall that the multiplications "$2 < 3$" (3 multiplying 2 from the right) and "$2 > 3$" (2 multiplying 3 from the left) yield the value 6 only for conventional mathematics. For the case of isomathematics, the multiplications "$2 < 3$" and "$2 > 3$" continue to yield the same result, but the value is now arbitrary, and depends on the assumed unit, e.g. "$2 < 3$" $=$ "$2 > 3$" $= 275.65$.

For the case of genomathematics, the multiplications "$2 < 3$" and "$2 > 3$" generally yield different results, each one being generally arbitrary and depending on the assumed unit per each multiplication, e.g., "$2 < 3$" $= 31.37 \neq$ "$2 > 3$" $= 27.81$.

For the case of hypermathematics, the two multiplications "$2\{<\}3$" and "$2\{>\}3$" not only yield different results, and each value is arbitrary, but also *each multiplication yield a set of values*, e.g., "$2 < 3$" $= \{5.23, 72.45, 0.67, \ldots\} \neq 2 > 3$" $= \{1.24, 45.43, 137.42, \ldots\}$.

Regrettably, we cannot outline hypermathematics to avoid a prohibitive length of this monograph. Therefore, we refer the interested reader to monograph [2e] for details.

An important property is that, as it was the case for isomathematics and genomathematics, *hypermathematics coincides with conventional mathematics at the abstract realization-free level.* For this reason, hypermathematics essentially consists of a new multivalued realization of conventional mathematical axioms.

It should be noted that *hypermathematics has been constructed for a quantitative representation of biological processes.* We recalled in Chapter 1, Fig. 1.6, that the correct representation of the *growth in time* of sea shells requires the doubling of the three Euclidean coordinates.

The necessity of hypermathematics in this case is due to the fact that, via our three Eustachian tubes, we perceive the growth of sea shells as occurring in a conventional *three-dimensional* space. The only known reconciliation of the three-dimensionality of our sensory perception with the doubling of the references axes is that permitted by hypermathematics, thanks to its abstract identity with the conventional mathematics.

More particularly, the hypereuclidean space coincides with the conventional Euclidean space at the abstract level. Therefore, our sensory perception of a 3×2-dimensional hyper-representation of sea shells growth is fully three-dimensional for our sensory perception. In other words, the representation space of Fig. 1.6 *is not* six dimensional. It is instead *three-dimensional multivalued.*

We should also indicate that hypermathematics in general, and hypernumbers in particular, have been submitted for a more adequate understanding of the *DNA code*, which continues to be studied to this day via the simplest possible numbers, those with the unit +1 dating back to biblical times. In reality, the complexity of the DNA code may result in being such to be beyond our comprehension, and may require hypernumbers whose unit is essentially an ordered set of some 10^{40} different values.

As a further illustration that science will never admit "final theories," we note that, by no means, hypermathematics is the most general possible form of mathematics. In fact, the subsequent step has already been identified, and consists of *hypermathematics* defined by the so-called *weak operations and identities* [8]. However, such a level of mathematical treatment is beyond the limited scope of this presentation.

5. Isodual Mathematics

One of the biggest scientific inbalances of the 20-th century relates to *antimatter*. In fact, throughout the 20-th century matter has been treated at all possible levels of study, from Newtonian mechanics to second quantization, while antimatter has been solely treated at the level of *second quantization*. The scientific inbalance is then evident if one meditates a moment on the fact that astrophysics of the 20-th century cannot even treat, let alone answer, the question whether a far away galaxy or quasar is made up of matter or of antimatter, evidently because of the complete lack of a *classical* description of antimatter.

Santilli [9] has studied for decades the above scientific inbalance, by first establishing that its origin rests on the *lack of appropriate mathematics*, rather than insufficiency of physical laws. In fact, the science of the 20-th century had only one mathematics, the conventional one, with consequential unique channel of quantization. Any classical treatment of antimatter, after quantization, yields *particles with the wrong sign of the charge*, rather than the needed antiparticles.

Note that exactly the same conclusion is reached via the use of iso-, geno-, and hyper-mathematics. In fact, each of these generalized mathematics has its own channel of quantization which can only yield particles, without resulting in antiparticles.

Santilli [*loc. cit.*] has established that a consistent treatment of antimatter at the classical level requires *yet new mathematics* which: 1) are anti-isomorphic to the conventional, iso-, geno-, and hyper-mathematics; 2) have their own channel of quantization; and 3) the operator images are indeed antiparticles, defined as charge conjugates of conventional particles on a Hilbert space.

The anti-isomorphic image of the *conventional* mathematics was proposed in Ref. [9a] of 1985. It is characterized by the map of the trivial unit +1 into its *negative* value,

$$I = +1 \rightarrow I^d = -1, \tag{2.78}$$

a map called *isoduality* for certain technical reasons and denoted with the upper symbol d, and then the reconstruction of the entire conventional mathematics in such a way to admit −1, rather than +1, as the correct left and rights units.

As an example, we have the *isodual fields* $\mathbb{F}^d(n^d, +^d, \times^d)$ [1b] which are rings of numbers $n^d = n \times I^d = -n$, where n is an ordinary number, equipped with the *isodual sum, isodual product, isodual norm, etc.*,

$$n^d +^d m^d = (n + m) \times I^d = -n - m, \tag{2.79a}$$

$$n^d \times^d m^d = n^d \times (-1) \times m^d = -n \times m, \tag{2.79b}$$

$$|n|^d = |n| \times I^d = -|n|, \tag{2.79c}$$

where +, × and |...| are the conventional operations for $\mathbb{F}(n, +, \times)$.

Note that isodual fields admit a *negative-definite norm.* It is then evident that *the isodual map reverses the sign of all quantities*, not only the charge, but also mass, time, energy, *etc.*, with *the understanding that these negative quantities are now referred to negative units.* For instance, the *isodual time* is indeed given by negative value of time $t^d = t \times I^d = -t$, but referred to *negative unit* $I^d = -1$ sec. As a result, *motion backward in time referred to a negative unit of time is as causal as motion forward in time referred to a positive unit of time.* Similar results hold for negative masses referred to negative units of mass, *etc.*

The *isodual mathematics* is given by *isodual numbers and fields, isodual spaces and geometries, isodual algebras and groups, isodual functional analysis*, and, inevitably, *isodual quantization* (see [9b] for brevity). It was then easy to prove in Refs. [9] that the isodual theory of antimatter yields a correct classical description which, after isodual quantization, yields genuine antiparticles.

It should be also indicated that map (2.78) is insufficient to construct anti-isomorphic images of the broader iso-, geno-, and hyper mathematics, which require the following *isodual map*

$$A = A(r, p, \psi, \dots) \rightarrow A^d = -A^\dagger(-r^\dagger, -p^\dagger, -\psi^\dagger, \dots), \tag{2.80}$$

with corresponding *isodual products*,

$$A > B = A \times Q \times B \rightarrow A >^d B = A \times (-Q^\dagger) \times B, \tag{2.81}$$

which must be applied to the *totality* of the iso-, geno-, and hyper-mathematics.

Regrettably, we cannot possibly review all these isodual mathematics, and must refer the interested reader to papers [9] or monograph [2e]. To illustrate the complexity and novelty of the resulting theories, we merely restrict ourselves to an indication in Fig. 2.1 of the emerging new notions of time. The reader should be aware that *isodual mathematics have resulted to be necessary for consistent quantitative representations of biological systems.* This *is not* surprising because the complexity of biological systems is such as to require the totality of our scientific knowledge.

THE NEW NOTIONS OF TIME

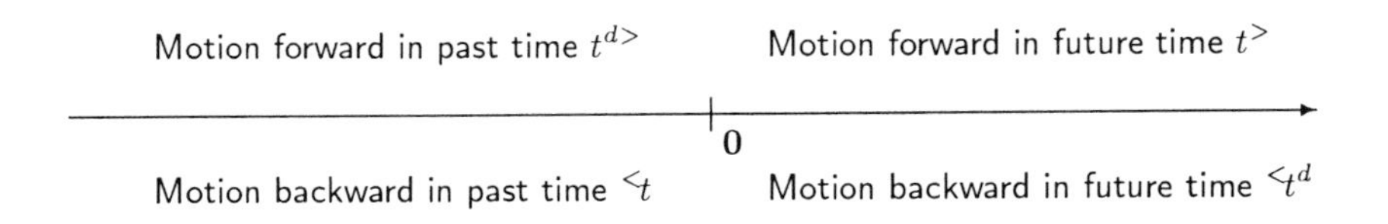

Figure 2.1. A summary view of the new notions of time introduced by the isotopic, genotopic, hyperstructural, and isodual methods for quantitative treatments of biological structures.

It has been generally believed until recently that Eddington's "time arrows" are *two*, motion forward and backward in time. The studies of Refs. [9] have established that *there exist four different notions of time of progressively increasing complexity and methodological needs.* The above picture depicts the generic four different times, which are given by *motion forward and backward in future and past times.* It is evident that these four times require two different conjugations, one for the map of the future into the past and the other for the map of motion forward into that backward, each conjugation being bi-injective (that is, such that, when applied twice it reproduces the original time). The map interconnecting Eddington "time arrows" (motion forward with that backward, or *vice versa*) is the familiar

$$\textit{time inversion: } t^{>} = t \rightarrow {}^{<}t = -t, \tag{2.82}$$

The map of the future into the past (or *vice versa*) must be anti-isomorphic to the preceding one for various reasons, thus being equivalent to the isodual map of Refs. [9], which, as such, *acts on the unit of time*,

$$\textit{time isoduality: } \hat{I}_t^{>} = t \rightarrow \hat{I}_t^{>d} = -\hat{I}_t^{>\dagger} = -t. \tag{2.83}$$

Since for conventional time its unit is $I_t = +1$, the above two maps are identical, i.e., $\hat{t}^{>} = t \times I^{>} = t, I^{>} = 1$, and $I^{>d}$ coincides with $-t$. However, the above trivial identity disappears when the unit of time is no longer +1 sec. It is easy to see that the conventional time constitutes the simplest conceivable realization of the notion of time *as perceived by our senses*, and, as such, it does not necessarily correspond to an intrinsic reality. Isotopies, genotopies, hyperstructures, and their isoduals permit *a chain of structural generalization of the above simple notion of time in a way compatible with our sensory perception*, which is achieved via the basic invariant of all geometries,

$$(\text{time})^2 \times (\text{unit})^2 = \text{ invariant}. \tag{2.84}$$

The first possible generalization of time meeting the above requirements is that provided by the isotopies and their isodualities, resulting in the following four different new times:

1) *Isotime*, representing motion forward to future time under a generalized unit solely treated via ordered multiplications to the right, and written $\hat{t}^{>} = t \times \hat{I}_t^{>}$, where $\hat{I}_t^{>} = \hat{I}_t^{>}(t, r, p, \ldots) = 1/\hat{T}_t > 0$ stands for the *forward time isounit*, i.e., the new numerical value of the unit of time (for instance $\hat{I}_t^{>} = 187$ sec, or, more properly a function of local variables), assumed to be positive-definite;

2) *Inverse isotime*, representing motion backward to past time under the same generalized unit for the forward motion, but solely applied to ordered products to the left, and written ${}^{<}\hat{t} = {}^{<}\hat{I}_t \times (-t), {}^{<}\hat{I}_t = \hat{I}_t^{>}$;

3) *Isodual isotime*, representing motion forward in past time solely applied ordered products to the right, which is written $\hat{t}^{>d} = -\hat{t}^{\dagger} = -\hat{I}^{>\dagger} \times t$; and

4) *Inverse isodual isotime*, representing motion backward in future time solely applied to ordered products to the left, and written ${}^{<}\hat{t}^{d} = -{}^{<}\hat{I}_t^{\dagger} \times (-t)$.

Despite the alteration of the unit of time, isotime is perceived by our senses as the conventional time because it satisfies the basic invariance,

$$(t_2 - t_1)^2 \times 1 \equiv [(t_2 - t_1) \times \hat{T}_t \times (t_2 - t_1)] \times \hat{I}_t,$$

thus establishing that *our perception of time is not necessarily the actual behavior of time in biological and other structures*. Moreover, the isodual isotime also satisfies the same basic invariance,

$$(t_2 - t_1)^2 \times 1 \equiv [(t_2 - t_1) \times \hat{T}^d \times (t_2 - t_1)] \times \hat{I}^d,$$

which establishes that *backward motions of time within biological structures is as causal as the conventional motion forward, and it is again compatible with our sensory perception.*

Note finally that the pace of time is inversely proportional to the unit, in the sense that *time slows down* for $\hat{I}_t^> > 1$ sec, while *time accelerates* for $\hat{I}_t^> < 1$ sec. As a consequence, *the pace of time perceived by our senses, by no means, is the actual pace of the structure considered.* In fact, if the internal isounit of time of a sea shell is $\hat{I}_t^> \ll 1$ sec, we appear motionless to its sensory perception, and sea shells perceive themselves as aging much faster than our perception of the same evolution. By no means the above four generalized notions of time exhaust all possible times, because we have *four additional genotimes*, in which the time genounit can assume complex or quaternionic values, and *four further hypertimes*, in which the time hyperunit is given by an ordered set of generally complex or quaternionic values.

The reader should be aware that the notion of time actually needed for scientific (that is mathematically rigorous and numerical) representations of biological structures and their evolutions, is that of *hypertime*. Consider, for instance, the problem of bifurcation of sea shells during their growth. It is easy to see that two notions of time are insufficient for their representation, which requires all four possible times (see monograph [2e] for details). But, as recalled in Sect. 2.4, the mathematics needed for the description of sea shells growth in time is hypermathematics. Consequently, the sole applicable notion of time is hypertime with a double multivalued structure in their four different realizations.

The need for generalized times is finally established by the fact that *conventional time in its four realizations solely permits the description of reversible systems (trivially, because the underlying mathematical axioms are reversible), thus being in dramatic disagreement with the irreversible character of nature.* The notorious irreversibility of biological structures can only be represented in a scientifically credible way, first, via time possessing a generalized unit, and, second, such generalized unit should be different for motions forward and backward in time.

It is hoped that the above results qualify any belief on the final character of quantum mechanics as being nonscientific, the sole scientific reality being the evidence that biological structures require theories of such a complexity to be beyond our comprehension.

References

[1] Santilli, R.M.: Hadronic J.**1**, 224, 574, and 1267 (1978) [1a]; Algebras, Groups and Geometries **10**, 273 (1993) [1b]; Lett. Nuovo Cimento **37**, 545, (1983) [1c]; Intern. J. Modern Phys. **D7**, 351 (1998) [1d]; Rendiconti Circolo Matematico Palermo, Suppl. **42**, 7 (1996) [1e]; Found. Phys. **27** 625 (1997) [1f]; Acta Appl. Math. **50**, 177 (1998) [1g].

[2] Santilli, R.M.: *Foundations of Theoretical Mechanics*, Vol. **I** (1978) [2a] and **II** (1983) [2b], Springer-Verlag, Heidelberg-New York; *Elements of Hadronic Mechanics*, Vol.*I* [2c] and *II* [2d] (1995), Ukraine Academy of Sciences, Kiev; *Isotopic, Genotopic and Hyperstructural Method in Theoretical Biology*, Ukraine Academy of Sciences, Kiev (1996) [2e]; *Isotopic Generalization of Galilei's and Einstein's Relativities*, Vols. **I** and bf II, Hadronic Press (1991) [2f].

[3] Myung, H.C., and Santilli, R.M.: Hadronic J. **5**, 1120 (1982) [3a]. Jannussis, A. and Mignani, R.: Physica **A187**, 575 (1992) [3b]. Kadeisvili, J.V.: Algebras, Groups and Geometries **9**, 283 and 319 (1992) [3c]. Tsagas, Gr.T. and Sourlas, D.S.: Algebras, Groups and Geometries **12**, 1 and 67 (1995) and **13**, 129 (1996) [3d]. Kamiya, N. and Santilli, R.M.: Hadronic J. **16**, 168 (1993) [3e]. Kadeisvili, J.V.: Math. Methods in Applied Sciences **19**, 1349 (1996) [3f]. Illert, C. and Santilli, R.M., *Foundations of Conchology*, Hadronic Press (1996) [3g].

[4] Jiang, C.X.: *An Introduction to Santilli's Isonumber Theory*, Hadronic Press, in press [4a]. Tsagas, Gr.R. and Sourlas, D.: *Mathematical Foundations of the Lie-Santilli Theory*, Ukraine Academy of Sciences, Kiev (1992) [4b]. Kadeisvili, J.V.: *Santilli's Isotopies of Contemporary Algebras, Geometries and Relativities*, Ukraine Academy of Sciences, Kiev, 2nd edition (1997) [4c].

[5] Santilli, R.M.: Nuovo Cimento **51**, 570 (1967) [5a]; Suppl. Nuovo Cimento **6**, 1225 (1968) [5b]; Meccanica **1**, 3 (1968) [5c]; Hadronic J. **3**, 440 (1979) [5d]; Found. Phys. **27**, 1159 (1997) [5e]; *Lie-Admissible Approach to the Hadronic Structure*, Vol. **I** (1978) and **II** (1982) [6d], Hadronic Press [5f].

[6] Gilmore, R.: *Lie Groups, Lie Algebras, and Some of their Representations*, Wiley, New York (1974) [6a]. Jacobson, N.: *Lie Algebras*, Interscience, New York (1962) [6b].

[7] Albert, A.A.: Trans. Amer. Math. Soc. **64**, 552 (1948).

[8] Santilli, R.M. and Vougiouklis, T.: in *Frontiers in Hyperstructures*, Vougiouklis, T,. Editor, Hadronic Press, Florida (1996).

[9] Santilli, R.M.: Hadronic J. **8**, 25 and 36 (1985) [9a]; Intern. J. Modern Phys., **A14**, 2205 (1999) [9b]; Hyperfine Interactions **109**, 63 (1997) [9c].

Chapter 3

FOUNDATIONS OF HADRONIC CHEMISTRY

1. Introduction

By following the historical argument by Einstein, Podolsky, and Rosen [1] on the "lack of completion" of quantum mechanics, a central assumption of this monograph is that *quantum chemistry* [2] is equally "incomplete."

In this Chapter we review a "completion" of quantum chemistry under the name of *hadronic chemistry* first proposed by Santilli and Shillady in Refs. [3]. The proposed "completion" is that via the addition of *nonlinear, nonlocal, and nonpotential interactions due to the deep wave-overlappings of the wavepackets of valence electrons*, which interactions are absent in quantum chemistry.

The invariant treatment of said interactions requires their representation via a generalization of the basic unit of the theory, resulting in the use of the novel iso-, geno-, and hyper-mathematics of the preceding chapter. Consequently, hadronic chemistry has three main branches called *iso-, geno- and hyper-chemistry* which are used for the representation of *molecular bonds, irreversible processes, and biological structures*, respectively.

As it was the case for the underlying mathematics, by conception and construction, hadronic and quantum chemistry coincide everywhere, with the sole general exception at distances of the order of 1 fm, in which the former theory presents novel contributions over the latter theory under the exact validity of all conventional quantum axioms and physical laws. As we shall see, the novel short range contributions are generally small, yet they permit the apparent resolution of at least some of the insufficiencies of quantum chemistry outlined in Sect. 1.3.

The main purpose of this Chapter is to outline the foundations of hadronic chemistry. Applications to a new model of the hydrogen, water, and other molecules with the essentially exact representation of molecular characteristics are presented in the subsequent chapters.

Predictably, our new model of molecular bonds also sees its roots in a number of pioneering studies by chemists. In fact, the notion of "electron correlation" [4a] is rather old in chemistry and some of the related studies did indeed reach notions close to that of our isochemical electron bonds of Chapter 4.

We also recall the studies by G.N. Lewis [4b], who originated the *octet structure* for light elements. I. Langmuir [4c] extended this notion to 2−He, 8−Ne, 8−Ar, 18−Kr, 18−Xe, 32−Ra. Additional contributions related to our work were made by D. Boyd *et. al.* [4d, 4e] with their "Coulomb hole" in Hartree-Fock-Roothaan calculations which achieves agreement with more accurate explicitly-correlated wavefunctions. This is in agreement with the work presented in this monograph, and clearly demonstrates the need for an attractive interaction among correlated electrons at very short distances (see Chapters 4 and 5).

While this is strong evidence for the reality of such a "Coulomb hole," Refs. [4] do not explain the nature of a "hole" in the electron correlation. As we shall see in the next chapters, hadronic chemistry implies not merely an absence of repulsion (the "Coulomb hole"), but also an *attraction* among the identical electrons, and the capability of its quantitative treatment.

Above all, preceding studies directly related to hadronic chemistry are those based on the various *screenings of the Coulomb law* [4] which have resulted in being necessary to improve the representation of molecular characteristics. By recalling that the Coulomb law is a central *invariant* of quantum mechanics and chemistry (that is, invariant under the unitary time evolution of the theory), said screenings can only be reached via *nonunitary* transforms of the Coulomb law, that is, by exiting the class of equivalence of quantum chemistry.

As a result, even though formulated on conventional Hilbert spaces over conventional fields, the use of the terms "quantum chemistry" is inappropriate for screened Coulomb laws on rigorous scientific grounds. It is easy to see that screened Coulomb laws do not admit the conventional notion of "quantum" of energy, they prohibit the existence of Bohr's quantized electron orbits, and have other features strictly outside quantum mechanics and chemistry in their original formulation.

Moreover, the reader should keep in mind the catastrophic inconsistencies of nonunitary theories reviewed in Sect. 1.7, most importantly, their *lack of invariance*, i.e., their general prediction of different numerical values for the same quantity under the same conditions but at different times.

A main purpose of hadronic chemistry is that of providing a general treatment for all possible nonunitary images of quantum chemistry, including nonunitary images of the Coulomb laws as a particular case, in a way invariant and compatible with the validity of quantum mechanics and chemistry at distances of the order of Bohr's orbits.

In fact, hadronic chemistry has the most general possible nonunitary structure, thus being *directly universal* for all possible screening, i.e., admitting all of them as particular cases (universality) in the fixed frame of the experimenter (direct universality). The representation is invariant because done via the basic nonunitary transform $U \times U^\dagger = \hat{I}$ assumed as the fundamental generalized unit of the theory. Compatibility with quantum mechanics at the distances of Bohr's orbits is easily achieved by assuming that $\lim_{r \gg 1 \text{ fm}}(U \times U^\dagger) = I$.

The above conception will permit the construction of a basically new model of molecular structures in which quantum mechanics and chemistry are not exactly valid for molecular structures, while preserving the validity of quantum mechanics for the structure of individual atoms, along the conceptual lines of Fig. 1.7.

As we shall see, the numerical value of the "completion" of quantum into hadronic chemistry via nonunitary effects is generally small. As a result, quantum mechanics and chemistry remain valid as a good approximation of molecular structures. Nevertheless, despite their small numerical values, the invariant addition of nonunitary interactions and effects have dramatic scientific implications, to such an extent to imply basically novel events and related technology which are simply unthinkable for quantum chemistry (see Chapters 7 and 8).

Another main objective of hadronic chemistry is to achieve a scientific representation of irreversible processes (such as chemical reactions), that is, a representation based on a truly irreversible structure predicting invariant numbers verified by measurements.

As well known, quantum chemistry is unable to represent irreversible processes. This is due to the fact that quantum chemistry can only represent systems via a Hamiltonian $H = p^2/2 \times m + V(r, p)$, while *all* potentials existing in nature (such as the electric, magnetic, gravitational, and other potentials) are strictly reversible with no known exception. Moreover, all geometric and other axioms of the mathematics underlying quantum chemistry are strictly reversible. It then follows that no attempt to represent irreversible processes via quantum chemistry can be expected to pass the test of time or of scientific rigor.

As treated in details in the preceding Chapter, our invariant representation of irreversibility initiates with the most fundamental notion, the lack of symmetric character of the basic unit of the theory. Then the same irreversibility propagates at the level of *all* subsequent mathematical structures. In this Chapter we shall show that physical and chemical theories based on such irreversible mathematics can indeed provide the first known axiomatically consistent and invariant representation of irreversibility from classical to quantum descriptions.

Yet another objective for the construction of hadronic chemistry is to provide methods for a scientific treatment of biological structures. By recalling that a central pillar of quantum mechanics is the *rotational symmetry of rigid bodies*, on strict scientific grounds, any attempt of studying biological structures via quantum chemistry necessarily implies that such structures are *perfectly rigid, perfectly reversible*, and *perfectly eternal*, (as it is the case of a crystal). Attempt at removing one of these inconsistencies for the intent of preserving the original theory can be easily dismissed as nonscientific because it is afflicted by a litany of inconsistencies.

As evident already in the mathematical structure presented in the preceding Chapter, hadronic chemistry does indeed permit the representation of biological structure as *deformable, irreversible and with a finite time*. However, in so doing, the ensuing methods become *multidimensional*, thus acquiring an unavoidable complexity. In the final analysis, the complexity of biological structures is dramatically distant from the elementary axioms of quantum mechanics and chemistry.

An understanding of this chapter requires some knowledge of analytic mechanics, quantum mechanics and chemistry, and the new mathematics outlined in the preceding chapter, which knowledge is herein assumed. Simple methods for the explicit construction of classical and operator hadronic chemistry will also be presented to assist readers less inclined toward advanced mathematics.

The reader should also be aware that quantum mechanics and chemistry have only *one* formulation, the conventional one [2]. On the contrary, hadronic theories have *two* different formulations, that on generalized spaces over generalized fields, and its *projection* on conventional spaces over conventional fields. To avoid excessive complexities, in this report we shall often consider the *projection* of hadronic theories on conventional spaces over conventional fields, and leave the mathematically correct formulation on generalized spaces to mathematical studies.

2. Classical Foundations of Hadronic Chemistry

2.1. The Historical Teaching of Lagrange and Hamilton

The central analytic equations of hadronic mechanics are the *original* Lagrange's and Hamilton's equations [5], which *are not* those used by chemistry throughout the 20-th century, bur rather the equations originally proposed, those with *external terms*

$$\frac{d}{dt}\frac{\partial L(t,\mathbf{r},\dot{\mathbf{r}})}{\partial \dot{\mathbf{r}}} - \frac{\partial L(t,\mathbf{r},\dot{\mathbf{r}})}{\partial \mathbf{r}} = \mathbf{F}(t,\mathbf{r},\dot{\mathbf{r}}), \tag{3.1a}$$

$$\frac{d\mathbf{r}}{dt} = \frac{\partial H(t,\mathbf{r},\mathbf{p})}{\partial \mathbf{p}}, \quad \frac{d\mathbf{p}}{dt} = -\frac{\partial H(t,\mathbf{r},\mathbf{p})}{\partial \mathbf{r}} + \mathbf{F}(t,\mathbf{r},\mathbf{p},\ldots), \tag{3.1b}$$

where $\mathbf{r} = \{r^k\}$, $\dot{\mathbf{r}} = \{\dot{r}^k\}$, $\mathbf{p} = \{p_k\}$, $\mathbf{F} = \{F^k\}$. All remaining original construction of analytic mechanics, such as the celebrated Jacobi theory [5c], was also formulated for the true analytic equations with external terms.

The external terms can be found in most studies up to the early part of the 20-th century, at which point they were eliminated following the success of the *truncated analytic equations*

$$\frac{d}{dt}\frac{\partial L(t,\mathbf{r},\dot{\mathbf{r}})}{\partial \dot{\mathbf{r}}} - \frac{\partial L(t,\mathbf{r},\dot{\mathbf{r}})}{\partial \mathbf{r}} = 0, \tag{3.2a}$$

$$\frac{d\mathbf{r}}{dt} = \frac{\partial H(t,\mathbf{r},\mathbf{p})}{\partial \mathbf{p}}, \quad \frac{d\mathbf{p}}{dt} = -\frac{\partial H(t,\mathbf{r},\mathbf{p})}{\partial \mathbf{r}}, \tag{3.2b}$$

for the representation of planetary and atomic structures. External terms were then generally avoided during the 20-th century because the addition implies evident departures from the very structure of Galilei's and Einstein's relativities.

However, Lagrange and Hamilton expressed quite clearly in their original writings the view that one single function, today called the *Lagrangian* or *Hamiltonian, cannot* represent the entire physical reality, because such a single function can represent only forces derivable from a potential. All remaining forces and effects of contact nonpotential type were represented with the external terms $\mathbf{F}(t,\mathbf{r},\mathbf{p},\ldots)$.

The classical foundations of hadronic chemistry is given precisely by the above historical teaching, that is, by following Lagrange's and Hamilton's forgotten legacy, we assume that a Hamiltonian H represents all possible action-at-a-distance interactions derivable from a potential, while the external terms $\mathbf{F}$ represent contact and other features or effects not representable by a potential or a Hamiltonian.

Moreover, an additional historical teaching of Lagrange and Hamilton, also generally forgotten in the 20-th century, is that *the origin of irreversibility rests with nonpotential forces represented with external terms*, since all potential interactions represented with a Lagrangian or a Hamiltonian are reversible, as recalled in Sect. 3.1.

Therefore, *the truncation of the external terms in the fundamental analytic equations implies the suppression of the very scientific means needed to represent irreversibility*. This is the reason why the original, historical external terms in the analytic equations have a truly central role for hadronic chemistry.

The primary scope of this section is to identify *methods for the invariant treatment of external terms*, as a necessary condition to have a viable analytic mechanics.

2.2. The Inevitability of the Historical Teaching

Comprehensive investigations conducted in monographs [6a] via the integrability conditions for the existence of a Lagrangian or a Hamiltonian (the so-called *conditions of variational selfadjointness*) have confirmed the historical teaching of Lagrange and Hamilton [5], to the effect that the truncated analytic equations can represent in the coordinates of the observer only a rather limited subclass of Newtonian systems.

The customary transformation of systems which are nonhamiltonian in a given frame to equivalent forms in different frames in which they assume a Hamiltonian character (i.e., the reduction of the true into the truncated analytic equations via coordinate transforms) will be strictly prohibited in our studies for a variety of reasons. The first is that the methods underlying the above transform (known as the Darboux theorem of the symplectic geometry or the Lie-Koening theorem in analytic mechanics [6a]) are solely applicable to *local-differential* systems while we are primarily interested in treating *nonlocal-integral* effects, for which the topological foundations of the methods are inapplicable.

Even assuming a local approximation via power series in the velocities, the reduction will be equally prohibited because the needed transformations are highly *nonlinear*, thus mapping the inertial frames of the observer into highly *noninertial frames*, with evident consequential violation of Galilei's and Einstein's relativities.

Finally, it is known that no experimental equipment can possibly be put on a Darboux transformed coordinate in which, e.g., $r' = \exp[Nrp]$.

Because of the above reasons, we shall solely admit in our studies *direct analytic representations* [6a, 6b], namely, representations of the equations of motion via an action functional, which hold in the *fixed* inertial frame of the observer. Only *after* achieving this direct representation, the use of the transformation theory may have practical relevance.

2.3. Problematic Aspects of External Terms

An inspection of Eqs. (3.1) reveals a number of methodological problems created by the presence of external terms. Recall that analytic equations (3.2) without external terms characterize the time evolution,

$$\frac{dA}{dt} = [A, H] = \frac{\partial A}{\partial r^k}\frac{\partial H}{\partial p_k} - \frac{\partial H}{\partial r^k}\frac{\partial A}{\partial p_k}, \tag{3.3}$$

via the celebrated *Poisson brackets* $[A, H]$ which, first, verify all conditions to characterize an algebra (right and left distributive and scalar laws), and, second, that algebra results in being Lie. By comparison, analytic equations (3.1) with external terms characterize the generalized time evolution,

$$\frac{dA}{dt} = (A, H) = \frac{\partial A}{\partial r^k}\frac{\partial H}{\partial p_k} - \frac{\partial H}{\partial r^k}\frac{\partial A}{\partial p_k} + \frac{\partial A}{\partial p_k}F^k, \tag{3.4}$$

whose brackets (A, H) *do not* characterize any algebra as commonly defined in mathematics (because the brackets violate the right distributive and scalar laws). As a result, Eqs. (3.1) must be rewritten in such a way as to admit a consistent algebraic structure, in which absence no effective covering of conventional analytic mechanics is possible.

Moreover, it is evident that time evolution (3.1b) is *noncanonical.* It is today known that, while the truncated Hamilton equations (3.2b) possess *invariant basic units of space and time* (because canonical transformations preserve, by definition, the units in the symplectic and Lie structures), this is not the case for Eqs. (3.1) when formulated on conventional spaces over conventional fields. As a result, Eqs. (3.1) do not admit unambiguous applications to actual measurement, as studied in details in Sect. 1.7.

The above occurrences have required the reformulation of Eqs. (3.1) in such a way to:

1) hold in the fixed inertial frame of the experimenter;
2) admit invariant units of space and time;
3) possess a consistent algebra in the brackets of the time evolution;
4) be derivable from an action principle; and, last but not least,
5) be axiom-preserving, i.e. preserving the axiomatic structure of conventional Hamiltonian mechanics (that without external terms).

It is today known that the above requirements are satisfied by the *iso-, geno-, and hyper-formulation of Eqs. (3.1)*, first presented in Ref. [7]

and known under the name of *Hamilton-Santilli iso-, geno-, and hyper-mechanics*. The latter mechanics are here assumed as the classical foundation of hadronic chemistry.

2.4. Classification of Hamilton's Equation with External Terms

During most of the 20-th century it has been believed that the ten total conservation laws underlying Galilean and special relativities (conservation of the total energy, linear momentum, angular momentum, and uniform motion of the center of mass) are solely applicable to a finite systems of isolated particles with *conservative* internal forces.

The above belief was disproved by the studies of monograph [6b] (see in particular Sect. 6.3, p. 236). In fact, a finite systems of particles with conservative as well as *nonpotential* internal forces represented by the external terms F^k in Eqs. (3.1) does indeed verify all ten *conventional* total conservation laws when the external forces verify the conditions

$$\sum_k F^k(t, r, p) = 0, \tag{3.5a}$$

$$\sum_k r_k \wedge F^k(t, r, p) = 0, \tag{3.5b}$$

$$\sum_k p_k \cdot F^k(t, r, p) = 0. \tag{3.5c}$$

The above condition constitute the classical foundation of a *new notion bound systems*, which are technically called *closed variationally nonselfadjoint* (or *closed nonhamiltonian*) [6b], in the sense that the systems verify all conventional total conservation laws (closure), yet they are variationally nonselfadjoint (not representable solely with a Hamiltonian).

The reader should be aware that the primitive classical structure of the new model of molecular bonds presented in the next chapters is precisely given by the above closed nonselfadjoint systems.

The broader class of *open nonselfadjoint systems* is evidently that when at least one of conditions (3.5) is violated by the external terms, resulting in exchange of energy, linear or angular momentum with an external system.

In summary, the classification in monograph [6b] of all possible systems characterized by the historical Hamilton's equations with external terms is given by:

1) *Closed selfadjoint systems*, in which all forces are conservative and the external terms are null. These systems are manifestly reversible in

time, and constitute the classical foundation of all models of conventional quantum chemistry. A representative example is given by planetary systems.

2) *Closed nonselfadjoint systems*, in which all potential forces are conservative and the external force are non-null, yet ratifying conditions (3.5). These systems constitute an evident generalization of the former, they are generally irreversible, and constitute the classical foundation of isochemistry. A representative example is given by Jupiter when isolated from the rest of the universe, which does indeed verify all ten conventional total conservation laws, yet the internal structure has vortices with *variable angular moments* visible with telescopes, entropy, and clear irreversibility.

3) *Open nonselfadjoint systems*, in which both potential and nonpotential forces are (sufficiently smooth, yet) arbitrary, and generally violate conditions (3.5). These systems are evidently *open*, in the sense of interacting with a system assumed as external, are generally irreversible, and constitute the foundations of genochemistry. A representative example is a space-ship during re-entry in Earth's atmosphere when considering the rest of the planet as external.

Another rather general belief of the physics of the 20-th century disproved by the studies via the conditions of variational selfadjointness [6] is that all open systems can be reduced to closed, conventional, Hamiltonian form via the addition of the external system. The achievement of closure (in the above sense of verifying all ten total conservation laws in our spacetime) via the addition of the external system is evident. However, the resulting new system is conventionally Hamiltonian only when the system is constituted by point particles under conservative, action-at-a-distance interactions without collisions. On the contrary, if the original constituents are extended particles under potential and nonpotential interactions, the resulting closed system *cannot* be conventionally Hamiltonian, thus mandating the above classification.

2.5. Hamilton-Santilli Isomechanics

When the external terms verify conditions (3.5), the resolution of the problematic aspects of Sect. 3.2.3 is permitted by isomathematics of Sect. 2.2, resulting in the so-called *Hamilton-Santilli isomechanics* [7].

The reformulation of Eqs. (3.1b) requires in this case the use of the 6-dimensional *isophase space* $T^*\hat{E}(\hat{r}, \hat{\delta}, \hat{\mathbb{R}})$ (Sect. 2.2.3, or more technically, the *isosymplectic geometry* [7]) with local coordinates $\hat{b} = \{b^\mu\} = \{b^\mu\} \times \hat{I}_6 = \{r^k, p_k\} \times \hat{I}_6$; defined with respect to the 6-dimensional genounit

[*loc. cit.*]

$$\hat{I}_6 = \text{Diag.}(\hat{I}_3, \hat{T}_3), \quad \hat{I}_3 = 1/\hat{T}_3 > 0. \tag{3.6}$$

The latter structure is due to the fact that $\hat{r}^k$ is contravariant while $\hat{p}_k$ is covariant, for which $\hat{d}\hat{r}^k = \hat{I}^k_i \times d\hat{r}^i$ and $\hat{d}\hat{p}_k = \hat{T}^i_k \times d\hat{p}_i$.

By ignoring the time isoderivative for simplicity, but using the isoderivatives in coordinates and momenta, Eqs. (3.1b) can then be *identically* rewritten in the *Hamilton-Santilli genoequations* [7]

$$(\omega) \times \begin{pmatrix} dr^k/dt \\ dp_k/dt \end{pmatrix} = \begin{pmatrix} 0_{3\times3} & -I_{3\times3} \\ I_{3\times3} & 0_{3\times3} \end{pmatrix} \times \begin{pmatrix} dr^k/dt \\ dp_k/dt \end{pmatrix} =$$
$$\begin{pmatrix} -dp_k/dt \\ dr^k/dt \end{pmatrix} = \begin{pmatrix} \hat{\partial}\hat{H}/\hat{\partial}r^k \\ \hat{\partial}\hat{H}/\hat{\partial}p_k \end{pmatrix} = \begin{pmatrix} \hat{T}^i_k \times \partial\hat{H}/\partial r^i \\ \hat{I}^k_i \times \partial\hat{H}/\partial p_i \end{pmatrix}, \tag{3.7a}$$

$$\hat{H} = \hat{p}_k \hat{\times} \hat{p}^k / \hat{2} \hat{\times} \hat{m} + \hat{V}(\hat{r}) = [p_k \times p^k/2 \times m + V(r)] \times \hat{I}_3, \tag{3.7b}$$

$$\hat{T}_3 = \text{Diag.}[I - F/(\partial H/\partial r)], \tag{3.7c}$$

where ω is the familiar canonical symplectic tensor, all generalized products are isotopic, e.g., $A\hat{\times}B = A \times \hat{T}_3 \times B$, and the contraction of the repeated k-indices occurs in isospace with isometric $\hat{\delta}^{>} = \hat{T}_3 \times \delta$; $r = \{r^k\}$, $\dot{r} = \{\dot{r}^k\}$, $p = \{p_k\}$, $F = \{F^k\}$.

As one can see, the main mechanism of Eqs. (3.7) is that of *transforming the external terms F into an explicit form of the isotopic element $\hat{T}_3$*. As a consequence, *reformulation (3.7) constitutes direct evidence on the capability to represent nonhamiltonian forces and effects with a generalization of the unit of the theory.*

Note in particular that *the external terms are embedded in the isoderivatives.* However, when written down explicitly, Eqs. (3.1) and (3.7) coincide. Note that $\hat{T}_3$ as in rule (3.7c) is fully symmetric, thus acceptable as the isotopic element of isomathematics. Note also that all nonlocal and nonhamiltonian effects are embedded in $\hat{T}$, thus permitting the use of the new Santilli-Sourlas-Tsagas topology [7] (a topology which is everywhere local-differential and conventional, except at the unit).

Note finally that *the Hamilton-Santilli isoequations are irreversible for all reversible Hamiltonians, whenever the external terms, or, equivalently their representation via the isounit, is not invariant under time reversal,*

$$F(t,\ldots) \neq F(-t,\ldots), \qquad \text{or} \tag{3.8a}$$

$$\begin{aligned} \hat{T}(t,\ldots) &= \text{Diag.}[I - F(t,\ldots)/(\partial H/\partial t)] \neq \\ &\neq \hat{T}(-t,\ldots) = [\text{Diag.}(I - F(t,\ldots)/(\partial H/\partial r)]. \end{aligned} \tag{3.8b}$$

$$(\hat{\partial}\hat{H}/\hat{\partial}\hat{r})|_t \neq (\hat{\partial}\hat{H}/\hat{\partial}\hat{r})|_{-t}, \quad (\hat{\partial}\hat{H}/\hat{\partial}\hat{p})|_t \neq (\hat{\partial}\hat{H}/\hat{\partial}\hat{p})|_{-t}, \tag{3.8c}$$

In particular, following our assumptions, we have irreversibility under the conservation of the total energy (see next subsection).

For future needs it is useful to represent Eqs. (3.7) in the unified notation

$$\hat{\omega}_{\mu\nu} \hat{\times} \frac{d\hat{b}^\nu}{dt} = \frac{\hat{\partial}\hat{H}(t,\hat{b})}{\hat{\partial}\hat{b}^\mu} = \hat{T}^\nu_{6\mu} \times \frac{\partial \hat{H}(t,b)}{\partial b^\nu}, \tag{3.9a}$$

$$\hat{\omega}_{\mu\nu} = \omega_{\mu\nu} \times \hat{I}_6, \quad \hat{b} = (\hat{r},\hat{p}), \quad \hat{T}_6 = \text{Diag.}(\hat{I}_3,\hat{T}_3). \tag{3.9b}$$

Another useful version of analytic equations (3.7) is characterized by the unified notation [6a, 6b]

$$R^\circ = \{R^\circ\} = (p_k, 0), \quad b = \{b^\mu\} = (r^k, p_k) \tag{3.10a}$$

$$\omega_{\mu\nu} = \frac{\partial R^\circ_\nu}{\partial b^\mu} - \frac{\partial R^\circ_\mu}{\partial b^\nu}. \tag{3.10b}$$

By recalling the properties of the isodifferential calculus, Eqs. (3.9) can then be written

$$\left(\frac{\hat{\partial}\hat{R}^\circ_\nu}{\hat{\partial}\hat{b}^\mu} - \frac{\hat{\partial}\hat{R}^\circ_\mu}{\hat{\partial}\hat{b}^\nu}\right) \hat{\times} \frac{d\hat{b}^\nu}{dt} = \frac{\hat{\partial}\hat{H}(b)}{\hat{\partial}\hat{b}^\mu} = \hat{T}^\alpha_{6\mu} \times \frac{\partial H(b)}{\partial b^\alpha}. \tag{3.11}$$

An important consequence is that *the Hamilton-Santilli isoequations coincide with the Hamilton equations without external terms at the abstract level.* In fact, as elaborated in Sect. 2.2, at the abstract level all differences between I and $\hat{I}$, $\times$ and $\hat{\times}$, ∂ and $\hat{\partial}$, *etc.*, disappear. This proves the achievement of a central objective of isomechanics, the property that the analytic equations with external terms can indeed be *identically* rewritten in a form equivalent to the analytic equations without external terms, provided, however, that the reformulation occurs via the broader isomathematics.

The "direct universality" of Eqs. (3.7), (3.9) should be noted and compared with the limited representational capabilities of the conventional equations (3.2b). In fact, Eqs. (3.7) can represent all possible (well behaved) systems with local and nonlocal as well a potential and nonpotential forces (universality), in the fixed inertial coordinates of the experimenter (direct universality). Note that this universality holds also for all possible *nonlocal-integral* realizations of the external force F, which are prohibited for the truncated Hamilton's equations. Note finally the simplicity of the representation of all possible external forces via *algebraic* Eqs. (3.7c). By comparison, the construction of a Lagrangian or a Hamiltonian for nonselfadjoint systems, if and when it exists, requires the solution of *nonlinear partial differential equations* [6a, 6b].

We should finally note that Eqs. (3.7) have been formulated, for simplicity, via the conventional time t and its derivative d/dt, rather than the isotime $\hat{t} = t \times \hat{I}$ (see Fig. 2.1) and its isoderivative $\hat{d}\hat{/}\hat{d}\hat{t}$. Such a formulation is sufficient for the study of classical and quantum isochemistry because the main systems to be studied (molecular structures) are reversible in time and the results will be referred to our conventional time. The extension of Eqs. (3.7) to include isotime and its isoderivative has been done in Ref. [7] resulting in remarkable additional degrees of freedom of the Hamilton-Santilli isomechanics.

2.6. Classical Lie-Santilli Brackets

It is important to verify that Eqs. (3.7), (3.9) resolve the problematic aspects of external terms indicated in Sect. 3.2.3. This is done by noting that the time evolution of Eqs. (3.7) has the structure [7]

$$\frac{d\hat{A}}{dt} = [\hat{A}\hat{,}\hat{H}] = \frac{\hat{\partial}\hat{A}}{\hat{\partial} r^k} \hat{\times} \frac{\hat{\partial}\hat{H}}{\hat{\partial}\hat{p}_k} - \frac{\hat{\partial}\hat{H}}{\hat{\partial} r^k} \hat{\times} \frac{\hat{\partial}\hat{A}}{\hat{\partial}\hat{p}_k} = \qquad (3.12)$$

$$= \left(\frac{\partial A}{\partial r^k} \times \frac{\partial H}{\partial p_k} - \frac{\partial H}{\partial r^k} \times \frac{\partial A}{\partial p_k} \right) \times \hat{I},$$

where the summation occurs in isospace. Note the conservation of the total energy H even in case the isotopic element $\hat{T}$ is not time reversal invariant,

$$\frac{d\hat{H}}{dt} = [\hat{H}\hat{,}\hat{H}] = \frac{\hat{\partial}\hat{H}}{\hat{\partial} r^k} \hat{\times} \frac{\hat{\partial}\hat{H}}{\hat{\partial}\hat{p}_k} - \frac{\hat{\partial}\hat{H}}{\hat{\partial} r^k} \hat{\times} \frac{\hat{\partial}\hat{H}}{\hat{\partial}\hat{p}_k} \equiv 0, \qquad (3.13)$$

Note that the projection of the above expression on isospaces into its corresponding form on conventional spaces reads

$$\frac{dH}{dt} = \frac{\partial H}{\partial r^k} \times \frac{\partial H}{\partial p_k} - \frac{\partial H}{\partial p_k} \times \frac{\partial H}{\partial r^k} + \frac{\partial H}{\partial p_k} \times F^k = \frac{\partial H}{\partial p_k} \times F^k \equiv 0, \quad (3.14)$$

where the last identity holds in view of the assumed condition (3.5c).

The conservation of the total energy as well as other conservation laws is then recommendable for the reformulation without ambiguities of Hamilton's equations (3.1b) in terms of their isotopic form (3.7).

The reader should however be aware that reformulation (3.7) is "universal", that is, applying for all possible external terms, including those for which conditions (3.5) are violated and the energy is no longer conserved. In this latter case property (3.13) continues to be valid, that is, the total energy is conserved on isospaces over isofields, while its *projection* on conventional spaces is not conserved.

As one can see, the latter occurrence may be cause of ambiguities to noninitiated readers, although the underlying mechanism is simple. In essence, the total energy is conserved on isospace while being nonconserved on ordinary space because in the former case the energy is referred to a generalized unit which is precisely the inverse of its rate of nonconservation. It is then evident that, under these conditions, the Hamiltonian is conserved. The same Hamiltonian is then nonconserved when projected on ordinary space because in this case it is referred to the trivial constant unit I.

The basic brackets of Hamilton-Santilli isoequations (3.7) are explicitly given by

$$[A\hat{,}B] = \frac{\partial A}{\partial r_j} \times \hat{T}_i^j(t,r,p,\ldots) \times \frac{\partial B}{\partial p_j} - \frac{\partial B}{\partial r_i} \times \hat{T}_i^j(t,r,p,\ldots) \times \frac{\partial A}{\partial p_j}, \quad (3.15)$$

and can be written from Eqs. (3.9) in the unified $\hat{b}$-notation [7]

$$\begin{gathered}[\hat{A}\hat{,}\hat{B}] = \frac{\hat{\partial}\hat{A}}{\hat{\partial}\hat{b}^\mu} \hat{\times} \hat{\omega}^{\mu\nu} \hat{\times} \frac{\hat{\partial}\hat{B}}{\hat{\partial}\hat{b}^\nu} = \\ = (\frac{\partial A}{\partial b^\alpha} \times \hat{T}_\mu^\alpha \times \omega^{\mu\nu} \times \hat{T}_\nu^\beta \times \frac{\partial B}{\partial b^\beta}) \times \hat{I} = (\frac{\partial A}{\partial b^\alpha} \times \omega^{\alpha\beta} \times \frac{\partial B}{\partial b^\beta}) \times \hat{I},\end{gathered} \quad (3.16)$$

where summations are, again, on isospaces, $\omega_{\mu\nu}$ is the canonical symplectic tensor of Eqs. (3.9),

$$\{\omega^{\mu\nu}\} = \{(|\omega_{\alpha\beta}|^{-1})^{\mu\nu}\} = \begin{pmatrix} 0_{3\times3} & I_{3\times3} \\ -I_{3\times3} & 0_{3\times3} \end{pmatrix}, \quad (3.17)$$

is the canonical Lie tensor, and we have used the invariance property (see [*loc. cit.*], Eq. (2.67))

$$\hat{T}_\mu^\alpha \times \omega^{\mu\nu} \times \hat{T}_\nu^\beta = \omega^{\alpha\beta}. \quad (3.18)$$

As one can see, the isotopic reformulation of Hamilton's historical equations with external terms does resolve the problematic aspects of Sect. 3.2.3. Unlike the case of brackets (3.4), the generalized brackets $[\hat{A}\hat{,}\hat{B}]$ do verify the left and right scalar and distributive laws, thus characterizing a perfectly acceptable (nonassociative) algebra.

Moreover, the algebra characterized by brackets $[\hat{A}\hat{,}\hat{B}]$ verifies the isotopic Lie axioms, thus being *a realization of the Lie-Santilli isotheory* of Sect. 2.2.9. An important property of isobrackets $[\hat{A}\hat{,}\hat{B}]$ is that *they coincide with the conventional Poisson brackets at the abstract, realization-free level.* In fact, the sole difference between the conventional and isotopic brackets is that the summations, e.g., in realization (3.13), are on isospaces, otherwise the two brackets formally coincide.

The geometric counterpart of Eqs. (3.9) is given by the *isosymplectic geometry* [7], which we cannot possibly review here for brevity. We merely mention for readers interested in geometry that the abstract identity of the conventional and isotopic brackets implies the corresponding abstract identity between the fundamental symplectic and isosymplectic forms. As a matter of fact, the isosymplectic geometry can be formulated via the use of all conventional symbols, and then merely subject them to the broader isotopic realization. These occurrences confirm that the isotopies are a new degree of freedom which is hidden in conventional geometric or algebraic axioms.

2.7. Isoaction Principle

It is easy to see that Eqs. (3.7) are derivable from the *isoaction principle* [7]

$$\begin{aligned} \hat{\delta}\hat{A} &= \hat{\delta}\int_{t_1}^{t_2}(\hat{p}_k\hat{\times}\hat{d}\hat{r}^k - \hat{H}\times dt) = \\ &= \hat{\delta}\int_{t_1}^{t_2}(p_k\times\hat{I}^k_i(t,r,p,dp/dt,\ldots)\times\hat{d}\hat{r}^i - \hat{H}\times dt) = 0. \end{aligned} \tag{3.19}$$

An equivalent variational principle is that in terms of notation (3.10) for which the isoaction principle,

$$\begin{aligned} \hat{A} &= \hat{\delta}\int_{t_1}^{t_2}(\hat{p}_k\hat{\times}\hat{d}\hat{r}^k - \hat{H}\times dt) \equiv \hat{\delta}\int_{t_1}^{t_2}(\hat{R}^\circ_\mu\hat{\times}\hat{d}\hat{b}^\mu - \hat{H}\hat{\times}dt) = \\ &= \hat{\delta}\int_{t_1}^{t_2}(R^\circ_\mu\times\hat{I}^\mu_{6\nu}\times db^\nu - H\times dt) = 0, \end{aligned} \tag{3.20}$$

yields unified version (3.11), where we have tacitly assumed the elimination of redundant isotopic products.

The derivation mainly requires the lifting of all partial derivatives and differentials of the conventional variational calculus according to the rules of the isodifferential calculus. Alternatively, one can use the conventional variation of a conventional action, and then subject the derivatives and differentials to their isotopic interpretation.

This establishes the *isocanonicity* of the theory (Sect. 2.2.8), namely, its capability to reconstruct a canonical structure on isospace over isofields, such as structures $\hat{p}_k\hat{\times}\hat{d}\hat{r}^k - H\times dt$. The extension to the isotopy of time adds a further degree of freedom [7], which is not essential for this work.

The reader can now begin to see that isomathematics is not a mere formal theory, because it carries rather deep implications. Recall that

Hamilton's equations with external terms are not derivable from a variational principle. In turn, such an occurrence has precluded the identification of the operator counterpart of Eqs. (3.1b) throughout the 20-th century.

We now learn that the *identical* reformulation of Eqs. (3.1b) into Hamilton-Santilli isoequation is fully derivable from a variational principle. In turn, this will soon permit the identification of the unique and unambiguous operator counterpart, which is at the foundation of the new model of molecular structure presented in the next chapters.

The direct universality of classical isochemistry can then be seen from the arbitrariness of the integrand of isoaction functional (3.6) once projected on conventional spaces over conventional fields.

An important property of isoaction principle (3.19) is that it possesses a functional dependence on *derivatives of arbitrary order*, which are evidently embedded in the isotopic element. By comparison, the conventional action can at most depend on *first-order* derivatives dr/dt. More generally, *the isotopies permit the reformulation of (well behaved) action functionals of arbitrary order into their identical isocanonical form (3.20).* This occurrence has important dynamical implications studied in the subsequent chapters.

Alternatively, we can say that *Lagrangians or variational actions of order higher than the first are inessential* because they can be all identically reduced to an isotopic first-order form by embedding all higher-order terms in the isoderivatives. This result is important because theories of higher than the first are known not to admit a meaningful phase space and Hamiltonian counterpart, the latter being typical of first-order theories. As a result, *the Hamilton-Santilli mechanics is directly universal for all possible orders.*

2.8. Hamilton-Jacobi-Santilli Isoequations

The first important consequence of isoaction principle (3.19) is the characterization of the following *Hamilton-Jacobi-Santilli equations* [7]

$$\frac{\partial \hat{A}}{\partial t} + \hat{H} = 0, \quad \frac{\hat{\partial} \hat{A}}{\hat{\partial} \hat{r}^k} - \hat{p}_k = 0, \quad \frac{\hat{\partial} \hat{A}}{\hat{\partial} \hat{p}_k} \equiv 0, \tag{3.21}$$

which will soon have basic relevance for isoquantization.

2.9. Examples of Classical Applications

A typical Newtonian application of the Hamilton-Santilli isomechanics is a space-ship during reentry in our atmosphere which experiences local potential forces applied at the center of gravity r represented by $V(r)$ as well as contact, nonlocal resistive forces depending on its shape, which can be written in terms of a suitable kernel $\mathcal{F}(\sigma, \mathbf{r}, \mathbf{p}, \ldots)$

$$m\frac{d\mathbf{p}}{dt} = -\frac{\partial H}{\partial \mathbf{r}} + \mathbf{F}^{\mathrm{NSA}}, \quad \mathbf{F}^{\mathrm{NSA}} = \int d\sigma\, \mathcal{F}(\sigma, \mathbf{r}, \mathbf{p}, \ldots), \tag{3.22}$$

where NSA stands for variational nonselfadjointness, namely, lack of derivability from a potential [6a].

The isoanalytic representation of the above system is then given by Eqs. (3.7) with $H = p^2/2m + V(r)$ and $\hat{T} = 1 - F/\ \partial V/\partial r$. The reader can easily construct an unlimited number of other direct representations of nonhamiltonian systems.

Note that the identification of an action functional permits the application of the *optimal control theory* for the first time to the optimization of shapes of extended objects moving within a resistive medium without approximation, which optimization was otherwise impossible due to the absence of an action principle.

Specific classical applications will be studied in future papers.

2.10. Connection Between Isotopic and Birkhoffian Mechanics

Monograph [6b] is devoted to a comprehensive presentation of a generalization of conventional Hamiltonian mechanics (that with truncated analytic equations) called by Santilli for certain historical reasons *Birkhoffian mechanics.* The action principle of the latter mechanics in unified notation is given by

$$\delta A = \delta \int_{t_1}^{t_2} (R_\mu \times db^\mu - H \times dt) = 0, \tag{3.23a}$$

$$R = R(b) = \{R_\mu\} = \{A_k(r,p), B^k(r,p)\}, \tag{3.23b}$$

$$k = 1, 2, 3, \quad \mu, \nu = 1, 2, 3, 4, 5, 6, \tag{3.23c}$$

namely, the main difference between conventional Hamiltonian and Birkhoffian mechanics is that the former is characterized by the particular realization $R^\circ = (p, 0)$, Eq. (3.10), while the latter is characterized by the general form $R = R(r,p)$, i.e., by arbitrary (nondegenerate) functions of r and p.

Birkhoff's equations can then be readily obtained from principle (3.23), and can be written in the following covariant and contravariant forms

$$\Omega_{\mu\nu}(b) \times \frac{db^\nu}{dt} = \frac{\partial H(b)}{\partial b^\mu}, \tag{3.24a}$$

$$\frac{db^\mu}{dt} = \Omega^{\mu\nu}(b) \times \frac{\partial H(b)}{\partial b^\nu}, \tag{3.24b}$$

$$\Omega_{\mu\nu} = \frac{\partial R_\nu}{\partial b^\mu} - \frac{\partial R_\mu}{\partial b^\nu}, \tag{3.24c}$$

$$\Omega^{\mu\nu} = |(\Omega_{\alpha\beta})^{-1}|^{\mu\nu}. \tag{3.24d}$$

An aspect important for hadronic chemistry is that Birkhoffian mechanics was also proved in Ref. [6b] to be "directly universal" for all possible Newtonian systems. As a result, all possible true Hamilton equations (3.1b) can always be represented with Eqs. (3.24) in the fixed reference frame $b = (r, p)$ of the experimenter.

However, *Birkhoffian mechanics has a noncanonical structure, thus suffering from the problematic aspects of Sect. 1.7, while it admits no known or otherwise consistent operator image despite considerable attempts.* Due to these insufficiencies identified years following the appearance of monograph [6b], research on the appropriate analytic mechanics for unrestricted Newtonian systems had to be initiated again from the very foundations, eventually resulting in the construction of isohamiltonian mechanics of Ref. [7] which does resolve all insufficiencies of Birkhoffian mechanics.

The reader may be interested in knowing that this laborious search can ultimately be reduced to the following identical reformulation of Birkhoff's into the Hamilton-Santilli equations

$$\frac{db^\mu}{dt} = \Omega^{\mu\mu}(b) \times \frac{\partial H(b)}{\partial b^\nu} = \\ = \omega^{\mu\nu} \times \hat{T}^\nu_{6\alpha}(b) \times \frac{\partial H(b)}{\partial b^\nu} = \omega^{\mu\nu} \times \frac{\hat{\partial} H(b)}{\hat{\partial} b^\nu}. \tag{3.24e}$$

More explicitly, the reformulation is based on: 1) the factorization of Birkhoff's tensor $\Omega^{\mu\nu}(b)$, Eqs. (3.23c), into the canonical Lie tensor $\omega^{\mu\nu}$, Eqs. (3.17),

$$\Omega^{\mu\nu}(b) = \omega^{\mu\nu} \times \hat{T}^\nu_{6\mu}(b); \tag{3.25}$$

2) the interpretation of the 6×6-dimensional factor matrix $\hat{T}$ as the isotopic element of the theory in phase space; and, most importantly, 3) the embedding of such isotopic element in the isoderivatives

$$\hat{T}^\nu_{6\mu}(b) \times \partial_\nu = \hat{\partial}_\mu. \tag{3.26}$$

The latter reformulation yields the isohamilton equations (3.9) which are formally identical to the conventional (truncated) equations, even though the underlying actual equations are those with unrestricted external terms. In turn, such a reformulation resolves all insufficiencies of

Birkhoffian mechanics because it achieves full invariance via the notion of isocanonicity (see Sect. 3.5), while admitting a unique and unambiguous operator image (see Sect. 3.3 and 3.4).

As an incidental note, the reader should be aware that the construction of an analytic representation via Birkhoff's equations of sufficiently smooth but otherwise arbitrary, generally nonconservative, Newtonian systems is rather complex, inasmuch as it requires the solution of nonlinear partial differential equations, or integral equations [6b].

By comparison, the construction of the same analytic equations via Hamilton-Santilli isoequations (3.7) or (3.9) is truly elementary, and merely requires the identification of the isotopic element according to rule (3.7c) for arbitrarily given external forces $F_k(t, r, p)$ without any solution of any equation.

2.11. Hamilton-Santilli Geno-, Hyper-, and Isodual- Mechanics

As indicated earlier, Hamilton-Santilli isoequations are always conservative on isospace due to law (3.13). As such, their physical interpretation is unambiguous for closed nonhamiltonian systems verifying conditions (3.5), while their use for open-nonconservative systems may be ambiguous for noninitiated readers.

Whenever the representation of the *time rate of variation* of the energy and other physical quantities is necessary, a more adequate mechanics is the *Hamilton-Santilli genomechanics* [7] which is based on the genomathematics of the preceding Chapter.

The main difference between the iso- and geno-mechanics is that in the former case the isotopic element $\hat{T}$ is real-valued and symmetric, while in the latter case is real-valued but nonsymmetric. This automatically implies the existence of *two* elements $\hat{T}^{>}$ and ${}^{<}\hat{T}$ interpreted as characterizing motion forward and backward in time with corresponding forward and backward genounits $\hat{I}^{>} = 1/\hat{T}^{>}$ and ${}^{<}\hat{I} = 1/{}^{<}\hat{T}$ and corresponding ordered products to the right ">" and to the left "<". We reach in this way an axiomatically consistent representation of irreversibility under nonconserved Hamiltonians which begins with the most fundamental elements of the theory, the basic units and products.

Genomechanics admits numerous realizations. A representative case is that of genoequations in which the symmetry of the isotopic element is broken at the geometric level of the phase-space (symplectic geometry). By ignoring again genoderivatives in time for simplicity, we have in this way the following realization of the *Hamilton-Santilli genoequations* [7]

$$\left(\frac{db^\mu}{dt}\right) = \left(\begin{array}{c} dr^k/dt \\ dp_k/dt \end{array}\right) = \hat{\omega}^> > \left(\begin{array}{c} \hat{\partial}^> \hat{H}/\hat{\partial}^> r^k \\ \hat{\partial}^> \hat{H}/\hat{\partial}^> p_k \end{array}\right) = \omega^{\mu\alpha} \times \hat{I}^{>\nu}_{6\alpha} > \frac{\partial H}{\partial b^\nu} =$$

$$= \left(\begin{array}{cc} 0_{3\times3} & I_{3\times3} \\ -I_{3\times3} & 0_{3\times3} \end{array}\right) \times \left(\begin{array}{cc} I_{3\times3} & K_{3\times3} \\ 0_{3\times3} & I_{3\times3} \end{array}\right) \times \left(\begin{array}{c} \partial H/\partial r^k \\ \partial H/\partial p_k \end{array}\right) = \tag{3.27a}$$

$$\left(\begin{array}{c} \partial H/\partial p_k \\ -\partial H/\partial r^k + F_k \end{array}\right),$$

$$\hat{I}^>_6 = \left(\begin{array}{cc} I_{3\times3} & K_{3\times3} \\ 0_{3\times3} & I_{3\times3} \end{array}\right), \tag{3.27b}$$

$$\begin{array}{c} b = \{r^k, p_k\}, \quad H = p^2/2 \times m + V(r,p), \\ K = -F/(\partial H/\partial p) = -F/(p/m). \end{array} \tag{3.27c}$$

In this case the time evolution is characterized by the following brackets [7]

$$\frac{dA}{dt} = (A, H) = \frac{\hat{\partial}^> A}{\hat{\partial}^> b^\mu} > \hat{\omega}^{>\mu\nu} > \frac{\hat{\partial}^> H}{\hat{\partial}^> b^\nu} \equiv$$

$$\equiv \frac{\partial A}{\partial r^k} \times \frac{\partial H}{\partial p_k} - \frac{\partial H}{\partial r^k} \times \frac{\partial A}{\partial p_k} - \frac{\partial H}{\partial p_k} K^i_k \frac{\partial A}{\partial p_i} \equiv \tag{3.28}$$

$$\equiv \frac{\partial A}{\partial r^k} \times \frac{\partial H}{\partial p_k} - \frac{\partial H}{\partial r^k} \times \frac{\partial A}{\partial p_k} + \frac{\partial A}{\partial p_k} \times F_k$$

As one can see, the new brackets (A, H) are no longer totally antisymmetric, thus permitting the representation of the *time rate of variation of the energy*

$$dH/dt = (H, H) = p_k \times F^k. \tag{3.29}$$

This confirms that genomechanics is indeed nonconservative as the original one. The irreversibility of Eqs. (3.27) is evident. Therefore, genomechanics permits a classical representation of irreversible and nonconservative systems, as needed for classical genochemistry,

Reformulation (3.27) of Hamilton's equations (3.1b) is not a mere mathematical curiosity, in view of its implications. First, the reformulation resolves the inconsistencies of brackets (3.4), by permitting the new brackets (A, B) to constitute a consistent algebra (verifying the right and left scalar and distributive laws). In particular, such algebra results to be Lie-admissible in the sense that the attached antisymmetric brackets $[A \hat{,} B] = (A, B) - (B, A)$ are Lie, although this classical algebra *is not* Jordan-admissible (see Ref. [7] for brevity).

Moreover, Hamilton's equations with external terms are notoriously *not* derivable from a variational principle. This insufficiency is readily resolved again by Eqs. (3.27) which are indeed derivable from the *genoaction principle* [7],

$$\begin{aligned}\delta^{>}\hat{A} = \delta^{>}\int_{t_1}^{t_2}(\hat{R}^{\circ}_{\mu} > \hat{d}^{>}\hat{b}^{\mu} - H \times dt) = \\ = \hat{\delta}^{>}\int_{t_1}^{t_2}(R^{\circ}_{\mu} \times \hat{I}^{>\mu}_{6\nu} \times db^{\nu} - H \times dt) = 0,\end{aligned} \tag{3.30}$$

where the term $H \times dt$ has not been lifted for simplicity. The above principle merely requires the lifting of all derivations in the conventional variation into genoderivatives.

In turn, the availability of an action principle permits the derivation of the following *Hamilton-Jacobi-Santilli genoequations* [7]

$$\frac{\partial \hat{A}}{\partial t} + \hat{H} = 0, \quad \frac{\hat{\partial}^{>}\hat{A}}{\hat{\partial}^{>}\hat{r}^{k}} - \hat{p}_k = 0, \quad \frac{\hat{\partial}^{>}\hat{A}}{\hat{\partial}^{>}\hat{p}_k} \equiv 0. \tag{3.31}$$

Still in turn, the availability of Hamilton-Jacobi equations permits the achievement of the first known unique and unambiguous operation image of Hamilton's equations with external terms studied in the next sections. By comparison, no operator image of Eqs. (3.1b) is known, or can be consistently formulated, e.g., because of the algebraic inconsistencies of the underlying brackets (3.4).

For additional realization of Hamilton-Santilli genoequations we refer the interested reader to monograph [6d] for brevity.

The *Hamilton-Santilli hypermechanics* [7] is formally the same as the genomechanics, with the exception that genotopic elements and genounits are now realized via sets of real-valued nonsymmetric matrices. Its explicit construction is left to the interested reader for brevity.

The *isodual geno- and hyper mechanics* [7] are the anti-isomorphic images constructed via the isodual map of Chapter 2. The latter mechanics were originally constructed for an invariant *classical* representation of antimatter [14]. Subsequently, they resulted in being necessary for an invariant classical representation of complex biological systems, such as bifurcation, which require time inversions (see Fig. 2.1).

2.12. Simple Construction of Classical Isochemistry

Classical isochemistry can be constructed via a simple method which does not need any advanced mathematics, yet it is sufficient and effective for practical applications.

In fact, *the Hamilton-Santilli isomechanics can be constructed via the systematic application of the following noncanonical transform to all quantities and operations of the conventional Hamiltonian mechanics*

$$U \times U^t = \hat{I}_6 = \begin{pmatrix} \hat{T} & 0 \\ 0 & \hat{T}^{-1} \end{pmatrix}, \quad (3.32a)$$

$$\hat{T} = I - \frac{F}{\partial H/\partial p} = I - \frac{F}{p/m}, \quad (3.32b)$$

where t stands for transpose and F represents the external terms of Eqs. (3.1b).

The success of the construction depends on the application of the above noncanonical transform to the *totality* of Hamiltonian mechanics, with no exceptions. We have in this way the lifting of: the 6-dimensional unit of the conventional phase space into the isounit

$$I_6 \rightarrow \hat{I}_6 = U \times I_6 \times U^t; \quad (3.33a)$$

numbers into the isonumbers,

$$n \rightarrow \hat{n} = U \times n \times U^t = n \times (U \times U^t) = n \times \hat{I}_6; \quad (3.33b)$$

associative product $A \times B$ among generic quantities A, B, into the isoassociative product with the correct expression and property for the isotopic element,

$$A \times B \rightarrow A \hat{\times} B = U \times (A \times B) \times U^t = A' \times \hat{T} \times B', \quad (3.34a)$$

$$A' = U \times A \times U^t, \quad B' = U \times B \times U^t, \quad \hat{T} = (U \times U^t)^{-1} = T^t; \quad (3.34b)$$

Euclidean into isoeuclidean spaces (where we use only the space component of the transform)

$$\begin{gathered} x^2 = x^t \times \delta \times x \rightarrow \hat{x}^{\hat{2}} = U \times x^2 \times U^t = \\ = [(x^t \times U^t) \times (U^{t-1} \times \delta \times U^{-1}) \times (U \times x) \times (U \times U^t) = \\ = [x'^t \times (\hat{T} \times \delta) \times x'] \times \hat{I}; \end{gathered} \quad (3.35)$$

and, finally, we have the following isotopic lifting of Hamilton's into Hamilton-Santilli equations,

$$\begin{gathered} db/dt - \omega \times \partial H/\partial b = 0 \rightarrow \\ \rightarrow U \times db/dt \times U^t - U \times \omega \times \partial/\partial b \times U^t = \\ = db/dt \times (U \times U^t) - (U \times \omega \times U^t) \times (U^{t\ -1} \times U^{-1}) \times \\ \times (U \times \partial H/\partial b \times U^t) \times (U \times U^t) = d\hat{b}/dt - \omega \times \hat{\partial} \hat{H}/\hat{\partial} \hat{b} = 0, \end{gathered} \quad (3.36)$$

where $\hat{b} = b \times \hat{I}, \partial H/\partial\hat{b} = \hat{T} \times \partial H/\partial b$, and we have used property $U \times \omega \times U^t \equiv \omega$.

As one can see, the seemingly complex isomathematics is reduced to a truly elementary construction. It is an instructive exercise for the reader interested in learning the new methods to work out the above construction in all details and prove its universality.

2.13. Invariance of Classical Isochemistry

A final requirement is necessary for physically and chemically acceptable theories, and that is *the invariance of the theory under the time evolution and the transformation theory*, which invariance assures the prediction of the same numbers for the same quantity under the same conditions at different times, thus avoiding the problematic aspects of noninvariant theories studied in Sect. 1.7.

Recall that conventional Hamiltonian mechanics is indeed physically and chemically consistent precisely because it verifies the above invariance requirement. In fact, the time evolution of the theory is a *canonical transformation*,

$$U \times \omega \times U^t \equiv \omega. \tag{3.37}$$

The transformation theory is also canonical, namely, it verifies the preceding law. The entire theory is then invariant, that is, Hamilton's equations are mapped into themselves, *etc.*

By comparison, *all analytic equations whose time evolution is not a canonical transformation are not invariant and, consequently, are not acceptable for valid physical or chemical applications* (Sect. 1.7). This is the case of the original Hamilton's equations with external terms, Birkhoff's equations, and all structural broadenings of the conventional analytic equations without external terms.

In fact, it is easy to see that a noncanonical transform, $U \times \omega \times U^t \neq \omega$, alters the isounit, besides changing the form of the isoequations, thus implying the transition from the originally system to a new one.

It is at this point that the use of the novel isomathematics becomes mandatory. In fact, *when formulated on an isospace over an isofield, the time evolution $\hat{U}$ of Hamilton-Santilli isomechanics is isocanonical*, i.e.

$$\hat{U} \hat{\times} \hat{\omega} \hat{\times} \hat{U}^t \equiv \hat{\omega}. \tag{3.38}$$

The invariance of the entire theory, let alone of the isounit and of the numerical results, then follows, as the reader is encouraged to verify.

2.14. Simple Construction of Classical Genochemistry

The broader genochemistry can be constructed in its entirety via the use of *two* noncanonical transforms [7]

$$U \times \omega \times U^t \neq \omega, \quad W \times \omega \times W^t \neq \omega, \tag{3.39}$$

and the identification of genounits

$$\hat{I}_6^> = U \times W^t, \quad {}^<\hat{I}_6 = W \times U^t = (\hat{I}_6^>)^t. \tag{3.40}$$

The application of the transform $U \times W^t$ to the *totality* of conventional Hamiltonian mechanics then yields genomechanics for motion forward in time, including forward genonumbers $\hat{n}^> = U \times n \times W^t = n \times \hat{I}_6^>$, genospaces, genoequations, *etc.*, as the reader is encouraged to verify [7].

Similarly, the use of the conjugate transform $W \times U^t$ to the totality of conventional Hamiltonian mechanics yields genomechanics for motion backward in time.

2.15. Invariance of Classical Genochemistry

It is easy to see that the genoequations are not invariant under a conventional canonical or noncanonical transform. However, when properly formulated on genospaces over genofields, the forward and backward time evolutions are individually invariant, because they constitute genocanonical transforms, e.g., verify the respective rules

$$\hat{U} > \hat{\omega}^> > \hat{W}^t \equiv \hat{\omega}^>, \tag{3.41a}$$

$$\hat{W} < {}^<\hat{\omega} < \hat{U}^t \equiv {}^<\hat{\omega}. \tag{3.41b}$$

The individual invariances of the theories then follows, as the reader is encouraged to verify.

2.16. Simple Construction and Invariance of Hyper- and Isodual Mechanics

The methods for the explicit construction of hyper- and isodual mechanics and their invariance are a simple extension of the preceding methods. Their explicit construction is suggested to readers interested in an axiomatically consistent and invariant study of irreversibility beginning at the classical level, and then continuing at the operator level (see Ref. [14] or monograph [6e] for details).

3. Operator Foundations of Hadronic Chemistry

3.1. Introduction

We are finally equipped to present the foundations of *nonrelativistic hadronic chemistry*, which will be used in the next chapters for the construction of a new model of molecular bonds.

These operator foundations of the new chemistry are given by hadronic mechanics [6c, 6d] when specifically formulated for chemistry, as done in this section. We shall provide primary attention to the *isotopic branch of hadronic chemistry*, called *operator isochemistry*, for short, and only outline the elements of the broader *operator genochemistry and hyperchemistry*, which are contemplated for applications at a later time.

A knowledge of Sect. 3.2 is requested for a technical understanding of this section. A mathematically simpler presentation is available in the next section. Unless otherwise specified, all quantities and operations represented with conventional symbols A, H, $\times$, *etc.*, denote quantities and operations on conventional Hilbert spaces over conventional fields. All quantities and symbols of the type $\hat{A}$, $\hat{H}$, $\hat{\times}$, *etc.*, are instead defined on isohilbert spaces over isofields. Similarly, all quantities and operations of the type $\hat{A}^{>}$, $\hat{H}^{>}$, $>$, *etc.*, are defined on forward genohilbert spaces and related genofields.

Note the use of the terms "operator" iso-, geno-, and hyper-chemistry, rather than "quantum," owing to the differences between quantum and hadronic mechanics.

3.2. Naive Iso-, Geno, Hyper-, and Isodual Quantization

An effective way to derive the basic dynamical equations of operator isochemistry is that via an isotopy of the conventional map of the classical Hamilton-Jacobi equations into their operator counterpart. More rigorous methods, such as those of symplectic quantization, essentially yields the same operator equations.

Recall that the *naive quantization* can be expressed via the following map of the canonical action functional

$$A = \int_{t_1}^{t_2} (p_k \times dr^k - H \times dt) \rightarrow -i \times \hbar \times \ln |\psi\rangle, \tag{3.42}$$

under which the conventional Hamilton-Jacobi equations are mapped into the Schrödinger equations,

$$-\partial_t A = H \rightarrow i \times \hbar \times \partial_t |\psi\rangle = H \times |\psi\rangle, \tag{3.43a}$$

$$\partial_k A = p_k \rightarrow -i \times \hbar \times \partial_k |\psi\rangle = p_k \times |\psi\rangle. \tag{3.43b}$$

Isocanonical action (3.19) is evidently different than the conventional canonical action, e.g., because it is of higher order derivatives. As such, the above naive quantization does not apply.

In its place we have the following *naive isoquantization* first introduced by Animalu and Santilli [8]

$$\hat{A} = \int_{t_1}^{t_2} [\hat{p}_k \hat{\times} d\hat{x}^k - \hat{H} \hat{\times} d\hat{t}] \rightarrow -i \times \hat{I} \times \ln |\hat{\psi}\rangle, \tag{3.44}$$

where $\hat{i} = i \times \hat{I}$ and we should note that $\hat{i} \hat{\times} \hat{I} \times \ln |\hat{\psi}\rangle = i \times \mathrm{isoln} |\hat{\psi}\rangle$.

The use of Hamilton-Jacobi-Santilli isoequations (3.21) yields the following operator equations (here written for the simpler case in which $\hat{T}$ has no dependence on r, but admits a dependence on velocities and higher derivatives)

$$-\partial_t \hat{A} = \hat{H} \rightarrow i \times \hbar \times \partial_t |\hat{\psi}\rangle = \hat{H} \hat{\times} |\hat{\psi}\rangle = \hat{H} \times \hat{T} \times |\hat{\psi}\rangle, \tag{3.45a}$$

$$\hat{\partial}_k \hat{A} = \hat{p}_k \rightarrow -i \times \hbar \times \hat{\partial}_k |\hat{\psi}\rangle = \hat{p}_k \hat{\times} |\hat{\psi}\rangle = \hat{p}_k \times \hat{T} \times |\hat{\psi}\rangle, \tag{3.45b}$$

which constitutes the fundamental equations of *operator isochemistry*, as we shall see in the next section.

The extension of the above results to the following *naive genoquantization* is elementary

$$\hat{A} = \int_{t_1}^{t_2} \left[\hat{p}_k > \hat{d}^{>} \hat{x}^k - \hat{H} \hat{\times} d\hat{t}\right] \rightarrow -i \times \hat{I}^{>} \times \ln |\hat{\psi}^{>}\rangle. \tag{3.46}$$

The use of Hamilton-Jacobi-Santilli genoequations (3.31) then yields the following operator equations:

$$-\partial_t \hat{A} = \hat{H} \rightarrow i \times \hbar \times \partial_t |\hat{\psi}^{>}\rangle = \hat{H} > |\hat{\psi}^{>}\rangle = \hat{H} \times \hat{T}^{>} \times |\hat{\psi}^{>}\rangle, \tag{3.47a}$$

$$\partial_k^{>} \hat{A} = \hat{p}_k \rightarrow -i \times \hbar \times \hat{\partial}_k^{>} |\hat{\psi}^{>}\rangle = \hat{p}_k > |\hat{\psi}^{>}\rangle = \hat{p}_k \times \hat{T}^{>} \times |\hat{\psi}^{>}\rangle, \tag{3.47b}$$

and their conjugate for motion backward in time, which constitute the fundamental equations of *operator genochemistry*, as we shall see in the next section.

The derivations of *naive hyperquantization* and *isodual quantization* [7] are left to the interested reader.

3.3. Structure of Operator Isochemistry

Note the crucial importance of the last Hamilton-Jacobi-Santilli isoequations (3.21), namely, $\hat{\partial}\hat{A}\hat{/}\hat{\partial}\hat{p}_k = 0$. In fact, these equations assure that *the isoeigenfunctions do not depend on the linear momentum*, $\hat{\partial}\hat{\psi}\hat{/}\hat{\partial}\hat{p} = 0$, $\hat{\psi} = \hat{\psi}(t, r)$, as it is the case for conventional quantum mechanics. By comparison, the impossibility of achieving a consistent operator image of Birkhoffian mechanics mentioned was due to the fact that the corresponding equations are of the type $\partial\hat{A}/\partial p_k = 0$, thus implying generalized wavefunctions with an explicit dependence on the linear momentum, $\psi = \psi(t, r, p)$ whose treatment is beyond our current knowledge for numerous reasons.

The structure of operator isochemistry is essentially given by (see Sect. 3.2 for treatments and references):

1) The description of chemical bonds, as well of closed-isolated systems in general, is done via *two* quantities, the Hamiltonian $H(t, r, p) = p^2/2m + V(r, p)$ used for the representation of all action-at-a-distance interactions derivable from a potential V, plus the isounit,

$$\hat{I} = \hat{I}(t, r, p, \psi, \nabla\psi, \ldots),$$

used for the representation of all nonlinear, nonlocal, and nonpotential-nonhamiltonian forces and effects. The explicit form of the Hamiltonian is that conventionally used in quantum chemistry [2]. The explicit form of the isounit depends on the case at hand. A generic expression for spinning particles with point-like change (such as the electrons) is given by the two quantities

$$H(t, r, p) = p^2/2m + V(r, p), \tag{3.48a}$$

$$\hat{I} = \exp\left[N \int dv\, \psi_{\downarrow}^{\dagger}(r)\psi_{\uparrow}(r)\right], \tag{3.48b}$$

the first one being used to represent conventional interactions, and the second to represent deep wave-overlappings of valence electrons. Additional explicit expressions of the isounit in chemistry will be presented in the subsequent chapters.

2) The lifting of the multiplicative unit $I > 0 \rightarrow \hat{I} = 1/\hat{T} > 0$, which requires the reconstruction of the entire theory into such a form to admit $\hat{I}$ as the correct left and right unit at all levels of study, including numbers and angles, conventional and special functions, differential and integral calculus, metric and Hilbert spaces, algebras and groups, *etc.*, without any exception known to the authors. This reconstruction is "isotopic" in the sense of being axiom-preserving. Particularly important is the preservation of all conventional quantum laws as shown below.

3) The mathematical structure characterized by [7]:

3a) The isofield $\hat{\mathbb{C}} = \hat{\mathbb{C}}(\hat{c}, +, \hat{\times})$ with *isounit* $\hat{I} = 1/\hat{T} > 0$, *isocomplex numbers* and related *isoproduct*

$$\begin{gathered} \hat{c} = c \times \hat{I} = (n_1 + i \times n_2) \times \hat{I}, \quad \hat{c}\hat{\times}\hat{d} = (c \times d) \times \hat{I}, \\ \hat{c}, \hat{d} \in \hat{\mathbb{C}}, \quad c, d \in \mathbb{C}, \end{gathered} \tag{3.49}$$

the isofield $\hat{\mathbb{R}}(\hat{n}, +, \hat{\times})$ of *isoreal numbers* $\hat{n} = n \times \hat{I}, n \in \mathbb{R}$, being a particular case;

3b) The isohilbert space $\hat{\mathcal{H}}$ with isostates $|\hat{\psi}\rangle, |\hat{\phi}\rangle, \ldots$, *isoinner product* and *isonormalization*

$$\langle\hat{\phi}|\hat{\times}|\hat{\psi}\rangle \times \hat{I} \in \hat{\mathbb{C}}, \quad \langle\hat{\psi}|\hat{\times}|\hat{\psi}\rangle = 1, \tag{3.50}$$

and related theory of isounitary operators;

3c) The isoeuclidean space $\hat{E}(\hat{r}, \hat{\delta}, \hat{\mathbb{R}})$ with *isocoordinates, isometric and isoinvariant*, respectively,

$$\hat{r} = \{r^k\} \times \hat{I}, \quad \hat{\delta} = \hat{T}(t, r, p, \psi, \nabla\psi, \ldots) \times \delta, \quad \delta = \text{Diag.}(1, 1, 1), \tag{3.51}$$

$$\hat{r}^{\hat{2}} = (r^i \times \hat{\delta}_{ij} \times r^j) \times \hat{I} \in \hat{\mathbb{R}};$$

3d) The isodifferential calculus and the isofunctional analysis (see Sect. 3.2) with particular reference to forms (2.13);

3e) The Lie-Santilli isotheory with enveloping isoassociative algebra $\hat{\xi}$ of operators $\hat{A}$, $\hat{B}$, ..., with isounit $\hat{I}$, isoassociative product $\hat{A}\hat{\times}\hat{B} = \hat{A} \times \hat{T} \times \hat{B}$, *Lie-Santilli isoalgebra with brackets and isoexponentiation*

$$[\hat{A}\hat{,}\hat{B}] = \hat{A}\hat{\times}\hat{B} - \hat{B}\hat{\times}\hat{A}, \tag{3.52a}$$

$$\hat{U} = \hat{e}^X = (e^{X \times \hat{T}}) \times \hat{I} = \hat{I} \times (e^{\hat{T} \times X}), \quad X = X^\dagger, \tag{3.52b}$$

and related isosymmetries with characterize groups of isounitary transforms on $\hat{\mathcal{H}}$ over $\hat{\mathbb{C}}$,

$$\hat{U}\hat{\times}\hat{U}^\dagger = \hat{U}^\dagger\hat{\times}\hat{U} = \hat{I}. \tag{3.53}$$

Note that quantum chemistry is admitted identically and in its entirety as a particular case whenever the isounit $\hat{I}$ recovers the conventional quantum unit I. In this sense, isochemistry is a *covering* of quantum chemistry,

Moreover, isochemistry solely applies at short distances whenever wave-overlappings are appreciable. As a result, all realizations of the isounit are restricted by the conditions of being positive-definite, admitting I as a particular case, and being such that $\lim_{r\to\infty} \hat{I} = I$.

Note that *composite isochemical systems*, such as *isomolecules*, are represented via the tensorial product of the above structures. This can be best done via the identification first of the *total isounit, total isofields, total isohilbert spaces, etc.*,

$$\hat{I}_{\text{tot}} = \hat{I}_1 \times \hat{I}_2 \times \ldots, \quad \hat{\mathbb{C}}_{\text{tot}} = \hat{\mathbb{C}}_1 \times \hat{\mathbb{C}}_2 \times \ldots, \quad \hat{\mathcal{H}}_{\text{tot}} = \hat{\mathcal{H}}_1 \times \hat{\mathcal{H}}_2 \times \ldots \quad (3.54)$$

Note also that some of the units, fields and Hilbert spaces in the above tensorial products can be *conventional*, namely, the composite structure may imply *local-potential long range interactions* (e.g. those of Coulomb type), which require the necessary treatment via *conventional* quantum mechanics, and *nonlocal-nonpotential short range interactions* (e.g., those in deep wave-overlappings), which require the use of operator isochemistry.

The above mixture of quantum and hadronic components occurs precisely in our isotopic model of chemical bonds presented in the next chapters. The mixture is permitted by the fact that both mechanics have the same axioms and physical laws, and can jointly apply since they represent long range and short range effects, respectively. Equivalently, the mixture is permitted by the arbitrariness of the isounit which, as such, is fully allowed to admit the conventional quantum unit I as a particular case.

3.4. Basic Equations of Operator Isochemistry

The use of naive isoquantization of Sect. 3.3 and the mathematical structure of the preceding subsection permit the formulation of the following basic *isoschrödinger equations* [7]

$$\hat{i}\hat{\times}\hat{\partial}_t|\hat{\psi}\rangle = \hat{H}\hat{\times}|\hat{\psi}\rangle = \hat{H} \times \hat{T} \times |\hat{\psi}\rangle = \hat{E}\hat{\times}|\hat{\psi}\rangle = E \times |\hat{\psi}\rangle, \quad (3.55a)$$

$$\hat{p}_k\hat{\times}|\hat{\psi}\rangle = \hat{p}_k \times \hat{T} \times |\hat{\psi}\rangle - \hat{i}\hat{\times}\hat{\partial}_k|\hat{\psi}\rangle = -i \times \hat{T}_k^i \times \nabla_i|\hat{\psi}\rangle, \quad (3.55b)$$

$$|\hat{\psi}(\hat{t})\rangle = \hat{U}\hat{\times}|\hat{\psi}(\hat{0})\rangle = \{\hat{e}^{i\hat{H}\hat{\times}\hat{t}}\}\hat{\times}|\hat{\psi}(\hat{0})\rangle, \quad (3.55c)$$

with corresponding basic *isoheisenberg equations* [7],

$$\hat{A}(\hat{t}) = \hat{U}\hat{\times}\hat{A}(\hat{0})\hat{\times}\hat{U}^{\dagger} = \{\hat{e}^{i\hat{H}\hat{\times}\hat{t}}\}\hat{\times}\hat{A}(\hat{0})\hat{\times}\{\hat{e}^{-i\hat{t}\hat{\times}\hat{H}}\}, \quad (3.56a)$$

$$\hat{i}\hat{\times}d\hat{A}\hat{/}d\hat{T} = [\hat{A}\hat{,}\hat{H}] = A\hat{\times}\hat{H} - \hat{H}\hat{\times}\hat{A} = \hat{A} \times \hat{T} \times \hat{H} - \hat{H} \times \hat{T} \times \hat{A}, \quad (3.56b)$$

$$[\hat{b}^{\mu}\hat{,}\hat{b}^{\nu}] = \hat{i}\hat{\times}\hat{\omega}^{\mu\nu} = i \times \omega^{\mu\nu} \times \hat{I}_6, \quad \hat{b} = (\hat{r}^k, \hat{p}_k). \quad (3.56c)$$

It is evident that the isoheisenberg equations in their infinitesimal and exponentiated forms are a realization of the Lie-Santilli isotheory of Sect. 3.2, which is therefore the algebraic and group theoretical structure of operator isochemistry.

Note that Eqs. (3.55) and (3.56) automatically bring into focus the general need for a *time isounit* $\hat{I}_t(t, r, \psi, \ldots) = \hat{T}_t > 0$ for the characterization of the time isodifferential $\hat{d}\hat{t} = \hat{I}_t \times dt$ and isoderivative $\hat{\partial}_t = \hat{T}_t \times \partial_t$. Note also that $\omega^{\mu\nu}$ in Eqs. (3.46b) is the *conventional* Lie tensor, namely, the same tensor appearing in the conventional canonical commutation rules, thus confirming the axiom-preserving character of isochemistry.

The "direct universality" of basic equations (3.55) and (3.56) should be kept in mind and compared to the limited capabilities of current models in chemistry. More specifically, the latter models are purely Hamiltonian and, as such, they can only represent systems which are linear, local and potential. By comparison, we can write Eq. (3.55a) in its explicit form

$$\begin{aligned} \hat{i}\hat{\times}\hat{\partial}_t\hat{\psi} = i \times \hat{I}_t \times \partial_t|\hat{\psi}\rangle = \hat{H}\hat{\times}|\hat{\psi}\rangle = \hat{H} \times \hat{T} \times |\hat{\psi}\rangle = \\ = \{\hat{p}_k \times \hat{p}_k/\hat{2}\hat{\times}\hat{m} + \hat{U}_k(\hat{t}, \hat{r})\hat{\times}\hat{v}^k + \\ +\hat{U}_0(\hat{t}, \hat{r})\} \times \hat{T}(\hat{t}, \hat{r}, \hat{p}, \hat{\psi}, \nabla\psi, \ldots) \times |\hat{\psi}(\hat{t}, \hat{r})\rangle = \\ = \hat{E}\hat{\times}|\hat{\psi}(t, \hat{x})\rangle = E \times |\hat{\psi}(\hat{t}, \hat{x})\rangle, \end{aligned} \tag{3.57}$$

thus showing the capability to represent all infinitely possible linear and nonlinear, local and nonlocal, and potential as well as nonpotential systems with conserved total energy (universality) directly in the frame of the observer (direct universality).

The isoheisenberg equations (4.56a) were first formulated by Santilli [9] as the primary structure of the proposed hadronic mechanics, jointly with the underlying isotopies of Lie's theory (Sect. 3.2). The isoschrödinger equations (3.55a) were formulated by Santilli [10a], Mignani [10b], and by Myung and Santilli [11] in isohilbert spaces.

A consistent formulation of the isolinear momentum (3.55b) escaped identification for two decades, thus delaying the completion of the construction of hadronic mechanics, as well as its practical applications. The consistent and invariant form (3.55b) with consequential isocanonical commutation rules (3.56b) were first identified by Santilli in report [7] thanks to the discovery of the isodifferential calculus in the same report.

3.5. Preservation of Quantum Physical Laws

As one can see, the fundamental assumption of isoquantization is the lifting of the basic unit of quantum chemistry, Planck's constant $\hbar$, into a matrix $\hat{I}$ with nonlinear, integro-differential elements which also depend on the wavefunction and its derivatives

$$\hbar = I > 0 \to \hat{I} = \hat{I}(t, r, p, \psi, \hat{\psi}, \ldots) = \hat{I}^\dagger > 0. \qquad (3.58)$$

It should be indicated that the above generalization is only *internal* in the system considered because, when measured from the outside, *the isoexpectation values and isoeigenvalues of the isounit recover Planck's constant identically*,

$$\langle \hat{I} \rangle = \frac{\langle \hat{\psi} | \hat{\times} \hat{I} \hat{\times} | \hat{\psi} \rangle}{\langle \hat{\psi} | \hat{\times} | \hat{\psi} \rangle} = 1 = \hbar, \qquad (3.59a)$$

$$\hat{I} \hat{\times} | \hat{\psi} \rangle = \hat{T}^{-1} \times \hat{T} \times | \hat{\psi} \rangle = 1 \times | \hat{\psi} \rangle = | \hat{\psi} \rangle. \qquad (3.59b)$$

Moreover, the isounit is the fundamental invariant of isochemistry, thus preserving all axioms of the conventional unit $I = \hbar$, e.g.,

$$\hat{I}^{\hat{n}} = \hat{I} \hat{\times} \hat{I} \hat{\times} \ldots \hat{\times} \hat{I} \equiv \hat{I}, \qquad (3.60a)$$

$$\hat{I}^{\hat{\frac{1}{2}}} \equiv \hat{I}, \qquad (3.60b)$$

$$\hat{i} \hat{\times} \hat{d} \hat{I} \hat{/} \hat{d}t = [\hat{I} \hat{,} \hat{H}] = \hat{I} \hat{\times} \hat{H} - \hat{H} \hat{\times} \hat{I} \equiv 0. \qquad (3.60c)$$

Despite their evident generalized structure, *Eqs. (3.55) and (3.56) preserve conventional quantum mechanical laws under nonlinear, nonlocal and nonpotential interactions* [7].

To begin an outline, the preservation of Heisenberg's uncertainties can be easily derived from isocommutation rules (3.56c):

$$\Delta x^k \times \Delta p_k \geq \frac{1}{2} \times \langle [\hat{x}^k \hat{,} \hat{p}_k] \rangle = \frac{1}{2}. \qquad (3.61)$$

To see the preservation of Pauli's exclusion principle, recall that the regular (two-dimensional) representation of $SU(2)$ is characterized by the conventional *Pauli matrices* σ_k with familiar commutation rules and eigenvalues on $\mathcal{H}$ over $\mathbb{C}$,

$$[\sigma_i, \sigma_j] = \sigma_i \times \sigma_j - \sigma_j \times \sigma_i = 2 \times i\varepsilon_{ijk} \times \sigma_k, \qquad (3.62a)$$

$$\sigma^2 \times |\psi\rangle = \sigma_k \times \sigma^k \times |\psi\rangle = 3 \times |\psi\rangle, \qquad (3.62b)$$

$$\sigma_3 \times |\psi\rangle = \pm 1 \times |\psi\rangle. \qquad (3.62c)$$

Isochemistry requires the construction of *nonunitary images of Pauli's matrices*, which, for diagonal isounits, can be written (see also Sect. 3.3.6)

$$\hat{\sigma}_k = U \times \sigma_k \times U^\dagger, \quad U \times U^\dagger = \hat{I} \neq I, \qquad (3.63a)$$

$$U = \begin{pmatrix} i \times m_1 & 0 \\ 0 & i \times m_2 \end{pmatrix}, \quad U^\dagger = \begin{pmatrix} -i \times m_1 & 0 \\ 0 & -i \times m_2 \end{pmatrix},$$
$$\hat{I} = \begin{pmatrix} m_1^2 & 0 \\ 0 & m_2^2 \end{pmatrix}, \quad \hat{T} = \begin{pmatrix} m_1^{-2} & 0 \\ 0 & m_2^{-2} \end{pmatrix}, \tag{3.63b}$$

where the m's are well behaved nowhere null functions, resulting in the *regular isopauli matrices*, first derived by Santilli [15]

$$\hat{\sigma}_1 = \begin{pmatrix} 0 & m_1^2 \\ m_2^2 & 0 \end{pmatrix}, \hat{\sigma}_2 = \begin{pmatrix} 0 & -i \times m_1^2 \\ i \times m_2^2 & 0 \end{pmatrix}, \hat{\sigma}_3 = \begin{pmatrix} m_1^2 & 0 \\ 0 & m_2^2 \end{pmatrix}. \tag{3.64}$$

Another realization is given by *nondiagonal unitary transforms* [*loc. cit.*],

$$U = \begin{pmatrix} 0 & m_1 \\ m_2 & 0 \end{pmatrix}, \quad U^\dagger = \begin{pmatrix} 0 & m_2 \\ m_1 & 0 \end{pmatrix},$$
$$\hat{I} = \begin{pmatrix} m_1^2 & 0 \\ 0 & m_2^2 \end{pmatrix}, \quad \hat{T} = \begin{pmatrix} m_1^{-2} & 0 \\ 0 & m_2^{-2} \end{pmatrix}, \tag{3.65}$$

with corresponding *regular isopauli matrices*,

$$\hat{\sigma}_1 = \begin{pmatrix} 0 & m_1 \times m_2 \\ m_1 \times m_2 & 0 \end{pmatrix}, \quad \hat{\sigma}_2 = \begin{pmatrix} 0 & -i \times m_1 \times m_2 \\ i \times m_1 \times m_2 & 0 \end{pmatrix},$$

$$\hat{\sigma}_3 = \begin{pmatrix} m_1^2 & 0 \\ 0 & m_2^2 \end{pmatrix}, \tag{3.66}$$

or by more general realizations with Hermitean nondiagonal isounits $\hat{I}$ [15].

All isopauli matrices of the above regular class verify the following isocommutation rules and isoeigenvalue equations on $\hat{\mathcal{H}}$ over $\hat{\mathbb{C}}$

$$[\hat{\sigma}_i \hat{,} \hat{\sigma}_j] = \hat{\sigma}_i \times \hat{T} \times \hat{\sigma}_j - \hat{\sigma}_j \times \hat{T} \times \hat{\sigma}_i = 2 \times i \times \varepsilon_{ijk} \times \hat{\sigma}_k, \tag{3.67a}$$

$$\hat{\sigma}^{\hat{2}} \hat{\times} |\hat{\psi}\rangle = (\hat{\sigma}_1 \hat{\times} \hat{\sigma}_1 + \hat{\sigma}_2 \hat{\times} \hat{\sigma}_2 + \hat{\sigma}_3 \hat{\times} \hat{\sigma}_3) \hat{\times} |\hat{\psi}\rangle = 3 \times |\hat{\psi}\rangle, \tag{3.67b}$$

$$\hat{\sigma}_3 \hat{\times} |\hat{\psi}\rangle = \pm 1 \times |\hat{\psi}\rangle, \tag{3.67c}$$

which preserve conventional spin $1/2$, thus establishing the preservation in isochemistry of the Fermi-Dirac statistics and Pauli's exclusion principle.

It should be indicated for completeness that the representation of the isotopic $S\hat{U}(2)$ also admit *irregular isorepresentations*, which no longer preserve conventional values of spin [15]. The latter structures are under study for the characterization of spin under the most extreme conditions, such as for protons and electrons in the core of collapsing stars and, as such, they have no known relevance for isochemistry.

The preservation of the superposition principle under nonlinear interactions occurs because of the reconstruction of linearity on isospace over isofields, thus regaining the applicability of the theory to composite systems.

Recall in this latter respect that conventionally nonlinear models,

$$H(t, x, p, \psi, \ldots) \times |\psi\rangle = E \times |\psi\rangle, \tag{3.68}$$

violate the superposition principle and have other shortcomings (see Sect. 1.7). As such, they cannot be applied to the study of composite systems such as molecules. All these models can be *identically* reformulated in terms of the isotopic techniques via the embedding of all nonlinear terms in the isotopic element,

$$H(t, x, p, \psi, \ldots) \times |\psi\rangle \equiv H_0(t, x, p) \times \hat{T}(\psi, \ldots) \times |\psi\rangle = E \times |\psi\rangle, \tag{3.69}$$

by regaining the full validity of the superposition principle in isospaces over isofields with consequential applicability to composite systems.

The preservation of causality follows from the one-dimensional isounitary group structure of the time evolution (3.56a) (which is isomorphic to the conventional one); the preservation of probability laws follows from the preservation of the axioms of the unit and its invariant decomposition as indicated earlier; the preservation of other quantum laws then follows.

The same results can be also seen from the fact that operator isochemistry coincides at the abstract level with quantum chemistry by conception and construction. As a result, hadronic and quantum versions of chemistry are *different realizations of the same abstract axioms and physical laws.*

Note that the preservation of conventional quantum laws under nonlinear, nonlocal and nonpotential interactions is crucially dependent on the capability of isomathematics to reconstruct linearity, locality and canonicity-unitarity on isospaces over isofields.

The preservation of conventional physical laws by the isotopic branch of hadronic mechanics was first identified by Santilli in report [12]. It should be indicated that the same quantum laws *are not* generally preserved by the broader genomechanics, evidently because the latter must represent by assumption *non*-conservation laws and other *departures* from conventional quantum settings.

With the understanding that the theory does not receive the classical determinism, it is evident that isochemistry provides a variety of "completions" of quantum chemistry according to the celebrated E-P-R argument [1], such as:

1) Isochemistry "completes" quantum chemistry via the addition of nonpotential-nonhamiltonian interactions represented by nonunitary transforms.

2) Isochemistry "completes" quantum chemistry via the broadest possible (non-oriented) realization of the associative product into the isoassociative form.

3) Isochemistry "completes" quantum chemistry in its classical image.

In fact, as proved by well known procedures based on *Bell's inequalities*, quantum chemistry does not admit direct classical images on a number of counts. On the contrary, as studied in details in Refs. [15], the nonunitary images of Bell's inequalities permit indeed direct and meaningful classical limits which do not exist for the conventional formulations.

Similarly, it is evident that isochemistry constitutes a specific and concrete realization of "hidden variables" λ which are explicitly realized by the isotopic element, $\lambda = \hat{T}$, and actually turned into an operator hidden variables. The "hidden" character of the realization is expressed by the fact that hidden variables are embedded in the unit and product of the theory. In fact, we can write the isoschrödinger equation $\hat{H}\hat{\times}|\hat{\psi}\rangle = \hat{H} \times \lambda \times |\hat{\psi}\rangle = E \times |\hat{\psi}\rangle$, $\lambda = \hat{T}$. As a result, the "variable" λ (now generalized into the operator $\hat{T}$) is "hidden" in the modular associative product of the Hamiltonian $\hat{H}$ and the state $|\hat{\psi}\rangle$.

For studies on the above and related issues, we refer the interested reader to Refs. [15] and quoted literature.

3.6. Simple Construction of Operator Isochemistry

Despite their *mathematical equivalence*, it should be indicated that chemistry and isochemistry are *physically inequivalent*, or, alternatively, isochemistry is outside the classes of equivalence of chemistry because the former is a *nonunitary image* of the latter.

As we shall see in the next chapters, the above property provides means for the explicit construction of the new model of isochemical bonds from the conventional model. The main requirement is that of identifying the *nonhamiltonian* effects one desires to represent, which as such, are necessarily *nonunitary*. The resulting nonunitary transform is then assumed as the fundamental space isounit of the new isochemistry [12]

$$U \times U^{\dagger} = \hat{I} \neq I, \tag{3.70}$$

under which transform we have the liftings of: the quantum unit into the isounit,

$$I \rightarrow \hat{I} = U \times I \times U^{\dagger}; \tag{3.71}$$

numbers into isonumbers,

$$a \rightarrow \hat{a} = U \times a \times U^{\dagger} = a \times (U \times U^{\dagger}) = a \times \hat{I}; \quad a = n, c, \tag{3.72}$$

associative products $A \times B$ into the isoassociative form with the correct isotopic element,

$$A \times B \rightarrow \hat{A} \hat{\times} \hat{B} = \hat{A} \times \hat{T} \times \hat{B}, \tag{3.73a}$$

$$\hat{A} = U \times A \times U^{\dagger}, \hat{B} = U \times B \times U^{\dagger}, \hat{T} = (U \times U^{\dagger})^{-1} = T^{\dagger}; \tag{3.73b}$$

Schrödinger's equation into the isoschrödinger's equations

$$\begin{aligned} H \times |\psi\rangle = E \times |\psi\rangle \rightarrow U(H \times |\psi\rangle) = \\ = (U \times H \times U^{\dagger}) \times (U \times U^{\dagger})^{-1} \times (U \times |\psi\rangle) = \\ = \hat{H} \times \hat{T} \times |\hat{\psi}\rangle = \hat{H} \hat{\times} |\hat{\psi}\rangle; \end{aligned} \tag{3.74a}$$

Heisenberg's equations into their isoheisenberg generalization

$$\begin{aligned} i \times dA/dt - A \times H - H \times A = 0 \rightarrow \\ \rightarrow U \times (i \times dA/dt) \times U^{\dagger} - U(A \times H - H \times A) \times U^{\dagger} = \\ = \hat{i} \hat{\times} d\hat{A}/dt - \hat{A} \hat{\times} \hat{H} - \hat{H} \hat{\times} \hat{A} = 0; \end{aligned} \tag{3.74b}$$

the Hilbert product into its isoinner form

$$\begin{aligned} \langle\psi| \times |\psi\rangle \rightarrow U \times \langle\psi| \times |\psi\rangle \times U^{\dagger} = \\ = (\langle\psi| \times U^{\dagger}) \times (U \times U)^{-1} \times (U \times |\psi\rangle) \times (U \times U)^{-1} = \langle\hat{\psi}| \hat{\times} |\hat{\psi}\rangle \times \hat{I}; \end{aligned} \tag{3.74c}$$

canonical power series expansions into their isotopic form

$$\begin{aligned} A(k) = A(0) + k \times [A, H] + k^2 \times [[A, H,], H] + \ldots \rightarrow U \times A(k) \times U^{\dagger} = \\ = U \times \Big[A(0) + k \times [A, H] + k^2 \times [[A, H], H] + \ldots\Big] \times U^{\dagger} = \\ = \hat{A}(\hat{k}) = \hat{A}(0) + \hat{k} \hat{\times} [\hat{A} \hat{,} \hat{H}] + \hat{k}^2 \hat{\times} [[\hat{A} \hat{,} \hat{H}] \hat{,} \hat{H}] + \ldots, \\ k > 1, \quad |\hat{T}| \ll 1; \end{aligned} \tag{3.74d}$$

Schrödinger's perturbation expansion into its isotopic covering (where the usual summation over states $p \neq n$ is assumed)

$$E(k) = E(0) + k \times \langle\psi| \times V \times |\psi\rangle + k^2 \frac{|\langle\psi|\times V\times|\psi\rangle|^2}{E_{0n} - E_{0p}} + \ldots \rightarrow$$

$$\rightarrow U \times E(k) \times U^\dagger = U \times \Big[E(0) + k \times \langle\psi| \times V \times |\psi\rangle + \ldots\Big] \times U^\dagger =$$

$$= \hat{E}(\hat{k}) = \hat{E}(0) + \hat{k} \hat{\times} \langle\hat{\psi}| \times \hat{T} \times \hat{V} \times \hat{T} \times |\hat{\psi}\rangle + \ldots,$$

$$k > 1, \quad |\hat{T}| \ll 1; \tag{3.74e}$$

etc. All remaining aspects of operator isochemistry can then be derived accordingly, including the isoexponent, isologarithm, isodeterminant, isotrace, isospecial functions and transforms, *etc.*

Note that the isotopic images of conventional perturbation expansions, as in the illustrative Eqs. (3.74d) and (3.74e), have far reaching implications for chemistry. Recall from Sect. 1.3 that one of the vexing problems of contemporary chemistry is the excessive computer time now needed for the calculation on complex molecular structures, despite the availability of the most modern and powerful computers. This limitation is resolved by isochemistry to such an extent that the rapidity of the calculations is essentially reduced to the appropriate selection of the isounit. As an explicit verification, in Chapters 4, 5 and 6 we shall present isochemical models of molecular structures whose computer use can be at least 1,000 times shorter than that requested by quantum chemistry for the calculation of the same quantity.

This remarkable advance is due to the fact illustrated by liftings (3.74d) and (3.74e), that any conventional series which is weakly (thus slow) convergent, can be turned by the isotopies into a series which are as strongly (thus fast) convergent as desired, under the sole condition that the absolute (or average) value of the isotopic element (isounit) is sufficiently smaller (bigger) than one, $|\hat{T}| \ll 1$ ($|\hat{I}| \gg 1$) (see monograph [6d] for details and references).

Moreover, even divergent series can be turned under the above isotopies into fast convergent forms. As a matter of fact, it was submitted since the first proposal to build hadronic mechanics [9] (see again [6d] for details) that the divergence of the perturbation theory of strong interactions may be due to the simplistic nature of the underlying mathematics, specifically, the use of an associative product $A \times B$ essentially learned since junior high school.

In fact, consider Eqs. (3.74d) and (3.74e) for $k > 1$, in which case the conventional perturbative series diverges as for the case of strong interactions. Consider then the isotopic series for the particular case in which $|\hat{T}| = 1/k$. It is then easy to see that the isotopic images of the same series (3.74d) and (3.74e) become convergent. This also illustrates that a perturbative series which is slowly convergent is easily turned into a series as fast convergent as desired under the appropriate selection of the value $|\hat{T}|$ as sufficiently smaller than one.

As we shall see, all realizations of the isounit result in verifying the above fundamental requirement, thus ensuring a convergence of the isoperturbation theory as fast as desired. Quite remarkably, as we shall see in Chapters 4 and 5, the value $|\hat{I}| \gg 1$ needed for fast convergence, rather than being imposed, emerges quite naturally from the characteristics of the molecular structures.

Note that the above construction via a nonunitary transform is the correct operator image of the derivability of the classical isohamiltonian mechanics from the conventional form via noncanonical transforms (Sect. 3.2.12).

The construction of hadronic mechanics via nonunitary transforms of quantum mechanics was first identified by Santilli in the original proposal [5e], and then worked out in subsequent contributions (see [12] for the latest presentation).

3.7. Invariance of Operator Isochemistry

It is important to see that, in a way fully parallel to the classical case (Sect. 3.2.13), operator isochemistry is indeed invariant under the most general possible nonlinear, nonlocal and nonhamiltonian-nonunitary transforms, provided that, again, the invariance is treated via the isomathematics. In fact, any given nonunitary transform $U \times U^\dagger \neq I$ can always be decomposed into the form [12]

$$U = \hat{U} \times \hat{T}^{1/2}, \tag{3.75}$$

under which nonunitary transforms on $\mathcal{H}$ over $\mathbb{C}$ are identically reformulated as isounitary transforms on the isohilbert space $\hat{\mathcal{H}}$ over the isofield $\hat{\mathbb{C}}$,

$$U \times U^\dagger \equiv \hat{U} \hat{\times} \hat{U}^\dagger = \hat{U}^\dagger \hat{\times} \hat{U} = \hat{I}. \tag{3.76}$$

The form-invariance of operator isochemistry under isounitary transforms then follows,

$$\hat{I} \rightarrow \hat{I}' = \hat{U} \hat{\times} \hat{I} \hat{\times} \hat{U}^\dagger \equiv \hat{I}, \hat{A} \hat{\times} \hat{B} \rightarrow \hat{U} \hat{\times} (\hat{A} \hat{\times} \hat{B}) \hat{\times} \hat{U}^\dagger = \hat{A}' \hat{\times} \hat{B}', \ etc., \tag{3.77a}$$

$$\hat{H}\hat{\times}|\hat{\psi}\rangle = \hat{E}\hat{\times}|\hat{\psi}\rangle \rightarrow \hat{U} \times \hat{H}\hat{\times}|\hat{\psi}\rangle =$$
$$= (\hat{U} \times \hat{H} \times \hat{U}^\dagger)\hat{\times}(\hat{U}\hat{\times}|\hat{\psi}\rangle) = \hat{H}'\hat{\times}|\hat{\psi}'\rangle = \tag{3.77b}$$
$$= \hat{U}\hat{\times}\hat{E}\hat{\times}|\hat{\psi}\rangle = \hat{E}\hat{\times}\hat{U}\hat{\times}|\hat{\psi}\rangle = \hat{E}\hat{\times}|\hat{\psi}'\rangle,$$

where one should note the preservation of the *numerical values* of the isounit, isoproducts and isoeigenvalues, as necessary for consistent applications.

Note that the invariance in quantum chemistry holds only for transformations $U \times U^\dagger = I$ *with fixed* I. Similarly, the invariance of isochemistry holds only for all nonunitary transforms such $\hat{U}\hat{\times}\hat{U}^\dagger = \hat{I}$ *with fixed* $\hat{I}$, and *not* for a transform $\hat{W}\hat{\times}\hat{W}^\dagger = \hat{I}' \neq \hat{I}$ because the change of the isounit $\hat{I}$ implies the transition to a *different physical system.*

The form-invariance of hadronic mechanics under isounitary transforms was first studied by Santilli in report [12].

3.8. Gaussian Screenings as Particular Cases of Isochemistry

In Sect. 1.3 we recalled that the Coulomb law is one of the fundamental invariants of quantum mechanics and chemistry, as a necessary condition for its physical consistency. In fact, the law is invariant under any possible transformation of quantum chemistry

$$F = 1/r^2 \rightarrow U \times (1/r^2) \times U^\dagger = 1/r^2, \quad U \times U^\dagger = I. \tag{3.78}$$

In Sect. 1.3 we then pointed out that the manipulations of the Coulomb law conducted in quantum chemistry, e.g., via the so-called "Gaussian screenings" in the hope of improving the representation of experimental data, requires a necessary exiting from the basic axioms of "quantum" chemistry. In fact, such screenings require the necessary use of *nonunitary transforms*, e.g., as in the simple case

$$F = 1/r \rightarrow W \times (1/r) \times W^\dagger = e^{A \times r}/r, \quad W \times W^\dagger = e^{A \times r} \neq I. \tag{3.79}$$

Independently from that, if a "screened Coulomb potential" is assumed as the potential of a Hamiltonian with a unitary time evolution, the preservation of "quantum" chemistry is only apparent because such a screened Coulomb potential prohibits the existence of the hydrogen atom, renders all electron orbits highly unstable, and prevents the very existence of the "quantum" of energy, which gives the name to "quantum" chemistry.

By using the analysis of the preceding sections, it is now easy to see that *all infinitely possible Gaussian screenings of the Coulomb law are particular cases of isochemistry.*

In fact, their derivation is a particular case of the construction of isochemistry via nonunitary transforms as outlined in the preceding sections. However, in order to avoid the catastrophic inconsistency of nonunitary theories (Sect. 1.7), it is mandatory to apply the selected nonunitary transform to the *totality* of the formalism of the real "quantum" chemistry, that is without any screening at all. The emerging theory is then, necessarily, operator isochemistry, as illustrated by the transformation of the Hamiltonian

$$U \times U^{\dagger} = \text{screening function} \neq I, \tag{3.80a}$$

$$\begin{aligned} U \times H \times U^{\dagger} &= U \times (p \times p/2 \times m + 1/r)U^{\dagger} = \\ &= \hat{p}\hat{\times}\hat{p}\hat{/}\hat{2}\hat{\times}\hat{m} + (1/r) \times (U \times U^{\dagger}) = \\ &= \hat{p}^{\hat{2}}\hat{/}\hat{2}\hat{\times}\hat{m} + \text{any desired function.} \end{aligned} \tag{3.80b}$$

In conclusion, rather than disproving current trends on screenings of the Coulomb law, operator isochemistry confirms their validity, although clarifying that, on true scientific grounds, they do not belong to "quantum" chemistry, but are instead particular cases of the structurally broader operator isochemistry.

A similar situation occurs for variational methods whose wavefunctions can be easily proved *not* to be solutions of purely Coulomb equations, as it is the case for the restricted three-body equation of the H_2^+ molecular ion. It then follows that, on rigorous scientific grounds, these wavefunctions belong to hadronic chemistry.

In different terms, the wavefunctions of variational models achieving an exact representation of molecular data via a number of *ad hoc* parameters can be easily proved *not* to be the solution of the Schrödinger's equations for the same structure under exact quantum chemical axioms, trivially, because the latter are not exact and miss up to 2% of the experimental values. However, as we shall see in Chapters 3 and 4, the isoschrödinger equations for the same structure are indeed exact to any desired digit. This proves that, as stated above, the solution of exact variational models are exact solutions of isochemistry and definitely not of quantum chemistry.

3.9. Elements of Operator Geno-, Hyper-, and Isodual- Chemistry

In Chapter 2 we have recalled that isomathematics is a particular case of the broader *genomathematics* which, in turn, is a particular case of the yet broader *hypermathematics* [7].

As indicated earlier, isomathematics is based on generalized units, which are nonsingular, real-valued, and *Hermitean*, thus implying a *unique* generalized product,

$$\hat{I} = 1/\hat{T} = \hat{I}^{\dagger}, \quad A\hat{\times}B = A \times \hat{T} \times B = A \times \hat{T}^{\dagger} \times B, \quad \hat{T} \text{ fixed.} \tag{3.81}$$

The broader genomathematics is characterized instead by generalized units, called *genounits*, which are also nonsingular and real-valued but *nonhermitean* (e.g., resulting from the addition of nondiagonal terms in a given real-valued, diagonal isounit). This implies the existence of *two* different genounits, with consequential need of *ordering the generalized products to the right and to the left*, usually denoted as follows:

$$\hat{I}^{>} = 1/\hat{T}^{>}, \quad A > B = A \times \hat{T}^{>} \times B, \quad \hat{T}^{>} \text{ fixed,} \tag{3.82a}$$

$$^{<}\hat{I} = 1/^{<}\hat{T}, \quad A < B = A \times {}^{<}\hat{T} \times B, \quad {}^{<}\hat{T} \text{ fixed,} \tag{3.82b}$$

$$\hat{I}^{>} = ({}^{<}\hat{I})^{\dagger}. \tag{3.82c}$$

As a result, genomathematics is characterized by two separate isotopies, one per each ordered direction of the product, which are interconnected by Hermitean conjugation, and assumed to represent the two directions of time.

The basic equations of operator genochemistry are the *genoschrödinger equations for motion forward in time*

$$\begin{aligned} \hat{i} > \hat{\partial}_t^{>} |\hat{\psi}^{>}\rangle = \hat{H} > |\hat{\psi}\rangle = \hat{H} \times \hat{T}^{>} \times |\hat{\psi}^{>}\rangle = \\ = \hat{E} > |\hat{\psi}^{>}\rangle = E \times |\hat{\psi}^{>}\rangle, \end{aligned} \tag{3.83a}$$

$$\begin{aligned} \hat{p}_k > |\hat{\psi}\rangle = \hat{p}_k \times \hat{T}^{>} \times |\hat{\psi}\rangle = \\ = -\hat{i} > \hat{\partial}_k > |\hat{\psi}^{>}\rangle = -i \times \hat{T}_k^{>i} \times \nabla_i |\hat{\psi}^{>}\rangle, \end{aligned} \tag{3.83b}$$

$$|\hat{\psi}(\hat{t})^{>}\rangle = \hat{U}^{>} > |\hat{\psi}(\hat{0})\rangle = \{\hat{e}^{i \times \hat{H} \times T^{>} \times t}\} \times |\hat{\psi}(\hat{0})^{>}\rangle, \tag{3.83c}$$

and their conjugate *genoequations for motion backward in time*

$$\langle {}^{<}\hat{\psi}(0)| {}^{<}\partial_t < \hat{i} = \langle {}^{<}\hat{\psi}| < \hat{H} = \langle {}^{<}\hat{\psi}| < \hat{E} = \langle {}^{<}\hat{\psi}| \times E', \tag{3.84a}$$

$$\begin{aligned} \langle {}^{<}\hat{\psi}(t)| = \langle {}^{<}\hat{\psi}(0)| < {}^{<}\hat{U} = \langle {}^{<}\hat{\psi}(0)| \times e^{t \times {}^{<}\hat{T} \times H}, \langle {}^{<}\hat{\psi}(t)| < \hat{p}_k = \\ = \langle {}^{<}\hat{\psi}(0)| \times {}^{<}\hat{T} < p_k = -\langle {}^{<}\hat{\psi}(0)| {}^{<}\hat{\partial}_k \times i = -\langle {}^{<}\hat{\psi}(0)| \partial_i \times {}^{<}\hat{T}_k^i \times i, \end{aligned} \tag{3.84b}$$

$$\hat{T}^{>} = ({}^{<}\hat{T})^{\dagger}, \tag{3.84c}$$

where quantities characterized by the symbol $>$ ($<$) refer to motion forward (backward) in time, and one should note the different genoeigenvalues $E \neq E'$ for motion forward and backward in times (due to the differences in the genotopic elements).

The preceding equations then characterize the following *genoheisenberg equations* in the finite and infinitesimal forms

$$\hat{A}(\hat{t}) = \hat{U}^{>} > \hat{A}(\hat{0}) < {}^{<}\hat{U}^{\dagger} = \{e^{i \times \hat{H} \times \hat{T}^{>} \times t}\} \times \hat{A}(\hat{0}) \times \{e^{-i \times \hat{t} \times {}^{<}\hat{T} \times \hat{H}}\}, \quad (3.85a)$$

$$i \times \hat{d}\hat{A}\hat{/}\hat{d}\hat{t} = (\hat{A}\hat{;}\hat{H}) = A < \hat{H} - \hat{H} > \hat{A} = = \hat{A} \times {}^{<}\hat{T} \times \hat{H} - \hat{H} \times \hat{T}^{>} \times \hat{A}, \quad (3.85b)$$

$$(\hat{b}^{\mu}\hat{;}\hat{b}^{\nu}) = i \times \hat{\omega}^{\mu\nu} = i \times \omega^{\mu\nu} \times \hat{I}_6^{>}, \quad \hat{b} = (\hat{r}^k, \hat{p}_k). \quad (3.85c)$$

Operator genochemistry must then be elaborated, for consistency, with the genomathematics of Chapter 2, including the use of genonumbers and genofields, genometric and genohilbert spaces, *etc.*

In this case, it is easy to see that *forward (or backward) genohermiticity* coincides with conventional Hermiticity. As a result, *all quantities which are observable for conventional quantum chemistry remain observable for genochemistry.* This property permits the achievement of *observable nonconserved physical quantities*, such as the nonconserved energy, nonconserved angular momentum, *etc.*, which is achieved for the first time by genomechanics, since all nonconserved physical quantities of quantum mechanics and chemistry are notoriously nonhermitean, thus nonobservable.

The reader interested in learning the new chemistry is encouraged to work out all remain details of genochemistry. For instance, *the genoexpectation values of genounits recover the conventional unit.* This implies that genomechanics is a second "completion" of quantum mechanics along the historical teaching by Einstein, Podolsky and Rosen.

The first axiomatically consistent formulation of genomechanics was achieved in Ref. [7], some 18 years following its original proposal [8].

As studied in Chapter 2, the yet broader *hypermathematics* occurs when the units are constituted by *ordered sets* of nonsingular, real-valued, and nonsymmetric matrices or operators, resulting in corresponding ordered multi-valued products to the right and to the left, also interconnected by a Hermitean conjugation,

$$\hat{I}^{>} = \{1/\hat{T}_1^{>}, 1/\hat{T}_2^{>}, \ldots\}, \quad A > B = A \times \hat{T}_1^{>} \times B + A \times \hat{T}_2^{>} \times B + \ldots, \quad (3.86a)$$

$$^{<}\hat{I} = \{1/{}^{<}\hat{T}_1, 1/{}^{<}\hat{T}_2, \ldots\}, \quad A < B = A \times {}^{<}\hat{T}_1 \times B + A \times {}^{<}\hat{T}_2 \times B + \ldots, \quad (3.86b)$$

$$\hat{I}^{>} = ({}^{<}\hat{I})^{\dagger}. \tag{3.86c}$$

The basic dynamical equations are then Eqs. (3.83)-(3.85) when formulated in terms of hypermathematics (see Refs. [6e, 7] for brevity).

The *isodual chemistry* is an image of conventional quantum chemistry under the anti-isomorphic *isodual map* of the basic unit and product

$$I \rightarrow I^d = -I, \tag{3.87a}$$

$$A \times B \rightarrow A \times^d B = A \times (-I) \times B, \tag{3.87b}$$

$$I \times A = A \times I = A \rightarrow I^d \times^d A = A \times^d I^d = A, \tag{3.87c}$$

and then the reconstruction of the entire mathematical formalism in such a way to admit $I^d = -I$, rather than I, as the correct left and right unit. This implies the use of *isodual numbers, isodual spaces, isodual algebras*, etc.

The *isodual iso-, geno-, and hyper chemistries* are the anti-isomorphic images of iso-, geno-, and hyper-chemistry, respectively under the *general isodual map* [14]

$$\hat{I}(r, p, \psi, \dots) \rightarrow \hat{I}^d = -I^{\dagger}(-r^{\dagger}, -p^{\dagger}, -\psi^{\dagger}, \dots), \tag{3.88a}$$

$$A \hat{\times} B = A \times \hat{T} \times B \rightarrow A \hat{\times}^d B = A \times \hat{T}^d \times B, \tag{3.88b}$$

$$\hat{I} \hat{\times} A = A \hat{\times} \hat{I} = A \rightarrow \hat{I}^d \hat{\times}^d A = A \hat{\times}^d \hat{I}^d = A. \tag{3.88c}$$

Readers interested in learning hadronic chemistry are suggested to construct the isodual iso-, geno-, and hyper-chemistry in all needed details via the systematic application of the above map to *all* aspects of the formalism.

3.10. Simple Construction of Operator Geno-, Hyper-, and Isodual Chemistry

As it is the case for isochemistry presented in Sect. 3.3.6, geno-, hyper-, and isodual chemistry can also be constructed in their entirety via a method which is simple, yet effective and sufficient for practical applications.

Recall from Sect. 3.3.6 that isochemistry can be constructed via the use of *one* nonunitary transform which characterizes the isounit, $U \times U^{\dagger} = \hat{I} \neq I$. The broader genochemistry can be constructed via the following *two* nonunitary transforms [12, 13]

$$U \times U^{\dagger} \neq I, W \times W^{\dagger} \neq I, \tag{3.89a}$$

$$U \times W^{\dagger} = \hat{I}^{>}, W \times U^{\dagger} = {}^{<}\hat{I}, \tag{3.89b}$$

provided that, again, the above transforms are applied to the totality of the theory for motion forward and, independently, for motion backward in time.

We have in this way the *forward genounit, forward genonumbers, forward genoproducts, forward genoschrödinger equations, etc.* [12, 13]

$$I \rightarrow \hat{I}^{>} = U \times I \times W^{\dagger}, \tag{3.90a}$$

$$n \rightarrow \hat{n}^{>} = U \times n \times W^{\dagger} = n \times \hat{I}^{>}, \tag{3.90b}$$

$$\begin{aligned} A \times B \rightarrow A' > B' = U \times (A \times B) \times W^{\dagger} = \\ = (U \times A \times W^{\dagger}) \times (U \times W^{\dagger})^{-1} \times (U \times B \times W^{\dagger}), \end{aligned} \tag{3.90c}$$

$$\begin{aligned} H \times |\hat{\psi}\rangle \rightarrow U \times (H \times |\hat{\psi}\rangle) = \\ = (U \times H \times W^{\dagger}) \times (U \times W^{\dagger})^{-1} \times (U \times |\hat{\psi}\rangle) = \\ = \hat{H} \times \hat{T}^{>} \times |\hat{\psi}^{>}\rangle = \hat{H} > |\hat{\psi}^{>}\rangle, \end{aligned} \tag{3.90d}$$

$$\begin{aligned} \langle\psi| \times |\psi\rangle \rightarrow U \times (\langle\psi| \times |\psi\rangle) \times W^{\dagger} = \\ = (\langle\psi| \times W^{\dagger}) \times (U \times W^{\dagger})^{-1} \times (U \times |\psi\rangle) \times (U \times W^{\dagger}) = \\ = \langle{}^{<}\hat{\psi}| > |\hat{\psi}^{>}\rangle \times \hat{I}^{>}. \end{aligned} \tag{3.90e}$$

All other aspects of the forward genochemistry can be constructed in a similar way. The construction of the backward genochemistry is done by the conjugate map $W \times U^{\dagger} = {}^{<}\hat{I}$.

Similarly, the construction of hyperchemistry occurs when the two nonunitary transforms (3.89) are multivalued.

The construction of the isodual chemistry can be easily done by using the following anti-unitary transform [14],

$$U \times U^{\dagger} = -I, \tag{3.91}$$

and applying it to the totality of the formalism of conventional quantum chemistry. We have in this way the *isodual unit, isodual numbers, isodual products, etc.* [14],

$$I \rightarrow I^{d} = U \times I \times U^{\dagger} = -I, \tag{3.92a}$$

$$n \rightarrow U \times n \times U^{\dagger} = n^{d} = -n, \tag{3.92b}$$

$$A \times B \rightarrow A^{d} \times^{d} B^{d} = U \times (A \times B) \times U^{\dagger}. \tag{3.92c}$$

The isodual iso-, geno-, and hyper- chemistry can be constructed via a combination of the above methods.

Note that the formalism resulting from the above map is *inconsistent* in the event even *one* quantity (say, a Fourier transform) or operation (say, power) is not transformed. The inconsistency is due to the evident mixing in this case of conventional and generalized mathematics, and would be the same as elaborating conventional quantum chemistry via iso-, or geno-mathematics.

3.11. Invariance of Operator Geno-, Hyper-, and Isodual Chemistry

Once the above geno-, hyper- and isodual formulations have been constructed via the simple methods provided above, their invariance can also be easily proved. It merely requires the formulation of the transformation theory via the appropriate mathematics, again, to avoid the mixing of different mathematics intended for different theories.

Consider, for instance, the problem of the invariance of the basic equations of genochemistry. Since this theory is structurally irreversible, the study of its invariance first requires the selection of *either* motion forward in time *or* that backward in time. Then, any given nonunitary transform must be reformulated in the appropriate *genounitary law* acting on genohilbert spaces over genofields [13]

$$Z \times Z^{\dagger} \neq I, Z = \hat{Z} \times \hat{T}^{>1/2}, \tag{3.93a}$$

$$Z \times Z^{\dagger} \equiv \hat{Z} > \hat{Z}^{\dagger} = \hat{Z}^{\dagger} > Z^{\dagger} = \hat{I}^{>}. \tag{3.93b}$$

It is easy to see that the forward genoequations are indeed invariant under the above genounitary transform. In fact, we have the invariance laws [13],

$$\hat{I}^{>} \rightarrow \hat{I}^{>\prime} = \hat{Z} > \hat{I}^{>} > \hat{Z}^{\dagger} \equiv \hat{I}^{>}, \tag{3.94a}$$

$$A > B \rightarrow \hat{Z} > (A > B) > Z^{\dagger} = \hat{A}' > \hat{B}', \tag{3.94b}$$

namely, genounits and genoproducts are *numerically preserved*. The invariance of the rest of the theory is then a consequence, as the interested reader is encouraged to verify.

Essentially the same reformulations are needed to prove the invariant of the remaining branches of hadronic chemistry.

3.12. Classification of Hadronic Chemistry

In summary, *hadronic chemistry has eight topologically and chemically different classical formulations which, following eight different maps, characterize eight different operator formulations, yield the following branches:*

- *quantum chemistry and its isodual;*
- *isochemistry and its isodual;*
- *genochemistry and its isodual;*
- *hyperchemistry and its isodual.*

These theories have been conceived for conditions of increasing complexity and methodological needs, all the way to attempting a deeper understanding of the DNA code, e.g., via a number theory whose unit is an infinite ordered set.

In particular, the covering of quantum chemistry we have called *isochemistry* has been constructed to achieve an invariant representation of *closed-isolated and reversible systems with nonlinear, nonlocal, and nonhamiltonian internal effects verifying conventional total conservation laws and conventional quantum principles*, as evident from the Heisenberg-Santilli isoequation [9] applied to the energy,

$$i \times \frac{d\hat{H}}{dt} = \hat{H}\hat{\times}\hat{H} - \hat{H}\hat{\times}\hat{H} = \hat{H} \times \hat{T} \times \hat{H} - \hat{H} \times \hat{T} \times \hat{H} \equiv 0. \tag{3.95}$$

The broader *genochemistry* has been built to achieve an invariant representation of *open, nonconservative, and irreversible chemical events under linear and nonlinear, local and nonlocal, and Hamiltonian as well as nonhamiltonian processes as occurring in chemical reactions and other events*, as evident from the Heisenberg-Santilli genoequations [9] applied to the energy,

$$i \times \frac{d\hat{H}}{dt} = \hat{H} < \hat{H} - \hat{H} > \hat{H} = \hat{H} \times {}^{<}\hat{T} \times \hat{H} - \hat{H} \times \hat{T}^{>}\hat{H} \neq 0. \tag{3.96}$$

The above law represents the *time-rate-of-variation of the energy*, and evidently admits the simpler conservative law (3.95) as a particular case.

The yet broader *hyperchemistry* has been constructed to achieve an invariant representation of *nonconservative, irreversible, and multivalued systems*, such as biological structures (see Fig. 1.6 for an illustration via sea shells), as shown by the *hyperheisenberg equation* applied to the energy

$$\begin{gathered} i \times \tfrac{d\hat{H}}{dt} = \hat{H} < \hat{H} - \hat{H} > \hat{H} = \\ = (\hat{H} \times {}^{<}\hat{T}_1 \times \hat{H} - \hat{H} \times \hat{T}_1^{>} \times \hat{H}) + \\ +(\hat{H} \times {}^{<}\hat{T}_2 \times \hat{H} - \hat{H} \times \hat{T}_2^{>} \times \hat{H}) + \cdots \neq 0. \end{gathered} \tag{3.97}$$

Isodual chemistry was constructed for the representation of complex biological systems implying *bifurcation* or other complex events requiring a mathematical representation with both motion forward and backward in time along the four directions of time of Fig. 2.1. In this case *all* quantities and their operations are subjected to the anti-isomorphic isodual map, e.g., for the case of the Heisenberg-Santilli isodual time evolution we have [15]

$$\begin{aligned} i^d \times^d d^d A^d /^d d^d t^d &= -i \times dA/dt = \\ = A^d \times^d H^d - H^d \times^d A^d &= -A \times H + H \times A. \end{aligned} \tag{3.98}$$

The joint use of chemistry and its isodual image then generally permit a quantitative representation of bifurcations and other similar events (Figs. 1.6 and 2.1). Essentially the same results hold for the combined use of iso-, geno-, and hyper-chemistry and their respective isoduals.

The above inclusive chains of generalizations confirms the historical teaching that, like all other quantitative sciences, chemistry is a discipline that will never admit a "final theory." No matter how effective any given theory may appear at any given point in time, its structural generalization for quantitative treatments of yet more advanced knowledge is simply inevitable.

References

[1] Einstein, A., Podolsky, B. and Rosen, N.: Phys. Rev. **47**, 777 (1935).

[2] Boyer, D.J.: *Bonding Theory*, McGraw Hill, New York (1968) [2a]. Hanna, M.W.:*Quantum Mechanical Chemistry*, Benjamin, New York (1965) [2b]. Pople, J.A. and Beveridge, D.L.: *Approximate Molecular Orbits*, McGraw Hill, New York (1970) [2c]. Linnet, J.W.: *The Electronic Structure of Molecules, A New Approach*, Methuen Press, London, and Wiley, New York (1964) [2d].

[3] Santilli, R.M. and Shillady, D.D.: Hadronic J. **21**, 633 (1998) [3a]; Intern. J. Hydrogen Energy **24**, 943 (1999) [3b]; Intern. J. Hydrogen Energy **25**, 173 (2000) [3c].

[4] Sinanoglu, O.: Proc. Nat. Acad. Sciences **47**, 1217 (1961) [4a]. Lewis, G.N.: J. Am. Chem. Soc. **38**, 762 (1916) [4b]. Langmuir, I.: J. Am. Chem. Soc. **41**, 868 (1919) [4c]. Boyd, R.J. and Yee, M.C.: J. Chem. Phys. *77*, 3578 (1982) [4d]. Cann, N.M. and Boyd, R.J.: J. Chem. Phys. *98*, 7132 (1992) [4e].

[5] Lagrange, L.: *Mechanique Analytique* (1788), reprinted by Gauthier-Villars, Paris (1888) [5a]; Hamilton, W.R.: *On a General Method in Dynamics* (1834), reprinted in *Hamilton's Collected Works*, Cambridge Univ. Press (1940) [5b]. Jacobi, C.G.: *Zur Theorie del Variationensrechnung under der Differentialgleichungen* (1937) [5c].

[6] Santilli, R.M.: *Foundations of Theoretical Mechanics*, Vol. **I** (1978) [6a] and **II** (1983) [6b], Springer-Verlag, Heidelberg-New York; *Elements of Hadronic Mechanics*, Vol. **I** [6c] and **II** [6d] (1995) and **III**, in preparation [6i], Ukraine Academy of Sciences, Kiev; *Isotopic, Genotopic and Hyperstructural Method in Theoretical Biology*, Ukraine Academy of Sciences, Kiev (1996) [6e].

[7] Santilli, R.M.: Rendiconti Circolo Matematico Palermo, Suppl. **42**, 7 (1996).

[8] Animalu, A.O.E. and Santilli, R.M.: in *Hadronic Mechanics and Nonpotential Interactions*, M. Mijatovic, Editor, Nova Science, New York (1990).

[9] Santilli, R.M.: Hadronic J. **1**, 574 (1978).

[10] Santilli, R.M.: Hadronic J. **3**, 440 (1979) [10a]. Mignani, R., Hadronic J. **5**, 1120 (1982) [10b].

[11] Myuing, H.C. and Santilli, R.M.: Hadronic J. **5**, 1277 and 1367 (1983).

[12] Santilli, R.M.: Found. Phys. **27**, 625 (1997).

[13] Santilli, R.M.: Found. Phys. **27**, 1159 (1997).

[14] Santilli, R.M.: Hyperfine Interactions **109**, 63 (1997); Intern. J. Modern Phys. **D7**, 351 (1998).

[3-15] antilli, R.M.: JINR Rapid Comm. **6**, 24 (1993); Acta Appl. Math. **50**, 177 (1998).

Chapter 4

ISOCHEMICAL MODEL OF THE HYDROGEN MOLECULE

1. Introduction

As it is well known, the primary structural characteristics of *quantum chemistry* (see, e.g., Refs. [1]) are those of being:

1) *linear*, in the sense that eigenvalue equations depend on wavefunctions only to the first power;

2) *local-differential*, in the sense of acting among a finite number of isolated points; and

3) *potential*, in the sense that all acting forces are derivable from a potential energy.

Therefore, quantum chemistry is a *Hamiltonian theory*, i.e., models are completely characterized by the *sole* knowledge of the Hamiltonian operator, with a *unitary structure*, i.e., the time evolution verifies the unitarity conditions

$$U = e^{iH\times t}, \quad U \times U^\dagger = U^\dagger \times U = I, \quad H = H^\dagger, \tag{4.1}$$

when formulated on conventional Hilbert spaces over the conventional fields of complex numbers.

Despite outstanding achievements throughout the 20-th century, quantum chemistry cannot be considered as "final" because of numerous insufficiencies identified in Chapter 1.

A most important insufficiency is *the inability to represent deep mutual penetrations of the wavepackets of valence electrons in molecular bonds.* The latter interactions are known to be:

$\hat{1}$) *nonlinear*, i.e., dependent on powers of the wavefunctions greater than one;

$\hat{2}$) *nonlocal-integral*, i.e., dependent on integrals over the volume of overlapping, which, as such, cannot be reduced to a finite set of isolated points; and

$\hat{3}$) *nonpotential*, i.e., consisting of "contact" interactions with consequential "zero range," for which the notion of potential energy has no mathematical or physical sense.

A representation of the latter features evidently requires a *nonhamiltonian theory*, i.e., a theory which cannot be solely characterized by the Hamiltonian, and requires additional terms. It then follows that the emerging theory is *nonunitary* i.e., its time evolution verifies the law,

$$U \times U^\dagger = U^\dagger \times U \neq I, \tag{4.2}$$

when formulated on conventional Hilbert spaces over conventional fields.

It is evident that the above features are beyond any hope of scientific-quantitative treatment via quantum mechanics and chemistry.

In the preceding Chapter 3 we have submitted the foundations of a generalization-covering of quantum chemistry under the name of *hadronic chemistry*, first submitted by Santilli and Shillady in Ref. [2], which is capable of providing an invariant representation of the above-mentioned nonlinear, nonlocal, nonpotential, nonhamiltonian, and nonunitary interactions in deep correlations of valence electrons.

In Chapter 3, we have also shown that the conventional "screenings" of the Coulomb potential (which are necessary for a better representation of experimental data) are outside the axiomatic structure of "quantum" chemistry because such screenings can only be reached via nonunitary maps of the Coulomb law, thus resulting in being particular cases of the broader hadronic chemistry.

The main purpose of this chapter is the application of hadronic chemistry to the construction of a new model of molecular bonds and its verification in the representation of experimental data of the hydrogen molecule.

Since molecular structures are considered as isolated, thus being closed, conservative, and reversible, the applicable branch of hadronic chemistry is *isochemistry*, which is characterized by the identification of the nonunitary time evolution with the *generalized unit* of the theory, called *isounit*,

$$U \times U^\dagger = \hat{I}(r, p, \psi, \partial\psi, ...) \neq I, \tag{4.3}$$

assumed hereon not to depend explicitly on time, and the reconstruction of the *totality* of the formalism of quantum chemistry into a new form admitting of $\hat{I}$, rather than I, as the correct right and left new unit.

The capability by the isounit to represent nonlinear, nonlocal, and nonhamiltonian interactions is evident. Its selection over other possible

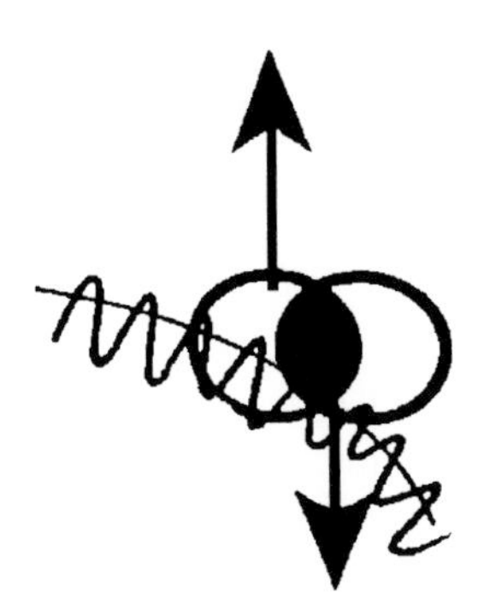

Figure 4.1. A schematic view of the central conditions studied in this chapter, the deep overlapping of the wavepackets of valence electrons in singlet coupling (to verify Pauli's exclusion principle). These conditions are known to be nonlinear, nonlocal, and nonpotential (due to the zero-range, contact character of the interactions), thus not being representable via a Hamiltonian, and, consequently, not being unitary. As a result, the ultimate nature of valence bonds is outside any credible representation via quantum chemistry. Hadronic chemistry (Chapter 3) has been built for the specific scope of representing the conditions herein considered of the bonding of valence electrons.

choices is mandated by the condition of *invariance*, that is, the prediction of the same numerical values for the same quantities under the same conditions, but at different times. In fact, whether generalized or not, the unit of any theory is the basic invariant.

A central assumption of this chapter is that *quantum mechanics and chemistry are exactly valid at all distances of the order of the Bohr radius ($\simeq 10^{-8}$ cm), and the covering hadronic chemistry only holds at distance of the order of the size of the wavepackets of valence electrons* (1 fm $= 10^{-13}$ cm).

This condition is evidently necessary, on one side, to admit the conventional quantum structure of the hydrogen atom, and, on the other side, to admit quantitative studies of the nonhamiltonian interactions of short range valence bonds.

The above condition is readily achieved by imposing that all isounits used in this chapter recover the conventional unit at distances greater than 1 fm,

$$\lim_{r \gg 1 \text{ fm}} \hat{I}(r, p, \psi, \partial\psi, \dots) = I, \tag{4.4a}$$

$$|\hat{I}| \ll 1, \quad |\hat{T}| \gg 1. \tag{4.4b}$$

In fact, under the above condition, hadronic chemistry recovers quantum chemistry everywhere identically.

The reader should keep in mind the crucial implications of conditions (4.4b) which, as shown in Sect. 3.6, Eqs. (3.74d) and (3.74e), permit a dramatic increase of the convergence of chemical series, with corresponding decrease of computer time, as verified in the models of this chapter and of the following ones.

The reader should also note that, quite remarkably, rather than being imposed, both conditions (4.4a) and (4.4b) are naturally verified by actual chemical models.

It should be recalled that, under the assumption of representing closed-isolated systems, *isochemistry verifies all conventional laws and principles of quantum mechanics* (Chapter 3). Therefore, there is no *a priori* mean for rejecting the validity of hadronic chemistry within the small region of space of valence bonds.

It then follows that the selection of which theory is valid is referred to the capability to represent experimental data. Quantum mechanics has been capable of achieving an exact representation of all experimental data for the structure of *one individual* hydrogen atom. Therefore, quantum mechanics is here assumed as being exactly valid within such a well defined physical system, any possible improvement being redundant at best.

By comparison, quantum mechanics and chemistry have not been able to achieve an exact representation of the experimental data of the *different* conditions of molecular structures, as discussed in detail in Chapter 1. As a result, these theories are *not* considered as being exactly valid for the different conditions of molecular bonds (see Fig. 1.7).

As we shall see in this chapter, hadronic chemistry can indeed provide an exact representation of molecular characteristics, and, therefore, it is consider as being exactly valid for the indicated conditions of applicability.

A knowledge of *isomathematics* of Chapter 2 and of *isochemistry* as presented in Chapter 3 is essential for a technical understanding of the content of this chapter (see also representative papers [3, 4]).

For mathematically less inclined readers, we recall from Sect. 3.3.6 that specific applications isochemistry can be constructed in their entirety via a simple nonunitary transform of conventional quantum chemical models. In fact such a transform adds precisely the desired short range, nonlinear, nonlocal, and nonhamiltonian effects.

2. Isochemical Model of Molecular Bonds

We now present the conceptual foundations of our *isochemical model of molecular bonds* for the simplest possible case of the H_2 molecule, which was first submitted by Santilli and Shillady in Ref. [5]. We shall

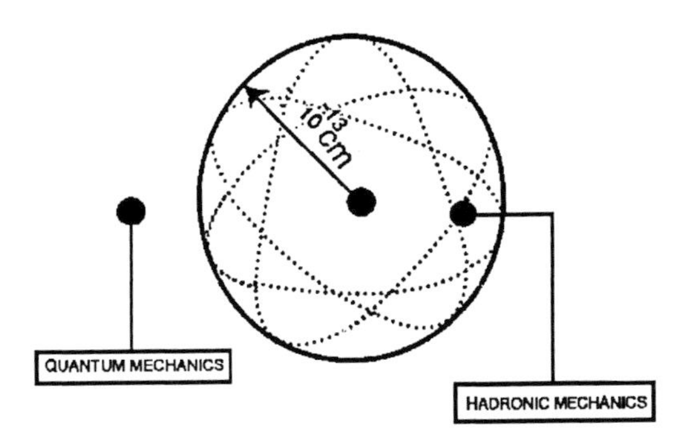

Figure 4.2. A schematic unit of the *hadronic horizon*, namely, of the sphere of radius 1 fm (= 10^{-13} cm) outside which quantum chemistry is assumed to be exactly valid, and inside which nonlinear, nonlocal, and nonpotential effects are no longer negligible, thus requesting the use of hadronic chemistry for their numerical and invariant treatment.

then extend the model to the water and to other molecules in the subsequent chapter.

Since the nuclei of the two H-atoms remain at large mutual distances, the bond of the H_2 molecule is evidently due to the bond of the peripheral valence electrons, as generally acknowledged [1].

Our main assumption [5] is that *pairs of valence electrons from two different atoms can bond themselves at short distances into a singlet quasi-particle state called "isoelectronium," which describes an oo-shaped orbit around the two nuclei similar to that of planets in binary star systems (Fig. 4.3).*

It is important to note that recent studies in pure mathematics [39] have established that the *oo*-shaped orbit, called the *figure eight* solution, is one of the most stable solutions of the N-body problem.

The primary binding force of the isoelectronium is assumed to be of nonlinear, nonlocal, and nonpotential type due to contact effects in deep overlappings of the wavepackets of the valence electrons, as studied in Sect. 4.3.

However, the reader should be aware that the isoelectronium is expected to have a component of the binding force of purely potential type because, when the electrons are in singlet coupling, the magnetostatic *attraction* may be conceivably bigger than the electrostatic *repulsion* at distances of the order of one fermi or less (see Fig. 4.4 for details).

It should be stressed, however, that a purely potential origin of the isoelectronium is not expected to be exactly valid for various reasons, the

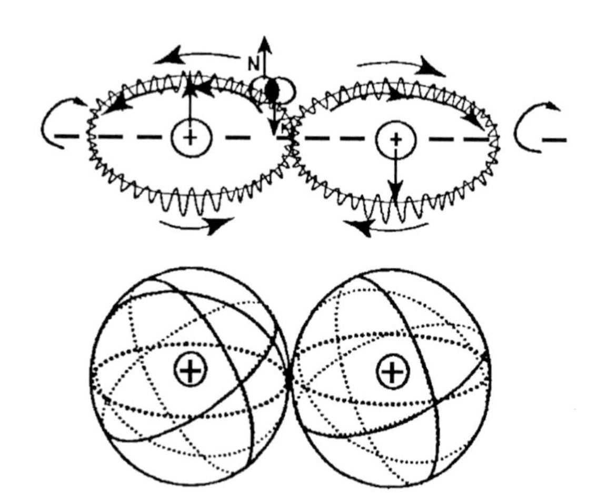

Figure 4.3. A schematic view of the proposed *isochemical model of the hydrogen molecule with fully stable isoelectronium*, where the top view refers to absolute zero degree temperature and in the absence of any motions, while the lower view includes rotations, thus recovering the conventional spherical distribution. The view is complementary to that of Fig. 4.7 for the unstable isoelectronium. The model is here depicted in terms of *orbits of the valence electrons*, rather than in terms of *orbitals, or density distributions*. The fundamental assumption is that the two valence electrons, one per each atom, correlate themselves into a bonded singlet state at short distance we have called *isoelectronium*, which is assumed in this figure to be stable. In this case the only orbit yielding a stable H-molecule is that in which the isoelectronium describes a *oo*-shaped orbit around the respective two nuclei, as it occurs for planets in certain systems of binary stars. The isoelectronium is then responsible for the *attractive force* between the two atoms. The *binding energy* is instead characterized by the *oo*-shaped orbit of the isoelectronium around the two nuclei, conceptually represented in this figure via a standing wave for a particle of spin 0, charge $-2e$, and null magnetic moment. As we shall see in this chapter, the model then permits a representation of: the reason why the H_2 and H_2O molecules have only two hydrogen atoms; the exact representation of the binding energy; the resolution of some of the inconsistencies of the conventional model; and other advances. Note finally that the model is easily extendable to dimers such as HO, HC, *etc.*, as studied in Chapter 3. The novelty in predictive character of the model can be seen from these preliminary lines. For instance, the model depicted in this figure predicts that *the hydrogen molecule becomes asymmetric, thus acquiring an infrared signature, under sufficient magnetic polarization, which removes its rotational motions.*

most visible one being the fact that, at the very small mutual distances here considered, magnetostatic and electrostatic laws diverge, thus prohibiting reliable quantitative studies.

Hadronic chemistry has been built to resolve all divergences in the study of the isoelectronium thanks to the isomathematics with product $A\hat{\times}B = A\times\hat{T}\times B$, and the isotopic element $\hat{T}$ restricted to have absolute

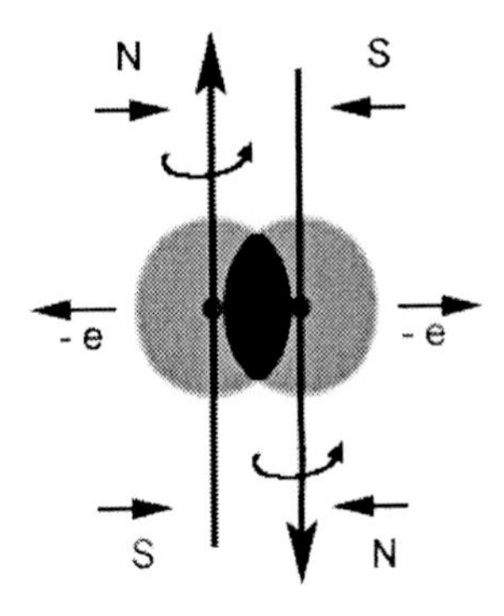

Figure 4.4. A schematic view of the conventional Coulomb forces of electrostatic and magnetostatic type in the structure of the isoelectronium. Since the charges are equal, they cause a *repulsion*. However, since the coupling is in singlet, the magnetic polarities are opposite, thus implying an *attraction*. Elementary calculations show that the magnetostatic attraction equals the electrostatic repulsion at a mutual distance of the order of 1 fm, while it becomes bigger at smaller distances, thus explaining the reason why the hadronic horizon has been set at 10^{-13} cm. This evidence establishes that the bonding force of the isoelectronium can also see its origin on purely Coulomb forces and, more particular, on the dominance of magnetic over electric effects at short distances, which is a rather general occurrence under the proper conditions (see the new chemical special of *magnecules* in Chapter 8). Despite this fully potential attractive total force, it should be stressed that the isoelectronium cannot be treated within a purely quantum mechanics context for various reasons. The first reason is that with the decrease of the distance, both electrostatic and magnetostatic effects diverge, thus preventing any serious scientific study. Hadronic mechanics and chemistry have been built precisely to remove these divergencies via the isotopies of generic products $A\hat{\times}B = A \times \hat{T} \times B$ with $|\hat{T}| \ll 1$ (Chapter 3). Therefore, the hadronic treatment of the isoelectronium permits convergent numerical predictions which would be otherwise impossible for quantum chemistry. Independently from that, the nonunitary lifting of quantum chemistry is mandated by the need to achieve an exact representation of experimental data on molecules which, as now established, requires screenings solely obtainable via nonunitary transforms of the Coulomb potential. Thus, any attempt to preserve old theories as exactly valid is doomed to failures. Despite that, the electrostatic and magnetostatic effects depicted in this figure illustrate that conventional potential effects should also be expected in the structure of the isoelectronium. In other words, rather than assuming either a purely quantum or a purely hadronic setting, we have *in media virtus*, i.e., the most plausible origin of the bonding force of the isoelectronium is that partially of potential and partially of nonpotential type. Still in turn, this implies the possibility of a significant (negative) binding energy for the isoelectronium, which is evidently that characterized by the potential component (Sect. 4.3).

values much smaller than 1. In this way, the hadronic component of the isoelectronium binding force will "absorb" all divergent or otherwise repulsive effects, resulting in convergent numerical values.

The reader is also discouraged to reduce the isoelectronium to a purely quantum structure because, in this way, the theory would preserve all the insufficiencies of chemistry studied in Chapter 1, most importantly, the inability to reach an exact representation of molecular characteristics from the strict application of first quantum principles without *ad hoc* adulterations. In fact, as now well established, such an exact representation requires screenings of the Coulomb law, which can only be obtained via nonunitary transforms. The same nonunitary broadening of quantum chemistry is requested on numerous other counts independent from the isoelectronium.

Despite these limitations, the purely magnetostatic-electrostatic structure of the isoelectronium remains important *in first approximation*, because it recovers in a very simple way the hadronic horizon (Fig. 4.2), as well as the prediction by hadronic mechanics dating back to 1978 that triplet couplings are highly unstable. In fact, in the latter case, both electrostatic and magnetostatic forces would be *repulsive*, thus prohibiting any possible bound state, in beautiful agreement with Pauli's exclusion principle.

It is easy to predict that *the isoelectronium cannot be permanently stable when interpreted as a quasi-particle of about* 1 *fm charge diameter.* In fact, the mere presence of exchange forces, which remain fully admitted by isochemistry, prevents the achievement of a complete stability under the indicated small mutual distances of the electrons. As we shall see in more details in Chapter 6, there are additional technical reasons which prevent the complete stability at short distances, and actually render the isoelectronium a short lived quasi-particles when the valence electrons are assumed at mutual distances of 1 fm.

However, it is easy to see that *the isoelectronium must be fully stable when the mutual distance of the two valence electrons is permitted to be of the order of molecular size. In fact, any instability under the latter long range conditions would imply a necessary violation of the fundamental Pauli's exclusion principle.*

In different words, *the isoelectronium is one of the first known quantitative representations of Pauli's principle*, in the sense that:

1) When assumed to be of potential type, the interaction responsible for Pauli's principle implies catastrophic inconsistencies, such as shifts of experimentally established energy levels, deviations from all spectroscopic lines, etc. As a result, a quantitative representation of Pauli's principle is *impossible* for quantum mechanics, evidently due

to its strictly potential character. For this reason, Pauli's principle is merely imposed in quantum mechanics without any explanations, as well known. By comparison, a quantitative representation is possible for hadronic mechanics precisely because of its admission of *nonpotential* interactions, that is, interactions which have no bearing on energy levels and spectroscopic lines.

2) Quantum mechanics admits, in general, both singlet and triplet couplings because particles are assumed to be point like as per the very topological structure of the theory. By comparison, hadronic mechanics represents particles as expended at mutual distances smaller than their wavepackets, and solely admits singlet couplings due to highly repulsive-unstable forces predicted for all triplet couplings. The latter repulsive forces originate from the drag experienced by one wavepackets when rotating within and against the rotation of the other wavepacket, as well as by the fact that in triplet couplings both magnetostatic and electrostatic effects are repulsive (Fig. 4.4); and

3) Quantum mechanics cannot provide an exact representation of an *attraction* between *identical* electrons at very short distances, as discussed earlier, in disagreement with the experimental evidence, e.g., that the two electrons of the helium are bonded most of the time, to such an extent that they are emitted in such a bonded form during photodisintegrations, and in other events. By comparison, hadronic mechanics has been built to represent precisely the *bonding* of identical electrons in *singlet* coupling under interactions *not* derivable from a potential.

The assumption of the isoelectronium as being unstable when its valence electrons are at mutual distances of molecular order, implies a violation of Pauli's principle, e.g., because of the automatic admission of triplet couplings for two electrons at the same energy level.

When assumed as being stable in the limit case of a quasi-particle of 1 fm charge radius, *the most stable trajectory of the isoelectronium is of oo-type, each o-branch occurring around each nucleus* (Fig. 4.3). As illustrated in Fig. 4.4 (see also Chapter 8), such a shape automatically prevents the inconsistent prediction of ferromagnetic character of all molecules.

When the correlation-bond is distributed over the entire molecular orbit, *the trajectory of the isoelectronium is also expected to be oo-shaped around the two nuclei with inverted direction of rotation from one o-branch to the other.* This is suggested by a variety of reasons, such as: the need of avoiding the inconsistent prediction of ferromagnetic character, the compatibility with the limit case of a fully stable particle at short distance (which, as we shall see, can describe several *oo*-shaped orbits prior to separation), and others.

It should be indicated that the assumption of a finite lifetime of the isoelectronium irrespective of size implies the possibility of adding several H-atoms to the H_2 molecule for the duration of the unbound valence electrons, as well as other inconsistencies, such as the capability by hydrogen and water to be paramagnetic (Chapter 8).

In this chapter, we apply the above hypothesis to the construction of a new model of the hydrogen molecule and prove its capability to:

1) provide an essentially exact representation of the binding energy and other characteristics of the hydrogen molecules;

2) said representation occurs from first axiomatic principles without exiting from the underlying class of equivalence as occurring for Coulomb screenings;

3) explain for the first time to our knowledge the reason why the hydrogen molecule has only two atoms;

4) introduce an actual "strongly" attractive molecular bond;

5) achieve a much faster convergence of power series with consequential large reduction in computer times;

6) prevent inconsistencies such as the prediction that the hydrogen is ferromagnetic. In fact, whatever magnetic polarity can be acquired by the orbit around one nucleus, the corresponding polarity around the second nucleus will necessarily be opposite, due to the opposite direction of the rotations in the two *o*-branches, thus preventing the acquisition of a net total polarity North-South of the molecule.

By recalling from Chapter 3 that Gaussian screenings of the Coulomb law are a particular case of the general nonunitary structure of hadronic chemistry, one can see from these introductory lines that our first achievement on scientific records of an essentially exact representation of molecular characteristics is reduced to the proper selection of the basic nonunitary transform, because the latter will permit dramatically more restrictive screenings.

The derivability of the essentially exact representation from first axioms of hadronic chemistry without adulterations is evident.

Equally evident is the first introduction of an actual, "strongly" attractive interatomic force (where the word "strongly" does not evidently refer to strong interactions in hadron physics), which is absent in current models due to the notorious "weak" nature of exchange and other forces of current used in molecular structures (where the word "weak" does not evidently refer to the weak interactions among leptons).

The representation of the reason why the hydrogen (or water) molecule has only two H-atoms is inherent in the very conception of the isoelectronium. Once the two valence electrons of the H-atoms couple themselves into a singlet quasi-particle state, there is no possibility for

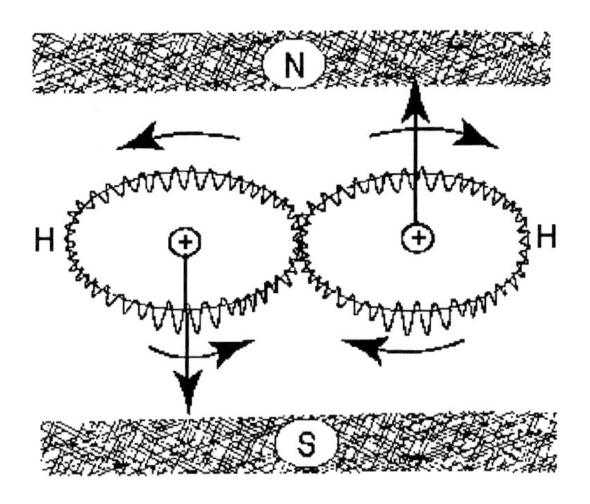

Figure 4.5. A schematic view of the impossibility for the isochemical model of the hydrogen molecule to acquire a net magnetic polarity, thus resolving a serious inconsistency of quantum chemistry. Recall from Chapter 1 that current molecular models are based on exchange, van der Waals, and other forces of nuclear origin, all implying the independence of the orbitals of the individual atoms. Under these assumptions, quantum electrodynamics demands that all molecules acquire a net total magnetic polarity North-South when exposed to an external magnetic field, in dramatic disagreement with reality. The isochemical model of molecular structure resolves this inconsistency because, as indicated in Fig. 4.3, the most stable trajectory for the isoelectronium is *oo*-shaped as it also occurs for the trajectory of planets in binary stars, with each *o*-branch around each nucleus. In this case, the rotation of the two *o*-branches are necessarily opposite to each other, thus resulting in *opposite* magnetic polarities, with the consequential impossibility to reach a *net* molecular magnetic polarity. As we shall see in Chapter 7, the above features have important industrial applications for new clean fuels and energies.

a third valence electron to participate in the bound state, e.g., because we would have an impossible bound state between a fermion (the third electron) and a boson (the isoelectronium).

The achievement of a much faster convergence of the power series, or, equivalently, a dramatic reduction of computer times for the same calculations, is evident from the structure of hadronic chemistry as discussed in Chapter 3.

The avoidance of the prediction of ferromagnetic features (acquisition of a total North-South polarity under an external magnetic field) is due to the nature of the orbit of the isoelectronium, as discussed in details below and in Chapter 8.

In this chapter, we shall study two realizations of the proposed new model of the hydrogen molecule, the first model is a limiting case in which the isoelectronium is assumed to be *stable* (with an infinite life-

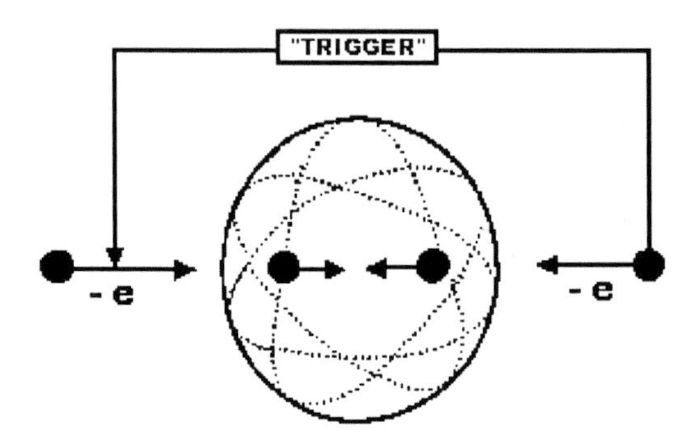

Figure 4.6. A schematic view of the *trigger*, namely, the external means suitable to force electrons with the same charge to penetrate the hadronic barrier (Fig. 4.2), in which attractive hadronic forces overcome the repulsive Coulomb barrier.

time) at ordinary conditions, and the second model in which the isoelectronium is assumed to be *unstable* (with a finite lifetime). The lifetime of the isoelectronium will then be computed in Chapter 6.

The hypothesis of the bonding of electrons at short distances was first introduced by Santilli [7a] for the structure of the π^0 meson as a hadronic bound state of one electron and one positron. Animalu [7b] and Animalu and Santilli [7c] extended the model to the Cooper pair in superconductivity as a hadronic bound state of two identical electrons.

A notion which is important for the very existence of the isoelectronium is that of a *trigger*, namely, *external (conventional) interactions, which cause the identical electrons to move one toward the other and to penetrate the hadronic horizon (Fig. 4.2) against their repulsive Coulomb interactions.* Once inside the above mentioned horizon, the attractive hadronic forces overcome the repulsive Coulomb interaction, resulting in a bound state.

In the case of the π^0 model as a bound state of an electron and a positron at short distances, there is no need for a trigger because the constituents naturally attract each other. On the contrary, the existence of the Cooper pair does indeed require a trigger, which was identified by Animalu [7b] and Animalu and Santilli [7c] as being provided by the Cuprate ions. For the case of an isolated hydrogen molecule, we conjecture that the trigger is constituted by the two H-nuclei, which do indeed attract the electrons. We essentially argue that the attraction of the electrons by the two nuclei is sufficient to cause the overlapping of the two wavepackets, thus triggering the electrons beyond the hadronic horizon.

It should be indicated that we cannot use the term "electronium" because it would imply a bound state of two identical electrons under *quantum* mechanics, which is known to be impossible. The term "electronium" would also be technically inappropriate because the constituents *are not* ordinary electrons, but rather "isoelectrons," i.e., the image of ordinary particles under *nonunitary* transforms or, more technically, irreducible isounitary representations of the covering of the Poincarè symmetry known as the *Poincarè-Santilli isosymmetry* [3c, 3d, 4a].

We cannot close this conceptual section without a few comments regarding the possibility of treating the isoelectronium via *quantum electrodynamics* (QED), since the latter appears to be the natural discipline for a valence bond of two identical electrons at short distance. This issue is compounded by the general belief of the *unrestricted* exact validity of QED all the way to very small distances of the order of 10^{-24} cm.

It is easy to see that, as it is the case for quantum mechanics, a quantitative treatment of the isoelectronium is beyond the technical capabilities of QED for numerous conceptual and technical reasons. In fact, QED is purely linear, local and potential, while the interactions we are interested in representing are nonlinear, nonlocal and nonpotential.

In any case, it is easy to prove via the use of the Feynman diagrams that QED *cannot* represent any *attraction* between identical electrons in singlet coupling at short distance, as it occurs in the physical reality for the two electrons of the Helium, the Cooper pair, the valence electrons, and other systems. On the contrary, the *isotopies of quantum electrodynamics* (ISOQED) are expected to provide such a representation, but their study here would be vastly beyond the limited scope of this monograph.

The reconciliation between the current belief of the unrestricted exact validity of QED and the bonding of identical electrons is permitted by the fact that all experimental verifications of QED at shorter and shorter distances have been conducted via the use of *higher and higher energies.* On the contrary, the experimental verification of QED for the conditions of the isoelectronium require *smaller and smaller energies* which experimental verifications have been absent in the physics of the 20-th century due to the notorious emphasis on high energies.

As a final comment, it should be noted that the limitations of QED for the study of the isoelectronium are purely classical, and rest on *the inability of classical electrodynamics to represent the physical evidence of the attraction of identical spinning charges at sufficiently small distances, evidence which is even visible to the naked eyes, e.g., in ball lighting as created by nature, in microwave ovens or other means.*

As a matter of fact, no classical theory of electromagnetism can possibly be considered as "final" until it achieves the capability of representing the attraction of identical charges under suitable conditions. As a result, no quantum theory of electromagnetism, including QED, can be considered as "final" unless it is based on the preceding classical theory. One of the objectives of classical and operator isochemistry is precisely that of achieving such a missing representation.

3. The Limit Case of Stable Isoelectronium

We are now equipped to conduct a nonrelativistic study of the isoelectronium (Fig. 4.3) in the limit case of full stability under the assumption that the binding force is of purely hadronic type without potential contributions (Fig. 4.4). This approach is evidently done to test the effectiveness of hadronic chemistry for the numerical studies of the problem considered, since corrections due to potential effects can be easily added.

The reader should be aware upfront that *the above assumptions imply that the isoelectronium has no binding energy*, trivially, because nonpotential forces have no potential energy by conception.

The reader should be aware that the actual hadronic treatment should be conducted within the context of isomathematics, that is, on isoeuclidean and isohilbert spaces defined over isofields. To avoid excessive mathematical complexity, in this section we study the *projection* of this isotopic treatment on conventional spaces over conventional fields. However, it should be stressed that the only correct formulation remains the isotopic one.

As we shall see, the hadronic treatment of the isoelectronium yields an attraction of the type of the Hulten potential which is so strong to "absorb" at short distances all other forces, whether attractive or repulsive. However, the direct interpretation of the Hulten potential as an actual potential would be erroneous, since it solely occurs in the *projection* of the model on conventional spaces, while being completely absent in the technically appropriate treatment on isospaces over isofields. The direct interpretation of the Hulten potential as an actual potential well of quantum mechanical nature would also be in direct contradiction with the absence of binding energy.

Therefore, the assumption of the projected model as the correct one leads to insidious inconsistencies and misrepresentations, such as the possible interpretation of the isoelectronium via a potential well, which treatment is very familiar in quantum mechanics, but the same treatment has no physical meaning for the isoelectronium. This is due to the fact that, as stressed earlier, a necessary condition to avoid inconsistencies in the interpretation of Pauli's principle is that its interaction *does not* admit a potential energy, thus rendering meaningless, or at best contradictory, conventional potential wells.

Note that the emergence of a "strong" Hulten potential eliminates the issue whether the isoelectronium is due to the dominance of the attractive magnetostatic forces over the repulsive electric ones (Fig. 4.4). This is due to the fact that the Hulten potential, as we shall review shortly in detail, behaves at short distances as $constant/r$, thus absorbing all Coulomb forces, irrespective of whether attractive or not. Moreover, the unified treatment via the Hulten potential presented below eliminates the divergent character of these forces at short distances, thus permitting meaningful numerical results.

We should finally indicate, to avoid inconsistencies, that the study of this section deals with the *limit* case of *a perfectly stable isoelectronium interpreted as a quasi-particle of about 1 fm charge diameter*, while in reality such form of the isoelectronium is unstable. Moreover, in this section we shall not study the expectation that the isoelectronium persists beyond the 1 fm mutual distance of the valence electrons, as necessary to prevent violations of Pauli's principle.

We begin our quantitative analysis with the nonrelativistic quantum mechanical equation of two ordinary electrons in singlet couplings, $e^-_\downarrow$ and $e^-_\uparrow$ represented by the wavefunction $\psi_{\uparrow\downarrow}(r) = \psi(r)$,

$$\left(\frac{p \times p}{m} + \frac{e^2}{r}\right) \times \psi(r) = E \times \psi(r). \tag{4.5}$$

To transform this state into the isoelectronium representing the bonding of the H-electron with a valence electron of another atom of generic charge ze, we need first to submit Eq. (4.5) to a nonunitary transform characterizing the short range hadronic effects, and then we must add the *trigger*, namely, the Coulomb attraction by the nuclei.

This procedure yields the *isoschrödinger equation for the isoelectronium* (Chapter 1),

$$U \times U^\dagger = \hat{I} = 1/\hat{T} > 0, \tag{4.6a}$$

$$\hat{A} = U \times A \times U^\dagger, \quad A = p,\ H, \ldots, \tag{4.6b}$$

$$U \times (A \times B) \times U^\dagger = \hat{A}\hat{\times}\hat{B} = \hat{A} \times \hat{T} \times \hat{B}, \quad \hat{\psi} = U \times \psi, \tag{4.6c}$$

$$\left(\frac{1}{m}\hat{p} \times \hat{T} \times \hat{p} \times \hat{T} + \frac{e^2}{r} \times \hat{T} - \frac{z \times e^2}{r}\right) \times \hat{\psi}(r) = E_0 \times \hat{\psi}(r), \tag{4.6d}$$

$$\hat{p}\hat{\times}\hat{\psi}(r) = -i \times \hat{T} \times \nabla\hat{\psi}(r), \tag{4.6e}$$

where the factor $\hat{T}$ in the first Coulomb term originates from the nonunitary transform of model (4.5), while the same factor is absent in the second Coulomb term because the latter is long range, thus being conventional. As a result, in the model here considered the trigger is merely added to the equation.

The angular component of model (4.6) is conventional [3], and it is hereon ignored. For the radial component $r = |\mathbf{r}|$, we assume the isounit [7]

$$\hat{I} = e^{N\times\psi/\hat{\psi}} \approx 1 + N \times \psi/\hat{\psi}, \quad N = \int dr^3\, \hat{\psi}^\dagger(r)_{1\downarrow} \times \hat{\psi}(r)_{2\uparrow}, \quad (4.7a)$$

$$\hat{T} \approx 1 - N \times \psi/\hat{\psi}, \quad (4.7b)$$

$$|\hat{I}| \gg 1, \quad |\hat{T}| \ll 1, \quad (4.7c)$$

$$\lim_{r\gg 1\mathrm{fm}} \hat{I} = 1, \quad (4.7d)$$

where one should note that Eqs. (4.7c) and (4.7d) are automatically verified by expressions (4.7a) and (4.7b).

Note that the explicit form of ψ is of Coulomb type, thus behaving like

$$\psi \approx N \times \exp(-b \times r), \quad (4.8)$$

with N approximately constant at distances near the hadronic horizon of radius

$$r_c = \frac{1}{b}, \quad (4.9)$$

while $\hat{\psi}$ behaves like

$$\hat{\psi} \approx M \times \left(1 - \frac{\exp(-b \times r)}{r}\right), \quad (4.10)$$

with M being also approximately constant under the same range [7a]. We then have

$$\hat{T} \approx 1 - \frac{V_{\mathrm{Hulten}}}{r} = 1 - V_0 \frac{e^{-b\times r}}{(1 - e^{-b\times r})/r}, \quad (4.11)$$

namely, we see the appearance of a Hulten potential in this local approximation. But the Hulten potential behaves at short distances like the Coulomb one,

$$V_{\mathrm{Hulten}} r \approx \frac{1}{b} \approx \frac{V_0}{b} \times \frac{1}{r}. \quad (4.12)$$

As a result, inside the hadronic horizon we can ignore the repulsive (or attractive) Coulomb forces altogether, and write

$$+\frac{e^2}{r}\times\hat{T} - \frac{e^2}{r} \approx +\frac{e^2}{r}\times\left(1 - \frac{V_{\mathrm{Hulten}}}{r}\right) - \frac{z\times e^2}{r} = -V\times\frac{e^{-b\times r}}{1 - e^{-b\times r}}, \quad (4.13)$$

by therefore resulting in the desired overall *attractive* force among the identical electrons inside the hadronic horizon.

By assuming in first approximation $|\hat{T}| = \rho \approx 1$, the radial equation of model reduces to the model of π^0 meson [7a] or of the Cooper pair [7b, 7c], although with different values of V and b.

$$\left[\frac{1}{r^2}\left(\frac{d}{dr}r^2\frac{d}{dr}\right) + \frac{m}{\rho^2 \times \hbar^2}\left(E_0 + V \times \frac{e^{-b\times r}}{1-e^{-b\times r}}\right)\right] \times \hat{\psi}(r) = E_0 \times \hat{\psi}(r). \tag{4.14}$$

The exact solution and related boundary conditions were first computed in Ref. [7a], Sect. 5, and remain fully applicable to the isoelectronium.

The resulting spectrum is the typical one of the Hulten potential,

$$|E_0| = \frac{\rho^2 \times \hbar^2 \times b^2}{4 \times m}\left(\frac{m \times V}{\rho^2 \times \hbar^2 \times b^2} \times \frac{1}{n} - n\right)^2, \tag{4.15}$$

which evidently possesses a *finite* spectrum, as well known.

To reach a numerical solution, we introduce the parametrization as in Ref. [7a],

$$k_1 = \frac{1}{\lambda \times b}, \tag{4.16a}$$

$$k_2 = \frac{m \times V}{\rho^2 \times \hbar^2 \times b^2}. \tag{4.16b}$$

We note again that, from the Eqs. (1.39) and related boundary conditions, k_2 must be bigger than but close to one, $k_2 \approx 1$ [7].

We therefore assume in first nonrelativistic approximation that

$$\frac{m \times V}{\rho^2 \times \hbar^2 \times b^2} = 1. \tag{4.17}$$

By assuming that V is of the order of magnitude of the total energy of the isoelectrons at rest as in the preceding models [7],

$$V \approx 2 \times \hbar \times \omega \approx 2 \times 0.5 \text{ MeV} = 1 \text{ MeV}, \tag{4.18}$$

and by recalling that $\rho \approx 1$, we reach the following estimate for the *radius of the isoelectronium*

$$r_c = b^{-1} \approx \left(\frac{\hbar^2}{m \times V}\right)^{1/2} = \left(\frac{\hbar}{m \times \omega_0}\right)^{1/2} =$$

$$= \left(\frac{1.054 \times 10^{-27} \text{erg} \cdot \text{sec}}{1.82 \times 10^{-27} \text{ g} \times 1.236 \times 10^{20} \text{ Hz}}\right)^{1/2} = \tag{4.19}$$

$$= 6.8432329 \times 10^{-11}\text{cm} = 0.015424288 \text{ bohrs} = 0.006843 \text{ Å},$$

It should be noted that: 1) the above values of r_c and V are only *upper boundary values* in the center-of-mass frame of the isoelectronium, i.e., the largest possible values under the assumptions of this section; 2) the values have been computed under the approximation of null relative kinetic energy of the isoelectrons with individual total energy equal to their rest energy; and 3) the values evidently *decrease* with the addition of the relative kinetic energy of the isoelectrons (because this implies the increase of m in the denominator).

The actual radius of the isoelectronium, when considered to be an quasi-particle as in this section, is also expected to vary with the trigger, that is, with the nuclear charges, as confirmed by the calculations presented in the next sections. This illustrates again the upper boundary character of value (4.19).

The value k_1 is then given by

$$k_1 = \frac{V}{2 \times k_2 \times b \times c_0} = 0.19, \quad k_2 \approx 1. \tag{4.20}$$

Intriguingly, the above two values for the isoelectronium are quite close to the corresponding values of the π^0 [7a] and of the Cooper pair [7b, 7c] (see also Sect. 1.9),

$$k_1 = 0.34, \quad k_2 = 1 + 8.54 \times 10^{-2}, \tag{4.21a}$$

$$k_1 = 1.3 \times \sqrt{z} \times 10^{-4}, \quad k_2 = 1.0 \times \sqrt{z}, \tag{4.21b}$$

It is important to see that, at this nonrelativistic approximation, *the binding energy of the isoelectronium is not only unique, but also identically null,*

$$|E_0| = \frac{\rho^2 \times \hbar^2 \times b^2}{4 \times m} \left(\frac{m \times V}{\rho^2 \times \hbar^2 \times b^2} - 1 \right)^2 = \frac{V}{4 \times k_2} \times (k_2 - 1)^2 = 0. \tag{4.22}$$

This result is crucial to prevent inconsistencies with Pauli's exclusion principle, which, as indicated earlier, *requires no potently energy between the two electrons for its interpretation in a way consistent with experimental data.*

The notion of a *bound state with only one allowed energy level* (called "hadronic suppression of the atomic spectrum" [7a]) is foreign to conventional quantum mechanics and chemistry, although it is of great importance for hadronic mechanics. In fact, any excitation of the constituents, whether the π^0, the Cooper pair or the isoelectronium, causes their exiting the hadronic horizon, by therefore re-acquiring the typical atomic spectrum. Each of the considered three hadronic states has, therefore, only one possible energy level.

The additional notion of a *bound state with null binding energy* is also foreign to quantum mechanics and chemistry, although it is another fundamental characteristic of hadronic mechanics and isochemistry. In fact, the hadronic interactions admit no potential energy, and as such, they cannot admit any appreciable binding energy, as typical for ordinary contact zero-range forces of our macroscopic Newtonian reality.

The null value of the binding energy can be confirmed from the expression of the meanlife of the isoelectronium, which can be written in this nonrelativistic approximation [7a]

$$\tau = \frac{\hbar}{4\times\pi\times\hbar^2}|\hat{\psi}(0)|\times\alpha\times E_{\hat{e}}^{\mathrm{Kin}} = 7.16\times10^4\times\frac{k_1}{(k_2-1)^3\times b\times c_0}. \quad (4.23)$$

The full stability of the isoelectronium, $\tau = \infty$, therefore, requires the *exact* value $k_2 \equiv 1$, which, in turn, implies $E_0 \equiv 0$.

The above derivation characterizes the *limiting assumption of a fully stable isoelectronium in nonrelativistic approximation.* By comparison, the Cooper pair under the same derivation *is not* permanently stable because its binding energy is very small, yet finite [7b], thus implying a large yet finite meanlife. Also by comparison, the π^0 *cannot* be stable, and actually has a very small meanlife, evidently because the constituents are a particle-antiparticle pair and, as such, they annihilate each other when bound at short distances.

Another important information of this section is that *the isoelectronium is sufficiently small in size to be treated as a single quasi-particle.* This property will permit rather important simplifications in the isochemical structure of molecules studied in the next sections.

By comparison, the Cooper pair has a size much bigger than that of the isoelectronium [7b, 7c]. This property is fundamental to prevent that the Cooper pair takes the role of the isoelectronium in molecular bonds, i.e., even though possessing the same constituents and similar physical origins, the isoelectronium and the Cooper pair are different, non-interchangeable, hadronic bound states.

The lack of binding energy of the isoelectronium is perhaps the most important information of this section. In fact, it transfers the representation of the binding energy of molecular bonds to the motion of the isoelectronium in a molecular structure, as studied in the next sections.

A novelty of isochemistry over quantum chemistry is that the mutual distance (charge diameter) between the two isoelectrons in the isoelectronium could, as a limited case, also be identically null, that is, the two isoelectrons could be superimposed in a singlet state. Rather than being far fetched, this limit case is intriguing because it yields the value $-2e$ for the charge of the isoelectronium, the null value of the relative kinetic energy, and an identically null magnetic field. This is a perfectly diamagnetic state, which evidently allows a better stability of the isochemical bond as compared to a quasi-particle with non-null size.

Note that, if conventionally treated (i.e., represented on conventional spaces over conventional fields), the nonunitary image of model (4.5) would yield *noninvariant numerical results* which, as such, are unacceptable (Sect. 1.7). This occurrence mandates the use of the covering isochemistry and related isomathematics which assures the achievement of invariant results.

Note also that the main physical idea of isounit (4.7) is the *representation of the overlapping of the wavepackets of the two electrons under the condition of recovering conventional quantum chemistry identically whenever such overlapping is no longer appreciable.* In fact, for sufficiently large relative distances, the volume integral of isounit (4.7a) is null, the exponential reduces to I, Eq. (4.7d), the nonunitary transform becomes conventionally unitary, and quantum chemistry is recovered identically.

It is also important to see that, under transform (4.7a), model (4.5) is implemented with interactions which are: *nonlinear*, due to the factor $\psi/\hat{\psi}$ in the exponent; *nonlocal*, because of the volume integral in (4.7a); and *nonpotential*, because not represented by a Hamiltonian.

We finally note that the explicit form of the isotopic element $\hat{T}$, Eq. (4.7b), emerges in a rather natural way as being *smaller than one* in absolute value, Eq. (4.7c), i.e.,

$$|\hat{T}| = |1 - N \times \psi/\hat{\psi}| \ll 1. \tag{4.24}$$

As pointed out in Chapter 3, this property alone is sufficient to *guarantee* that all slowly convergent series of quantum chemistry converge faster for isochemistry.

4. Isochemical Model of the Hydrogen Molecule with Stable Isoelectronium

We are now sufficiently equipped to initiate the study of the *isochemical model of the hydrogen molecule*, first submitted by Santilli and Shillady in Ref. [5] (see Figs. 4.3, 4.4 and 4.5). In this Section we shall begin the study by identifying the equation of structure of the H-molecule under the limit assumption that the isoelectronium is perfectly stable at short distances, namely, that the two valence electrons are permanently trapped inside the hadronic horizon, resulting in the main features derived in the preceding section

$$\begin{gathered}\text{mass} \approx 1\ \text{MeV}, \quad \text{spin} = 0,\\ \text{charge} = 2 \times e, \quad \text{magnetic moment} \approx 0,\end{gathered} \tag{4.25a}$$

$$\begin{gathered}\text{radius} = r_c = b^{-1} = 6.8432329 \times 10^{-11}\text{cm} =\\ = 0.015424288\ \text{bohrs} = 0.006843\ \text{Å}.\end{gathered} \tag{4.25b}$$

The more realistic case when the isoelectronium is unstable at such small distances is studied later on in this chapter, where we shall also reach an essentially exact representation of the characteristics of the hydrogen molecule.

The main reason for assuming the isoelectronium to be stable at short distances with characteristics (4.25) is that such an approximation permits rather major structural simplifications, most notably, the transition, from the conventional hydrogen molecule (which is a *four-body system*), to the isochemical model of this section (which is a *three-body system*, Fig. 4.3). By recalling that four-body systems do not admit an exact solution, while restricted three-body systems do admit an exact analytic solution, the implications of the approximate model of this section are sufficient to warrant an inspection.

Our foundation is the conventional quantum model of H_2 molecule [1],

$$\begin{gathered}\left(\frac{1}{2\mu_1} p_1 \times p_1 + \frac{1}{2\mu_2} p_2 \times p_2 + \right.\\ \left. + \frac{e^2}{r_{12}} - \frac{e^2}{r_{1a}} - \frac{e^2}{r_{2a}} - \frac{e^2}{r_{1b}} - \frac{e^2}{r_{2b}} + \frac{e^2}{R}\right) \times |\psi\rangle = E \times |\psi\rangle.\end{gathered} \tag{4.26}$$

Our task is that of subjecting the above model to a transform

$$U \times U^\dagger|_{r \approx r_c} = \hat{I} = 1/\hat{T} \neq I, \tag{4.27}$$

which is nonunitary only at the short mutual distances

$$r_c = b^{-1} = r_{12} \approx 6.8 \times 10^{-11}\text{cm}, \tag{4.28}$$

and becomes unitary at bigger distances,

$$U \times U^\dagger|_{r \leq 10^{-10}\text{cm}} \neq I, \quad I_{r \gg 10^{-10}\text{cm}} = I. \tag{4.29}$$

This guarantees that our isochemical model coincides with the conventional model everywhere except for small contributions at small distances.

Assumption (4.29) also guarantees that *the conventional energy level of the individual hydrogen atoms are not altered.* In other words, assumption (4.29) realizes the main conception of this monograph, the exact character of quantum mechanics for the structure of *one* hydrogen atom, and its insufficiency for *two* hydrogen atoms bounded into the hydrogen molecule (Chapter 1).

The Hilbert space of systems (4.26) can be factorized in the familiar form (in which each term is duly symmetrized or antisymmetrized) as in Refs. [1]

$$|\psi\rangle = |\psi_{12}\rangle \times |\psi_{1a}\rangle \times |\psi_{1b}\rangle \times |\psi_{2a}\rangle \times |\psi_{2b}\rangle \times |\psi_R\rangle, \tag{4.30a}$$

$$\mathcal{H}_{\mathrm{Tot}} = \mathcal{H}_{12} \times \mathcal{H}_{1a} \times \mathcal{H}_{1b} \times \mathcal{H}_{2a} \times \mathcal{H}_{2b} \times \mathcal{H}_R. \tag{4.30b}$$

The nonunitary transform we are looking for shall act only on the r_{12} variable while leaving all others unchanged. The simplest possible solution is given by

$$U(r_{12})\times U^{\dagger}(r_{12}) = \hat{I} = \exp\left[\frac{\psi(r_{12})}{\hat{\psi}(r_{12})}\int dr_{12}\hat{\psi}^{\dagger}(r_{12})_{1\downarrow}\times\hat{\psi}(r_{12})_{2\uparrow}\right], \tag{4.31}$$

where the ψ's represent conventional wavefunctions and the $\hat{\psi}$'s represent isowavefunctions.

As an alternative yielding the same results, one can transform short-range terms (isochemistry), and add un-transformed long-range terms (quantum chemistry), resulting in the radial equation

$$\left(-\frac{\hbar^2}{2\times\mu_1}\hat{T}\times\nabla_1\times\hat{T}\times\nabla_1 - \frac{\hbar^2}{2\times\mu_2}\hat{T}\times\nabla_2\times\hat{T}\times\nabla_2 + \right.$$
$$\left. + \frac{e^2}{r_{12}} - \frac{e^2}{r_{1a}} - \frac{e^2}{r_{2a}} - \frac{e^2}{r_{1b}} - \frac{e^2}{r_{2b}} + \frac{e^2}{R}\right)\times|\hat{\psi}\rangle = E\times|\hat{\psi}\rangle. \tag{4.32}$$

By recalling that the Hulten potential behaves at small distances like the Coulomb one, Eq. (4.32) becomes

$$\left(-\frac{\hbar^2}{2\times\mu_1}\times\nabla_1^2 - \frac{\hbar^2}{2\times\mu_2}\times\nabla_2^2 - V\times\frac{e^{-r_{12}\times b}}{1-e^{-r_{12}\times b}} - \right.$$
$$\left. - \frac{e^2}{r_{1a}} - \frac{e^2}{r_{2a}} - \frac{e^2}{r_{1b}} - \frac{e^2}{r_{2b}} + \frac{e^2}{R}\right)\times|\hat{\psi}\rangle = E\times|\hat{\psi}\rangle. \tag{4.33}$$

The above equation does indeed achieve our objectives. In fact, it exhibits a *new explicitly attractive force between the neutral atoms of the hydrogen molecule, which force is absent in conventional quantum chemistry.* The equation also explains the reasons why the H_2 molecule admits only *two* H-atoms. As we shall see in the remaining sections, Eq. (4.33) also permits essentially exact representations of the binding energy and other molecular characteristics, yields much faster convergence of series with much reduced computer times, and resolves other insufficiencies of conventional models.

5. Exactly Solvable, Three-Body, Isochemical Model of the Hydrogen Molecule

Our isochemical model of the hydrogen molecule, Eqs. (4.33), can be subjected to an additional simplification, which is impossible for quantum chemistry. In our isotopic model, the two isoelectrons are bonded together into a single state we have called isoelectronium. In particular, the charge radius of the latter is sufficiently small to permit the values (see Fig. 4.3)

$$r_{12} \leq r_{1a}, \text{ and } r_{1b}, \quad r_{12} \approx 0, \tag{4.34a}$$

$$r_{1a} \approx r_{2a} = r_a, \quad r_{1b} \approx r_{2b} = r_b. \tag{4.34b}$$

Moreover, the H-nuclei are about 2,000 times heavier than the isoelectronium. Therefore, our model (4.33) can be reduced to a *restricted three body problem* similar to that possible for the conventional H_2^+ *ion* [1], but *not* for the conventional H_2 molecule.

Such a restricted model essentially consists of two H-protons at rest at a fixed mutual distance plus the isoelectronium moving around them in the *oo*-shaped orbit of Fig. 4.4, according to the structural equation

$$\left(-\frac{\hbar^2}{2\mu_1} \times \nabla_1^2 - \frac{\hbar^2}{2\mu_2} \times \nabla_1^2 - V \times \frac{e^{-r_{12}b}}{1-e^{-r_{12}b}} - \right.$$
$$\left. - \frac{2e^2}{r_a} - \frac{2e^2}{r_b} + \frac{e^2}{R} \right) \times |\hat{\psi}\rangle = E \times |\hat{\psi}\rangle. \tag{4.35}$$

Under the latter approximation, the model permits, for the first time, the achievement of an *exacts solution for the structure of the H_2 molecule*, as it is the case for the H_2^+ ion or for all restricted three-body problems. This solution will be studied in Chapter 6 via variational methods. The exact analytic solution has not been studied at this writing, and its study is here solicited by interested colleagues. At this introductory level we only limit ourselves to a few comments.

Note that *the above exact solution of the hydrogen molecule is only possible for the case of the isoelectronium fully stable at short mutual distances.* In fact, for the case of the mutual distance of the valence electrons no longer restricted to 1 fm, the model is a full *four-body structure*, which, as such, admits no exact solution.

Note also that model (4.35) is the isochemical model of the H_2 molecule inside the hadronic horizon. The matching representation *outside* the hadronic horizon is presented in the next section.

Note also that the above restricted three-body model can be used for the study of the bonding of an H-atom to another generic atom, such as HO, thus permitting, again for the first time, novel exact calculations on the water as HOH, namely, as two intersecting isotopic bonds HO and OH, each admitting an exact solution, with possible extension to molecular chains, and evident extensions to other molecules.

Readers interested in studying model (4.35) should keep in mind that *the rest energy of the isoelectronium is unknown at this writing, thus being a free parameter suitable for fitting experimental data.* More specifically, in Eq. (4.35) we have *assumed* from Sect. 4.3 that

$$m_{\text{isoelectronium}} = 2 \times m_{\text{electron}}. \tag{4.36}$$

However, the results of Sect. 4.3 are approximate. In particular, they hold under *the assumption that the isoelectronium has no internal binding energy.* Such an assumption was made for the specific purpose of proving that nonpotential forces represented with the isounit can indeed yield a bound state. In particular, the assumption was suggested by the need to represent Pauli's exclusion principle without the introduction of a potential.

However, such a view may be solely valid at molecular distances of valence electrons, and not necessarily at short distances. As a result, the isoelectronium may indeed have an internal binding energy, that is, it can have internal forces derivable from a potential in addition to the nonpotential forces without binding energy of hadronic chemistry, as outlined in Fig. 4.4.

This is due to the fact that the structure of the isoelectronium implies *three* acting forces: *one repulsive* Coulomb force due to the same changes, plus *two attractive* forces due to the two pairs of opposite magnetic polarities in singlet configuration. The latter two attractive forces may overcome the repulsion due to the change beginning at distances of the order of one Fermi, resulting in a conceivable net attractive force derivable from a potential.

Under the latter conditions, the isoelectronium would indeed have a *negative* binding energy, resulting in the unknown value

$$m_{\text{isoelectronium}} < 2 \times m_{\text{electron}}. \tag{4.37}$$

The understanding is that the case $m_{\text{isoelectronium}} > 2 \times m_{\text{electron}}$ is impossible.

The unknown character of the isoelectronium mass alters considerably the perspective of restricted model (4.35). As we shall see in Chapter 6, it is possible to prove via variational and other methods that *model (4.35) under assumption (4.36) does not admit exact solutions accurately representing the binding energy of the hydrogen molecule.* However, under the use of the isoelectronium mass free for fitting experimental data, the situation may be different.

Another information which should not be assumed to be exact is the *size of the isoelectronium*, Eq. (4.19). In fact, as stressed in Sect. 4.3, such a value too must be assumed to be an *upper boundary value.* In model (4.35) the isoelectronium is assumed to be point-like. However, the model can be first extended via Eq. (4.35) for a stable isoelectronium with a *fixed* unknown radius

$$r_c = b^{-1} \leq 6.8 \times 10^{-11}\text{cm}. \tag{4.38}$$

A second extension of model (4.35) should also be taken into consideration, that in which

$$r_c = b^{-1} \geq 6.8 \times 10^{-11}\text{cm}, \tag{4.39}$$

because, as stressed in Sects. 4.2 and 4.3, any assumption that the isoelectronium ceases to exist at distances bigger than 10^{-11} cm would imply a violation of Pauli's exclusion principle.

As a matter of fact, the assumed mass (4.36) is more in line with assumption (4.39), than with assumption (4.38), again, to prevent the existence at large mutual distances of the valence electrons of attractive internal potential forces with a binding energy which would alter conventional atomic structures.

Even though, admittedly, *the size of the isoelectronium is variable in the physical reality*, its average into a constant value may have meaning, of course, as a first approximation.

A third quantity of model (4.35) deserving inspection is the experimental value of the bond length, which is generally referred to the distance between the two nuclei R. In principle, such a distance is expected to be altered by a fully stable isoelectronium. Therefore, a solution of model (4.35) in which R is fitted from the experimental data is indeed meaningful, of course, as a first approximation.

In conclusion, in both, the four-body model (4.33) and the restricted three-body model (4.35), we have *three quantities which, in principle, can be assumed to be unknown and, therefore, should be derived from the fit of experimental data: 1) the mass of the isoelectronium; 2) the size of the isoelectronium; and 3) the bond length.*

There is no doubt that an exact analytic solution of model (4.35) suitable to represent the binding energy of the hydrogen is permitted by the above three free fits with intriguing implications for all H-bonds whose study is left to interested researchers.

6. Isochemical Model of the Hydrogen Molecule with Unstable Isoelectronium

In this section we review the study of Ref. [5] on the solution of the restricted isochemical model of the hydrogen molecule, Eq. (4.35) and Fig. 4.3, via conventional variational methods used in chemistry, under the assumption that the isoelectronium has characteristics (4.25). As we shall see, these studies have achieved an essentially exact representation of experimental data on the hydrogen molecule, including its binding energy and bond length, for the first time from exact first principles without ad hoc adulterations.

For historical papers in chemistry connected to our model, see Refs. [6]. Representative, more recent papers with technical connections to our study as outlined below are given by Refs. [8 – 38].

The possibilities that the mass of the isoelectronium be smaller than 2×mass of electron and its radius be bigger than 6.8×10^{-11} cm will not be considered in this section.

For this purpose we first note that the solution of the full model with the Hulten potential $e^{-rb}/(1 - e^{-rb})$ where $r_c = b^{-1}$ is the size of the isoelectronium, implies rather considerable technical difficulties. Therefore, we shall study model (4.35) under an *approximation* of the Hulten potential given by one Gaussian of the type

$$\frac{e^{-rb}}{1 - e^{-rb}} \approx \frac{1 - Ae^{-br}}{r}, \tag{4.40}$$

with A a constant identified below.

It is known that a linear combination of sufficient number of Gaussians can approximate any function. Therefore, the achievement of an essentially exact representation of molecular data via approximation (4.40) will evidently persist under the full use of the Hulten potential.

Recall from Sect. 4.3 that the *stable* character of the isoelectronium is crucially dependent on the use of the attractive Hulten potential, which "absorbs" repulsive Coulomb forces at short distances resulting in attraction. Therefore, the weakening of the Hulten potential into the above Gaussian form has the direct consequence of turning the isoelectronium into an *unstable* state.

In this and in the following sections, we shall therefore study an isochemical model of the hydrogen molecule which is somewhat intermediary between the conventional chemical model and the isochemical model with a fully stable isoelectronium.

It should be indicated that the terms "unstable isoelectronium" should be referred as the period of time in which the two valence electrons remain within the hadronic horizon of 6.8×10^{-11} cm. The same terms *should not* be interpreted to the fact that the isoelectronium does not exist outside the hadronic horizon, because the latter view implies a number of inconsistencies, such as possible violation of Pauli's exclusion principle, acquisition by molecules of ferromagnetic character, *etc.*

The main objective of this section is to show the achievement of the exact representation of molecular characteristics even for the case of one Gaussian approximation (4.40). The question whether the isoelectronium is stable or unstable evidently depends on the amount of instability and its confrontation with experimental data, e.g., on magnetic susceptibility. As such, the issue will be addressed theoretically and experimentally in a future paper.

Under the above assumption, our first step is the study of model (4.35) in an exemplified Coulomb form characterized by the following equation, hereon expressed in atomic units (a.u.)

$$H \times \Psi = \left(-\frac{1}{2}\nabla^2 - \frac{2}{r_a} - \frac{2}{r_b} + \frac{1}{R} \right) \times \Psi, \tag{4.41}$$

where the differences from the corresponding equation for the H_2^+ ion [1] are the replacement of the reduced mass $\mu = 1$ with $\mu = 2$, and the increase in the electric charge from $e = 1$ to $e = 2$.

The standard method for solving the above equation is the following. The variational calculation is set up in matrix algebra form in a nonorthogonal basis set S which has been normalized to 1. The metric of this non-orthogonal system of equations S is used to set up the orthogonal eigenvalue problem and the eigenvalues are sorted to find the lowest value. H and S are Hermitean matrices. E is a diagonal matrix with the energy eigenvalues

$$HC = ESC; \quad \text{define } C = S^{-\frac{1}{2}}C', \text{ then } HS^{-\frac{1}{2}}C' = ES^{-\frac{1}{2}}C', \tag{4.42a}$$

$$(S^{-\frac{1}{2}}HS^{+\frac{1}{2}}C') = H'C' = E(S^{-\frac{1}{2}}SS^{-\frac{1}{2}})C' = EC', \tag{4.42b}$$

where the last equation is obtained by multiplying the first equation from the left by $S^{-\frac{1}{2}}$, and use the unitary property that $S^{-\frac{1}{2}} = S^{+\frac{1}{2}}$ to form an orthogonal eigenvalue problem. Finally we solve for C' by diagonalizing H' and obtain $C = S^{-\frac{1}{2}}C'$.

Here the basis is formed from contracted basis sets Φ, which are fixed linear combinations of Gaussian spheres χ fitted to real shapes of spherical harmonic functions. The eigenvector column in C gives the basis coefficients of the molecular orbitals according to the expression

$$\Psi_i = \sum_{c_{i,j}} (j : \Psi_j = \sum_{a_{j,k}} \Psi_k; \; \chi = \left(\frac{2\alpha}{\pi}\right)^{3/4} \exp[-(\alpha - A)^2] = /\alpha, A). \tag{4.43}$$

The problem of how to form a sharp cusp on a $1s$ orbital is solved to a practical extent by using up to six Gaussians; here we use the very best "least-energy" 1s orbital from Pople's group [18]. In this problem the s-, p-, d- and f-orbitals are polarization functions that merely serve to evaluate the effect of other angular components on the $1s$ orbitals which are the main terms of the $1s$-sigma bond in H_2.

Gaussian orbitals can easily be scaled to screened nuclear charge values by multiplying the Gaussian exponents by the square of the scaling factor (in effect, shrinking the space of the H-atom model) followed by renormalization of the linear combination of Gaussians. In this work the scaling constant of the $1s$ orbitals was optimized to 1.191 and the 2-, 3- and 4-shell orbitals optimized as scaled shells rather than optimizing each orbital individually.

As the orbitals were optimized using parabolic fitting to three energy values as a function of the scaling value, it became apparent that the bond length of the three-body model is much shorter than the usual value of 1.4011 Bohr (= 0.74143 Angströms). Thus, the bond length was re-optimized after optimization of the scaling for each principle shell. The scaling constants and the orbital contractions are Angströms at an energy of -7.61509174 Hartrees (= -207.2051232 eV) where the achievement of an exact representation of the binding energy is studied in detail.

Although a large basis set of $1s$, $2s$, $2p$, $3s$, $3p$, $3d$, $4s$, $4p$, and $4f$ orbitals was used, this variational energy is probably higher than the exact solution of the type used by Bates, Ledsham and Stewart [12]. However, the energy of the 6-gaussian ($6G$) Least-Energy $1s$ function [18] is -0.499946 Hartrees for the H atom so the energy quoted here should be within 0.001 Hartrees of the exact solution.

While it is expected that a collapsed isoelectronium pair would be even more unstable than a collapsed positronium quasi-particle due to the repulsive interaction of the electrons, this three-body model of H_2 predicts over 6 Hartrees added molecular stability and a substantial decrease in bond length. The $E(1)$ value of the electronium-pair of some

-11.473164 Hartrees is lower than the total energy of the molecule due to the repulsion of the proton-nuclei from the $1/R$ term of the Hamiltonian.

7. Gaussian Approximation of the Isochemical Model of the Hydrogen Molecule as a Four-Body System

As indicated earlier, it is possible that the valence electrons bond themselves into the isoelectronium not in a permanent fashion, but rather in a statistical fashion, with only a percentage of their time in a bonded state, in which case the restricted three-body model is evidently insufficient. In this section we review the studies of Ref. [5] on the full four-body isochemical model of H_2, which model also permits the achievement of an exact representation of the binding energy from first principles without adulterations (see Fig. 4.7).

A considerable effort has been made since the time of Hylleraas [20] in the 1930's to find a way to calculate the last 2% of the binding energy of molecules. Boys [22] derived a form of "configuration interaction" which offered exact variational solutions, but this proved to be very slowly convergent and only applicable to small molecules. Moller-Plesset perturbation developed by Pople *et. al.* [23] is popular today, but studies up to eighth order have been shown not to converge after huge expense in computer time. Linked-cluster diagrams by Bartlett *et. al.* [24]) multiconfiguration-self-consistent-field (MCSCF) calculations by Schaefer *et. al.* [25] and Goldstone-Bruekner-Feynman diagrammatic perturbation by Kelly [26] have all been shown to require very large computer resources, are limited to small molecules and sometimes fail to give even negative binding energies as shown by Goddard [27] for Cr_2.

All these slow and expensive methods seem to share one common feature, the use of high energy empty "virtual" orbitals from a ground state calculation, usually of Hartree-Fock-Roothaan type, to improve the representation of the ground state.

One might ask how it is possible to lower the energy by using higher energy wavefunctions. The fact that some energy lowering is found suggests electron dynamics is indeed complicated, and the rate of convergence of this method is quite slow.

The method adopted here is to use the usual Hartree-Fock-Roothaan self-consistent-field equations [1] (which also has some formal flaws such as the self-interaction terms [27]), and question the form of the Coulomb interaction of the electron.

Note that reducing the values of the Coulomb integrals will lower the energy by reducing the electron-electron repulsion while reducing the exchange terms will raise the energy, but the 1/2 factor reduces the effect

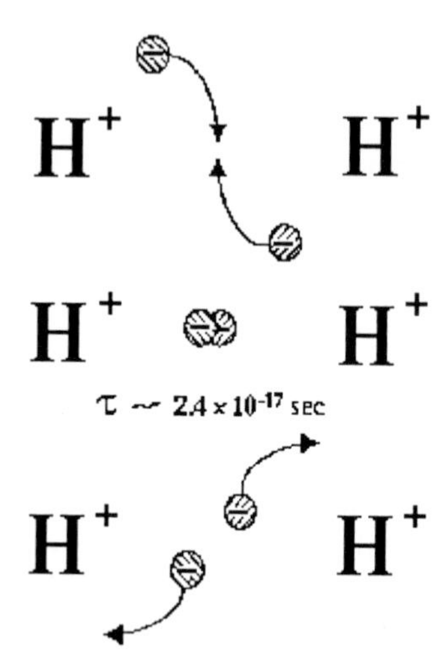

Figure 4.7. A schematic view of the *isochemical model of the hydrogen molecule with unstable isoelectronium* due to the weakening of the Hulten potential and consequential relaxation of the infinite lifetime, as a view complementary to that of Fig. 4.4 with stable isoelectronium. In this case the notion of isoelectronium essentially represents a tendency of pairs of valence electrons to correlate-bond in singlet states at short distances. The use of isochemistry, rather than conventional chemistry, is necessary because even this weaker form of the isoelectronium, as well as all screenings of the Coulomb potential at large, are nonunitary images of conventional Coulomb settings, as established in Chapter 1. Therefore, all these models require a representation on isospaces over isofields for their invariant formulation. In the text, we present the projection of such an invariant formulation on conventional spaces over conventional fields for simplicity.

of the exchange terms. Thus, a reduction of the value of the integrals will lower the energy. Note that Goddard [27] has already recommended reducing the atomic self-energy by subtracting 1.39 eV from Hartree-Fock exchange integrals in the cases of Cr_2 and Mo_2. In this way, we reach the expressions

$$FC = ESC; \quad F_{i,j} = H_{i,j} + \sum_{k,l} P_{k,l}[(i,j|k,l) - \frac{1}{2}(i,k|j,l)], \tag{4.44a}$$

$$(i,j|k,l) = \iint \chi_i(1)\chi_j(1)\frac{1}{r_{12}}\chi_k(2)\chi_l(2)\, d\tau_1 d\tau_2, \tag{4.44b}$$

$$P_{i,j} = 2\sum_n c_{n,i}c_{n,j} \quad \text{(sum } n \text{ only over occupied orbitals).} \tag{4.44c}$$

The 1995 paper on Cooper pairs by Santilli and Animalu [7c] invokes the non-local hadronic attractive force first identified in the π^0-meson

[7a] as applied to singlet-paired electrons which form a boson particle. After using a non-local isotopic nonlinear transformation, the hadronic attraction was transformed back to real-space and modeled resulting into an attractive force which overcomes the repulsive Coulomb force. In turn, the latter occurrence constitutes the physical-chemical origin in the use of a suitably screened Coulomb potential for the binding energy.

Examination of the original 1978 paper on positronium collapse by Santilli [7a], reveals that the Hulten potential is not necessarily a unique representation of the hadronic force; since a linear combination of similar potentials could be used to represent the same hadronic bound state, provided that they characterize an attractive force among the electrons capable of overcoming their Coulomb repulsion.

This work assumes that until matrix elements of a two-Gaussian-screened-Coulomb potential can be used to approximate the real-space form of the hadronic attraction. This form has the important property that it can be merged with the general case of the four-center Coulomb or exchange integral derived by Shavitt [16] using the famous Gaussian transform technique.

$$\frac{1}{r_{12}} = \sqrt{\frac{1}{\pi}} \int_0^\infty s^{-\frac{1}{2}} \exp[-s r_{12}^2] ds, \tag{4.45a}$$

$$1 + \frac{p+q}{pq} s = \frac{1}{1-t^2}. \tag{4.45b}$$

For future reference, note that this transform already has a pole at the lower limit where $s = 0$. This pole was removed at the last step by a change in variable, given as Eqs. (4.45). Shavitt was able (as a former graduate student of S.F. Boys) to show that the Gaussian transform technique reproduced the formula previously derived by Boys [6] in 1950 using electrostatic arguments. The Gaussian-lobe basis SCF programs by Shillady [8, 28] and others.

It is important to note that the formula is completely general in orientation of four Gaussian sphere lobe-orbitals as well as the distance between two electrons. As modified for description of correlation of two electrons, *such a general formula can describe angular correlation as well as distance interaction.* Thus, matrix Coulomb repulsion to model the real-space form of the hadronic attraction of two electrons.

Well-founded admiration for Shavitt's work in deriving the Coulomb interaction was rekindled as his derivation was checked. This work added the Gaussian screening as $\exp[-r^2]/r$ so that the special properties of Gaussians could be used, especially the property that polar coordinates readily separate into factorable x, y, z components. The goal is to evaluate the two-electron four-center matrix elements of the Gaussian-screened-Coulomb potential as shown below,

$$Y(r) = \frac{1 - 2\exp[-\alpha r^2]}{r}. \tag{4.46}$$

Intriguingly, the Gaussian exponent carried through the original derivation of the Coulomb interaction by resorting to a well known auxiliary function F_0 which has been studied by Shillady [8, 28] and others. Since both $s^{\frac{1}{2}}$ and $(s+\alpha)^{\frac{1}{2}}$ occur in the denominator of the screened-Coulomb form, two poles occur in the integral. A change of variable absorbs the term

$$1 + \frac{p+q}{pq}(s+\alpha) = \frac{1}{1-t^2}, \tag{4.47}$$

while the pole due to $(s+\alpha)^{-1/2}$ shifts the other pole at $s^{-\frac{1}{2}}$ to the lower limit of the integral. A smooth spike is evident at the lower value of the integration using a 70 point Simpson's Rule Integration (two ranges are used with 20 points more closely spaced near the pole and 50 points for the remaining range.)

The above work was carried out using 64 bit double precision arithmetic which provides 14 significant figures. A simple offset (δ) of 1.0×10^{-15} has provided useful results with this simple offset to avoid numerical overflow. While this pole is a problem in need of a continuous function to integrate, numerical integration seems to handle this well to 14 significant figures, particularly since the routines used for the Coulomb integrals are known to be accurate only to 12 significant figures [28].

The area under the pole-spike is estimated as a narrow triangle upon a rectangle 1.0×10^{-15} wide with the height of the triangle set at 1.79940×10^{13} times the height of the point set 1.0×10^{-15} into the range of integration (the first Simpson point).

The present code for this screened-Coulomb integral is presently slower than the corresponding function used for the Coulomb integrals due to the 70 point Simpson integration [28], but the integrand is nearly flat after the spike at $s = 0.0$ so that portion of the integrand can be evaluated more rapidly with fewer points. The simple offset of the lower limit by 1.0×10^{-15} is adequate for this monograph.

$$\left(aA(1), bB(1) \left| \frac{\exp(-\alpha(r_{12})^2)}{r_{12}} \right| cC(2), dD(2) \right) = \tag{4.48a}$$

$$= \frac{2\pi^{5/2}}{pq\sqrt{p+q}} e^{[ab/(a+b)]\overline{AB}^2 - [cd/(c+d)]\overline{CD}^2} \times$$

$$\times \int_{\rho}^{1} e^{-[pq/(p+q)]\overline{PQ}^2 t^2} \left(\frac{[pq/(p+q)]t^2}{[pq/(p+q)]t^2 + \alpha(t^2-1)} \right)^{1/2} dt, \quad (4.48b)$$

$$\rho = \delta + \frac{\alpha\sqrt{p+q}}{pq+(p+q)\alpha}, \quad p = (a+b), \quad q = c+d, \quad \delta = 1.0 \times 10^{-15}, \quad (4.48c)$$

$$\text{pole} = (1.79940 \times 10^{13}) e^{-[pq/(p+q)]\overline{PQ}^2 \rho^2}. \quad (4.48d)$$

The new integral was incorporated into the same routine used to evaluate the usual Hartree-Fock-Roothaan SCF scheme except F_0 was supplemented by the new auxiliary function (4.46). The H_2 molecule was treated using the same fixed-nuclei method with a bond distance of 1.4011 Bohrs. A simple basis set of just one Least-Energy $6G$-$1s$ orbital [18] centered on each H-nucleus was used to test the new program "Santilli-Animalu-Shillady-Lobe" (SASLOBE), which is set to handle up to 512 contracted orbitals.

It must be stated that the energies given are now parametrically dependent on the Gaussian-screening constant as $E(\alpha)$. The energy is variationally bound to be above the true energy in a narrow range around the optimum value. It is extremely important to note that the energy is lowered using the new attractive hadronic term, but the optimum value is difficult to locate and "variational collapse" occurs when r_c is extended or reduced away from a shallow minimum in the energy.

In order to minimize the number of parameters in the model (only one, the Coulomb screening constant A) two equations were imposed on the Gaussian-function. First, the function was required to be equal to zero at some radial cutoff value r_c which is assumed to be the inverse of the b-variable of Sect. 5,

$$b = \frac{1}{r_c} = A\frac{\exp[-\alpha r_c^2]}{r_c}, \quad A = \exp[+\alpha r_c^2]. \quad (4.49)$$

Second, this radial cutoff value was used as sigma of the inverted Gaussian (radius at half-height),

$$A \exp\left[-\frac{\ln 2}{r_c^2} r_c^2\right] = \frac{A}{2}, \quad \alpha = \frac{\ln 2}{r_c}, \quad A = 2. \quad (4.50)$$

The upper boundary of the radius of the isoelectronium has been estimated in Sect. 4.3 to be about 0.6843×10^{-10} cm, which corresponds to 0.012931401 Bohrs. This radius does lower the Hartree-Fock-Roothaan energy noticeably for H_2, and further optimization of the pole-spike produced an SCF energy of -1.17446875 Hartrees with a cutoff radius of

0.0118447 Bohrs or 1.18447×10^{-10} cm using the minimum $1s$ basis. In conclusion, the fitted value of $b \equiv 1/r_c$ is reasonably close to the estimate value for the H_2 molecule. The minimum basis was later extended to $6G$-$1s$, $1G$-$2s$, $1G$-$2p$ for pole calibration.

Details of the exact representation of the binding energy via the above second method are presented in Appendix 4.B.

8. Summary of the Results

In order to demonstrate the advantage of the isochemical model using a Gaussian-screened-Coulomb attraction between electrons, a standard Boys-Reeves [22] calculation was carried out in Ref. [2]. This included all single- and double-excitations CISD from the ground state Hartree-Fock-Roothaan SCF orbitals for a 99×99 "codetor" [6] interaction. Only the 1s orbitals were optimized with a scaling of 1.191 for the Least-Energy $6G$-$1s$ orbitals, but the basis also included $1G$-$2s$, $2G$-$2p$, $1G$-$3s$, $1G$-$3p$, $3G$-$3d$, Thd $1G$-$4sp$ (tetrahedral array of four Gaussian spheres), and $4G$-$4f$ orbitals scaled to hydrogenic values as previously optimized [17].

Table 4.1. Summary of results for the hydrogen molecule.

Species	H_2	H_2 [a]	H_2
Basis screening			
$1s$	1.191	6.103	1.191
$2s$	0.50	24.35	0.50
$2p$	0.50	24.35	2.36
$3s$	0.34	16.23	*
$3p$	0.34	16.23	*
$3d$	0.34	-16.2^b	*
$4sp$	0.25	12.18	*
$4f$	0.25	12.18	*
Variational energy (a.u.)	*	−7.61509174	*
SCF energy (a.u.)	−1.12822497	*	−1.13291228
CI energy (a.u.)	−1.14231305	*	*
CINO energy (a.u.)	−1.14241312	*	*
SAS energy (a.u.)	*	*	−1.174444
Exact energy (a.u.) [30]	−1.174474	*	−1.174474
Bond length (bohr)	1.4011	0.2592	1.4011
Isoelectronium radius (bohr)	*	*	0.01124995

[a] Three-body Hamiltonian (5.1).
[b] The negative $3d$ scaling indicates five equivalent three-sphere scaled to 16.20 rather than "canonical" $3d$ shapes.

The additional basis functions provide opportunity to excite electrons to higher orbitals as is the standard technique in configuration interaction, somewhat contrary to the main hypothesis of this work, which is that there is an attractive hadronic force between electron pairs inside the r_c critical radius. The results of the above calculations are summarized in Table 4.1.

The Boys-Reeves C.I. achieved an energy of -1.14241305 Hartrees based on an SCF energy of -1.12822497 Hartrees. This was followed by one additional iteration of "natural orbitals" (CINO), in which the first order density matrix is diagonalized to improve the electron pairing to first order [29]. The fact that this procedure lowered the energy only slightly to -1.14241312 Hartrees (i.e., -7.0×10^{-7} Hartrees), indicates the 99-configuration representation is close to the lower energy bound using this basis set while the isochemistry calculation produced the exact energy with a comparatively much smaller basis set.

Since SASLOBE has only a n^7 routine for the necessary integral transformation instead of the most efficient n^5 algorithm ($\simeq n$ is the number of basis functions), the SASLOBE C.I. runs are somewhat slow and required about 20 hours on a 300 MFLOPS Silicon Graphics computer.

With more efficient routines, this time can be reduced to about three hours. However, the screened-Coulomb attraction method used a smaller basis and achieved lower energies in a few seconds. It is also estimated that careful spacing of fewer quadrature points in the new integral routine can certainly reduce the SASLOBE run times by a factor of 2 at least.

Therefore it is clear that calculations in hadronic chemistry are, conservatively, at least 1,000 times faster than a C.I. calculation, an occurrence fully similar to the corresponding case in hadronic *vs.* quantum mechanics.

Another estimate is that, since the new integral corrections require a little more time than the usual Coulomb integrals (but do not take any additional storage space), the computer run-times for an isochemistry calculation should only be about three times the run-times for the corresponding Hartree-Fock-Roothaan calculation in any given basis set.

The extension of the isochemical model of the H_2 molecule to other molecules is conceptually straightforward. In particular, the notion of isoelectronium essentially restricts all possible bonds to the established ones, as it is the case for the water molecule (see next chapter).

In order to generalize the underlying quantitative treatment to molecules containing H–to–F, the pole-spike was re-optimized to obtain 100% of the correlation energy below the SCF energy in the given basis set since the SCF energy here was not quite at the Hartree-Fock limit.

Table 4.2. Isoelectronium results for selected molecules.

Species	H_2	H_2O	HF
SCF-energy (DH) (a.u.)	-1.132800^a	-76.051524	-100.057186
Hartree-Fock^d (a.u.)			-100.07185^d
Iso-energy (a.u.)	-1.174441^c	-76.398229^c	-100.459500^c
Horizon R_c (Å)	0.00671	0.00038	0.00030
QMC energy^{d,e} (a.u.)	-1.17447	-76.430020^e	-100.44296^d
Exact non-rel. (a.u.)	-1.174474^f		-100.4595^d
Corellation (%)	99.9^b	91.6^b	103.8
SCF-dipole (D)	0.0	1.996828	1.946698
Iso-dipole (D)	0.0	1.847437	1.841378
Exp. dipole (D)	0.0	1.85^g	1.82^g
Time^h (min:s)	0:15.49	10:08.31	6:28.48

(DH^+) Dunning-Huzinaga (10S/6P), [6,2,1,1,1/4,1,1]+H_2P_1+3D1.
^{a}LEAO-6G1S + optimized GLO-2S and GLO-2P.
^{b}Relative to the basis set used here, not quite HF-limit.
^{c}Iso-energy calibrated to give exact energy for HF.
^{d}Hartree-Fock and QMC energies from Luchow and Anderson [33].
^{e}QMC energies from Hammond *et al.* [30].
^{f}First 7 sig. fig. from Kolos and Wolniewicz [34].
^{g}Data from Chemical Rubber Handbook, 61st ed., p. E60.
^{h}Run times on an O2 Silicon Graphics workstation (100 MFLOPS max.).

The energy obtained here results from the calibration of the pole-spike to the experimental value of HF, and is below the Quantum Monte Carlo (QMC) energy of Luchow and Anderson [33], which requires hours on a much larger computer, as compared to less than 10 minutes for this work. In fact, the run times for HF were about 8 CPU minutes on a 100 MFLOP Silicon Graphics O2 workstation.

The principal value of the pole (4.48d) was calibrated for 100% energy of HF, H_2O has two tight sigma bonds and two diffuse lone-pairs so a single compromise value is a good test of the method. In HF the F^- is nearly spherical so an average r_c value does a better job of describing the "correlation hole" of transient isoelectronium. The computed dipole moments are in excellent agreement with the experimental values. The use of the same pole value for H_2O and HF degrades the H_2 energy slightly. The results of our studies for H–to–F based molecules are summarized in Table 4.2.

A comparison of the above date (particularly those on computer times) with corresponding data obtained via conventional approaches is instructive.

9. Concluding Remarks

The fundamental notion of the new model of molecular bonds studied in this chapter [5] is the bonding at short distances of pairs of valence electrons from two different atoms into a singlet quasi-particle state we have called *isoelectronium*, which travels as an individual particle on an oo-shaped orbit around the two respective nuclei.

The isoelectronium and related methodology are then characterized by a covering of contemporary chemistry called *isochemistry*, which is the branch of the more general *hadronic chemistry* specifically constructed to represent closed-isolated systems with linear and nonlinear, local and nonlocal, and potential as well as nonpotential internal forces.

A main assumption is that linear, local, and potential interactions are sufficient for atomic structures since the atomic distances are much bigger than the size of the wavepackets of the electrons. However, in the transition to molecular structures we have the additional presence of nonlinear, nonlocal, and nonpotential effects due to the deep penetration of the wavepackets of valence electrons, which is essentially absent in atomic structures (Fig. 1.7).

The attractive short-range interactions needed to overcome the repulsive Coulomb force in the isoelectronium structure originate precisely from nonlinear, nonlocal, and nonhamiltonian effects in deep wave-overlappings; they are described by *hadronic mechanics* [3b]; and their invariant formulation is permitted by the recently achieved broadening of conventional mathematics called *isomathematics*.

Specific experimental studies are needed to confirm the existence of the isoelectronium, by keeping in mind that the state may not be stable outside a molecule in which the nuclear attraction terms bring the electron density to some critical threshold for binding, a feature we have called the "*trigger*."

Nonrelativistic studies yield a radius of the isoelectronium of 0.69×10^{-10} cm. This "horizon" is particularly important for isochemical applications and developments because outside the horizon the electrons repel one-another while inside the horizon there is a hadronic attraction.

The same nonrelativistic studies also predict that, as a limit case, the isoelectronium is stable within a molecule, although partially stable configurations also yield acceptable results. The question of the stability vs. instability of the isoelectronium inside the hadronic horizon must therefore also be left to experimental resolutions.

The understanding is that, when the restriction to the hadronic horizon is lifted, and molecular dimensions are admitted for the inter-electron distance, the isoelectronium must be stable, otherwise violations of Pauli's exclusion principle could occur. In this sense, the isoelectronium is a direct representation of Pauli's exclusion principle.

The foundations of the isoelectronium can be seen in a paper by Santilli [7a] of 1978 on the structure of the π^0-meson as a bound state of one electron and one positron. The latter model also illustrates the capability of hadronic mechanics vs. quantum mechanics. In fact, quantum mechanics *cannot* represent the π^0 as the indicated bound state of one electron and one positron because of numerous inconsistencies, such as: the inability to represent the rest energy of the π^0, which would require a "positive" binding energy, since the sum of the rest energies of the constituents is much smaller than the rest energy of the bound state; the impossibility to represent the charge radius of the π^0, which can only be that of the positronium for quantum mechanics; the lack of representation of the meanlife of the π^0; and other insufficiencies.

By comparison, *all* the above insufficiencies are resolved by hadronic mechanics, which permits the first quantitative, numerical representation of *all* characteristics of the π^0 as a bound state of one electron and one positron at short distances, including its spontaneous decay with the lowest mode $\pi^0 \rightarrow e^- + e^+$, which results in being the hadronic tunnel effect of the constituents [7a].

In particular, the indicated model of the π^0 contains the first identification of the attractive character of nonlinear, nonlocal, and nonhamiltonian interactions due to deep wave-overlappings in singlet coupling (and their repulsive character in triplet coupling).

The isoelectronium also sees its foundations in subsequent studies by Animalu [7b] of 1994 and Animalu and Santilli [7c] of 1995 on the construction of *hadronic superconductivity* for a quantitative representation of the structure of the Cooper pair. We have in this case an occurrence similar to the preceding one for the structure of the π^0. In fact, quantum mechanics can indeed represent superconductivity, but only via an *ensemble of Cooper pairs*, all assumed to be point-like. In particular, quantum mechanics simply cannot represent the structure of *one* Cooper pair, due to the divergent character of the Coulomb repulsion between the identical electrons of the pair.

Again Animalu-Santilli hadronic superconductivity did indeed resolve this insufficiency and permitted, for the first time, the achievement of a structure model of *one* Cooper pair in remarkable agreement with experimental data (Sect. 1.9). Hadronic superconductivity also shows predictive capacities simply absent in quantum mechanics, such as *the prediction of a new electric current mostly given by the motion of electron pairs, rather than the conventional electric current composed of individual electrons* (patent pending). Such a new hadronic current implies an evident reduction of the electric resistance due to the essentially null magnetic moment of the pair, as compared to the large magnetic moment of individual electrons, and its interactions with atomic electrons when moving within a conductor.

Note finally that the preceding hadronic model of the π^0 and of the Cooper pair are ultimately due to the capability of hadronic mechanics to eliminate divergencies at short distances, which is technically realized via the isotopies of the unit and related associative products of quantum mechanics

$$I \rightarrow \hat{I} = 1/\hat{T}, \tag{4.51a}$$

$$|\hat{I}| \gg 1, \quad |\hat{T}| \ll 1, \tag{4.51b}$$

$$A \times B \rightarrow A \times \hat{T} \times B, \tag{4.51c}$$

under which divergent or slowly convergent series can be evidently turned into rapidly convergent forms.

The tendency of identical valence electrons to bond into the isoelectronium is additionally confirmed by other evidence, such as ball lighting, which are composed by a very large number of electrons bonded together into a small region.

In summary, incontrovertible experimental evidence establishes that *electrons have the capability of bonding themselves at short distances contrary to their Coulomb repulsion.* Quantum mechanics simply cannot provide a scientific study of this physical reality. Hadronic mechanics resolved this impasse, by first identifying the conditions needed to achieve attraction, called "trigger," and then permitting quantitative numerical study of the bond.

The isoelectronium results in having deep connections with a variety of studies in chemistry conducted throughout the 20-th century [6, 8 – 38], and actually provides the physical-chemical foundations for these studies as well as their appropriate mathematical formulation for the invariance of the results.

In summary, *the isochemical model of molecular bonds submitted by Santilli and Shillady [5] is supported by the following conceptual, theoretical and experimental evidence:*

1) The isoelectronium introduces a new *attractive* force among the *neutral* atoms of a molecular structure which is absent in quantum chemistry and permits a quantitative understanding of the *strength* and *stability* of molecular bonds.

2) The isoelectronium permits an immediate interpretation of the reasons why the H_2 and H_2O molecules only admit *two* H-atoms.

3) The isoelectronium permits the achievement of a representation of the binding energy of the hydrogen molecule which is accurate to the *seventh digit*, thus allowing meaningful thermodynamical calculations.

4) The isoelectronium provides an explanation of the long known, yet little understood Pauli's exclusion principle, according to which electrons correlate themselves in singlet when on the same orbital without any exchange of energy, thus via a process essentially outside the representational capabilities of quantum mechanics and chemistry.

5) The isoelectronium is consistent with the known existence of superconducting electron-pairs which bond themselves so strongly to tunnel together through a potential barrier.

6) The isoelectronium provides a quantitative model for the explanation of electron correlations. Instead of a complicated "dance of electrons" described by positive energy excitations, the isochemistry explanation is that electrons are energetically just outside the horizon of a deep attractive potential well due to their wavefunctions overlapping beyond the critical threshold of the hadronic horizon.

7) The isoelectronium is consistent with the "Coulomb hole" studied by Boyd and Yee [35] as found from subtracting accurate explicitly-correlated wavefunctions from self-consistent-field wavefunctions. In our studies the "Coulomb hole" is re-interpreted as a "hadronic attraction."

8) The isoelectronium is also in agreement with the "bipolaron" calculated for anion vacancies in KCl by Fois, Selloni, Parinello and Car [36] and bipolaron spectra reported by Xia and Bloomfield [37].

9) The isoelectronium permits an increase of the speed in computer calculations conservatively estimated at least 1,000-fold, and prevents the inconsistent prediction that all molecules are ferromagnetic (see Chapter 7).

Moreover, *another remarkable result of this study is that the value of the radius of the isoelectronium,* 0.69×10^{-10} *cm, computed via dynamical equations in Sect. 4.3 has been fully confirmed by the independent calculations conducted in Sects. 4.6 and 4.7 via the Gaussian-lobe basis set, yielding 0.00671* Å.

We should also mention *preliminary yet direct experimental verifications of the isoelectronium offered by the ongoing experiments on photoproduction of the valence electrons in the helium indicating that electrons are emitted in pairs* [38]. The studies of this monograph warrant the systematic conduction of these experiments *specifically for the hydrogen molecule*, and the experimental finalization as to whether electrons are emitted in an isolated form or in pairs, including relative percentages of both emissions. If conducted below the threshold of disintegration of the isoelectronium, the proposed experiments can evidently provide final proof of the existence of the isoelectronium.

We should finally note that *the representation of the binding energy and other characteristics of the hydrogen molecule exact to the seventh digit first achieved in Ref. [5] constitutes the strongest experimental evidence to date on the insufficiency of quantum mechanics and the validity of the covering hadronic mechanics for the representation of nonlinear, nonlocal, and nonpotential, thus nonhamiltonian and nonunitary effects due to deep overlappings of the "extended wavepackets" of electrons with a "point-like charge structure."*

It is evident that all the above results provide scientific credibility for the isoelectronium, the related isochemical model of molecular bonds, and the underlying hadronic chemistry, sufficient to warrant systematic theoretical and experimental studies.

As shown in Chapter 7, a significant feature of the proposed novel isochemistry is not only the capability to provide accurate representations of experimental data in shorter computer times, but also the capability to predict and quantitatively treat *new industrial applications.*

APPENDIX 4.A: Isochemical Calculations for the Three-Body H_2 Molecule

This appendix contains a summary of the computer calculations conducted in Ref. [5] for the restricted three-body model of the hydrogen molecule according to isochemistry, Eq. (4.35), showing an exact representation of the binding energy. The calculations are based on the isoelectronium as per characteristics (4.25).

```
Gaussian-Lobe Program for Large Molecules
  set up by D. Shillady and S. Baldwin
     Richmond Virginia 1978-1997
       3 BODY H2 (Electronium)
```

ipear = 1, dt = 0.0, tk = 0.0, imd = 0, ntime = 60, mul = 1, iqd = 0, icor = 3, mdtim = 0, idb = 0.

ELECTRONIUM-PAIR CALCULATION

Atomic Core				Nuclear		Coordinates
		X		Y		Z
1.		0.000000		0.000000		0.000000
	Z1s =	6.103	Z2s =	24.350	Z2p =	24.350
	Z3s =	16.230	Z3p =	16.230	Z3d =	-16.200
	Z4p =	12.180	Z4f =	12.180		
1.		0.000000		0.000000		0.259200
	Z1s =	6.103	Z2s =	24.350	Z2p =	24.350
	Z4p =	12.180	Z4f =	12.180	Z3d =	-16.200

Basis Size = 50 and Number of Spheres = 142 for 2 Electrons.

Distance Matrix in Angströms:

	H	H
H	0.00000	0.13716
H	0.13716	0.00000

A-B-C Arcs in Degrees for 2 Atoms.
The Center of Mass is at Xm = 0.000000, Ym = 0.000000, Zm = 0.129600.

One-Electron Energy Levels:

E(1) =	-11.473116428176	E(26) =	28.974399759209
E(2) =	-4.103304982059	E(27) =	28.974400079775
E(3) =	-1.621066945385	E(28) =	31.002613061833
E(4) =	-1.621066909587	E(29) =	31.002614578175
E(5) =	0.735166320188	E(30) =	35.201145239721
E(6) =	3.760295564718	E(31) =	38.003259639003
E(7) =	3.760295673022	E(32) =	44.948398097510
E(8) =	4.206194459198	E(33) =	44.94839B118458
E(9) =	4.813241859203	E(34) =	52.259825531212
E(10) =	11.2330B0571453	E(35) =	57.732587951875
E(11) =	15.70B645318078	E(36) =	57.732589021798
E(12) =	15.708645469273	E(37) =	68.743644612501
E(13) =	18.535761604401	E(38) =	68.743644649428
E(14) =	18.535761951543	E(39) =	73.195648957615
E(15) =	19.329445299735	E(40) =	79.303486379907
E(16) =	19.329445306194	E(41) =	85.865499885249
E(17) =	19.644048052034	E(42) =	85.865531919077
E(18) =	24.002368034839	E(43) =	127.196518644932
E(19) =	24.002368621986	E(45) =	130.602186113463
E(20) =	24.076849036707	E(46) =	130.602190550265
E(21) =	24.076853269415	E(47) =	137.484863078186
E(22) =	24.574406183060	E(48) =	158.452350229845
E(23) =	26.836031180463	E(49) =	205.158233049979
E(25) =	27.860752485358	E(50) =	446.152984041077

epair Energy = -7.615091736818.

APPENDIX 4.B: Isochemical Calculations for the Four-Body H_2 Molecule

In this appendix we present a summary of the computer calculations conducted in Ref. [5] for the four-body model of the hydrogen molecule, Eq. (4.33), according to isochemistry by using only $6G$-$1s$ orbitals for brevity. The calculations are also based on the characteristics of the isoelectronium in Eqs. (4.25). Note, again, the exact representation of the binding energy at -1.174447 Hartrees.

```
Gaussian-Lobe Program for Large Molecules
set up by D. Shillady and S. Baldwin
Virginia Commonwealth University
Richmond Virginia
1978-1997
```

Test of SASLOBE on H2

SANTILLI-RADIUS = 0.01184470000000.
Cutoff = (A/r)*(exp(-alp*r*r)), A = 0.20E+01, alp = 0.49405731E+04.

Atomic Core		Nuclear				Coordinates
		X		Y		Z
1.		0.000000		0.000000		0.000000
	Z1s =	1.200	Z2s =	0.000	Z2p =	0.000
	Z3s =	0.000	Z3p =	0.000	Z3d =	0.000
	Z4p =	0.000	Z4f =	0.000		
1.		0.000000		0.000000		1.401100
	Z1s =	1.200	Z2s =	0.000	Z2p =	0.000
	Z3s =	0.000	Z3p =	0.000	Z3d =	0.000
	Z4p =	0.000	Z4f =	0.000		

Basis Size = 2 and Number of Spheres = 12 for 2 Electrons.

Distance Matrix in Angströms:

	H	H
H	0.00000	0.74143
H	0.74143	0.00000

The center of Mass is at: Xm = 0.000000, Ym = 0.000000, Zm = 0.700550.

Spherical Gaussian Basis Set:

No. 1	alpha =	0.944598E+03	at X = 0.0000	Y = 0.0000	Z = 0.0000 a.u.
No. 2	alpha =	0.934768E+02	at X = 0.0000	Y = 0.0000	Z = 0.0000 a.u.
No. 3	alpha =	0.798123E+01	at X = 0.0000	Y = 0.0000	Z = 0.0000 a.u.
No. 4	alpha =	0.519961E+01	at X = 0.0000	Y = 0.0000	Z = 0.0000 a.u.
No. 5	alpha =	0.235477E+00	at X = 0.0000	Y = 0.0000	Z = 0.0000 a.u.
No. 6	alpha =	0.954756E+00	at X = 0.0000	Y = 0.0000	Z = 0.0000 a.u.
No. 7	alpha =	0.1944598E+03	at X = 0.0000	Y = 0.0000	Z = 1.4011 a.u.
No. 8	alpha =	0.7934768E+02	at X = 0.0000	Y = 0.0000	Z = 1.4011 a.u.
No. 9	alpha =	0.40798123E+01	at X = 0.0000	Y = 0.0000	Z = 1.4011 a.u.
No. 10	alpha =	0.11519961E+01	at X = 0.0000	Y = 0.0000	Z = 1.4011 a.u.
No. 11	alpha =	0.37235477E+00	at X = 0.0000	Y = 0.0000	2 = 1.4011 a.u.
No. 12	alpha =	0.12954756E+00	at X = 0.0000	Y = 0.0000	Z = 1.4011 a.u.

Contracted Orbital No. 1:
0.051420*(1), 0.094904*(2), 0.154071*(3), 0.203148*(4), 0.169063*(5), 0.045667*(6).
Contracted Orbital No.2:
0.051420*(7), 0.094904*(8), 0.154071*(9), 0.203148*(10), 0.169063*(11), 0.045667*(12).

***** Nuclear Repulsion Energy in au = 0.71372493041182. *****

Overlap Matrix:

#		at-orb	1	2
1	H	1s	1.000	0.674
2	H	1s	0.674	1.000

S(-1/2) Matrix:

#		at-orb	1	2
1	H	1s	1.263	-0.490
2	H	1s	-0.490	1.263

H-Core Matrix:

#		at-orb	1	2
1	H	1s	-1.127	-0.965
2	H	1s	-0.965	-1.127

Initial-Guess-Eigenvectors by Column:

#		at-orb	1	2
1	H	1s	0.546	1.239
2	H	1s	0.546	-1.239

One-Electron Energy Levels: E(1) = -1.249428797385, E(2) = -0.499825553916.

(1,1/1,1) =	0.75003658795676
minus (1,1/1,1) =	0.08506647783478
total (1,1/1,1) =	0.66497011012199
(1,1/1,2) =	0.44259146066210
minus (1,1/1,2) =	0.02960554295227
total (1,1/1,2) =	0.41298591770983
(1,1/2,2) =	0.55987025041920
minus (1,1/2,2) =	0.01857331166211
total (1,1/2,2) =	0.54129693875709
(1,2/1,2) =	0.30238141375547
minus (1,2/1,2) =	0.01938180841827
total (1,2/1,2) =	0.28299960533720
(1,2/2,2) =	0.44259146066210
minus (1,2/2,2) =	0.02960554295227
total (1,2/2,2) =	0.41298591770983
(2,2/2,2) =	0.75003658795676
minus (2,2/2,2) =	0.08506647783478
total (2,2/2,2) =	0.66497011012199

Block No. 1 Transferred to Disk/Memory. The Two-Electron Integrals Have Been Computed.
Electronic Energy = -1.88819368266525 a.u., Dif. = 1.8881936827,
Electronic Energy = -1.88819368266525 a.u., Dif. = 0.0000000000.
Energy Second Derivative = 0.00000000000000.

e1a =	-2.499	e1b =	-2.499		
e2a =	0.611	e2b =	0.611	e2ab =	0.611

Iteration No. = 2, alpha = 0.950000.
Electronic Energy = -1.88819368266525 a.u., Dif. = 0.0000000000.
Total Energy = -1.17446875 a.u.
One-Electron Energy Levels: E(1) = -0.638764885280, E(2) = 0.561205833046

Reference State Orbitals for 1 Filled Orbitals by Column:

#		at-orb	1	2
1	H	1s	0.546	1.239
2	H	1s	0.546	-1.239

Dipole Moment Components in Debyes:
Dx = 0.0000000, Dy = 0.0000000, Dz = 0.0000000.
Resultant Dipole Moment in Debyes = 0.0000000.
Computed Atom Charges: Q(1) = 0.000, Q(2)= 0.000.
Orbital Charges: 1.000000, 1.000000.

Mulliken Overlap Populations:

#		at-orb	1	2
1	H	1s	0.597	0.403
2	H	1s	0.403	0.597

Total Overlap Populations by Atom:

	H	H
H	0.597222	0.402778
H	0.402778	0.597222

Orthogonalized Molecular Orbitals by Column:

#		at-orb	1	2
1	H	1s	0.422	2.172
2	H	1s	0.422	-2.172

Wiberg-Trindie Bond Indices:

#		at-orb	1	2
1	H	1s	0.127	0.127
2	H	1s	0.127	0.127

Wiberg-Trindie Total Bond Indices by Atoms:

	H	H
H	0.127217	0.127217
H	0.127217	0.127217

References

[1] Boyer, D.J.: *Bonding Theory*, McGraw Hill, New York (1968) [1a]. Hanna, M.W.: *Quantum Mechanical Chemistry*, Benjamin, New York (1965) [1b]. Pople, J.A. and Beveridge, D.L.: *Approximate Molecular Orbits*, McGraw Hill, New York (1970) [1c]. Schaefer, H.F.: *The Electronic Structure of Atoms and Molecules*, Addison-Wesley, Reading, Mass. (1972) [1d].

[2] Santilli, R.M. and Shillady, D.D.: *Ab Initio Hadronic Chemistry: Basic Methods*, Hadronic J. **21**, 633 (1998).

[3] Santilli, R.M.: Rendiconti Circolo Matematico Palermo, Suppl. **42**, 7 (1996) [3a]; Found. Phys. **27**, 625 (1997) [3b]; Nuovo Cimento Lett. **37**,545 (1983) [3c]; J. Moscow Phys. Soc. **3**, 255 (1993) [3d]; *Elements of Hadronic Mechanics*, Vol. **I** and **II**, Ukraine Academy of Sciences, Kiev, 2-nd Edition (1995) [3e]; *Isotopic, Genotopic and Hyperstructural Methods in Theoretical Biology,* Ukraine Academy of Sciences, Kiev (1996) [3f]; Intern. J. Modern Phys. **A14**, 3157 (1999) [3g].

[4] Kadeisvili, J.V.: Math. Methods Applied Sciences **19**, 1349 (1996) [4a]. Tsagas, Gr. and Sourlas, D.S.: Algebras, Groups and Geometries **12**, 1 and 67 (1995) [4b]. Tsagas, Gr. and Sourlas, D.: *Mathematical Foundations of the Lie-Santilli Theory*, Ukraine Academy of Sciences, Kiev (1992) [4c]. Lôhmus, J., Paal, E. and Sorgsepp, L.: *Nonassociative Algebras in Physics*, Hadronic Press (1995) [4d]. Vacaru, S.: *Interactions, Strings and Isotopies on Higher order Superspaces*, Hadronic Press (1998) [4c].

[5] Santilli, R.M. and Shillady, D.D.: Intern. J. Hydrogen Energy, **24**, 943 (1999).

[6] Lewis, G.N.: J. Am. Chem. Soc., **38**, 762 (1916) [6a]. Langmuir, I.: J. Am. Chem. Soc., **41**, 868 (1919) [6b]. Frost, A.A.: J. Chem. Phys., **47**, 3707 (1967) [6c]. Bates, D.R., Ledsham, K. and Stewart, A.L.: Phil. Tran. Roy. Soc. (London) **A246**, 215 (1954) [6e]. Wind, H.: J. Chem. Phys. **42**, 2371 (1965) [6f]. Boys, S.F.: Proc. Roy. Soc. (London) **A200**, 542 (1950) [6g].

[7] Santilli, R.M.: Hadronic J. **1**, 574 (1978) [7a]. Animalu, A.O.E.: Hadronic J. **17**, 379 (1994) [7b]. Animalu, A.O.E. and Santilli, R.M.: Intern. J. Quantum Chemistry **29**, 175 (1995) [7c].

[8] Whitten, J.L.: J. Chem. Phys. **39**, 349 (1963); Sambe, H.: J. Chem. Phys. **42**, 1732 (1965); Preuss, H.: Z. Naturforsch. **11a**, 823 (1956); Whitten, J.L. and Allen, L.C.: J. Chem. Phys. **43**, S170 (1965); Harrison, J.F.: J. Chem. Phys. **46**, 1115 (1967); Frost, A.A.: J. Chem. Phys. **47**, 3707 (1967).

[9] Le Rouzo, H. and Silvi, B.: Int. J. Quantum Chem. **13**, 297, 311 (1978); Nguyen, T.T., Raychowdhury, P.N. and Shillady, D.D.: J. Comput. Chem. **5**, 640 (1984).

[10] Shavitt, I.: *Methods in Computational Physics*, B. Alder (ed.), Academic Press, New York (1963).

[11] Born, M. and Oppenheimer, J.R.: Ann. Physik **84**, 457 (1927).

[12] Bates, D.R., Ledsham, K. and Stewart, A.L.: Phil. Tran. Roy. Soc. (London), **A246**, 215 (1954); Wind, H.: J. Chem. Phys. **42**, 2371 (1965).

[13] Boys, S.F.: Proc. Roy. Soc. (London) **A200**, 542 (1950).

[14] Whitten, J.L.: J. Chem. Phys. **39**, 349 (1963). Sambe, H.: J. Chem. Phys. **42**, 1732 (1965). Preuss, H.: Z. Naturforsch. **11a**, 823 (1956). Whitten, J.L. and Allen, L.C.: J. Chem. Phys. **43**, S170 (1965). Harrison, J.F.: J. Chem. Phys. **46**, 1115 (1967). Frost, A.A.: J. Chem. Phys. **47**, 3707 (1967).

[15] Le Rouzo, H. and Silvi, B.: Intern. J. Quantum Chem. **13**, 297, 311 (1978). Nguyen, T.T., Raychowdhury, P.N. and Shillady, D.D.: J. Comput. Chem. **5**, 640 (1984).

[16] Shavitt, I.: *Methods in Computational Physics*, B. Alder, (ed.), Academic Press, New York, (1963).

[17] Shillady, D.D. and Talley, D.B.: J. Computational Chem. **3**, 130 (1982). Shillady, D.D. and Richardson, F.S.: Chem. Phys. Lett. **6**, 359 (1970).

[18] Ditchfield, R., Hehre, W.J. and Pople, J.A.: J. Chem. Phys. **52**, 5001 (1970).

[19] Davis, K.B., Mewes, M.O., Andrews, M.R., van Druten, N.J., Durfee, D.S., Kurn, D.M. and Ketterle, W.: Phys. Rev. Lett. **75**, 3969 (27 November 1995). Ketterle, W. and van Drutten, N.J.: Phys. Rev. **A54** 656 (1996).

[20] Hylleras, E.: Z. Physik **54**, 347 (1929).

[21] Yip S.K. and Sauls, J.A.: Phys. Rev. Lett. **69** 2264 (1992). Xu, D., Yip, S.K. and Sauls, J.A.: Phys. Rev. **B51** 16233 (1955); see also the levels of "normal" H_2 in *Symmetry and Spectroscopy*, by D.C. Harris and M.D. Bertolucci, p. 345, Oxford Univ. Press. New York (1978).

[22] Boys, S.F. and Cook, G.B.: Rev. Mod. Phys. **45**, 226 (1960). Reeves, C.M.: Commun. Assoc. Comput. Mach. **9**, 276 (1966); Lengsfield, B. and Liu, B.: J. Chem. Phys. **75**, 478 (1981); Walch, S.P., Bauschlicher, C.W. Jr., Roos, B.O. and Nelin, C.J.: Chem. Phys. **103**, 175 (1983).

[23] Pople, J.A., Krishnan, R., Schlegel, H.B. and Binkly, J.S.: Int. J. Quantum Chem. **14**, 545 (1978). Krishnan R. and Pople, J.A.: Int. J. Quantum Chem. **14**, 91 (1978).

[24] Cizek, J.: J. Chem. Phys. **45**, 4256 (1966); Paldus, J., Cizek, J. and Shavitt, I.: Phys. Rev. **A5**, 50 (1972). Bartlett, R.J. and Purvis, G.D.: Int. J. Quantum Chem. **14**, 561 (1978). Purvis, G.D. and Bartlett, R.J.: J. Chem. Phys. **76**, 1910 (1982).

[25] Schaefer, H.F.:*The Electronic Structure of Atoms and Molecules,* Addison-Wesley, Reading, Mass. (1972).

[26] Kelly, H.P.: Phys. Rev. **131**, 684 (1963).

[27] Goodgame, M.M. and Goddard, W.A.: Phys. Rev. Lett. **54**, 661 (1985).

[28] Mosier, C. and Shillady, D.D.: Math. of Computation **26**, 1022 (1972); cf. program LOBE140 described by D.D. Shillady and Sheryl Baldwin, Int. J. Quantum Chem., Quantum Biology Symposium **6**, 105 (1979).

[29] Bender, C.F. and Davidson, E.R.: Phys. Rev. **183**, 23 (1969). Alston, P.V., Shillady, D.D. and Trindle, C.: J. Am. Chem. Soc. **97**, 469 (1975).

[30] Hammond, B.L., Lester, W.A. Jr., and Reynolds, P.J.: Monte Carlo Methods, in: *Ab Initio Quantum Chemistry,* World Scientific Lecture and Course Notes in Chemistry, Vol. **1**, World Scientific, New Jersey, p. 67 (1994).

[31] Miehlich, B., Savin, A., Stoll, H. and Preuss, H.: Chem. Phys. Lett. **157**, 200 (1989). Becke, A.D.: J. Chem. Phys. **88**, 1053 (1988). Lee, C.L., Yang, W. and Parr, R.G.: Phys. Rev. **B37**, 785 (1988).

[32] Dunning, T.H.: J. Chem. Phys. **55**, 716 (1971).

[33] Luchow, A. and Anderson, J.B.: J. Chem. Phys. **105**, 7573 (1996).

[34] Kolos, W. and Wolniewicz, L.: J. Chem. Phys. **49**, 404 (1968).

[35] Boyd, R.J. and Yee, M.C.: J. Chem. Phys. **77**, 3578 (1982).

[36] Fois, E.S., Selloni, A., Parinello, M. and Car, R.: J. Phys. Chem. **92**, 3268 (1988).

[37] Xia, P. and Bloomfield, L.A.: Phys. Rev. Lett. **70**, 1779 (1993).

[38] Burnett Collaboration, http://eve.physics.ox.ac.uk/intense/

[39] Moore, C.: Phys. Rev. Lett. **70**, 3675 (1993). Chenciner, A. and Montgomery, R.: Annals of Math., 2001, in press. Montgomery, R.: Notices of the Am. Math. Soc. **48**, 471 (2001).

Chapter 5

ISOCHEMICAL MODEL OF THE WATER MOLECULE

1. Introduction

Water is an extremely important compound on Planet Earth in a biological as well as geophysical sense. As a consequence, comprehensive studies on water have been conducted since the beginning of quantitative science with outstanding scientific achievements (see, e.g., Ref. [1]). Nevertheless, despite all these efforts, a number of fundamental issues on the structure of the water molecule remain still open, such as:

1) The total electrostatic force among the atomic constituents of a water molecule is null in semiclassical approximation, while the currently used forces (exchange, van der Waals and other forces [1]) are known from nuclear physics to be "weak," thus insufficient to fully explain the "strong" bond among the constituents (where the words "weak" and "strong" do not refer hereon to the corresponding interactions in particle physics). In different words, the representation of the nuclear structure required the introduction of the "strong nuclear force" because of the insufficient strength of the exchange, van der Waals and other forces. It appears that current models on the water molecule lack the equivalent of the "strong nuclear force" to achieve a full representation of molecular structures.

2) Quantum chemistry has not provided a rigorous explanation of the reason why the water molecules only has *two* hydrogen atoms. This is an evident consequence of the assumption of exchange and other nuclear-type forces which were built in nuclear physics for an *arbitrary number of constituents*, a feature which evidently persists in its entirety in molecular structures.

3) Quantum chemistry has been unable to achieve an exact representation of the binding energy of the water molecule under the rigorous implementation of its basic axioms, such as the Coulomb law. In fact, there is a historical 2% still missing despite efforts conducted throughout the 20-th century.

4) More accurate representations have been recently achieved although via the use of Gaussian screenings of the Coulomb law, which, however, are outside the class of equivalence of quantum chemistry, since they are *nonunitarily* connected to the Coulomb law.

5) Quantum chemistry cannot provide a meaningful representation of thermodynamical properties related to water. In fact, the value of 2% missing in the representation of binding energy corresponds to about 950 Kcal/mole while an ordinary thermodynamical reaction takes about 50 Kcal/mole. The use of quantum chemistry in thermodynamical calculations would, therefore, imply an error of the order of 20 times the value considered.

6) Quantum chemistry has been unable to reach an exact representation of the electric and magnetic dipole and multipole moments of the water molecule to such an extent that, sometimes, the models result in having even the *wrong sign* (see, e.g., Ref. [1a], p. 22). This insufficiency is generally assumed to be due to the incompleteness of the assumed basis, although one should not keep adding terms without deeper analysis.

7) Computer usages in quantum chemical calculations require excessively long periods of time. This occurrence, which is due to the slow convergence of conventional quantum series, has persisted to this day, despite the availability of more powerful computers.

8) Quantum chemistry has been unable to explain the "correlation energy" which is advocated for the missing percentages of the binding energies. Orbital theories work well at qualitative and semi-empirical levels, but they remain afflicted by yet unresolved problems, such as the currently used correlation among many electrons as compared to the evidence that the *correlation solely occurs for electron pairs.*

9) Quantum chemistry predicts that the water molecule is ferromagnetic, in dramatic disagreement with experimental evidence. This prediction is a consequence quantum electrodynamics, which establishes that, under an external magnetic field, the orbits of valence electrons must be polarized in such a way as to offer a magnetic polarity opposite to that of the external homogeneous field. As it is well known, the individual atoms of a water molecule preserve their individuality in the current model of chemical bonds. As a result, quantum electrodynamics predicts that all valence electrons of the individual atoms of a water molecule acquire the same magnetic polarization under a sufficiently strong

external magnetic field, resulting in a total net magnetic polarity North-South.

Particularly insidious are variational methods because they give the impression of achieving exact representations within the context of quantum chemistry, while this can be easily proved *not* to be the case. To begin, representations of 100% of the experimental data occur with the introduction of a number of *empirical parameters* which lack a physical or chemical meaning. Moreover, it is easy to prove that *variational solutions cannot be the solution of quantum chemical equations*, trivially, because the former provide 100% representations, while the latter do not. In reality, the arbitrary parameters introduced in variational and other calculations are a measure of the *deviation from the basic axioms of quantum chemistry.*

When passing from the structure of one water molecule to more general molecular structures the number of open, basic, unsolved issues increases. For instance, it is generally admitted that quantum chemistry has been unable to provide a systematic theory of the liquid state in general, let alone that of liquid water in particular [1].

Also, chemical reactions in general are *irreversible*, while the axiomatic structure of quantum chemistry is *strictly reversible* because the theory is strictly Hamiltonian and all known potential forces are reversible. This results in an irreconcilable incompatibility between the very axiomatic structure of quantum chemistry and chemical reactions in general, and those involving water in particular. In fact, an axiomatically consistent representation of irreversibility is expected to imply effects which are simply inconceivable for quantum chemistry, evidently because they are outside its structure.

When passing to water as a constituent of biological entities, the limitations of quantum chemistry reach their climax. In fact, biological structures (such as a cell) are not only irreversible (because they grow, age and die), but have such a complex structure to require multi-valued theories (also known in mathematics as hyperstructures). The expectation that quantum chemistry, with its reversible and single-valued structure, can effectively represent biological systems and their evolution is beyond the boundaries of science.

In view of the above numerous and basic limitations, in the preceding papers [2] Santilli and other scientists have constructed a covering of quantum mechanics under the name of *hadronic mechanics* (Sect. 1.8). By conception and construction, quantum and hadronic mechanics coincide everywhere, except inside a small sphere of radius of the order of 1 fm ($= 10^{-13}$ cm) called *hadronic horizon*, in which interior (only) the broader theory holds.

Hadronic mechanics results in being a form of "completion" of quantum mechanics much along the historical Einstein-Podolsky-Rosen argument, although achieved via the addition of contact, nonhamiltonian, nonlinear, nonlocal, and nonpotential forces due to deep overlappings of the wavepackets of particles.

On more technical grounds, hadronic mechanics is based on *new mathematics*, called *iso-, geno- and hyper-mathematics* [2c] (see Chapter 2) for the characterization of reversible, irreversible, and multivalued systems, respectively, possessing features not representable via the Hamiltonian.

These new mathematics are characterized by a progressive generalization of the trivial unit I of quantum mechanics into generalized units $\hat{I}$ of Hermitean single-valued, nonhermitean single-value, and nonhermitean multi-valued character, respectively, first proposed by Santilli in 1978 [2],

$$I \to \hat{I} = \hat{I}^\dagger, \quad I \to \hat{I} \neq \hat{I}^\dagger, \quad I \to \{\hat{I}\} = \{\hat{I}_1, \hat{I}_2, \hat{I}_3, ...\} \neq \{\hat{I}\}^\dagger. \tag{5.1}$$

The new mathematics then emerge from the reconstruction of the conventional mathematics of quantum mechanics in such a way as to admit $\hat{I}$, rather than I, as the correct left and right unit at *all* levels.

The iso-, geno-, and hyper-mathematics characterize corresponding branches of hadronic mechanics, called *iso-, geno-, and hyper-mechanics*, which have been constructed for the corresponding representation of:

1) closed-isolated, reversible, single-valued systems with Hamiltonian and nonhamiltonian internal forces;

2) open-nonconservative, irreversible, single valued systems with unrestricted interactions with an external system; and

3) open-nonconservative, irreversible, multi-valued systems of arbitrary structure.

Subsequently, Animalu and Santilli [6] constructed *hadronic superconductivity*, with corresponding iso-, geno-, and hyper-branches (Sect. 1.9) for the representation of the *structure* (rather than an ensemble) of the Cooper pairs, in a way remarkably in agreement with experimental data.

In 1999, Santilli and Shillady [3a] constructed *hadronic chemistry* (Chapter 3) with corresponding branches called *iso-, geno-, and hyper-chemistry.* Since molecules are considered as isolated from the rest of the universe, and are reversible in time, they are studied via *isochemistry.*

Santilli and Shillady [3a] also constructed a *new isochemical model of the hydrogen molecule* (Chapter 4) based on the assumption that *pairs of valence electrons from different atoms couple themselves into a singlet quasi-particle state called isoelectronium.*

As shown in Chapter 4, the new model was proved to resolve at least the major insufficiencies of the quantum chemical model of the hydrogen molecule, such as: explain why the molecule has only two H-atoms; represent the binding energy to the seventh digit; achieve computer calculations which converge at least 1,000 times faster than those of quantum chemistry; and permit other advances.

The main scope of this chapter is that of studying the *new isochemical model of the water molecule* first submitted by Santilli and Shillady in Ref. [3b] via a suitable expansion of the results obtained for the hydrogen molecule.

The main assumption is that, when the valence electrons of the water molecule correlate-bond themselves into singlet pairs in accordance with Pauli's exclusion principle, there is the emergence of *new interactions structurally beyond any hope of representation by quantum mechanics and chemistry, trivially, because they are nonhamiltonian.*

In particular, the new interactions are strongly attractive, thus introducing, for the first time, a molecular bond sufficiently "strong" to represent reality. These and other features of the model, such as the sole possible correlation-bond being in pairs, will resolve all insufficiencies 1)–9) indicated earlier, as we shall see.

To provide introductory guidelines, let us recall that the main function of the isounit $\hat{I}$ (hereon assumed to be Hermitean, single-valued and positive-definite) is that of representing all interactions, characteristics and effects outside the representational capabilities of a Hamiltonian. This includes the representation of contact, nonpotential and nonhamiltonian interactions in deep overlapping of the wavepackets of valence electrons.

By recalling that, whether conventional or generalized, the unit is the fundamental invariant of any theory, the representation of the new interactions via the generalized unit assures invariance, that is, the prediction of the same numbers for the same quantities under the same conditions but at different times.

Representation of nonhamiltonian effects via quantities other than the generalized unit are encouraged, *provided that they achieve the indicated invariance*, as a necessary condition to avoid the catastrophic inconsistencies of Sect. 1.7.

The most fundamental mathematical, physical, and chemical notion of the new model of structure of the water molecule studied in this Chapter is, therefore, the generalization of the trivial unit +1 of current models into the isounit.

On pragmatic grounds, isochemistry can be easily constructed via a step-by-step application of the nonunitary transform

$$\hat{I} = 1/\hat{T} = U \times U^{\dagger} > 0, \tag{5.2}$$

to *all* aspects of quantum chemistry (Sect. 3.6). In particular, we shall assume that the above isounit recovers the conventional unit outside the hadronic horizon, and its average value is much bigger than 1,

$$\lim_{r \gg 1 \text{ fm}} \hat{I} = I. \tag{5.3a}$$

$$|\hat{I}| \gg 1. \tag{5.3b}$$

Assumption (5.3a) will assure the compatibility of a generalized discipline for the bonding of valence electrons, while preserving conventional quantum mechanics identically for the structure of the individual atoms composing the water molecule. Assumption (5.3b) will assure a much faster convergence of perturbative expansions, and other features.

In summary, the new isochemical model of the water molecule [3b] studied in this chapter can be constructed via the following steps:

1) Select a nonunitary transforms according to rules (5.2) and (5.3) which is representative of contact, nonlinear, nonlocal, and nonpotential effects in deep wave-overlapping, essentially similar to that used for the hydrogen molecule [3a] of the preceding Chapter;

2) Submitting to the selected nonunitary transform the totality of the notions, equations, and operations of the conventional quantum chemical model of the water molecule; and

3) Reconstructing the entire mathematics of the conventional model in such a way as to admit $\hat{I}$, rather than I, as the correct left and right unit at *all* levels, with no known exceptions. This lifting is necessary to avoid the catastrophic inconsistencies of Sect. 1.7 (e.g., to achieve invariance), thus requiring the isotopic lifting of numbers and fields, Cartesian and trigonometric functions, ordinary and partial differential equations, *etc.*

The axiomatically correct isochemical model of the water molecule is that formulated on isospaces over isofields. However, on pragmatic grounds, one can study its *projection* on ordinary spaces over ordinary numbers, *provided* that the results are interpreted with care.

For instance, in the indicated projection there is the general emergence of a *potential*, which, as such, may lead to imply that the model carries a *potential energy* and/or it can be treated via a conventional *potential well.* Such interpretations are correct if and only if the potential is well defined on isospaces over isofields. On the contrary, if said potential solely emerges in the projection, then it has a purely mathematical meaning without any associated energy.

The best illustration of the above seemingly contradictory occurrences was that for the isoelectronium of Sect. 4.3, whose structure did indeed exhibit the appearance of the *Hulten potential*, yet the quasi-particle had *no binding energy*. The reason is that binding energies are indeed well defined on isospaces over isofields via the isoschrödinger's equation and related isoeigenvalues, while the Hulten potential does not exist on isospaces, and solely occurs in the projection of the isoschrödinger's equation on ordinary Hilbert spaces.

To illustrate this important point, consider the isotopies of the conventional Schrödinger's equation via the nonunitary transform indicated above,

$$\begin{aligned} U\times(H\times|\psi\rangle) &= (U\times H\times U^{\dagger})\times(U\times U^{\dagger})^{-1}\times(U\times|\psi\rangle) = \\ &= \hat{H}\times\hat{T}\times|\hat{\psi}\rangle = \hat{H}\hat{\times}|\hat{\psi}\rangle = U\times(E\times|\psi\rangle) = \\ &= [E\times(U\times U^{\dagger})]\times(U\times U^{\dagger})^{-1}\times(U\times|\psi\rangle) = \hat{E}\times\hat{T}\times|\hat{\psi}\rangle = \\ &= E\times|\hat{\psi}\rangle, \end{aligned} \tag{5.4}$$

with corresponding liftings of numbers and Hilbert spaces,

$$U\times n\times U^{\dagger} = n\times(U\times U^{\dagger}) = n\times\hat{I}, \quad \hat{n}\in\hat{\mathbb{R}}, \tag{5.5a}$$

$$U\times\langle\psi|\times|\psi\rangle\times U^{\dagger} = \langle\hat{\psi}|\times\hat{T}\times|\hat{\psi}\rangle\times(U\times U^{\dagger}) = \langle\hat{\psi}|\hat{\times}|\hat{\psi}\rangle\times\hat{I}\in\hat{\mathbb{C}}. \tag{5.5b}$$

As one can see, binding energies $\hat{E} = E\times\hat{I}$ are fully defined on isohilbert spaces $\hat{\mathcal{H}}$ over isofields $\hat{\mathbb{R}}$, and actually acquire the conventional value E following the simplification $\hat{E}\times\hat{T}\times|\hat{\psi}\rangle = (E/\hat{T})\times\hat{T}\times|\hat{\psi}\rangle = E\times|\hat{\psi}\rangle$.

However, *the Hulten potential does not exist on isospaces over isofields*, trivially, because it does not exist in the Hamiltonian $\hat{H}$ which is fully conventional.

The Hulten potential of the isoelectronium of Sect. 4.3 emerge only when we project the real system, that on isohilbert spaces with equation $\hat{H}\hat{\times}|\hat{\psi}\rangle = E\times|\hat{\psi}\rangle$, on conventional Hilbert spaces. As such, one should not expect that the Hulten potential necessarily carries an actual binding energy.

The reader should equally exercise caution for other aspects, and generally abstain from formulating opinions for hadronic chemistry essentially dependent on quantum chemical concepts and notions.

2. Main Characteristics of Water

Water is a mixture of several different molecules in different percentages and molecular weights. In fact, we know *three* different isotopes of the hydrogen, ^{1}H, ^{2}H and ^{3}H, and *six* different isotopes of the oxygen ranging from ^{14}O to ^{19}O. In this monograph, we shall solely study the molecule $^1H_2 - ^{16}O$, and denote it H_2O = H-O-H where the symbol "-" is referred to the molecular bond. Such a water molecule will be studied hereon under the following conditions: 1) at absolute zero degrees °K; 2) in the absence of any rotational, vibrational, translational, or other motions; and 3) with all atoms in their ground state (see Ref. [1] for all details contained in this section).

The electrons of the individual H-atoms are assumed to be in the ground state $1s$. Of the eight electrons of the oxygen, two electrons with opposite spin orientation are in the lowest $1s$ state which is tightly bound to the nucleus; two electrons are in the next possible state $2s$; and the remaining four electrons are in the $2p$ state.

By using a three-dimensional reference frame with the y-z plane containing the nuclei of the H and O-atoms with origin in the latter, the $1s$ and $2s$ electrons have a spherical distribution while the $2p$ electrons are in orbitals perpendicular to the yz plane denoted $2p_x$; the remaining two electrons have orbitals perpendicular to the xz and xy planes denoted $2p_y$ and $2p_z$, respectively.

Also, the energy of formation of the water molecule from hydrogen and oxygen is -9.511 eV; the binding energy is -10.086 eV; the sum of the ground state energies of the three separate atoms is $-2,070.46$ eV; the total molecular energy at 0°K is $-2,080.55$ eV as a result of kinetic energy +2,080.6 and potential energy $-4,411.4$; the nuclear repulsion energy is +250.2 eV; the total electrostatic energy is $-2,330.8$ eV; the dissociation energy of O-H is 5.11 eV and that of H alone is 4.40 eV.

Again at 0°K and for all atoms in their ground states, the bond length of the H-O dimer is 0.95718×10^{-8} cm, while the two dimers H-O and O-H form a characteristic angle of 104.523°. Therefore, by no means the scripture H-O-H denotes that the water has a linear structure because of the indicated characteristic angles in between the two dimers H-O and O-H.

It is evident that when the individual atoms are in their excited states, the bond length and the characteristic angle change. In fact, increases of up to 8.5° have been measured for the characteristic H-O-H angle for excited states. The same characteristic angle is expected to be altered by the application of sufficiently strong electric and magnetic fields, although we are unaware of accurate measurements under the indicated conditions.

The water molecule possesses an electric dipole moment of 1.83×10^{-8} e.s.u. cm and a mean quadrupole moment of -5.6×10^{-26} e.s.u. cm. It should be recalled that the very existence of a non-null value of electric dipole and quadruple moments excludes the linear structure of the water

H-O-H in ordinary isolated conditions (that with a characteristic angle of 180°).

Water is a *diamagnetic* substance with a magnetic polarization (also called susceptibility) of (2.46, 0.77 and 1.42)$\times 10^{-6}$ e.m.u./mole for the corresponding three space-dimension xx, yy and zz, respectively.

In first approximation, the water molecule can be represented via two individual H-O dimers with wavefunction of the molecular orbitals (m.o.'s),

$$\psi_1 = \lambda\phi(\mathrm{H}', 1s) + \mu\phi(\mathrm{O}, 2p_z), \tag{5.6a}$$

$$\psi_2 = \lambda\phi(\mathrm{H}'', 1s) + \mu\phi(\mathrm{O}, 2p_y), \tag{5.6b}$$

where λ and μ are parameters.

However, the above simple model predicts a characteristic angle of 90°. As a consequence, the model is generally modified with a mixture of electrons from the $2p$ and $2s$ states also called *hybridization*. The occurrence confirms that any model of the water with charge distributions of the valence electrons in the H-O-H plane is insufficient to represent the experimental data. In turn, this mixing creates the known two *lobes* on the side of the oxygen atom, away from the hydrogen atoms, above and below the molecular plane. This results in models of the type

$$\psi_1 = \lambda[\cos\varepsilon\psi(\mathrm{O}, 2s) + \sin\varepsilon\phi((\mathrm{O}, 2p)] + \mu\phi(\mathrm{H}', 1s), \tag{5.7a}$$

$$\psi_2 = \lambda[\cos\varepsilon\psi(\mathrm{O}, 2s) + \sin\varepsilon\phi((\mathrm{O}, 2p)] + \mu\phi(\mathrm{H}'', 1s), \tag{5.7b}$$

where ε is the hybridization parameter with generic value of the order of $\cos\varepsilon = 0.093$ confirming that the valence electrons are mainly from $2p$ states.

It should be indicated that the exact configuration, location and function of the two lone-pair electron lobes are unsettled at this writing, since they are evidently dependent on the selected theoretical model. Also, the individual electric and/or magnetic dipoles of the lobes cannot be measured (only their total values is measurable), thus implying lack of direct experimental evidence on the individual lobes.

We should also recall that the individual H-O and O-H bonds are not independent from each other, as confirmed by the different values of the dissociation energies.

Water is both an acid and a base due to dissociation of H_2O into H^+ and O-H^- to the extent that the product of the concentrations $[H^+]$[O-H^-] sets up an equilibrium whose constant value is 1.0×10^{-14}, which is the well known pH scale of the equations

$$\mathrm{pH} = -\log_{10}[\mathrm{H}^+], \quad \mathrm{pOH} = 14 - \mathrm{pH}. \tag{5.8}$$

In neutral water the ion concentrations are $[H^+] = [O\text{-}H^-] = 1.004\times10^{-7}$ mole/liter.

Water is quite polar with a dipole moment of 1.84 to 1.834 Debye and a bulk dielectric constant of 80 at 20 °C. This implies that pure water is not a good conductor, with a direct current conductivity of only 5.7×10^{-8} $ohm^{-1}\cdot cm^{-1}$.

However, it is well known that small amounts of strong acids such as HCl or H_2SO_4 can make water highly conducting due to the ease with which H^+ can attach to H_2O to form H_3O^+ which then offers a *domino effect* for one H^+ to successively "bump" an H^+ off the other side of H_3O^+ and so produce a very effective conduction mechanism [4]. In fact it is well known that in aqueous solutions the transport numbers for the anions and cations are not equal, because up to 70% of the current is carried by H^+. Although OH^- typically carries much less current than H^+ in aqueous conduction of electricity (due to its larger size and lack of the domino-effect cited earlier for H^+), once a current flow is initiated additional ions are created due to collisions in solution.

An important aspect is the known existence of an equilibrium between H-O-O-H and HO^- around pH 11.63 [5] with a voltage dependence of 1.363 ± 0.0293 pH as given by M. Pourbaix for aqueous equilibria involving H^+, $O\text{-}H^-$, H^-, H-O-O-H and $H\text{-}O^-$. Thus, there is no doubt of the existence of small amounts of H-O-O-H in water at high pH.

In a high current process the flow of H^+ will be much greater than that of OH^- so that as H_2O is electrolyzed to $2H_2$ and $1O_2$, local concentrations/fluctuations will slightly favor higher pH (local depletion of H^+) and hence favor the existence of H-O-O-H.

We should finally mention the inability of quantum chemistry to achieve a scientific-quantitative representation (or at least an understanding) of the *different types of water when exposed to magnetic fields*, as established by the evidence, e.g., that plants grow faster when irrigated with water exposed to one type of magnetic field, while they die rapidly when exposed to a different type of magnetic field. In fact, quantum chemistry admits only *one* type of water, H_2O.

It is easy to see that this additional insufficiency of quantum chemistry is a direct consequence of the current use of exchange, van der Waals and other forces of nuclear origin under which the individual H and O atoms in the H_2O molecule preserve their individuality, thus resulting in one single configuration.

On the contrary, isochemistry introduces a real, strong bond for the valence electrons via the notion of isoelectronium. In this latter case different types of water, that is, water molecules with different physical characteristics, are indeed readily possible, as we shall see.

3. Exactly Solvable Model of the Water Molecule with Stable Isoelectronium

In the preceding Chapter 3 [3a], we have introduced the main hypothesis of the *isochemical molecular model*, according to which two electrons from two different atoms bond themselves into a singlet quasi-stable and quasi-particle state called *isoelectronium*, which describes an *oo*-shaped orbit around the nuclei, as it is the case for planets in certain binary stars (Fig. 4.3). The main characteristics of the isoelectronium in first nonrelativistic approximation were calculated in Sect. 4.3 and resulted in being:

$$\begin{gathered}\text{charge} - 2e, \quad \text{spin } 0, \quad \text{magnetic dipole moment } 0, \\ \text{mass } 1.022 \text{ MeV}, \quad \text{radius} = r_c = b^{-1} \approx \\ \approx (\hbar^2/m \times V)^{1/2} = (\hbar/m \times \omega)^{1/2} = 6.8432329 \times 10^{-11}\text{cm} = \\ = 0.015424288 \text{ bohrs} = 0.006843 \text{ Å}.\end{gathered} \tag{5.9}$$

In the above nonrelativistic approximation, the meanlife resulted in being infinite (full stability, with the understanding that relativistic corrections are expected to render such a meanlife finite (partial stability). All conventional forces of current use in chemistry (exchange, van der Waals and other forces) then hold when the valence electrons are at mutual distances bigger than the hadronic horizon.

In this Chapter, we study the *isochemical model of the water molecule* H_2O=H-O-H first introduced by Santilli and Shillady [3b], under the assumption that the molecule is considered at °C and in the absence of any rotational, oscillation or other motion. The main hypothesis is that each electron from the two H-atoms couples in singlet with one $2p$ electron from the O-atom, resulting in *two isoelectronia*, one per each H-O dimer as in Fig. 5.1.

In this Section we shall study a hadronic/isoschrödinger equation for the water molecule under the above assumptions, which equation evidently approximate, yet *exactly solvable* for the first time to our knowledge. We shall then show that the model is extendable to all other dimers comprising one hydrogen atom, such as H-C.

For this purpose, we approximate the H-O-H molecule as being composed of two intersecting identical dimers H-O with evidently only one oxygen atom. This requires a first correction due to the lack of independence of said dimers reviewed in Sect. 5.2. Moreover, in each H-O dimer we shall assume that the oxygen appears to the isoelectronium as having only one net positive charge $+e$ located in the nucleus. This evidently requires a second correction which essentially represents the screening

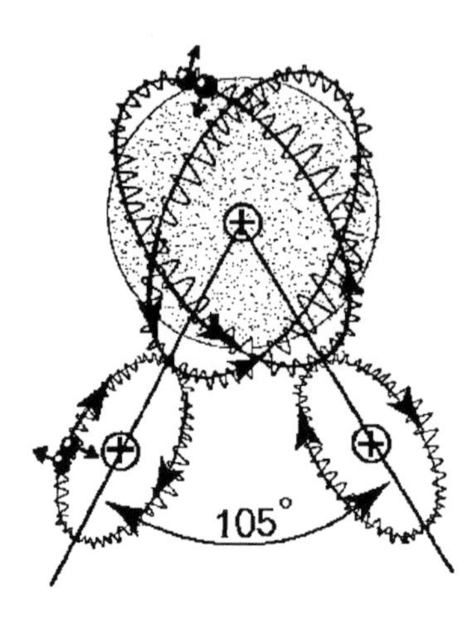

Figure 5.1. A schematic view of the proposed *isochemical model of the water molecule* here depicted at absolute zero degrees temperature and in the absence of any motion for the case of fully stable isoelectronium. It should be stressed that *at ordinary temperature rotational motions recover the conventional space distribution*, thus recovering the conventional "Mickey Mouse" configuration of the water. Also, the model is presented in terms of the *orbits* of the valence electrons (rather than in terms of *density distributions*). The fundamental assumption is that the two valence electrons, one per each pair of atoms, correlate themselves into two bonded singlet states at short distance we have called *isoelectronia*, one per each dimer H-O, which states are assumed to be mostly stable (see the text for the mostly unstable case). The water molecule is then reducible to two intersecting H-O dimers with a common O-atom. The only orbits yielding a stable water molecule are those in which each isoelectronium describes a *oo*-shaped orbit around the respective two nuclei of the H- and O-atoms. The isoelectronia are then responsible for the *attractive force* between the atoms. The *binding energy* is instead characterized by the *oo*-shaped orbits of the isoelectronia around the respective two nuclei, conceptually represented in this figure via a standing wave for a particle of spin 0 and charge $-2e$. Note that, in the absence of molecular motions, the orbits of the two isoelectronia are perpendicular to the H-O-H plane, thus confirming a characteristic of the water molecule reviewed in Sect. 5.2. Conventional exchange, van der Waals and other forces remain admitted by the model when the isoelectronia are mostly unstable. The model permits a representation of: 1) the "strong" value of the molecular bond; 2) the reason why the H_2O molecule has only two hydrogen atoms and one oxygen atom; 3) a representation of the binding energy, electric and magnetic moments accurate to several digits; and other advances studied in the text. The above model of the H-O dimer is then extendable to other H-based dimers, such as H-C.

of the various electrons of the oxygen. Additional corrections are also in order along conventional lines [1].

A study of these corrections has indicated that they can all be represented via one single Gaussian screening of the Coulomb law of the type

[3b]

$$\frac{+e}{r} \rightarrow \frac{+e(1 \pm e^{-\alpha r^2})}{r}, \tag{5.10}$$

where α is a positive parameters to be determined from experimental data, the sign "$-$" applies for the screened O-nucleus as seen from an *electron* (because of the *repulsion* caused by the electron clouds of the oxygen), while the sign "$+$" applies for the screened O-nucleus as seen from the H-*nucleus* (because of the *attraction* this time caused by said electron clouds).

The resulting model is structurally equivalent to the isochemical model of the hydrogen molecule of Chapter 3 [3a], except for the modifications indicated about, and can be outlined as follows.

By denoting with the sub-indices 1 and a the hydrogen, 2 and b the oxygen, prior to the indicated screening and in the absence of all hadronic effects, the conventional Schrödinger equation of the H-O dimer with the oxygen assumed to have only one elementary charge $+e$ in the nucleus is given by

$$\left(\frac{1}{2\mu_1} p_1 \times p_1 + \frac{1}{2\mu_2} p_2 \times p_2 - \right.$$
$$\left. - \frac{e^2}{r_{1a}} - \frac{e^2}{r_{2a}} - \frac{e^2}{r_{1b}} - \frac{e^2}{r_{2b}} + \frac{e^2}{r_R} + \frac{e^2}{r_{12}} \right) \times |\psi\rangle = E \times |\psi\rangle, \tag{5.11}$$

As it was the case for the H_2-molecule, our task is that of subjecting the above model to a transform, which is nonunitary only at the short mutual distances $r_c = b^{-1} = r_{12}$ of the two valence electrons (here assumed to be hadronic horizon), and becomes unitary at bigger distances $\hat{I}_{r \leq 10^{-10} \mathrm{cm}} \neq I$, $I_{r \gg 10^{-10} \mathrm{cm}} = I$.

We assume that the state and related Hilbert space of systems (5.11) can be factorized in the familiar form (in which each term is duly symmetrized or antisymmetrized)

$$|\psi\rangle = |\psi_{12}\rangle \times |\psi_{1a}\rangle \times |\psi_{1b}\rangle \times |\psi_{2a}\rangle \times |\psi_{2b}\rangle \times |\psi_R\rangle, \tag{5.12a}$$

$$\mathcal{H}_{\mathrm{Tot}} = \mathcal{H}_{12} \times \mathcal{H}_{1a} \times \mathcal{H}_{1b} \times \mathcal{H}_{2a} \times \mathcal{H}_{2b} \times \mathcal{H}_R. \tag{5.12b}$$

The nonunitary transform we are looking for shall act only on the r_{12} variable characterizing the isoelectronium while leaving all other variables unchanged. The simplest possible solution is given by

$$U(r_{12}) \times U^\dagger(r_{12}) = \hat{I} = e^{[\psi(r_{12})/\hat{\psi}(r_{12})] \int dr_{12} \hat{\psi}^\dagger(r_{12})_{1\downarrow} \times \hat{\psi}(r_{12})_{2\uparrow}}, \tag{5.13}$$

where the ψ's represents conventional wavefunction and the $\hat{\psi}$'s represent isowavefunctions, for which we have, again the fundamental condition of fast convergence

$$|\hat{T}| = |(U \times U^\dagger)^{-1}| \ll 1. \tag{5.14}$$

We now construct the isochemical model by transforming short-range terms (isochemistry) and adding un-transformed long range ones (chemistry), thus resulting in the radial equation

$$\left(-\frac{\hbar^2}{2 \times \mu_1} \times \hat{T} \times \nabla_1 \times \hat{T} \times \nabla_1 - \frac{\hbar^2}{2 \times \mu_2} \times \hat{T} \times \nabla_2 \times \hat{T} \times \nabla_2 + \right.$$

$$\left. + \frac{e^2}{r_{12}} \times \hat{T} - \frac{e^2}{r_{1a}} - \frac{e^2}{r_{2a}} - \frac{e^2}{r_{1b}} - \frac{e^2}{r_{2b}} + \frac{e^2}{R} \right) \times |\hat{\psi}\rangle = E|\hat{\psi}\rangle. \tag{5.15}$$

By recalling that the Hulten potential behaves at small distances like the Coulomb one, Eq. (5.15) becomes

$$\left(-\frac{\hbar^2}{2 \times \mu_1} \times \nabla_1^2 - \frac{\hbar^2}{2 \times \mu_2} \times \nabla_2^2 - V \times \frac{e^{-r_{12} \times b}}{1 - e^{-r_{12} \times b}} - \right.$$

$$\left. - \frac{e^2}{r_{1a}} - \frac{e^2}{r_{2a}} - \frac{e^2}{r_{1b}} - \frac{e^2}{r_{2b}} + \frac{e^2}{R} \right) \times |\hat{\psi}\rangle = E \times |\hat{\psi}\rangle. \tag{5.16}$$

The above model can be subjected to an important simplification. In first approximation under the assumption herein considered, the H-O dimer (5.16) can be reduced to a *restricted three body problem* similar to that possible for the conventional H_2^+ *ion* [1], but *not* for the conventional H_2 molecule, according to the equation

$$\left(-\frac{\hbar^2}{2\mu_1} \times \nabla_1^2 - \frac{\hbar^2}{2\mu_2} \times \nabla_1^2 - V \times \frac{e^{-r_{12}b}}{1 - e^{-r_{12}b}} - \right.$$

$$\left. - \frac{2e^2}{r_a} - \frac{2e^2}{r_b} + \frac{e^2}{R} \right) \times |\hat{\psi}\rangle = E \times |\hat{\psi}\rangle. \tag{5.17}$$

The indicated corrections due to the screening of the various electrons in the oxygen and other corrections are needed in the "sensing" of the O-nucleus by the isoelectronium as well as by the H-nucleus, yielding in this way our final model

$$\left(-\frac{\hbar^2}{2\mu_1} \times \nabla_1^2 - \frac{\hbar^2}{2\mu_2} \times \nabla_2^2 - V \times \frac{e^{-r_{12}b}}{1 - e^{-r_{12}b}} - \right.$$

$$\left. - \frac{2e^2}{r_a} - \frac{2e^2(1 - e^{-\alpha r_b})}{r_b} + \frac{e^2(1 + e^{-\alpha R})}{R} \right) \times |\hat{\psi}\rangle = E \times |\hat{\psi}\rangle, \tag{5.18}$$

where: α is a positive parameter; E is half of the binding energy of the water molecule; and, as it was the case for model (4.35), the mass of the isoelectronium, the internuclear distance, and the size of the isoelectronium can be fitted from the value of the binding energy and other data.

Under the latter approximation, the model admits an exact analytic solution, for the first time to our knowledge, which solution however exists only for the case of the *fully stable isoelectronium.* In fact, for the unstable isoelectronium, the model becomes *a four-body structure*, which as such admits no exact solution.

Besides being exactly solvable, model (5.18) exhibits a *new explicitly attractive "strong" force among the neutral atoms of the H-O dimer*, which is absent in conventional quantum chemistry; the equation also explains the reasons why the water molecule admits only *two* H-atoms.

As we shall see in the remaining sections, the model permits an essentially exact representation of the binding energy, electric and magnetic moments; the model yields much faster convergence of series with much reduced computer times, and resolves other insufficiencies of conventional models.

Finally, the model is evidently extendable with simple adjustment to an exact solution of other dimers involving the hydrogen, such as H-C. In addition, it permits the identification of electric and magnetic polarizations, which are not predictable with quantum chemistry (Chapter 8).

4. Gaussian Approximation of the Isochemical Model of the Water Molecule with Unstable Isoelectronium

The solution of the exactly solvable model (5.18) are not available at this writing, and its study is here encouraged. In this section, we review the studies of Ref. [3b] on a Gaussian *approximation* of the isochemical model of the water molecules, with the Hulten potential approximated to a certain Gaussian form.

It should be indicated from the onset that such an approximation implies an evident weakening of the Hulten attraction among the two isoelectrons of the isoelectronium, which, in turn, implies the instability of the isoelectronium itself, thus reaching a model which is somewhat intermediate between the full isochemical model and the conventional quantum chemical model of the water.

Despite this approximate character, the results of this section are significant because they show the capability of isochemistry to achieve an essentially exact representation of the binding energy, electric and magnetic moments and other characteristics of the water molecule.

The results of this study can be outlined as follows. Since HOOH will be slightly more allowed under the assumed conditions, collisions of HO^- with neutral H_2O and the internal repulsion within the anion could favor the release of a quasiparticle with charge $-2e$ to form OH^+. Collisions of OH^+ with OH^- will then further enhance the concentration of HOOH, and transport of $-2e$ will contribute to the current.

The question here is whether under extreme cases of forced conduction a singlet-pair of electrons (isoelectronium), can be "triggered" (Fig. 4.6) within a water molecule to form and release a $-2e$ charged isoelectronium particle which will provide an additional conduction mechanism analogous to Cooper-pairs of electrons in superconducting solids.

Since the energy depth of the V_0 parameter in the isoelectronium Hulten potential of the original 1978 derivation by Santilli [6] is not known, nor how closely the Gaussian representation fits the Hulten form, we can only match the radius of the two potentials and calculate the energy differences caused by the "sticky-electrons" model in which a transient form of isoelectronium can occur (the Gaussian potential well may not be deep enough to ensure a permanent bound state for the isoelectronium).

The "sticky-electron" model is a parametric model which includes the magnetic dipole attraction between singlet-paired electrons, as well as the nonlocal merging of the wave-packets of each electron at short distance. The radius of the Gaussian screening is then determined empirically by fitting the calculated energy as nearly as possible to the most accurate energy values available.

As used here, it should be emphasized that the off-axis positions of the Gaussian-lobe basis sets [7-9] ensure that angular correlation is included as well as radial dependence, and can include the magnetic dipole attraction of opposite electron spins as well as merger of wavepackets.

One radial screening parameter used with off-axis basis sets parametrically covers all forms of short range attraction which may include angular dependence. Thus the present model can give us an approximate energy difference required to release an electron-pair from OH^-.

$$OH^- \rightarrow OH^+ + (2e)^{-2}. \tag{5.19}$$

It will be seen below that the energy difference between OH^- and OH^+ as calculated, allowing a transient form of isoelectronium, is well within the voltage accessible using capacitive discharge through water. Such a mechanism which would allow $-2e$ particles to flow through water would not be superconductivity as conventionally understood, since the freely moving molecules and ions are not constrained to lattice positions as in solids, so that resistive I^2R heating will still occur.

This is mainly due to the fact that conduction in liquids occurs by mobility of both anions and cations along with size differences, polarizability differences and special mechanisms such as the hydrogen-bonding "domino effect" for H^+ transport. In solid-state conduction, only the electrons move by ignoring in-place phonon oscillations because the atoms do not travel from one electrode to the other.

Despite the indicated lack of superconducting character, it should be indicated that yet, the essentially null magnetic moment of the $-2e$ particle would imply indeed a reduction of the resistivity.

The apparent motion of positively charges "holes" is also due to motion of electrons while the atoms merely oscillate about mean lattice positions. In solutions there is a two-way traffic with positive and negative ions traveling in opposite directions and with differing velocities, thus leading to resistive heat even up to the vaporization of the water as well as a high probability of ion collisions.

It should be noted that in recent work Ashoori *et. al.* [10] have measured migration of paired-electrons to quantum dot wells in GaAs, while Boyd and Yee [11] have observed "bipolaron" electron pairs in alkali halide lattice vacancies. Calculations leading to unexpected bipolarons in crystal lattice vacancies have also been observed by using the method of Car and Parinello [12]. These findings in solids lend support to the concept of an electron pair as an individual particle, called by the authors isoelectronium.

The calculations given here do not prove the presence of isoelectronium particles in high current aqueous electrical conduction; they only indicate the energy threshold necessary to form the isoelectronium within the conducting solution by double-ionization of OH^-.

It is not easy to envision an experiment that would be able to analyze components of a given current, due to multiple ion species in terms of the amount of current due to $-2e$ particles, and none is proposed here. However, there may be a chemical test for such a mechanism. Once OH^- is doubly-ionized to form OH^+, collisions with $-2e$ particles would regenerate OH^- ions just as collisions of H^+ with OH^- will reform H_2O,and no new species will be evident.

However, if OH^- collides with OH^+ a new chemical species HO-OH will be formed that may last long enough in the liquid to behave as a strong oxidizing agent. Thus, organic compounds with double bonds (alkenes), which have negligible conductance, could be added to water undergoing a high current flow to cause hydroxylation of such compounds [13], (i.e., conversion of alkenes to epoxide, which are then readily hydrolyzed in the presence of H^+ to diols). Enhanced concentrations of epoxides and diols would be indirect evidence of double ionization of

OH^-, according to the expressions

$$OH^+ + OH^- \rightarrow HO-OH, \tag{5.20}$$

$$CH_2{=}CHR + HO-OH \rightarrow \begin{array}{c} CH = CHR \\ \backslash \;\; / \\ O \end{array} + H_2O.$$

If the isoelectronium can be detected indirectly by a chemical method, this would in itself be an important inference on the existence of a two-electrons, spin-zero particle. More importantly, "isochemical reactions" could be driven by high conduction "liquid plasma" environments where the isoelectronium is at an enhanced concentration.

Another case of interest is that of aqueous mixtures of insoluble organic compounds forming a separate oil layer over water in an intense magnetic field of several Tesla. At normal thermal energy of room temperature $kT \cong RT$ per mole the main energy form would be random Brownian motion.

However, in the presence of a strong magnetic field HO^+ and HO^- would be constrained to favor circular motions in the magnetic field by the "cyclotron effect," but there is no obvious source of HO^+.

Since two ions of opposite sign charges would be favored to collide by both electrical attraction and by opposite path curvature in a magnetic field, there is an enhancement that when created as a normal result of H^+, OH^-,H^-, H_2O_2, HO_2^-, H_2O equilibrium system studied by Pourbaix [5], any natural concentration of HOOH would be augmented by collision of H^+ with HOO^-.

In addition there is some slight chance that H^+ would collide with OH^- with sufficient excess energy to produce OH^+ and H^-. Thus the presence of an intense magnetic field would cause positive and negative ions to collide more easily while traveling in opposite rotational arcs in such a way as to enhance the concentration of HOOH,

$$H^+ + HOO^- \rightarrow HOOH, \tag{5.21a}$$

$$H^+ + OH^- \rightarrow H^- + OH^+, \tag{5.21b}$$

$$OH^+ + OH^- \rightarrow HOOH, \tag{5.21c}$$

which could then epoxidate alkenes and upon hydrolysis would lead to diols.

A direct measure of this effect would be to determine the enhanced solubility of alkenes in water. The alkenes are only slightly soluble in water ("oil and water do not mix") but alkenes converted to diols will have a measurably greater solubility in water due to the attached OH-groups. Again. If such enhanced solubility of alkenes in water can be caused by intense magnetic fields, this would be indirect evidence of the existence of an electron-pair particle with charge $-2e$.

In the description of the calculations below the key to the above possibility is that it is easy to calculate the energy of OH^- when one subtracts a small amount from the two-electron repulsion terms in the usual HFR-SCF treatment, due to the attraction of singlet-paired electrons at close range within 1.0 picometer.

In the recent *Handbook of Computational Quantum Chemistry* by Cook, Ref. [14], p. 438, it is noted that solutions to the HFR-SCF scheme may not always exist for anions. However, in the method used here convergence of the HFR-SCF method was normal for an SCF process, because the so called "self-energy" error of the Hartree-Fock method [15] (in which each electron repels all electrons including itself) is largely cancelled by the new attractive terms used here. In effect, this description of OH^- is possible because of the easy convergence of the "correlated-SCF process."

5. The Method

The model adopted in Ref. [3b] is to use the usual Hartree-Fock-Roothan self-consistent-field equations [16] (which also has some formal flaws such as the self-interaction terms [15]) and lift in a nonunitary way the form of the Coulomb interaction of the electrons.

Note that reducing the values of the Coulomb integrals will lower the energy by reducing the electron-electron repulsion while reducing the exchange terms will raise the energy, but the factor 1/2 reduces the effect of the exchange terms. Thus a reduction of the value of the integrals will lower the energy.

Note that Goddard [15] has already recommended reducing the atomic self-energy by subtracting 1.39 eV from Hartree-Fock exchange integrals in the cases of Cr_2 and Mo_2.

$$\mathrm{FC} = \mathrm{ESC}; \quad \mathrm{F}_{i,j} = H_{i,j} + \sum_{k,l} P_{k,l}[(i,j|k,l) - 1/2(i,k|j,l)], \tag{5.22a}$$

$$(i,j|k,l) = \iint \chi_i(l)\chi_j(l)\frac{1}{r_{12}}\chi_k(2)\chi_l(2)d\tau_1 d\tau_2, \tag{5.22b}$$

$$P_{i,j} = 2\sum_n c_{n,i}c_{n,j} \quad \text{(sum } n \text{ only over occupied orbitals).} \tag{5.22c}$$

The 1995 paper on electron-electron pairs by Animalu and Santilli [6b] invokes the non-local hadronic attractive force evident in the π^0-meson by Santilli [6a] applied to a pair of singlet-paired electrons which form a boson quasi-particle. However, the "collapsed positronium" rapidly decays since the particle-antiparticle annihilation takes place in less than a picosecond.

In the electron-electron case it is believed that there may be a stable quasi-particle singlet bond we have called the isoelectronium. After using a non-local isotopic nonlinear transformation, the hadronic attraction was projected into real-space, and modeled with a Hulten potential.

Considerable effort was made to evaluate the matrix elements for the Hulten potential without success. Examination of the original 1978 paper on positronium collapse by Santilli [6a] revealed that the Hulten potential is not necessarily a unique representation of the hadronic force. In fact, a linear combination of similar potentials could be used to represent the Hulten potential if matrix elements of such other potentials could be evaluated.

The depth of the screened Gaussian approximation is determined by requiring that the width at half height of the Gaussian is equal to the b value of the Hulten horizon (the radius at which the Coulomb repulsion is annulled by other attractive forces). Thus, the screened Gaussian potential probably has a depth which is too shallow although the V_0 depth parameter for the Hulten potential is not known at present.

This work assumes that until matrix elements of a two-electron interaction for singlet-pairs can be found for the Hulten potential, a Gaussian-screened-Coulomb potential can be used to describe the real-space form of the hadronic attraction and as a parameter fitted to experimental energies the screening exponent probably includes other effects such as the magnetic dipole interaction of two electrons with opposite spin-magnetic-moments. This form has the important property that it can be merged with the general case of the four-center Coulomb or exchange integral derived by Shavitt [17] using the famous Gaussian transform technique.

The Gaussian transform two-electron integral for four Gaussian spheres has been used in a number of Gaussian-lobe basis SCF programs written by Shillady [18, 19] and others. It is important to note that the formula is completely general in orientation of four Gaussian sphere lobe-orbitals as well as the distance between two electrons.

As modified for a description of the correlation of two electrons, such a general formula can describe angular correlation as well as distance interaction. Thus matrix elements of a screened-Coulomb interaction were subtracted from the usual $1/r$ Coulomb repulsion to model the real-space form of the hadronic attraction of two electrons. The work outlined in this section, first presented in Ref. [3b], added the Gaussian screening as $\exp[-\alpha r^2]/r$ so that the special properties of Gaussians could be used, especially the properties that the product of two Gaussians form another

Gaussian (times a re-centering factor), and that polar coordinates readily separate into factorable x, y, z components.

The goal was to evaluate the two-electron four-center matrix elements of the Gaussian-screened Coulomb potential in the expression

$$Y(r) = \frac{1 - 2\exp[-\alpha r^2]}{r}. \tag{5.23}$$

Amazingly, the Gaussian-Gaussian exponent and carried through the original derivation until the last step when integration over "s" is required. α is usually a very high number, this work used 0.13441885×10^7. At this point the usual Coulomb interaction resorts to a well known auxiliary function F_0 which has been studied by Shillady [18] and others.

Since both $s^{1/2}$ and $(s+\alpha)^{1/2}$ occur in the denominator of the screened-Coulomb form, two poles occur in the integral. A change of variable absorbs the pole due to $(s + \alpha)^{-1/2}$ and shifts the other pole due to $s^{-1/2}$ to the lower limit of the integral. A smooth spike is evident at the lower limit of the numerical integration using a 70 point Simpson's Rule integration (two ranges are used with 20 points more closely spaced near the pole and 50 points for the remaining range).

This work was carried out using 64 bit double precision arithmetic, which provides 14 significant figures. A simple offset δ of 1.0×10^{-15} has provided useful results with this simple offset to avoid numerical overflow.

While this pole is formally a problem in needing a continuous function to integrate, numerical integration seems to handle these Coulomb integrals are known to be accurate only to 12 significant figures. The area under the pole-spike is estimated as a narrow triangle upon a rectangle 1.0×10^{-15} wide with the height of the triangle set at 3.43207×10^8 times the height of the point set 1.0×10^{-15} into the range of integration (the first Simpson point).

The present code for this screened-Coulomb integral is presently slower that the corresponding F_0 function [17] used for the Coulomb integrals due to the 70 point Simpson integration, but the integrand is nearly flat after the spike at $s = 0.0$ so that portion of the integrand can be evaluated more rapidly with fewer points. For results presented here, the simple offset of the lower limit by 1.0×10^{-15} is adequate for this monograph. Further details on the auxiliary integral can be found in a previous paper on the H_2 molecule [20].

Work in progress indicates it may be possible to express the new auxiliary integral to an analytical expression involving the erf(x) function (see Chapter 6), but until further checks are completed this work used the Simpson integration. Note the integral is a result of a simplification of a twelve-fold integration over the volume elements of two electrons, and has been reduced to a one-dimensional integration multiplied by appropriate factors.

6. The Main Results

The geometry given for H_2O by Dunning [21] was used to carry out the usual HFR-SCF calculation after an additional $3d$ orbital mimic [19] was optimized in Ref. [3b] for the O atom and ($2s$,$2p$) orbitals were added for the H atoms. The exponent for the O$3d$ orbitals was optimized to three significant figures and the (O$3d$,H$2s$,H$2p$) exponents were (2.498, 0.500, 1.000). These polarization orbitals were added to the Dunning-Huzinaga ($10s6p$) [20] basis with the H$1s$ orbitals scaled to 1.2 which produced a lower energy than that of a 6-31G^{**} basis using the GAMESS program. The bond length of OH^+ was Angströms.

The same bond length was used for OH^- since the anion calculation using the usual HFR-SCF process was not feasible, and, in any case, the bond length is only slightly longer than that in water. The horizon cutoff value of 0.00038 Angströms optimized for H_2O was also used for OH^+ and OH^-.

The spike in the numerical integral routine was optimized by fitting the R_c cut-off value so as to obtain as near as possible the non-relativistic energy of the HF molecule as determined from Quantum Monte Carlo calculations [21]. The dipole moments for the ions are not very useful since ion dipoles are origin dependent, but they were calculated using the center-of-mass as the origin.

Table 5.1. Isoelectronium results for selected molecules [3b].

	OH^+	OH^-	H_2O	HF
SCF-Energy[a]	-74.860377	-75.396624	-76.058000	-100.060379
Hartree-Fock[b]				-100.07185[b]
Iso-Energy[c]	-75.056678	-75.554299	-76.388340	-100.448029
Horizon R_c (Å)	0.00038	0.00038	0.00038	0.00030
QMC Energy[b,d]	-76.430020[d]			-100.44296[b]
Exact non-rel.				-100.4595
Iso-Dipole (D)	5.552581	8.638473	1.847437	1.8413778
Exper. Dipole			1.84	1.82

[a] Dunning-Huzinaga ($10s/6p$), (6,2,1,1,1/4,1,1)+H$2s$1+H$2p$1+$3d$1.
[b] Iso-Energy calibrated to give maximum correlation for HF.
[c] Hartree-Fock and QMC energies from Luchow and Anderson [22].
[d] QMC energies from Hammond, Lester and Reynolds [21].

As we see in Table 5.1, the difference in energy between OH^- and OH^+ is 0.497621 Hartrees (13.54 eV) according to the Correlated-SCF calculations. It is clear from the standard SCF energy value for H_2O that this basis is very good, but not quite at the Hartree-Fock limit of energy. In addition, the fitting of the numerical integration spike so as to most nearly reproduce the total energy of HF is not exact.

These two artifacts introduce an energy uncertainty of about 0.0115 Hartrees, but this is less uncertainty than that of the Quantum Monte Carlo (QMC) energy of Luchow and Anderson [22]. Note that the Iso-Dipoles for H_2O and HF are very close to the experimental values which indicates that the calculated wavefunctions are of high quality.

Since the ionization energy of a neutral H atom is 13.60 eV and the energy difference of 13.54 eV would convert OH^- to OH^+, a threshold of about 13.7 eV should maintain H^+ in solution as well as transfer $(2e)^{-2}$ through an aqueous solution to or from the OH^-/OH^+ system.

These calculations indicate that there may be an enhancement of current flow with a potential above 13.7 volts across an aqueous cell and that the enhanced concentration of HOOH may be measurable above a potential of 13.7 volts. It is worth repeating that this estimate is possible largely due to the easy convergence of the Correlated-SCF process for a negative ion species; a process which is formally not defined under the usual Hartree-Fock-Roothan process (14), and most quantum chemists are familiar with the difficulty in treating negative ions using the standard Hartree-Fock-Roothaan method.

Admitting that the Correlated-SCF equations are a parametrized approximation to the Santilli derivation of the Hulten potential [6a] for a bound electron-pair, the method has the advantage of easy incorporation into an existing Hartree-Fock-Roothaan Gaussian basis program merely by subtracting a small "correlation integral" from the usual two-electron integrals.

With some thought, one should realize that fitting the single parameter (Gaussian screening exponent, α) to experimental energies, and/or Quantum Monte Carlo results will incorporate another attraction in the form of a magnetic dipole interaction between the spin moments of paired electrons. Including the magnetic dipole interaction and substituting a Gaussian form for the Hulten exponential potential leaves only a simulation of the bound electron-pair Isoelectronium. Thus, these results are for a model in which the usual HFR-SCF method is corrected for at least two attractive interactions of electrons causing them to approach each other as if they were "sticky"; hence the term "sticky-electron-pair model."

7. Conclusions

In Chapter 3, we have presented a covering of quantum chemistry under the name of *hadronic chemistry*. In Chapter 4, we have applied the new discipline to the construction of a new model of molecular structures based on the bonding of a pair of valence electrons from different atoms into a singlet quasi-particle state called *isoelectronium.*

We have then applied the model to the structure of the hydrogen molecule, by achieving results manifestly not possible with quantum chemistry, such as: a representation of the binding energy and other features of the hydrogen molecule accurate to the *seventh digit*; an explanation of the reason why the hydrogen molecule has only *two* hydrogen atoms; a reduction of computer usage at least 1,000 fold; and other advances.

In this chapter, we have applied the isochemical model of molecular bonds to the water and other molecules with similar results. In fact, *the isochemical model of the water and other molecules is supported by the following conceptual, theoretical and experimental evidence:*

1) It introduces a new strong binding force (which is absent in current models) capable of explaining the strength and stability of molecules;

2) It explains the reason why the water molecule has only *two* hydrogen atoms and one oxygen;

3) It permits a representation of the binding energy of the water and other molecules, which are accurate to *several digits*;

4) It represents electric and dipole moments and other features of the water and other molecules, also accurate to *several digits*;

5) It permits a reduction of computer usages in calculations at least 1,000 fold; as well as it permits other achievements similar to those obtained for the hydrogen molecule.

Moreover, as it happened for the hydrogen molecule in Chapter 4, *the value of the radius of the isoelectronium, Eqs. (5.9) computed via dynamical equations has been fully confirmed by independent calculations for the water and other molecules conducted via the Gaussian-lobe basis set.*

The emission of electron pairs in superconductivity has been emphasized in Chapter 3. In Chapter 4 we also indicated *preliminary, yet direct experimental verifications of the isochemical model of molecular bonds offered by the ongoing experiments on photoproduction of the valence electrons in the helium indicating that electrons are emitted in pairs* [23]. The systematic repetition of these experiments *specifically for water* is here recommended. The statistical percentages of electron pairs over the total number of emitted electrons would then establish whether the isoelectronium is fully or only partially stable.

We should finally note that *the representation of the binding energy, electric and magnetic moments and other characteristics of the water and other molecules exact to the several digits, as first achieved in Refs. [3] constitutes the strongest experimental evidence to date on the insufficiency of quantum mechanics and the validity of the covering hadronic mechanics for the representation of nonlinear, nonlocal and nonpotential-nonunitary effects, due to deep overlappings of the "extended wavepackets" of electrons with a point-like charge structure.*

The new isochemical model of the water molecule outlined in this chapter has a number of intriguing new applications. For instance, the correlated-SCF method is used to easily obtain an energy for the OH-anion in water, while the OH^+ ion is easily treated in either the standard or modified method. The difference in energy between the 8-electron OH^+ system and the 10-electron OH^- system is found to be 13.54 eV. This represents the energy needed to remove $(2e)^{-2}$ from OH^-. This indicates there may be a threshold for current flow in terms of $(2e)^{-2}$ as a quasi-particle in aqueous media at 13.6 eV. This voltage will also maintain H^+ in solution to some extent. Organic alkenes in solution should undergo epoxidation followed by solvolysis to diols under the conditions of abundant $(2e)^{-2}$.

Another interesting result is that the natural trace amounts of HOOH in water may be increased in water by merely placing the sample in an intense magnetic field. Positive and negative ions will traverse short arc segment paths driven by simple thermal Brownian motion in a way which will lead to an increase in collisions of oppositely charged ions. In particular, OH^- and OH^+ may undergo collisions more frequently leading to an increase in HOOH.

This additional HOOH should then be available to react with alkenes to form epoxides which will then hydrolyze in water to form diols. Such diols would be much more soluble in water than the original alkenes. This leads to the important possibility that merely exposing water-insoluble alkenes to water in a magnetic field will lead to a chemical reaction of the alkenes to form modified compounds which are more soluble in water. In other words, organic oils containing some double bonds may be made somewhat more soluble in water just by mechanical emulsification of the oils in water in an environment of a high magnetic field.

Thus, mixtures of oils and water could be mechanically agitated in a magnetic field of several Tesla to produce new oils which are chemically similar to the original oils (assuming a large organic structure) but more soluble in water after exposure to the magnetic field (see Chapter 8 for details).

Similarly, it is easy to see that, while the conventional quantum chemical model of the water molecule predicts one and only configuration, our isochemical model predicts various physically inequivalent configurations depending on the relative orientation of the two *oo*-shaped orbits and other properties, which are under separate study.

The industrial significance of the studies outlined in this chapter will be presented in Chapters 7 and 8.

References

[1] Eisenberg, D. and Kauzmann, W.: *The Structure and Properties of Water*, Oxford University Press, New York (1969).

[2] Santilli, R.M.: *Elements of Hadronic Mechanics*, Vols. **I** and **II**, Ukraine Academy of Sciences, Kiev, 2-nd ed. (1995); Found. Phys. **27**, 625 (1998); Rendiconti Circolo Matematico Palermo, Suppl. **42**, 7 (1996); Algebras, Groups and Geometries **10**, 273 (1993); *Isotopic, Genotopic and Hyperstructural Methods in Theoretical Biology*, Ukraine Academy of Sciences, Kiev (1996). Sourlas, D.S. and Tsagas, G.T.: *Mathematical Foundations of the Lie-Santilli Theory*, Ukraine Academy of Sciences, Kiev, (1993). Lôhmus, J., Paal, E. and Sorgsepp, L.:*Nonassociative Algebras in Physics*, Hadronic Press, Palm Harbor, FL (1994).

[3] Santilli, R.M. and Shillady, D.D.: Intern. J. Hydrogen Energy **24**, 943 (1999) [3a]; Intern. J. Hadrogen Energy **25**, 173 (2000) [3b].

[4] Barrow, G.M.: *Physical Chemistry*, 6th Edition, McGraw-Hill, New York (1996), Chapter 8.

[5] Pourbaix, M.: *Atlas of Electrochemical Equilibria in Aqueous Solutions*, National Assoc. Corrosion Engr., Houston, Texas, USA, p. 99. (1974).

[6] Santilli, R.M.: Hadronic J. **1**, 574 (1978); Animalu, A.O.E. and Santilli, R.M.: Int. J. Quantum Chem. **29**, 175 (1995).

[7] Boys, S.F.: Proc. Roy. Soc. (London), **A200**, 542 (1950).

[8] Whitten, J.L.: J. Chem. Phys. **39**, 349 (1963); Sambe, H.: J. Chem. Phys. **42**, 1732 (1965); Preuss, H.: Z. Naturforsch. **11a**, 823 (1956); Whitten, J.L. and Allen, L.C.: J. Chem. Phys. **43**, S170 (1965); Harrison, J.F.: J. Chem. Phys. **46**, 1115 (1967); Frost, A.A.: J. Chem. Phys. **47**, 3707 (1967).

[9] Le Rouzo, H. and Silvi, B.: Int. J. Quantum Chem. **13**, 297, 311 (1978); Nguyen, T.T., Raychowdhury, P.N. and Shillady, D.D.: J. Comput. Chem. **5**, 640 (1984).

[10] Zhitenev, N.B., Ashoori, R.C., Pfeiffer, L.N. and West, K.W.: Phys. Rev. Lett. **79**, 2308 (1997).

[11] Boyd, R.J. and Yee, M.C.: J. Chem. Phys. **77**, 3578 (1982).

[12] Fois, E.S., Selloni, A., Parinello, M. and Car, R.: J. Phys. Chem. **92**, 3268 (1988).

[13] March, J.: *Advanced Organic Chemistry*, 3rd ed., John Wiley and Sons, New York, p. 733 (1985),

[14] Cook, D.B.: *Handbook of Computational Quantum Chemistry*, Oxford Science Publications, Oxford New York, pp. 285-295 and 438-441 (1998).

[15] Goodgame, M.M. and Goddard, W.A.: Phys. Rev. Lett. **54**, 661 (1985).

[16] Schaefer, H.F.: *The Electronic Structure of Atoms and Molecules*, Addison-Wesley, Reading, Mass. (1972).

[17] Shavitt, I.: *Methods in Computational Physics*, B. Alder (ed.), Academic Press, New York (1963).

[18] Mosier, C. and Shillady, D.D.: Math. of Computation, **26**, 1022 (1972): Program LOBE140 described by D.D. Shillady and Sheryl Baldwin, Int. J. Quantum Chem., Quantum Biology Symposium **6**, 105 (1979).

[19] Shillady D.D. and Talley, D.B.: J. Computational Chem. **3**, 130 (1982); Shillady D.D. and Richardson, F.S.: Chem. Phys. Lett. **6**, 359 (1970).

[20] Dunning, T.H.: J. Chem. Phys. **53**, 2823 (1970), J. Chem. Phys. **55**, 716 (1971), J. Chem. Phys. **55**, 3958 (1971).

[21] Hammond, B.L., Lester, W.A. Jr., and Reynolds, P.J.: Monte Carlo Methods, in: Ab Initio Quantum Chemistry, World Scientific Lecture and Course Notes in Chemistry, Vol. **1**, World Scientific, New Jersey, p. 67 (1994).

[22] Luchow, A. and Anderson, J.B.: J. Chem. Phys. **105**, 7573 (1996).

[23] Burnett Collaboration, http://eve.physics.ox.ac.uk/INTENSE/zz.

[24] USMagnegas, Inc., http://www.santillimagnegas.com.

Chapter 6

VARIATIONAL CALCULATIONS OF ISOCHEMICAL MOLECULAR MODELS

1. Introduction

In Ref. [1a] outlined in Chapter 4, Santilli and Shillady introduced a restricted isochemical three-body model of the hydrogen molecule admitting an exact solution, and a full four-body isochemical model of the hydrogen molecule which no longer admits an exact solution.

In Ref. [1b] outlined in Chapter 5, Santilli and Shillady introduced two corresponding isochemical models of the water and other molecules, one based on a restricted three-body model of the HO dimer admitting exact solutions, and a second fully isochemical four-body model.

As also reviewed in Chapters 4 and 5, Shillady's SASLOBE variational method [1] showed the capability of the isochemical models to reach an essentially exact representation of experimental data on the hydrogen, water and other molecules, as well as resolving other shortcomings or inconsistencies of conventional quantum chemical molecular models.

A greatly detailed, independent verification of models [1a,1b] was conducted by A.K. Aringazin and M.G. Kucherenko [2a] via exact solution and by A.K. Aringazin [2b] via Ritz's variational method, by confirming all numerical results of Refs. [1].

In this chapter we outline Refs. [2] since they achieve new important insights and results on isochemistry of rather general character, and possible application to a variety of other molecules and applications of isochemistry.

2. Aringazin-Kucherenko Study of the Restricted, Three-Body Isochemical Model of the Hydrogen Molecule

In this section we outline the studies by Aringazin and Kucherenko [2a] of Santilli-Shillady exactly solvable, restricted three-body isochemical model of the H_2 molecule [1a], Eq. (4.35), $r_{12} \simeq 0$, i.e.,

$$-\frac{\hbar^2}{2M}\nabla^2_{ab}\psi + \left(-\frac{2e^2}{r_a} - \frac{2e^2}{r_b} + \frac{2e^2}{R}\right)\psi = E\psi. \tag{6.1}$$

As the reader will recall from Chapter 4, model (6.1) constitutes a limit case in which the two valence electrons are assumed to be permanently bonded together into the stable singlet quasi-particle state with features (4.25), i.e.,

$$\text{mass} \simeq 1 \text{ MeV}, \quad \text{spin} = 0, \quad \text{charge} = 2e, \quad \text{magnetic moment} \simeq 0,$$

$$\text{radius} = r_c = b^{-1} = 6.8432329 \times 10^{-11} \text{ cm} = \tag{6.2}$$

$$= 0.015424288 \text{ bohrs} = 0.006843 \text{ Å},$$

which we have called *isoelectronium.*

The assumption of stationary nuclei (or, equivalently, nuclei with infinite inertia), then turns the four-body hydrogen molecule H_2 into a restricted three-body system which, as such, admits exact solution.

The reader should also recall that, the assumption of the rest energy of the isoelectronium as given by twice the electron mass is merely an upper boundary occurring when the internal forces are of purely non-potential type. In reality, a total attractive force of purely potential type is possible because the magnetostatic attraction is bigger than the electrostatic repulsion as illustrated in Fig. 4.4. It is evident that the latter bond implies a negative binding energy resulting in a value of the isoelectronium mass

$$M_{\text{isoelectronium}} < 2m_{\text{electron}}, \tag{6.3}$$

which is unknown, and should be derived from fitting experimental data.

As one can see, the above restricted isochemical model of the H_2 molecule is similar to the conventional restricted three-body H_2^+ ion. To avoid confusion, we shall denote the three-body isochemical model with the "hat", $\hat{H}_2$, and the conventional (four-body) model without the "hat," H_2.

More specifically, studies [2a] were conducted under the following conditions:

1) the isoelectronium is stable;
2) the effective size of the isoelectronium is ignorable, in comparison to internuclear distance of H_2;
3) the two nuclei of H_2 are at rest;
4) the rest energy of the isoelectronium is assumed to be unknown and to be determined by the fit of the binding energy of the molecule;
5) the internuclear distance R of H_2 is also assumed to be unknown and to be fitted from the stability condition of the solution, and then compared with its experimental value.

A main result of Ref. [2a] is that the restricted three-body Santilli-Shillady model $\hat{H}_2$ is capable to fit the experimental binding energy for the following value of the isoelectronium mass,

$$M = 0.308381 m_e, \tag{6.4}$$

although its stability condition is reached for the following internuclear distance

$$R = 1.675828 \text{ a.u.}, \tag{6.5}$$

which is about 19.6% bigger than the conventional experimental value $R[H_2] = 1.4011$ a.u. $= 0.742$ Å.

These results confirm that the isochemical model (6.1) is indeed valid, but only in first approximation, in accordance with the intent of the original proposal [1a].

In Born-Oppenheimer approximation, i.e., at fixed nuclei, the equation for the H_2^+ ion-type system for a particle of mass M and charge q is given by

$$\nabla^2\psi + 2M\left(E + \frac{q}{r_a} + \frac{q}{r_b}\right)\psi = 0. \tag{6.6}$$

In spheroidal coordinates,

$$x = \frac{r_a + r_b}{R}, \quad 1 \le x \le \infty; \quad y = \frac{r_a - r_b}{R}, \quad -1 \le y \le 1; \quad 0 \le \varphi \le 2\pi, \tag{6.7}$$

where R is the separation distance between the two nuclei a and b, we have

$$\nabla^2 = \frac{4}{R^2(x^2 - y^2)}\left(\frac{\partial}{\partial x}(x^2 - 1)\frac{\partial}{\partial x} + \frac{\partial}{\partial y}(1 - y^2)\frac{\partial}{\partial y}\right) + \tag{6.8}$$

$$+ \frac{1}{R^2(x^2 - 1)(1 - y^2)}\frac{\partial^2}{\partial \varphi^2}.$$

Eq. (6.1) then becomes

$$\left[\frac{\partial}{\partial x}(x^2 - 1)\frac{\partial}{\partial x} + \frac{\partial}{\partial y}(1 - y^2)\frac{\partial}{\partial y} + \frac{x^2 - y^2}{4(x^2 - 1)(1 - y^2)}\frac{\partial^2}{\partial \varphi^2} + \right. \tag{6.9}$$

$$+\frac{MER^2}{2}(x^2-y^2)+2MqRx\Big]\,\psi=0,$$

where

$$\frac{1}{r_a}+\frac{1}{r_b}=\frac{4}{R}\frac{x}{x^2-y^2}. \tag{6.10}$$

The use of the expression

$$\psi=f(x)g(y)e^{im\varphi}, \tag{6.11}$$

then allows the separation

$$\frac{d}{dx}\left((x^2-1)\frac{d}{dx}f\right)-\left(\lambda-2MqRx-\frac{MER^2}{2}x^2+\frac{m^2}{x^2-1}\right)f=0,$$

$$\frac{d}{dy}\left((1-y^2)\frac{d}{dy}g\right)+\left(\lambda-\frac{MER^2}{2}y^2-\frac{m^2}{1-y^2}\right)g=0, \tag{6.12}$$

where λ is the separation constant. The exact solutions for $f(x)$ and $g(y)$ are given by the angular and radial Coulomb spheroidal functions (CSF) containing infinite recurrence relations.

Aringazin and Kucherenko [2a] calculated the energy levels via the use of recurrence relations of the type

$$Q_{k+1}=Q_k\bar{\kappa}_{N-k}-Q_{k-1}\bar{\rho}_{N-k}\bar{\delta}_{N-k+1},\quad Q_{-1}=0,\quad Q_0=1, \tag{6.13}$$

where the coefficients are

$$\rho_s=\frac{(s+2m+1)[b-2p(s+m+1)]}{2(s+m)+3},$$

$$\kappa_s=(s+m)(s+m+1)-\lambda, \tag{6.14}$$

$$\delta_s=\frac{s[b+2p(s+m)]}{2(s+m)-1}.$$

Ref. [2a] then used the value $N=16$ for the power degree approximation of both the radial and angular components. The two polynomials have 16 roots for λ from which only one root is appropriate due for its asymptotic behavior at $R\to 0$. Numerical solution of the equation,

$$\lambda^{(x)}(p,a)=\lambda^{(y)}(p,b), \tag{6.15}$$

gives the list of values of the electronic ground state energy,

$$E(R)=E_{1s\sigma}(R), \tag{6.16}$$

which corresponds to $1s\sigma_g$ term of the H_2^+ ion-like system, as a function of the distance R between the nuclei. Note that the term "exact solution" refers to the fact that by taking greater values of N, for example $N = 50$, one can achieve higher accuracy, up to a desired one (for example, twelve decimals).

Also, the scaling method based on the Schrödinger equation has been developed which enables one to relate the final $E(R)$ dependence of different H_2^+ ion-like systems to each other.

Table 6.1 presents result of the calculations of the *minimal total energy* and the corresponding *optimal distance*, at various values of the isoelectronium mass parameter

$$M = \eta m_e, \tag{6.17}$$

where $M = \eta$, in atomic units.

Table 6.1. The minimal total energy E_{min} and the optimal internuclear distance R_{opt} of Santilli-Shillady restricted three-body isochemical model $\hat{H}_2$ as functions of the mass M of the isoelectronium[a].

M, *a.u.*	E_{min}, *a.u.*	R_{opt}, *a.u.*	M, *a.u.*	E_{min}, *a.u.*	R_{opt}, *a.u.*
0.10	-0.380852	5.167928	0.32	-1.218726	1.614977
0.15	-0.571278	3.445291	0.33	-1.256811	1.566041
0.20	-0.761704	2.583964	0.34	-1.294896	1.519981
0.25	-0.952130	2.067171	0.35	-1.332982	1.476553
0.26	-0.990215	1.987664	0.40	-1.523408	1.291982
0.27	-1.028300	1.914050	0.45	-1.713834	1.148428
0.28	-1.066385	1.845688	0.50	-1.904260	1.033585
0.29	-1.104470	1.782044	0.75	-2.856390	0.689058
0.30	-1.142556	1.722645	1.00	-3.808520	0.516792
0.307	-1.169215	1.683367	1.25	-4.760650	0.413434
0.308	-1.173024	1.677899	1.50	-5.712780	0.344529
0.308381	-1.174475	1.675828	1.75	-6.664910	0.295310
0.309	-1.176832	1.672471	2.00	-7.617040	0.258396
0.31	-1.180641	1.667073			

[a] See also Figs. 6.1 and 6.2.

Aringazin and Kucherenko [2a] computed some 27 tables, each with the identification of the minimum of the total energy, together with the corresponding optimal distance R. Then, they collected all the obtained energy minima and optimal distances in Table 6.1.

With the fourth order interpolation/extrapolation, the graphical representations of Table 6.1 (see Figs. 6.1 and 6.2) show that the minimal

Table 6.2. Summary of main data and results on the ground state energy E and the internuclear distance R.

	E, *a.u.*	R, *a.u.*
H_2^+ ion, exact theory, N=16 [2a]	-0.6026346	1.9971579
H_2^+ ion, experiment [3]	-0.6017	2.00
3-body $\hat{H}_2$, M=$2m_e$, exact theory [2a]	-7.617041	0.258399
3-body $\hat{H}_2$, M=$2m_e$, var. theory [1a]	-7.61509174	0.2592
3-body $\hat{H}_2$, M=$0.381m_e$, exact theory [2a]	-1.174475	1.675828
4-body H_2, r_c=0.01125 a.u., V_g var. theory [1a]	-1.174474	1.4011
4-body H_2, r_c=0.01154 a.u., V_e var. theory [2b]	-1.144	1.4011
4-body H_2, r_c=0.08330 a.u., V_e var. theory [2b]	-1.173	1.3184
H_2, experiment	-1.174474	1.4011

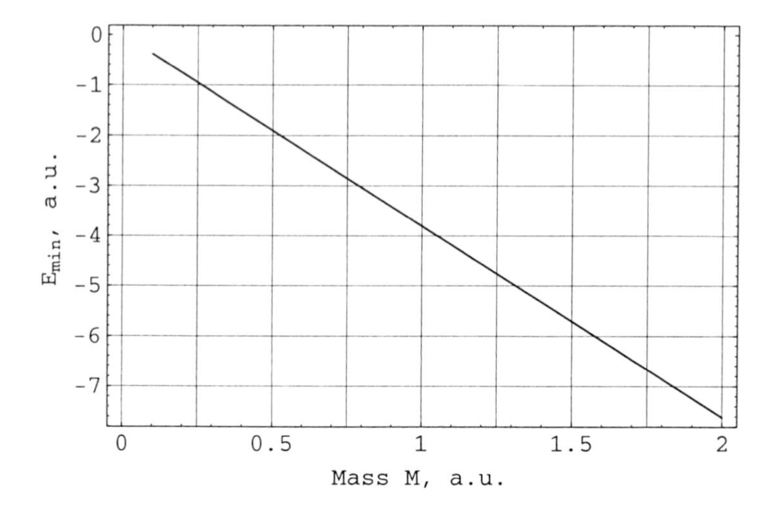

Figure 6.1. The minimal total energy $E_{min}(M)$ of the $\hat{H}_2$ system as a function of the isoelectronium mass M.

total energy behaves as

$$E_{min}(M) \simeq -3.808M, \tag{6.18}$$

and the optimal distance behaves as

$$R_{opt}(M) \simeq 0.517/M. \tag{6.19}$$

One can see that at $M = 2m_e$ we have

$$E_{min}(M) = -7.617041 \text{ a.u.}, \quad R_{opt}(M) = 0.258399 \text{ a.u.}, \tag{6.20}$$

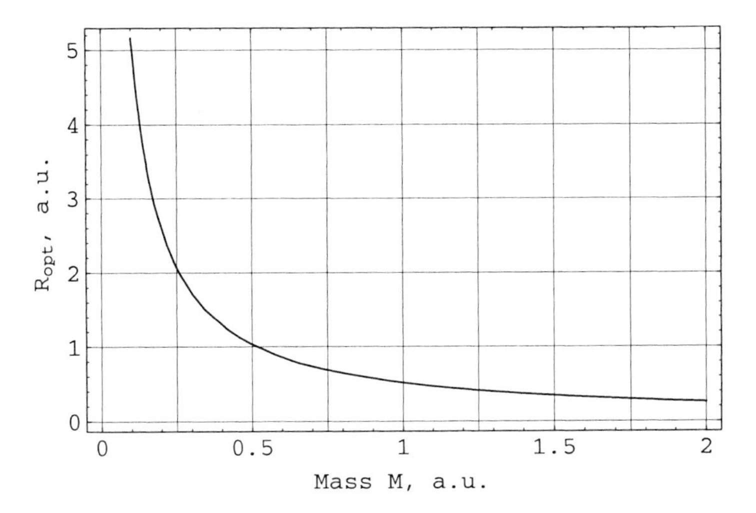

Figure 6.2. The optimal internuclear distance $R_{opt}(M)$ of the $\hat{H}_2$ system as a function of the isoelectronium mass M.

which recover the values obtained in Ref. [1a]

$$E_{min} = -7.61509174 \text{ a.u.}, \quad R_{opt} = 0.2592 \text{ a.u.}, \tag{6.21}$$

to a remarkable accuracy.

The conclusion by Aringazin and Kucherenko is that the Santilli-Shillady restricted three-body isochemical model of the hydrogen molecule is indeed valid as suggested, that is, as in first approximation. The main data and results on E_{min} and R_{opt} are collected in Table 6.2.

An important conclusion of Ref. [2a] is, therefore, that *the two valence electrons of the hydrogen molecule cannot be permanently bound inside the hadronic horizon with radius of one Fermi.*

The clear understanding, stressed in Chapter 4, is that the isoelectronium must continue to exist beyond the hadronic horizon, otherwise, in its absence, we would have a violation of Pauli's exclusion principle.

3. Aringazin Variational Study of the Four-Body Isochemical Model of the Hydrogen Molecule

In the subsequent Ref. [2b] Aringazin applied Ritz variational method to Santilli-Shillady four-body isochemical model of molecule of the hydrogen molecule (4.33), i.e.

$$\left(-\frac{\hbar^2}{2m_1}\nabla_1^2 - \frac{\hbar^2}{2m_2}\nabla_2^2 - V_0\frac{e^{-r_{12}/r_c}}{1-e^{-r_{12}/r_c}} + \frac{e^2}{r_{12}} - \right. \tag{6.22}$$

$$-\frac{e^2}{r_{1a}} - \frac{e^2}{r_{2a}} - \frac{e^2}{r_{1b}} - \frac{e^2}{r_{2b}} + \frac{e^2}{R}\Bigg) |\psi\rangle = E|\psi\rangle,$$

without restriction that the isoelectronium has the permanent dimension of about one Fermi.

In particular, Aringazin's objective was to identify the ground state energy and bond length of the H_2 molecule, in Born-Oppenheimer approximation, via a Gaussian screening of the Coulomb potential, V_g, the exponential screening of the Coulomb potential,

$$V_e = -\frac{Ae^{-r_{12}/r_c}}{r_{12}}, \tag{6.23}$$

as well as the original Hulten potential V_h of the model (6.22). The resulting analysis is quite sophisticated, and cannot be reviewed herein the necessary detail. Readers seriously interested in this verification of the new isochemical model of the hydrogen molecule are suggested to study Aringazin's original memoir [2b].

The Coulomb and exchange integrals were calculated only for V_e while for V_g and V_h Aringazin achieved analytical results only for the Coulomb integrals because of the absence of Gegenbauer-type expansions for the latter potentials.

A conclusion is that *the Ritz's variational treatment of model (6.22) with the potential (6.23) is capable to provide an exact fit of the experimental data of the hydrogen molecule in confirmation of the results obtained by Santilli and Shillady [1a] via the SASLOBE variational approach to Gaussian V_g-type model.* The main data and results on the ground state energy E_{min} and internuclear distance R_{opt} are collected in Table 6.2.

Note that in the variational approach of Ref. [2b] Aringazin used a *discrete* variation of the hadronic horizon r_c and approximate exchange integral (6.26) that resulted in approximate fittings of the energy and distance, as shown in Table 6.2.

In addition, Ref. [2b] computed the weight of the isoelectronium phase which results to be of the order of 1% to 6% that for the case of V_e model. However, we note that this is it not the result corresponding to the original Santilli-Shillady model, which is based on the Hulten potential V_h.

An interesting result is that in order to prevent divergency of the Coulomb integral for V_h *the correlation length parameter r_c should run discrete values* due to Eq. (6.27). This condition has been used in the V_e model, although it is not a necessary one within the framework of this model.

As recalled earlier, Aringazin [*loc. cit.*] assumes that the isoelectronium undergoes an increase of length beyond the hadronic horizon, and the resulting two electrons are separated by sufficiently large distance. This leads us to problem of how to compute *the effective life-time of isoelectronium.*

To estimate the order of magnitude of such a life-time, Aringazin uses the ordinary formula for radioactive α-decay since the total potential $V(r)$ is of the same shape as that here considered, with very sharp decrease at $r < r_{max}$ and Coulomb repulsion at $r > r_{max}$, where r_{max} corresponds to a maximum of the potential.

This quasiclassical model is a crude approximation because in reality the electrons do not leave the molecule. Moreover, the two asymptotic regimes act simultaneously, with some distribution of probability, and it would be more justified to treat the frequency of the decay process (i.e., the tunneling outside the hadronic horizon), rather than the life-time of the isoelectronium.

However, due to the assumption of the small size of isoelectronium in comparison to the molecule size, we can study an elementary process of decay separately, and use the notion of life-time. The results of Aringazin's calculations are presented in Table 6.3.

Table 6.3. Summary of Aringazin's calculations [2b] on the lifetime of the isoelectronium, where E is relative kinetic energy of the electrons, at large distance, $r \gg r_{max}$, in the center of mass system.

Energy E, a.u.	*eV*	*Lifetime, D_0·sec*
2	54.4	$2.6 \cdot 10^{-18}$
1	27.2	$1.6 \cdot 10^{-17}$
0.5	13.6	$2.2 \cdot 10^{-16}$
0.037	1	$5.1 \cdot 10^{-6}$
0.018	0.5	4.0
0.0018	0.1	$3.1 \cdot 10^{+25}$

In Ritz's variational approach, the main problem is to calculate analytically the so-called *molecular integrals.* The variational molecular energy in which we are interested, is expressed in terms of these integrals. These integrals arise when using some wave function, usually a simple hydrogen-like ground state wave function, as an infinite separation asymptotic solution, in the Schrödinger equation for the diatomic molecule. The main idea of Ritz's approach is to introduce parameters into the wave function, and vary them together with the separation parameter R, to achieve a minimum of the total molecular energy, which is treated as the resulting ground state energy.

In the case under study, Aringazin [*loc. cit.*] uses two parameters, γ and ρ, where γ enters hydrogen-like ground state wave function

$$\psi(r) = \sqrt{\frac{\gamma^3}{\pi}}\, e^{-\gamma r}, \tag{6.24}$$

and $\rho = \gamma R$ measures internuclear distance. These parameters should be varied *analytically or numerically* in the final expression of the molecular energy, after the calculation is made for the associated molecular integrals.

However, the four-body Santilli-Shillady model H_2 suggests an additional Hulten potential interaction between the electrons, which potential contains two parameters V_0 and r_c, where V_0 is a general factor, and r_c is a correlation length parameter characterizing the hadronic horizon. Thus, four parameters should be varied, γ, ρ, V_0, and r_c.

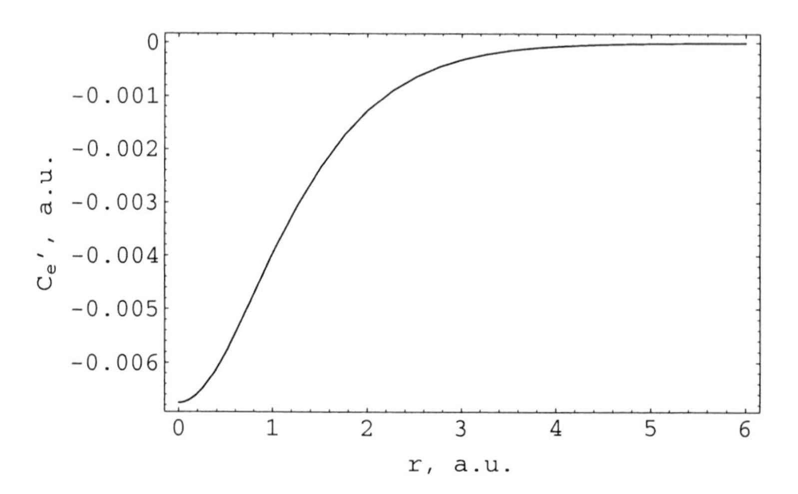

Figure 6.3. The Coloumb integral C'_e as a function of ρ, at $\lambda = 1/37$, where $\rho = \gamma R$, R is the internuclear distance, $\lambda = 2\gamma r_c$, and r_c is the hadronic horizon.

The introducing of Hulten potential leads to a modification of some molecular integrals, namely, of the Coulomb and exchange integrals. The other molecular integrals remain the same as in the case of the usual model of H_2, with well-known analytic results. Normally, the Coulomb integral, which can be computed in bispherical coordinates, is much easier to resolve than the exchange integral, which is computed in bispheroidal coordinates.

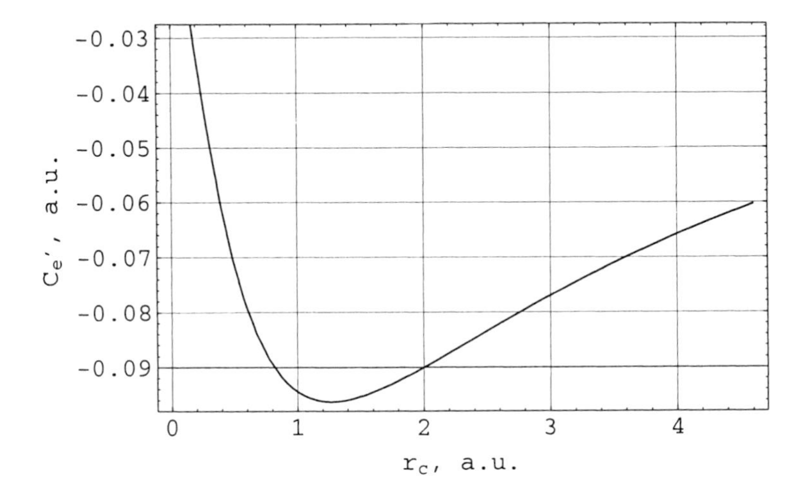

Figure 6.4. The Coloumb integral C'_e as a function of r_c, at $\rho = 1.67$. For $r_c > 0.2$ a.u., the regularized values are presented.

Calculations of the Coulomb integral for Hulten potential V_h appeared to be quite nontrivial [2b]. Namely, in the used bispherical coordinates, several special functions, such as polylogarithmic function, Riemann zeta-function, digamma function, and Lerch function, appeared during the calculation.

In order to proceed with the Santilli-Shillady approach, Aringazin [2b] invoked two different *simplified* potentials, the exponential screened Coulomb potential V_e, and the Gaussian screened Coulomb potential V_g, instead of the Hulten potential V_h. The former potentials both approximate well the Hulten potential at short and long range asymptotics, and each contains two parameters denoted A and r_c.

In order to reproduce the short range asymptotics of the Hulten potential, the parameter A should have the value $A = V_0 r_c$, for both potentials. The Coulomb integrals for these two potentials have been calculated *exactly* owing to the fact that they are much simpler than the Hulten potential.

In particular, we note that the final exact expression of the Coulomb integral for V_g contains only one special function, the error function erf(z), while for V_e it contains no special functions at all. In this way, Aringazin [2b] reaches the exact expression

$$C'_e = -\frac{A\lambda^2}{8(1-\lambda^2)^4}\frac{\gamma e^{-2\rho}}{\rho}\Big[-(\rho + 2\rho^2 + \frac{4}{3}\rho^3) + 3\lambda^2(5\rho + 10\rho^2 + 4\rho^3) -$$

$$-\lambda^4(15\rho + 14\rho^2 + 4\rho^3) + \lambda^6(8 + 11\rho + 6\rho^2 + \frac{4}{3}\rho^3 - 8e^{2\rho - \frac{2\rho}{\lambda}})\Big], \quad (6.25)$$

where $\lambda = 2\gamma r_c$. This Coloumb integral is plotted in Figs. 6.3 and 6.4.

The most difficult part of calculations [2b] is the exchange integral. Usually, to calculate it one has to use bispheroidal coordinates, and needs in an expansion of the potential in some orthogonal polynomials, such as Legendre polynomials in bispheroidal coordinates. In Ref. [2b], only the exponential screened potential V_e is known to have such an expansion but it is formulated, however, in terms of bispherical coordinates (so called Gegenbauer expansion). Accordingly, the exchange integral E'_e for V_e at *null* internuclear separation, $R = 0$ (in which case one can use bispherical coordinates) was calculated exactly. After that, the R-dependence using the standard result for the exchange integral for Coulomb potential E'_C (celebrated Sugiura's result) was partially recovered,

$$E'_e \simeq \frac{A\lambda^2}{(1+\lambda)^4}\left(\frac{1}{8} + \frac{1}{2}\lambda + \frac{5}{8}\lambda^2\right)\frac{8}{5}E'_C, \quad (6.26)$$

where $\lambda = 2\gamma r_c$ (see Fig. 6.5). Thus, only some approximate expression of the exchange integral for the case of V_e has been achieved. In this way, all subsequent results apply to the approximate V_e-based model.

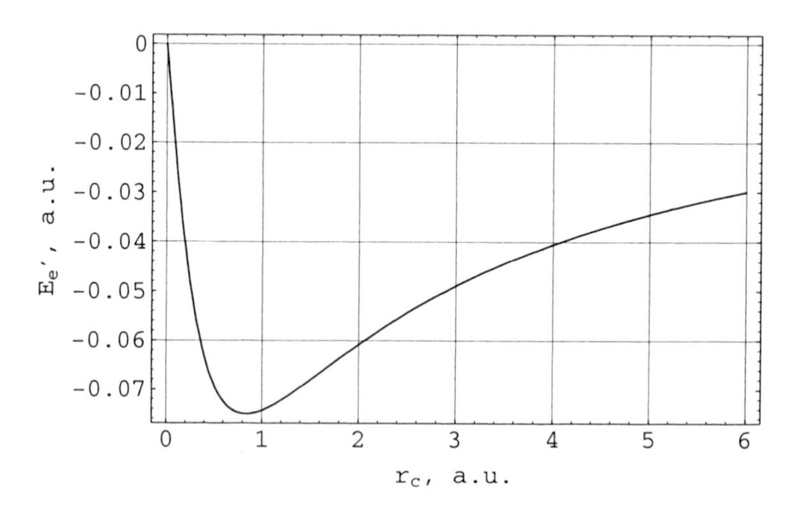

Figure 6.5. The exchange integral E'_e as a function of r_c, at $\rho = 1.67$.

Inserting the so-obtained V_e-based Coulomb and exchange integrals into the total molecular energy expression, the final analytical expression containing four parameters, γ, ρ, A, and r_c, was obtained. From a

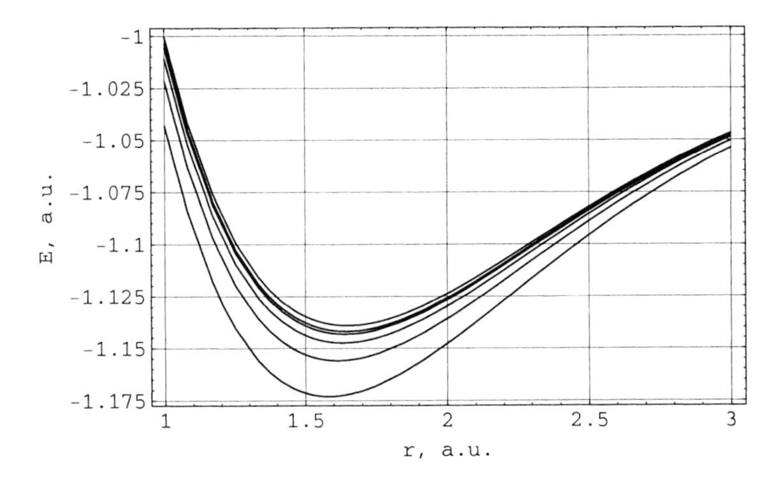

Figure 6.6. The total energy $E = E_{mol}$ as a function of ρ, at $\lambda = 1/60$, $1/40$, $1/20$, $1/10$, $1/5$. The lowest plot corresponds to $\lambda = 1/5$.

separate consideration of the Hulten potential case, the existence of a bound state of two electrons (which is the proper isoelectronium) leads to the following relationship between the parameters for the case of one energy level of the electron-electron system, $V_0 = \hbar^2/(2mr_c^2)$. Thus, using the relation $A = V_0 r_c$ Aringazin has $A = 1/r_c \equiv 2\gamma/\lambda$, in atomic units ($\hbar = m_e = c = 1$).

Note that Aringazin [2b] introduces the *one-level isoelectronium* characterized by the fact that the condition,

$$\lambda^{-1} = \text{integer numbers} > 0, \tag{6.27}$$

follows from the analysis of the Coulomb integral for Hulten potential.

With the above set up, minimization of the total molecular energy of the V_e-based model can be made. Numerical analysis shows that the λ-dependence does not reveal any minimum in the interval of interest,

$$4 \leq \lambda^{-1} \leq 60, \tag{6.28}$$

while there is a minimum of the energy for some values of γ and ρ, at fixed λ.

Therefore, 56 tables have been calculated to identify the energy minima and optimal distances for different values of λ, in the interval (6.28).

Aringazin's results are collected in Tables 6.4, 6.5, and Fig. 6.6. One can see that the binding energy decreases with the increase of the parameter r_c, which corresponds to an effective radius of the isoelectronium.

In conclusion, the calculation by Aringazin [2b] reviewed in this Chapter have not identified the meanlife of the isoelectronium assumed as a

Table 6.4. The total minimal energy E_{min} and the optimal internuclear distance R_{opt} as functions of the correlation length r_c for the exponential screened Coloumb potential V_e.

λ^{-1}	r_c, $a.u.$	R_{opt}, $a.u.$	E_{min}, $a.u.$
4	0.10337035071618050	1.297162129235449	-1.181516949656805
5	0.08329699109108888	1.318393698326879	-1.172984902150024
6	0.06975270534273319	1.333205576478603	-1.167271240301846
7	0.05999677404817234	1.344092354783681	-1.163188554065554
8	0.05263465942162049	1.352417789644028	-1.160130284706318
9	0.04688158804756491	1.358984317233049	-1.157755960428922
10	0.04226204990365446	1.364292909163710	-1.155860292450436
11	0.03847110142927672	1.368671725082009	-1.154312372623724
12	0.03530417706681329	1.372344384866235	-1.153024886026671
13	0.03261892720535206	1.375468373051375	-1.151937408039373
14	0.03031323689615631	1.378157728092548	-1.151006817317425
15	0.02831194904031777	1.380497017045902	-1.150201529091051
16	0.02655851947236431	1.382550255552670	-1.149497886394651
17	0.02500959113834722	1.384366780045693	-1.148877823925501
18	0.02363136168905809	1.385985219224291	-1.148327310762828
19	0.02239708901865092	1.387436244558651	-1.147835285349041
20	0.02128533948435381	1.388744515712491	-1.147392910500336
21	0.02027873303335994	1.389930082626193	-1.146993041730378
22	0.01936302821907175	1.391009413196452	-1.146629840949675
23	0.01852644434336641	1.391996158084790	-1.146298491232105
24	0.01775915199935013	1.392901727808297	-1.145994983116511
25	0.01705288514774330	1.393735733699196	-1.145715952370148
26	0.01640064219648127	1.394506328745493	-1.145458555325045
27	0.01579645313764336	1.395220473843219	-1.145220372020229
28	0.01523519631632570	1.395884147817973	-1.144999330178493
29	0.01471245291356761	1.396502514589167	-1.144793644973560
30	0.01422439038752817	1.397080057337240	-1.144601770891686

quasiparticle of charge radius r_c of about 1 fm. As one can see in Table 6.3, the predicted meanlife varies over a rather large range of values.

The achievement of an accurate meanlife of the isoelectronium of 1 fm charge radius can be reached only after reaching a more accurate knowledge of its rest energy. As the reader will recall from Chapter 4, the value of 1 MeV should be solely considered as an upper boundary value of the rest energy of the isoelectronium, since it holds only in the absence of internal potential forces while the latter cannot be excluded. Therefore, the actual value of the rest energy of the isoelectronium is today basically unknown.

Table 6.5. A continuation of Table 6.4.

λ^{-1}	r_c, *a.u.*	R_{opt}, *a.u.*	E_{min}, *a.u.*
31	0.01376766836566138	1.397620687025853	-1.144422362947838
32	0.01333936209977966	1.398127830817745	-1.144254245203342
33	0.01293689977547854	1.398604504597664	-1.144096385030938
34	0.01255801083612469	1.399053372836414	-1.143947871939897
35	0.01220068312791624	1.399476798299823	-1.143807900045981
36	0.01186312715793131	1.399876883556063	-1.143675753475045
37	0.01154374612489787	1.400255505817128	-1.143550794143290
39	0.01095393745919852	1.400954915288619	-1.143320213707519
40	0.01068107105944273	1.401278573036792	-1.143213620508321
41	0.01042146833640030	1.401586548200467	-1.143112256673494
42	0.01017418516195214	1.401879953246168	-1.143015746732479
43	0.00993836493541500	1.402159797887369	-1.142923750307661
44	0.00971322867044429	1.402427000676349	-1.142835958109381
45	0.00949806639934841	1.402682399061957	-1.142752088467028
46	0.00929222969498477	1.402926758144872	-1.142671884314343
47	0.00909512514431396	1.403160778323019	-1.142595110561057
48	0.00890620863525624	1.403385101987775	-1.142521551794315
49	0.00872498034101540	1.403600319405678	-1.142451010262626
50	0.00855098030451296	1.403806973898863	-1.142383304102633
51	0.00838378454080327	1.404005566419838	-1.142318265775268
52	0.00822300158793934	1.404196559601683	-1.142255740683024
53	0.00806826944722482	1.404380381352424	-1.142195585944305
54	0.00791925286251402	1.404557428052374	-1.142137669304475
55	0.00777564089552400	1.404728067404676	-1.142081868166104
56	0.00763714476025456	1.404892640982100	-1.142028068723488
57	0.00750349588477794	1.405051466507240	-1.141976165188595
58	0.00737444417302681	1.405204839898059	-1.141926059097351
59	0.00724975644291090	1.405353037106507	-1.141877658686723
60	0.00712921502024112	1.405496315774223	-1.141830878334298

The reader should also recall that the terms "meanlife of the isoelectronium when of charge radius of about 1 fm" are referring to the duration of time spent by two valence electrons at a mutual distance of 1 fm which is expected to be small. The understanding explained in Chapter 3 is that, *when the restriction of the charge radius to 1 fm is removed, and orbital mutual distances are admitted, the isoelectronium must have an infinite life (for the unperturbed molecule), because any finite meanlife under the latter conditions would imply the admission of two electrons with identical features in the same orbit, and a consequential violation of Pauli's exclusion principle.*

An interesting result of the Ritz variational approach to the Hulten potential studied by Aringazin [2b] is that *the charge radius of the isoelectronium r_c entering the Hulten potential and the variational energy, should run discrete set of values during the variation.*

In other words, this means that *only some fixed values of the effective radius of the one-level isoelectronium are admitted in the Santilli-Shillady model when treated via the Ritz approach.*

This result was completely unexpected and may indicate a kind of "hadronic fine structure" of the isoelectronium whose origin and meaning are unknown at this writing. It should be indicated that such a "hadronic fine structure" of the isoelectronium is solely referred to the case when r_c is restricted to be about 1 fm or less. The problem whether such a "hadronic fine structure" persists for values of r_c up to orbital distances is also unknown at this writing. It should be also indicated that this remarkable property is specific to the Hulten potential V_h, while it is *absent* in the V_e, or V_g models.

Moreover, Aringazin [2b] has achieved an estimation of *the weight of the isoelectronium phase* for the case of V_e model which appears to be of the order of 1% to 6%. This weight has been estimated from the energy contribution related to the exponentially screened potential V_e, in comparison to the contribution related to the usual Coulomb interelectron repulsive potential.

Finally, an important result of the Ritz variational four-body model studied by Aringazin [2b] is its fit to the experimental data of both the binding energy E and the bond length R of the hydrogen molecule thus providing an excellent independent confirmation of the results obtained by Santilli and Shillady [1].

References

[1] Santilli, R.M. and Shillady, D.D.: Intern. J. Hydrogen Energy **24**, 943 (1999) [1a]; Intern. J. Hadronic Energy **25**, 173 (2000) [1b].

[2] Aringazin, A.K. and Kucherenko, M.G.: Hadronic J. **23**, 1 (2000). e-print physics/0001056 [2a]. Aringazin, A.K.: Hadronic J. **23**, 57 (2000). e-print physics/0001057 [2b].

[3] Flugge, Z.: *Practical Quantum Mechanics*, Vols. **1**, **2**, Springer-Verlag, Berlin (1971).

Chapter 7

APPLICATION OF HADRONIC CHEMISTRY TO NEW CLEAN ENERGIES AND FUELS

1. Introduction

Undoubtedly, the most serious problem of contemporary society is the alarming decay of our environment due to the disproportional combustion of about 74 million barrels of fossil fuels per day (see Sect. 7.2 for details).

A primary objective for which hadronic chemistry was built is the theoretical prediction and industrial development of basically new, environmentally acceptable energies and fuels.

This chapter reports the research initiated by Santilli in 1998 at *Toups Technologies Licensing, Inc.*, a U.S. public company in Largo, Florida, under the presidency of Leon Toups. The research was then continued at *USMagnegas, Inc.*, a subsidiary of *EarthFirst Technologies, Inc.*, a public company under the presidency of John Stanton.

Note that there is simply no natural resource which can provide an environmentally acceptable replacement of fossil fuels. As a consequence, new clean fuels must be synthesized.

To be acceptable, new fuels must meet the following main requirements:

1) Resolve the environmental problems caused by fossil fuel combustion;

2) Be usable in existing internal combustion engines; and

3) Be price-competitive with respect to currently available fossil fuels.

As it will be soon evident, quantum chemistry is insufficient to fulfill all these rather difficult requirements, thus mandating the study of a structurally novel chemistry.

It should be indicated from these introductory lines that we shall report hereon *corporate*, rather than academic research along the indicated objectives. This is requested by the need that the results have direct *industrial* significance. It is, therefore, best to outline the differences between academic, corporate and military research, and then pass to the study of new energies and fuels as needed from an industrial profile.

As it is well known, the U.S. Military essentially halted in the 1970's all funding of advanced research in the U.S. academia. A primary reason for this decision was the impossibility of permitting the security of the United States of America be subject to preferred academic theories.

It is today generally admitted that the physical or chemical research conducted at military centers, such as *Scandia Laboratories*, is dramatically more advanced than the research conducted at academic institutions. This disparity is due to the notorious *restrictions* of academic research to be aligned with preferred theories, such as Einstein's special relativity, relativistic quantum mechanics and quantum chemistry, as compared to the general *lack* of such constraints in military research, with consequential advances that would be otherwise impossible.

It is lesser known that a similar divergence in advanced research is now well under way between the U.S. industry and academia for essentially the same reasons, thus resulting in an essentially similar gap.

First, basic novelty is a notorious enemy to oppose in academia, while basic novelty is a necessary prerequisite for interest in the industry, as it should be. Therefore, academic research is aimed at more and more sophisticated advances in established doctrines, with consequential lesser and lesser industrial relevance, while industrial research is aimed at basically novel advances, as a necessary condition to fulfill new societal needs or create new markets.

As a consequence of this scenario, there is a growing gap between the U.S. industry and academia in the very conception of advanced research which appears to be as irreconcilable as the divergence between military and academic research.

Moreover, the industry tends more and more to avoid the propagation of its results to academic circles, as an understandable protective measure to avoid predictable damage, thus resulting in a secrecy somewhat reminiscent of military research.

When passing to specific technical issues, the divergences in approaches, methods and values between industrial and academic research become more visible.

As an illustration, it will be evident in this and the remaining Chapters that clusters play a rather fundamental role in the synthesis and industrial production of new fuels meeting the indicated requirements. In fact, academic and industrial research on clusters has increased considerably in the past decades (see representative Refs. [1] and literature quoted therein).

However, an inspection of these studies soon reveals dramatic differences in the conception and realization of the research. Academic studies are generally conducted without any quantitative identification of the truly fundamental notion, the *attractive force permitting the very existence of clusters.* As a result, these studies have no clear industrial relevance evidently because the lack of a precise identification of the bonding force prevents the industrial reproduction and optimization of said clusters. For this reason, industrial studies on clusters have been primarily centered in the identification of the attractive force creating the clusters, and then on the description of their properties.

At first, the above divergence of views does not appear to be important. The divergence appears in all its dimension when entering into the technical identification of said bonding force, which is generally treated in academic circles under the constraint of being compatible with existing doctrines, thus resulting in terminologies without an actual underlying model, such as the assumption of "covalence bonds" (when the bond is known *not* to be of valence type), or "ionic bonds" (when it is well known that ions *repel* and do not attract each other because they have charges of the same sign) and other abstract assumptions compatible with pre-existing lines of research.

As we shall see in this and in the next chapter, a quantitative, numerical and plausible identification of the *attractive* force responsible for clusters of direct relevance for new clean fuels requires the admission of *a basically new chemical species.* It is at that point where the divergence between corporate and academic research becomes visible, again, because contemporary academia generally *opposes* the pursuit of true novelty, of course, with due exceptions.

Dramatically bigger divergencies exist between industrial and academic research in an area of chemistry where one would expect them the least: experimental measurements via Gas Chromatographic Mass Spectrometers (GC-MS). In fact, the latter equipment is generally used in academic research without an associated InfraRed Detector (IRD), while only measurements from GC-MS equipped with IRD are generally admitted in industry.

The reason for this additional disparity is the general dismissal in academia of a possible new chemical species, thus implying the expectation that all peaks detected by a GC-MS must be one or another molecule.

By comparison, industry is aware of the fact that, with the sole exception of rare cases of full spherical symmetry (such as the hydrogen), molecules must admit an IR signature as a *necessary* condition for their existence. Therefore, for the industry, a true experimental identification of a molecule beyond the level of personal opinions generally requires both, its identification as a peak in the MS, *as well as* the independent verification of the IR signature of that particular peak considered (rather than others).

As we shall see, basically new, environmentally acceptable energies and fuels are crucially dependent on the existence of a new chemical species consisting of heavy clusters detected via large MS peaks with high atomic weight *which cannot be identified in the MS scan and have no IR signature.* Since these large clusters cannot possibly all be perfectly symmetric, the joint use of the IRD prevents personal opinions on conventional interpretations, thus permitting a really scientific measurement.

In view of the above aspects, these final Chapters report ongoing *industrial* efforts to fulfill pressing societal needs for new, environmentally acceptable energies and fuels, without any consideration for currently preferred lines of research in academia.

2. Alarming Environmental Problems Caused by Gasoline and Coal Combustion

As it is well known, gasoline combustion requires atmospheric oxygen which is then turned into CO_2 and various hydrocarbons (HC). In turn, CO_2 is recycled by plants via the known reaction

$$H_2O + CO_2 + h\nu \rightarrow O_2 + CH_2O, \tag{7.1}$$

which restores oxygen in the atmosphere. This was essentially the environmental scenario at the beginning of the 20-th century.

The same scenario at the beginning of the 21-st century is dramatically different, because forests have rapidly diminished while we have reached the following unreassuring *daily consumption of crude oil* according to official releases by the U.S. Department of Energy (see Appendix 7.A for units and their conversions)

$$\begin{aligned} &74.18 \times 10^6 \text{ barrels/day} = 4.08 \times 10^9 \text{ gallons/day} = \\ &= 9.4984 \times 10^{10} \text{ moles n-octane/day}, \end{aligned} \tag{7.2}$$

(see, e.g. [2]), where we have replaced, for simplicity, crude oil with a straight chain of n-octanes $CH_3\text{-}(CH_2)_6\text{-}CH_3$ possessing the known density of 0.7028 g/cc at 20° C.

It should be indicated that data (7.2) *do not* include the additional large use of coals and natural gas, which would bring the daily combustion of fossil fuels up to the equivalent of about *one hundred and twenty millions barrels of fossil oil per day*. The primary environmental problems caused by the above disproportionate consumption of fossil fuel per day are the following:

1) *Excessive emission in our atmosphere of carcinogenic and other toxic substances.* It is well known by experts that gasoline combustion releases in our atmosphere the largest percentage of carcinogenic and other toxic substances as compared to any other source. The term "atmospheric pollution" is an euphemism for very carcinogenic breathing. As a known yet generally unspoken illustration, the recent alarming increase of breast and other types of cancer have been directly linked by medical research to the breathing of carcinogenic substances from fossil fuel combustion.

2) *Excessive release of carbon dioxide in our atmosphere.* It is evident that, under the very large daily combustion (7.2), plants cannot possibly recycle the entire production of CO_2, thus resulting in an alarming increase of CO_2 in our atmosphere, an occurrence known as the *green house effect*. In fact, by using the known reaction

$$C_8H_{18} + \frac{25}{2}O_2 \rightarrow 8CO_2 + 9H_2O, \tag{7.3}$$

we have the following *alarming daily production of CO_2 from fossil fuel combustion*:

$$7.5987 \times 10^{11} \text{ moles } CO_2/\text{day} = 3.3434 \times 10^7 \text{ metric tons/day}. \tag{7.4}$$

It is evident that plants cannot possibly recycle such a disproportionate amount of daily production of CO_2. This has implied a considerable increase of CO_2 in our atmosphere which can be measured by any person seriously interested in the environment via the mere purchase of a CO_2 meter, now available from numerious industries, at low cost, and then the comparison of current readings of CO_2 with standard values on record, e.g.,

$$\begin{array}{c}\text{Percentage of } CO_2 \text{ in our atmosphere}\\ \text{at sea level in 1950: } 0.033 \pm 0.01\%\end{array} \tag{7.5}$$

(see, e.g., the *Encyclopedia Britannica* of that period).

Along these lines, in the laboratory of *USMagnegas, Inc.*, in Florida we measure a *thirty fold* increase of CO_2 in our atmosphere over the above indicated standard. We assume the reader is aware of recent reports of the detection this past summer of *small lakes of fresh water in the North Pole*, an occurrence which has never been observed before. Increasingly catastrophic climactic events are known to everybody.

3) *Excessive removal of directly usable oxygen from our atmosphere*, an environmental problem of fossil fuel combustion, here called, apparently for the first time, *oxygen depletion*, which is lesser known than the green house effect even among environmentalists, but potentially more catastrophic. The problem refers to the difference between the oxygen needed for the combustion of fossil fuels less that recycled by plants and expelled in the exhaust.

By using reaction (7.3) and data (7.4), it is easy to obtain an estimate of the following additionally alarming *daily removal of oxygen from our atmosphere for the combustion of fossil fuels*:

$$1.1873 \times 10^{12} \text{ moles of } O_2 \text{ per day } = 3.7994 \times 10^7 \text{ metric tons per day.} \tag{7.6}$$

Again, this large volume of oxygen is turned by the combustion into CO_2 of which only an *unknown part* is recycled by plants into usable oxygen.

It should be stressed that *the very existence of the green house effect is unquestionable evidence of oxygen depletion*, because we are dealing precisely with the quantity of O_2 in the CO_2 of the green house effect which has not been re-converted by plants.

Besides the deaths due to cancer originated from the breathing of excessive carcinogenic substances of fossil fuels exhaust, and the additional deaths due to catastrophic climactic episodes caused by the green house effect, the combustion of fossil fuels causes additional deaths due to heart attacks linked by medical research to locally insufficient oxygen.

Oxygen depletion is today measurable by any person seriously interested in the environment via the mere purchase of an oxygen meter, which is available today at low cost even in small portable versions, measure the local percentage of oxygen, and then compare the result to standards on record, e.g.,

$$\begin{array}{c}\text{Oxygen percentage in our atmosphere} \\ \text{at sea level in 1950: } 20.946 \pm 0.02\%\end{array} \tag{7.7}$$

(see, again, the *Encyclopedia Britannica* of that period). Along these lines, in the laboratory of *USMagnegas, Inc.*, in Florida we measure a *local oxygen depletion* of 3% to 5%. Evidently, bigger oxygen depletions are expected for densely populated areas, such as Manhattan, London, and Tokyo, or at high elevation.

We assume the reader is aware of the recent decision by U.S. airlines to *lower* the altitude of their flights despite the evident increase of cost. Reportedly, this decision has been motivated by oxygen depletion causing fainting spells due to insufficient oxygen suffered by passengers during flights at previous higher altitudes.

The above environmental problem caused by the disproportional daily consumption (7.2) of fossil fuels is so serious to be potentially catastrophic. As such, they mandate their study without any consideration to personal interests, whether academic or not.

3. Alarming Environmental Problems Caused by Natural Gas Combustion

It is generally believed that the replacement of gasoline with natural gas implies an improvement of the combustion exhaust with consequential environmental advantages.

It is important for researchers and industrialists seriously interested in the environment to know that this *is not* the case. In fact, current gasoline fueled vehicles with catalytic converters essentially emit the same hydrocarbon and carbon monoxide as vehicles operated by Compressed Natural Gas (CNG). However, it is little known that *CNG vehicles emit more carbon dioxide in the exhaust then gasoline fueled car, thus causing a serious further deterioration of the already alarming green house problem.*

As an illustration, comparative EPA tests indicate that a Honda Civic Natural Gas Vehicle (with 1,600 cc engine) emits (see later on Sect. 7.9)

$$646.35 \text{ grams of } CO_2 \text{ per mile,} \tag{7.8}$$

while a similar gasoline operated Honda emits

$$458.65 \text{ grams of } CO_2 \text{ per mile.} \tag{7.9}$$

As a result, *the automotive combustion of natural gas for the same performance implies an increase in the emission of carbon dioxide of about 40% which renders the combustion of natural gas environmentally unacceptable.*

It should be indicated that emissions (7.8) and (7.9) *are not* related to the running of engines at idle, and refer instead to measurements following the formal EPA testing routine which include city driving as well as hill climbing.

It is evidently true that the combustion of natural gas in electric power plants in replacement of coal or other fossil fuels without catalytic converters does imply an improvement of the combustion exhaust with respect to carcinogenic and other toxic emissions, the understanding being that the emission of carbon dioxide as well as the oxygen depletion remain alarming.

4. Alarming Environmental Problems Caused by Hydrogen Combustion, Fuel Cells and Electric Cars

Hydrogen is generally perceived as the cleanest and most inextinguishable fuel available to mankind, evidently because it merely produces water vapor in the combustion exhaust, and can be produced from water.

It is important to note that, whether used for direct combustion or in fuel cells, hydrogen produced from regeneration methods (e.g., from natural gas) causes environmental problems *bigger* then those due to fossil fuels combustion, as recently studied, e.g., by P. Spath and M. Mann of the *U.S. National Renewable Energy Laboratory* [3].

This is evidently due to the fact that the production of hydrogen from regenerating methods requires the use of *electric energy generally produced by fossil fuels combustion.* Since the efficiency of hydrogen regeneration plants is evidently smaller than one, and fossil powered electric plants are *more* polluting than gasoline exhaust, it is then evident that the combustion of hydrogen of the indicated origin is more polluting than gasoline combustion on a global scale, since the electricity for the production of the latter is ignorable.

Moreover, as indicated in the preceding Section, CO_2 produced from fossil fuel is partially recycled by plants into O_2, while *hydrogen combustion permanently removes directly breathable oxygen from our planet, thus causing an alarming oxygen depletion*, since the separation of water produced in the exhaust to restore the original oxygen balance (again, for hydrogen from regenerating plants) is excessively expensive.

By assuming that gasoline is solely composed of one octane C_8H_{18}, thus ignoring other isomers, the combustion of one mole of H_2 gives 68.32 Kcal, while the combustion of one mole of octane produces 1,302.7 Kcal. Thus, we need $19.07 = 1302.7/68.32$ moles of H_2 to produce the same energy of one mole of octane.

In turn, the combustion of 19.07 moles of H_2 requires 9.535 moles of O_2, while the combustion of one mole of octane requires 12.5 moles of O_2. Therefore, on grounds of the same energy release, the combustion of hydrogen requires less oxygen than gasoline (about 76% of the oxygen consumed by the octane).

The alarming oxygen depletion occurs, again, because of the fact that the combustion of hydrogen turns oxygen into water, by therefore permanently removing breathable oxygen from our planet. When used in modest amounts, the combustion of hydrogen constitutes no appreciable environmental problem. However, *when used in large amounts, the combustion of hydrogen produced via regenerative methods is potentially catastrophic on environmental grounds, because oxygen is the foundation of life.*

At the limit, a global combustion of hydrogen of regenerating origin in complete replacement of fossil fuels would imply the permanent removal from our atmosphere of 76% of the oxygen currently consumed to burn fossil fuels, i.e., from Eq. (7.6), we would have the following *permanent oxygen depletion due to global hydrogen combustion*:

$$\begin{aligned} &76\% \text{ oxygen used for fossil fuel combustion} = \\ &= 2.8875 \times 10^7 \text{ metric tons of } O_2 \text{ depleted per day.} \end{aligned} \tag{7.10}$$

In addition, one should take into account the quantitatively similar oxygen depletion caused by the production of electricity, resulting in a truly catastrophic oxygen depletion which could imply the termination of any life on Earth within a few years.

Predictably, the above feature of hydrogen combustion has alarmed environmental groups, labor unions, and other concerned people. As an illustration, calculations show that, in the event all fuels in Manhattan were replaced by hydrogen, the local oxygen depletion would cause heart failures, with evident large financial liabilities and legal implications for hydrogen suppliers and users alike.

It should be noted that, in addition to carbon dioxide, plants also recycle water, as clearly shown by the main reaction creating the chlorophyl, Eq. (7.1). However, by no means, this resolves the oxygen depletion for the combustion of hydrogen from reformation processes, because: 1) the oxygen belonging to the water in Eq. (7.1) is only one third of the total original oxygen; 2) a large oxygen depletion occurs in the carbon dioxide emitted by reformation processes; and 3) the largest oxygen depletion is created by the fossil fuel combustion of the power plant producing the needed electricity.

Another potentially misleading aspect is that the combustion of one mole of hydrogen requires only 78% of the oxygen needed for the combustion of one mole of gasoline. This occurrence is insignificant for the problem of the oxygen depletion because comparison have to be done for same performances, since about 400 cf of hydrogen are needed to reach the energy content of one gallon of gasoline.

In conclusion, under the strict requirement of comparable performances between gasoline-fueled and hydrogen-fueled vehicles, and in view of the fact that hydrogen is the fuel with the lowest energy content, the dramatic increase of oxygen depletion in the combustion of hydrogen originating from reformulating processes is simply beyond any possible or otherwise credible doubt. Unfortunately, the numerical value of said increase cannot be accurately identified at this moment due to a number of yet unsolved issues, beginning with the unknown daily amount of CO2 which is no longer recycled by plants, whose O2 content is precisely the oxygen depletion.

The combustion of hydrogen produced from the electrolytic separation of water via electricity originating from conventional power plants, has similar environmental problems. In fact, the original separation of the water, and its subsequent recombination in the combustion does indeed preserve the original oxygen balance in our atmosphere. However, an oxygen depletion greater than that of Eq. (7.10) is caused by the combustion of fossil fuels to produce the electricity needed for the separation of water.

Recall that the combustion of fossil fuels in electric power plants implies the emission of large amounts of carcinogenic substances and carbon dioxide, plus oxygen depletion. As a result, *the automotive use of hydrogen whose production requires electricity originating from conventional power plants is more polluting than contemporary gasoline vehicles.*

Along similar lines, and contrary to popular beliefs, *the large use in fuel cells of hydrogen from regenerating plants is also environmentally unacceptable because of the disproportionate oxygen depletion caused by fuel cells, as well as the disproportionate environmental problems caused by the power plants for the production of the electricity needed for the hydrogen production.*

In other words, the sole restriction of the analysis to the exhaust itself is an approach belonging to the past century because the emphasis now is in *global environmental problems*, as it should be. Therefore, *a fuel cell vehicle operating with hydrogen from regeneration or electrolytical sources is more polluting than a contemporary gasoline vehicle*, again, because the fossil fuels burned to generate the electricity needed to produce hydrogen is more polluting than gasoline exhaust for the same energy output.

Along similar lines, *battery powered electric cars recharged by fossil fueled power plants are considered nowadays much more polluting than gasoline operated vehicles*, again, because power plants are more polluting than gasoline cars.

We should also indicate that the use of hydrogen for automotive fuel, e.g., as in the BMW hydrogen car [4], has the following primary insufficiencies:

1) Hydrogen does not possess sufficient energy density to be used in a compressed form for a practically meaningful range, thus requiring its liquefaction. All readers know the technical difficulties in liquefying hydrogen down to the requested minus 253° C, plus the additional difficulties in transporting such liquefied hydrogen, and the further problems in maintaining hydrogen in an automotive tank.

2) When used as fuel for an internal combustion engine, hydrogen implies the loss of about 35% in power as compared to the use of gasoline in the same engine. This problem is evidently linked to the preceding one and it is due to insufficient energy output.

3) The current use of hydrogen as fuel for vehicles is excessively more expensive than gasoline. This is due to: the need to produce hydrogen with the low (under-unity) efficiencies of electrolytical plants; the need to liquefy hydrogen; the need to transport and store hydrogen in a liquefied form; and other costs.

Therefore, in addition to the achievement of an environmentally acceptable method for hydrogen production, the practical, competitive use of hydrogen as fuel requires basic advances in chemistry capable of resolving the above three shortcomings.

It should be indicated that manufacturers of hydrogen powered vehicles do recommend the production of hydrogen via renewable energy sources, such as solar, wind or hydroelectric sources. However, such sources are dramatically insufficient to produce hydrogen in any amount which could be appreciable when compared to the daily need of the equivalent of 120 millions barrels of crude oil per day.

All conceivable studies based on quantum chemistry have been exhausted long ago. At any rate, the chances of discovering really new energies or fuels via quantum chemistry after studies for about one century by so many scientists, said chances are virtually null. The need for a generalization of quantum chemistry to reach the needed advances is then evident, as we shall see in more detail in the rest of this chapter.

5. The Need for New, Environmentally Acceptable Primary Sources of Electricity

The above scenario identifies quite clearly the fact that the most serious environmental problems exist nowadays in the *primary energy sources*, those used for the production of electricity, while *secondary energy sources*, the fuels used for locomotion and other needs, have a secondary relevance.

As indicated earlier, the environmentally ideal selection of energies is the use of renewable primary energy sources, such as solar, wind, and other energies, and the use for fuel of hydrogen originating from the separation of the water. Within such assumed scenario, the production of electricity would leave the environment unaffected; the separation of water to create hydrogen releases in the atmosphere breathable oxygen; while the combustion of hydrogen re-uses exactly the same amount of oxygen emitted in the separation of the water, while producing no other toxic substance, thus resulting in no alteration of our atmosphere.

As a result of the gross insufficiency of the indicated renewable primary energy, primary human and financial resources should be devoted to the most serious environmental problems, those pertaining to the production of electricity, and only secondary efforts should be devoted to secondary energy sources.

Unfortunately, the current scenario is exactly the opposite, inasmuch as primary efforts are devoted to fuel production, under the generally myopic ignorance that the production of electricity to synthesize fuels is more polluting than current uses of gasoline.

6. Insufficiencies of Quantum Mechanics, Superconductivity, and Chemistry for the Solution of Current Environmental Problems

The research that led to the construction of hadronic mechanics, hadronic superconductivity, and hadronic chemistry outlined in this monograph have established that *quantum mechanics, superconductivity and chemistry are structurally insufficient to provide a real solution of our alarming environmental problems.*

As an illustration, recall from Chapter 1 that the current 2% error by quantum chemistry in the representation of molecular binding energies is deceptively small, because in actuality it corresponds to about 950 Kcal/mol, thus implying a *20-fold error* for thermochemical calculations, since the latter are of the order of 50 Kcal/mole. Therefore, any study of environmental problems via quantum chemistry is deceptive, let alone ineffective, unless the basic insufficiencies of quantum chemistry are identified with clarity and resolved via a broader theory.

A plethora of additional insufficiencies of pre-established disciplines has been identified in Chapter 1 and in the quoted literature. Irrespective of that, it is well known that the study of all conceivable energies predicted or permitted by pre-established disciplines has been exhausted in the 20-th century. In any case, the chances of identifying really new sources of clean energies compatible with pre-established doctrines is virtually null, thus not worth serious considerations.

It is easy to see that, even assuming the availability of a primary clean source of energy for the production of hydrogen, quantum chemistry is unable to resolve insufficiencies 1), 2) and 3) of Sect. 7.5 for its practical use in a way effective and cost competitive over gasoline.

More specifically, despite large investments by the industry (see, e.g., Ref. [4]), quantum chemistry has been unable to identify a form of hydrogen permitting its automotive use in a compressed form, with an energy output comparable to that of gasoline, and at a competitive cost. To see the latter occurrence, suppose that one attempts to upgrade hydrogen with other combustible gases so as to achieve environmental acceptability. It is easy to see that this question does not admit an industrially and environmentally acceptable answer via the use of quantum chemistry.

For instance, the addition of CO to H_2 in a 50%–50% mixture would leave the oxygen depletion unchanged. In fact, each of the two reactions $H_2 + \frac{1}{2}O_2 \rightarrow H_2O$, and $CO + \frac{1}{2}O_2 \rightarrow CO_2$, requires 1/2 mole of O_2. Therefore, the 50%-50% mixture of H_2 and CO would also require 1/2 mole of O_2, exactly as it is the case for the pure H_2.

After studying the issue for years, the only solution known to the author to resolve the indicated three intrinsic insufficiencies of hydrogen as fuel is the theoretical prediction, quantitative study, and industrial realization of *new chemical species beyond that of molecules*, as outlined later on in this Chapter and Chapter 8.

7. The New Clean Primary Energies Predicted by Hadronic Mechanics, Superconductivity and Chemistry

Since its inception by R.M. Santilli in 1978 under support by the U.S. Department of Energy, *hadronic mechanics has been conceived, constructed, and verified for the specific purpose of predicting new clean sources of primary energy, as summarized in memoir* [5].

All energies predicted by hadronic mechanics are suitable for the clean production of electricity which, in turn, can be used for the environmentally acceptable production of clean fuels.

This monograph is devoted to molecular aspects and, as such, we cannot enter into a technical review of new energies at the deeper nuclear and particle levels. Nevertheless, it may be useful for readers seriously interested in the environment to provide at least a conceptual outline.

As shown in the preceding chapter, hadronic mechanics, superconductivity and chemistry coincide with conventional theories everywhere, except at short distances of the order of 10^{-13} cm in which the former theories permit axiomatically rigorous and invariant treatments of new forces and effect which simply do not exist in the latter, the nonpotential-nonhamiltonian interactions due to mutual overlapping of the wavepackets of particles.

Since these effects are nonpotential by definition, they should be represented with *anything except the Hamiltonian.* Santilli suggested in 1978 to represent these new effects via a *generalization of the basic unit of quantum mechanics* which is written in the form

$$I = \text{Diag.}(1,1,1,1) \rightarrow \hat{I}(t,r,p,\psi,\partial\psi,\dots) = 1/\hat{T} = \\ = \text{Diag.}(n_1^2,n_2^2,n_3^2,n_4^2) \times \exp\left\{N \int dv\, \psi_\downarrow^\dagger(r) \times \psi_\uparrow(r)\right\}, \tag{7.11}$$

where n_1^2, n_2^2, n_3^2, represent the *shape* of the charge distribution considered, n_4^2 represents its *density*, and the exponential represents nonlinear, nonlocal and nonpotential interactions.

In particular, the representation of nonhamiltonian effects via a generalized unit is the only known representation assuring invariance, that is the prediction of the same numbers for the same effects under all the same conditions but at different times.

As the reader will recall from Chapters 1, 2, and 3, specific applications of hadronic mechanics can be constructed by submitting any given quantum model to a nonunitary transform which, for consistency, must be applied to the totality of the original formulation. This simple method then yields *all formulations* of hadronic mechanics, e.g.,

$$U \times U^\dagger \neq I, \quad U \times U^\dagger > 0, \tag{7.12a}$$

$$I \rightarrow \hat{I} = U \times I \times U = 1/\hat{T}, \tag{7.12b}$$

$$A \times B \rightarrow A \hat{\times} B = A' \times \hat{T} \times B' = U \times (A \times B) \times U^\dagger, \\ A' = U \times A \times U^\dagger, \quad B' = U \times B \times U^\dagger, \tag{7.12c}$$

$$<\psi| \times |\psi> \times I \rightarrow <\hat{\psi}|\hat{\times}|\hat{\psi}> \times \hat{I} = U \times (<\psi| \times |\psi>), \tag{7.12d}$$

$$H \times |\hat{\psi}> = E \times |\psi> \rightarrow H' \hat{\times} |\hat{\psi}> = U \times (H \times |\psi>) = \\ = E' \times |\hat{\psi}> = U \times (E \times |\psi>), \quad |\hat{\psi}> = U \times |\psi>. \tag{7.12e}$$

The resulting theories are called hadronic mechanics, superconductivity and chemistry. The assumption that $U \times U^\dagger$ is different than one only at short distances permits the recovering of quantum mechanics at large distances. Whether in particle physics, nuclear physics or molecular structures, this type of nonunitary completion of quantum mechanics essentially adds characteristics, features and effects which are beyond any representational capability of quantum mechanics. New clean energies and fuels are permitted precisely by said nonunitary short range effects.

A feature of the new short range effects which is of paramount importance for new clean energies and fuels, yet it is absent in conventional disciplines, is that *nonhamiltonian interactions due to the deep mutual overlappings of the wavepackets of particles in singlet coupling, not only are attractive, but they actually are so attractive to overcome repulsive Coulomb interactions.*

Another important feature of the new interactions is that *they occur by conception without any binding energy*, since the latter are possible only for potential interactions.

Yet another feature of the new interactions is that they imply, in general, a mutation of the intrinsic characteristics of particles, which is characterized by irreducible representations of the Poincaré-Santilli isosymmetry. As an example, a deformation of the charge distribution of protons and neutrons, which is impossible for quantum mechanics (since these particles are points), is readily representable by hadronic mechanics via isounit (7.11), and permits the first known exact-numerical representation of nuclear magnetic moments [5].

The attractive character of the new interactions, combined with their lack of energy exchange, and the possibility of altering the intrinsic characteristics of particles, permit the prediction, quantitative treatment, and experimental verification of truly new and clean energies.

Readers should be aware that all the above novel features are beyond not only by quantum mechanics, but also Einstein's special relativity.

In fact, special relativity is strictly linear, local and potential in conception and axiomatic structure. Therefore, all nonlinear, nonlocal, and nonpotential effects depart from special relativity.

Above all, the *intrinsic* characteristics of particles are perennial and immutable according to special relativity. Therefore, the alteration of the intrinsic characteristics which is inherent in nonpotential interactions (because they provide a new form of renormalization) are incompatible with special relativity.

Because of these reasons, a generalization of special relativity via the isomathematics has been worked out in details by Santilli under the name of *isospecial relativity* [6] and subjected to a number of verifications in various fields (see, for brevity, memoir [5]).

Systematic and protracted studies based on isomechanics and isorelativity have permitted the identification of the *basic physical laws* which have to be verified by the new energies, as well as the *geometries* and *embodiments* for their practical realization.

These systematic studies conducted for over two decades have resulted in the identification of various basically new, clean energies all conceived for environmentally acceptable production of electricity.

These new clean energies can be classified and outlined as follows [5]:

1) New clean energies of Class I, those occurring at the level of elementary particles.

The neutron is, by far, the largest and most inexhaustible reservoir of clean energy available to mankind. In fact, the neutron is *naturally unstable* with spontaneous decay

$$n \rightarrow p^{+} + e^{-} + \bar{\nu}, \tag{7.13}$$

which releases electrons with a kinetic energy up to 0.8 MeV, i.e., an energy which is at least 100,000 times bigger than the energy of the electrons hitting a computer screen. Studies on the production and use of this energy were conducted in Refs.[7].

The capture of electrons from the decay of the neutron via a conducting screen provides a *dual* source of energy, called *hadronic energy* [7d], the first being the creation of a difference of electric potential, and the second being the creation of a large amount of heat (see Fig. 7.1).

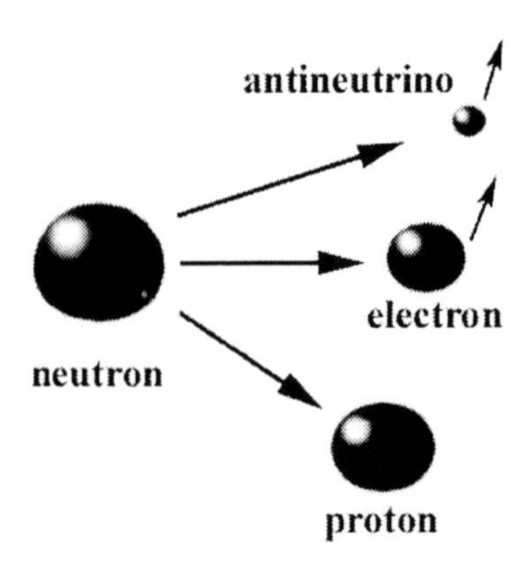

Figure 7.1. [7] A schematic illustration of the fact that an isolated neutron at rest is unstable and decays *spontaneously* into an at rest proton, a highly energetic electron and an antineutrino. The spontaneous character of the decay permits the study of mechanisms stimulating its decay, with the consequential production of clean energy. It should be recalled that, contrary to popular beliefs, the neutron has a *variable* meanlife depending on the characteristics of its environment. In fact, for certain nuclei the neutron has a meanlife of only nanoseconds; when isolated in vacuum the neutron has a meanlife of about 15 minutes; when member of other nuclei, the neutron has a meanlife of several years; and when a member of light, natural stable nuclei the neutron has an infinite meanlife. The variable character of the meanlife of the neutron then provides additional support for the existence of a mechanism stimulating its decay. Note that the reconciliation of the finite meanlife in certain nuclei with perfect stability in other is best resolved by assuming that the neutron is a bound state of a proton and an electron as originally conceived by Rutherford, and as first permitted by hadronic mechanics (see Fig. 1.18).

The above source of energy is clean because it does not emit dangerous radiations (since electrons can be easily captured by a metal screen while the neutrinos are notoriously harmless) and does not release harmful waste, as shown below.

Hadronic energy is essentially based on the following predictions [7d]:

1) A peripheral neutron belonging to a certain group of light, natural, stable, elements $N(A, Z)$, called *hadronic fuels*, can be *stimulated* to decay via a flux of photons $\hat{\gamma}$ with the resonating frequency of 1.294 MeV,

$$\hat{\gamma} + n \rightarrow p^+ + e^- + \bar{\nu}; \tag{7.14}$$

2) The resulting nuclei N(A, Z + 1) are naturally unstable with spontaneous beta decay

$$\hat{\gamma}+N(A, Z) \rightarrow N(A, Z+1)+e_1^- +\bar{\nu}_1 \rightarrow N(A, Z+2)+e_2^- +\bar{\nu}_1+\bar{\nu}_2; \tag{7.15}$$

3) The final nuclei $N(A, Z+2)$ of the class of hadronic fuels are light, natural, and stable elements, thus leaving no harmful waste.

Representative example of hadronic fuels are Zn(70, 30), Mo(100, 42), and several others [7a]. Note that the energy of 1.294 MeV of the original resonating photon *is not lost*, but remains available in the final usable energy; for each resonating photon there is the production of *two* electrons and related kinetic energy; and the process essentially transforms the original nuclei into nuclei with smaller mass, while producing an amount of energy at least seven times the energy of the resonating photon, thus assuring the positive energy output even after including the energy needed for the production of $\hat{\gamma}$.

Processes (7.15) are extremely exoenergetic. In fact, calculations show that a few grams of hadronic fuels under an efficiency rate of one reaction (7.15) per 10^{10} nuclei could produce energy sufficient to power a car for years.

The environmental acceptability of the hadronic energy is evident. Note, in particular, the occurrence of *two* nuclear transmutations *without* the emission of neutrons or other harmful radiations.

It should be indicated that the hadronic energy *is not* possible for quantum mechanics, because the cross section of reaction (7.15) is very small, thus having no practical value. Therefore, academicians supportive of quantum mechanics as a final theory dismiss the existence of hadronic energy, despite the availability of a first experimental verification.

However, *the cross section of reaction (7.14) has not been measured at 1.294 MeV. Therefore, the claim that the cross section of reaction (7.15) is very small at 1.294 MeV is a pure personal opinion of individual scientists and absolutely not a physical truth.*

The hadronic covering of the γ-n scattering with a nonunitary transform which is different than one only inside the neutron plus other conditions (under patent pending) recovers the prediction that the cross

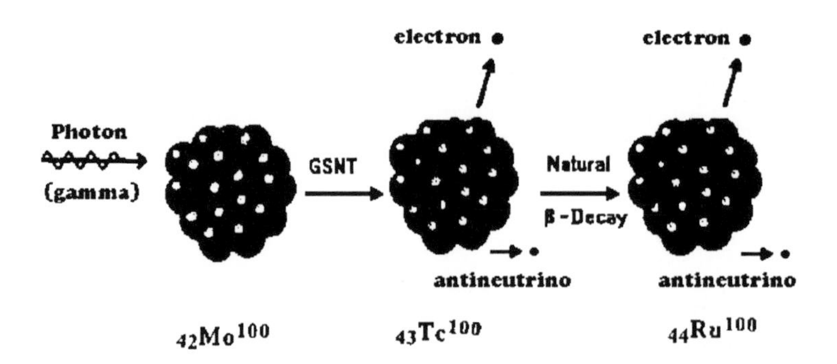

Figure 7.2. [7] A schematic view of the *hadronic energy* submitted by Santilli [7d] (patent pending). At least one peripheral neutron belonging to certain nuclei (called "hadronic fuels") can be stimulated to decay via a photon with the resonating frequency of 1.294 MeV, plus adjustments varying from nucleus to nucleus. The figure illustrates the case of the hadronic fuel Mo(42,100) which is transformed by one resonating photon into the unstable Tc(43, 100) which, in turn, beta decays into the stable Ru(44, 100) resulting in the emission of two electrons with large energies ranging from 5 to 7 MeV. Other cases of hadronic fuels are given by Zn(30, 70), various "mirror nuclei," and others. Note that the reactions of this figure are *prohibited* by hadronic mechanics. Note also that the resulting hadronic energy is clean because it does not emit harmful radiations (since the electrons can be easily trapped and the antineutrino are harmless), and does not leave harmful waste.

section for photons is very low at small energies, *except the emergence of a sharp peak at* 1.294 MeV, which is somewhat reminiscent of the sharp peak in the 1960s predicting the Ω^- particle.

Another more insidious objection against hadronic energy is that quark theories do not permit the stimulated decay of the neutron, and, in particular, photons with energy 1.294 MeV have no resonating feature at all for quarks.

The above occurrence illustrates the *necessity of new structure model of hadrons with physical constituents for the prediction of new clean energies.* In fact, hadronic energy can exist only under the new structure model of the neutron as a hadronic bound state of a proton and electrons (Fig. 1.18). Only under the assumption that the electron is an actual constituent of the neutron (although in a form mutated with respect to that in vacuum solely detected until now), there exist a resonating frequency or other mechanisms capable of exiting said structural electron, with evident consequential decay.

On real scientific grounds, the use of quark conjectures to dismiss the hadronic energy has no credibility because of the well known unresolved

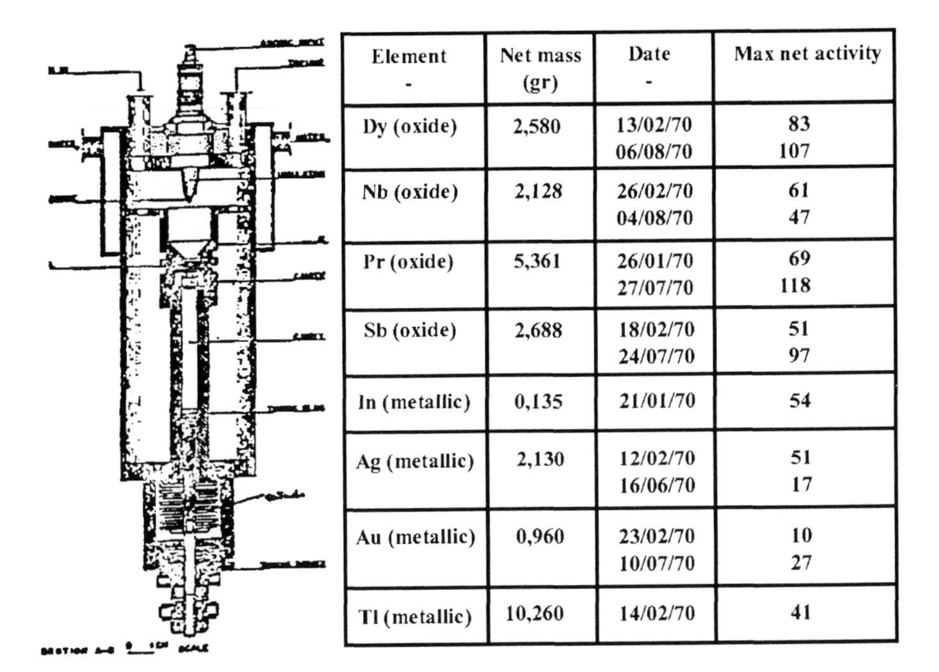

Element -	Net mass (gr)	Date -	Max net activity
Dy (oxide)	2,580	13/02/70 06/08/70	83 107
Nb (oxide)	2,128	26/02/70 04/08/70	61 47
Pr (oxide)	5,361	26/01/70 27/07/70	69 118
Sb (oxide)	2,688	18/02/70 24/07/70	51 97
In (metallic)	0,135	21/01/70	54
Ag (metallic)	2,130	12/02/70 16/06/70	51 17
Au (metallic)	0,960	23/02/70 10/07/70	10 27
Tl (metallic)	10,260	14/02/70	41

Figure 7.3. [7] A schematic view of the experiment conducted in Brasil by the late Italian physicist-priest don Borghi and his associates [7e] on the synthesis of the neutron from protons and electrons only. The experiment, which has a remarkable simplicity, consisted in maintaining via electric discharges a gas of protons and electrons at ordinary temperature inside a metal chamber called klystron and shown in the l.h.s. Various substances put in the outside of the chamber showed clear nuclear transmutation which can only be due to a flux of neutrons emanating from the interior of the klystron. Quantum mechanics prohibits the indicated synthesis on numerous grounds. By contrast, hadronic mechanics permits the consistent and invariant representation of the totality of the characteristics of the neutron as a bound state of a proton and an electron [7a, 7b, 7c], thus permitting the synthesis here considered. It should be recalled that the synthesis of the neutron from protons and electrons is possible only when the electrons have the sharp threshold energy of 0.80 MeV, and verify other conditions such as singlet coupling. Needless to say, the experiment is in need of a number of independent verifications.

problems of quark theories, such as the lack of a rigorous confinement with a proved identically null probability of free quarks. Moreover, the assumption of quarks as composites indicated in Sect. 1.8 removes any objection, since the new structure model of hadrons with physical constituents does recover quark structures as a first approximation. Intriguingly, composite quarks are recovered via the use of *hypermathematics*. In fact, according to hadronic mechanics *hadrons have different isounits*.

Triplets, octets and other SU(3) multiplets are *isomorphically* recovered from corresponding isounits multiplets.

The experimental foundations of hadronic energy are the following. First, there is the need to verify experimentally that the electron is indeed a physical constituent of the neutron. This objective can be achieved via the *synthesis of neutrons from protons and electrons only.* Once this synthesis is experimentally proved, hadronic energy is a mere consequence. Note that quark theories imply the belief that a *permanently stable particle* such as the electron can disappear and be replaced by hypothetical quarks which is outside the boundaries of science. The reader should be aware that the synthesis of the neutron has indeed been experimentally verified by a group of Italian experimentalists in Brazil [7e]. This test now awaits independent confirmation or denial.

Secondly, the inverse process of the synthesis, neutron's stimulated decay, has also been experimentally verified at a nuclear physics laboratory in Greece [7f]. This second experiment too is in need of independent verifications or denials.

Note the need for the experimental verification of both, the synthesis of the neutron from protons and electrons, and its stimulated decay in hadronic fuels. Note also that a new clean energy at the particle level mandates a new structure model of hadrons with physical constituents freely emitted in the spontaneous decay, precisely as it was the case for atomic and nuclear energies. In fact, the latter energies are permitted precisely by the capability of extracting the constituents.

Contrary to such a historical teaching at the nuclear and atomic levels, quarks theories have instead being developed under the *assumption* that the hadronic constituents cannot be produced free.

Therefore, as indicated in Sect. 1.8, one of the biggest *obstacle* against new clean energies at the particle and nuclear level is given precisely by the conjecture that the hypothetical elementary quarks under a hypothetical permanent confinement are the physical constituents of hadrons. On the contrary, the assumption that quarks are composites, by therefore *not* being the ultimate elementary constituents of hadrons, does indeed permit new clean energies.

Despite this evident scientific scenario, and despite the large environmental implications here at stake, it has been impossible for Santilli to locate a nuclear physics laboratory in the U.S.A. and Europe willing to repeat the tests on the synthesis and stimulated decay of the neutron for the confirmation or denial of tests [7e, 7f].

This opposition in academic circles, *but not in corporate research*, is evidently due to the fact that the verification of the synthesis of the neutron and its stimulated decay would imply the termination of the

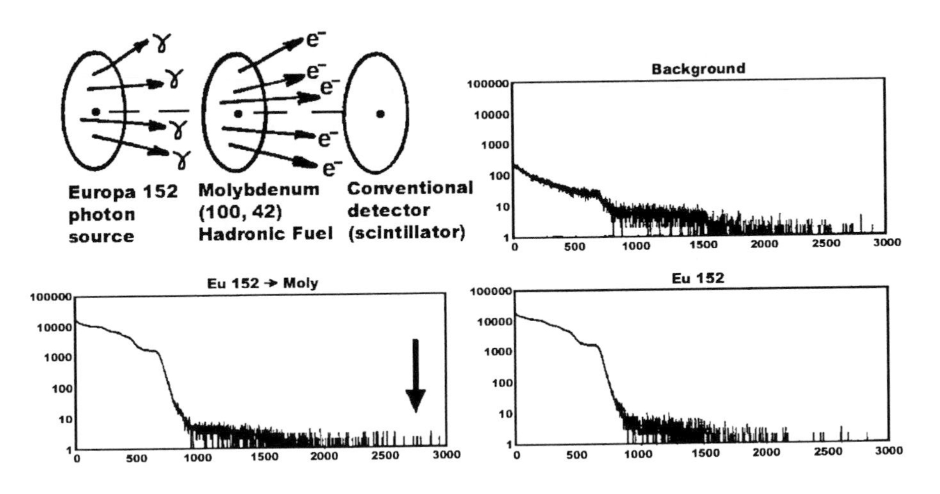

Figure 7.4. [7] A schematic view of the experiment on the stimulated decay of the neutron conducted by Tsagas and his associates [7f] at the University of Thrace in Xanthi, Greece. This experiment is evidently complementary to that on the synthesis of the neutron illustrated in Fig. 7.3, and it is also of a remarkable simplicity. The experiment consisted in the use of disk of Europa 52 as a source of photons with the needed resonating frequency of 1.294 MeV, which is placed next to a disk of Molybdenum. The two combined disks are then placed within a scintillator suitable to measure the energy of the emitted electrons. As one can see, the comparison of the measurements of the background, the Europa source alone, and of the combined disks Eu-Mo does indeed show the emission of electrons with energy above 2 MeV which can only be of nuclear origin, thus confirming the stimulated decay of the neutron, since the maximal possible energy of electrons due to Compton scattering between photons and peripheral atomic electrons is 1 MeV. Needless to say, this experiment too requires a number of independent verifications or disproofs. It should be noted in this respect that the experimenters used natural Molybdenum rather than its isotope Mo(42, 100), which is contained in the natural element for about 0.6%. All other isotopes of the Molybdenum have been proved not to allow the stimulated decay of the neutron because of the lack of verification of one or another conservation laws and/or superselection rules. As a result, the repetition of the same experiment is recommended with the use of the pure isotope Mo(42, 100) to yield results better than those reached by Tsagas [7f].

final character of quantum mechanics and quark conjectures in favor of new covering theories.

After decades of fruitless efforts, the author has no hope for self-corrective measures within academic circles, thus leaving industry as the only possible conduit for basic advances in the identification of the true constituents of the neutron.

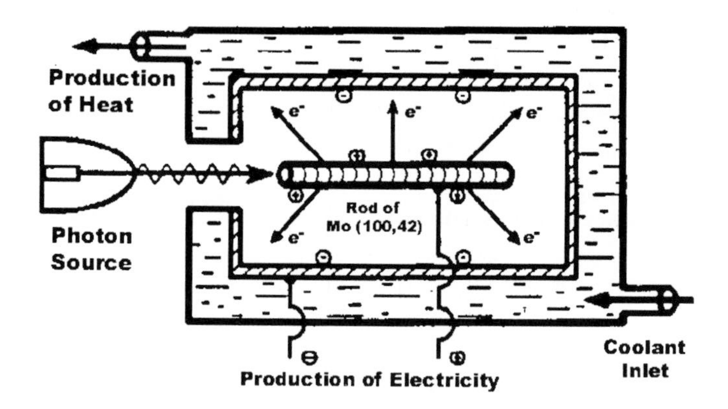

Figure 7.5. [7] An example of industrial application of hadronic mechanics currently under development at *Clean Energies Technologies, Inc.*, of Palm Harbor, Florida. *The utilization of the inexhaustible clean energy contained in the neutron via its stimulated decay* (patents pending). The predicted energy source is dual, in the sense of constituting a source of electricity (due to the difference of potential between the screen trapping the electrons and the original, conducting, hadronic fuel), and heat (which can be utilized via heat exchangers).

2) New Clean Energies of Class II, those occurring at the nuclear level with contributions of new energies of Class I.

As it is well known, quantum mechanics does not predict the possibility of low energy stimulated nuclear transmutations, and, when these transmutations are admitted, they are predicted to occur with the emission of harmful neutrons. As a matter of fact, the *lack* of emission of neutrons is generally assumed in academia as "evidence" for the lack of existence of the transmutations themselves.

It should be stressed that these views are nothing more than *personal opinions* by individual scientists with vested interests in quantum mechanics and, *absolutely they do not constitute a physical truth.*

The experimental verification of stimulated nuclear transmutations at low energy without the emission of neutrons would, therefore, constitute direct experimental support for the termination of the final character of quantum mechanics in favor of a covering theory.

In view of the above, organized interests in academic nuclear physics have strenuously opposed scientific research in the field. A vivid illustration is the publicly released opposition by nuclear physicists against the announcement by Fleishmann and Pons one decade ago on the possible existence of the "cold fusion." As a result of this public opposition, studies on the "cold fusion" are still considered today a "fraud" by the newsmedia and other uninformed sectors.

In reality, it should be recalled that studies on the "cold fusion" were *privately* funded, while the studies preferred by opposing nuclear physicists are those on the "hot fusion" which have been and continue to be funded by large amounts of *public* money. Moreover, the "cold fusion" remains at this writing an open scientific possibility not yet sufficiently explored, while the practical impossibility of the "hot fusion" has been proved beyond doubt decades ago. This identifies where problems of scientific accountability really rest.

Hadronic mechanics and isospecial relativity predict numerous new clean forms of primary energy of Class II, all within current experimental and industrial capabilities, some of which are reported in memoir [5] (see the following figures).

A necessary condition for these predictions is the new structure model of nuclei which are reduced by hadronic mechanics to protons and electrons only, although in generalized states called *isoprotons* and *isoelectrons*, while recovering conventional nuclear structures in terms of protons and neutrons in first approximation.

These new nuclear models are an evident consequence of the structure model of the neutron as a hadronic bound state of a proton and an electron as originally conceived by Rutherford (Fig. 1.18). As a result, the new clean energies of Classes I and II are deeply interconnected, to such an extent that the experimental evidence of one of them is direct experimental evidence of the other.

Three types of new clean energies of Class II are predicted by hadronic mechanics. They can all be conceptually identified by inspecting and, above all, admitting physical reality.

The first type can be identified by recalling that the neutron is synthesized in nature from protons and electrons in the core of stars. Objections to the effect that such a synthesis requires the very high pressures and temperatures in the core of stars are soon dismissed by the experimental evidence of Electron Capture (EC), namely, the spontaneous capture of electrons by certain nuclei under ordinary conditions on Earth. In fact, the EC implies precisely the synthesis of neutrons from protons and electrons, resulting in the low energy nuclear transmutations of the type

$$N(A, Z) + e^- \rightarrow N(A, Z-1) + \nu. \tag{7.16}$$

By recalling that the above EC is *spontaneous*, hadronic mechanics has permitted the quantitative identification of the physical laws and conditions under which nuclear transmutations (7.16) can be *stimulated* with the release of energy (i.e., the mass of $N(A, Z-1)$ is *smaller* than that of $N(Z, A)$) via the release of heat without the release of any radiation (i.e., $N(A, Z-1)$ is stable) [5],

$$N(A, Z) + e^- + \text{TR} \rightarrow N(A, Z-1) + \text{ heat without radiations.} \quad (7.17)$$

These conditions are numerous and cannot be technically reviewed here (see memoir [5] for brevity). We merely mention that transmutations (7.16) require singlet couplings, the electron must be near the threshold energy for the synthesis of the neutron (0.80 MeV), and the transmutation requires a "trigger" (TR) as it is the case for isochemistry (Chapter 4).

A second group of new clean energies of Class II can be derived by observing and admitting that Earth should have reached a state with a completely frozen core billions of years ago, as proved by using conventional laws for the dissipation of heat. The fact that Earth's core is instead as hot as it was billions of years ago establishes the existence of an internal source of heat which can only be due to nuclear transmutations. This conclusion is independently confirmed by known geophysical evidence that Earth is actually *expanding*.

The observation and admission of the geometry of Earth's core then permits the identification of specific new energies of Class II (patents pending). In fact, Earth's core is exposed to a locally intense *magnetic field* which implies the alignment of nuclear spins which, in certain cases, is necessary to verify the conservation of spin. Additional observations, confirmed by hadronic mechanics, establishes that the nuclear transmutation originating heat in the interior of our Earth can be reproduced in our atmospheric conditions because, contrary to popular belief, they occur only under a *threshold energy* which varies from case to case.

A third group of new clean energies of Class II can be identified by recalling that chemical analyses of bubbles of air trapped in amber has revealed that *Earth's atmosphere one hundred million years ago contained about 40% nitrogen.*

This finding establishes the existence of an evidently slow natural process which has caused the doubling of nitrogen in out atmosphere during the past 100 millions years.

The geological origin of the process is soon ruled out, since volcanic activities are known not to release nitrogen in the needed amount. The astrophysical origin of the process is equally ruled out for various reasons, such as the slowly progressive nature of the change over a long period of time which rules collisions of our planet with comets and other bodies.

Therefore, the most probable, thus most credible, origin of the increase of nitrogen in our atmosphere is that *nitrogen is synthesized in our atmosphere from other natural elements via a low energy nuclear process without the emission of radiations*, that is, precisely via the process prohibited by quantum mechanics but permitted by the covering hadronic mechanics.

An inspection of all possible natural events reveals that the synthesis of nitrogen on Earth is caused by *lighting*. The synthesis must evidently occur from a natural elements available on Earth. The most plausible candidates are carbon (contained in all living organisms) and deuteron (contained in ordinary water although in small amounts), resulting in the *stimulated synthesis of nitrogen predicted by hadronic mechanics* [5]

$$\text{TR} + {}_1\text{H}^2 + {}_6\text{C}^{12} \rightarrow {}_7\text{N}^{14} + \text{ heat}, \tag{7.18}$$

where the trigger TR is in this case given by lighting and the related events (extreme magnetic fields, consequential alignment of nuclei, and other features). The low rate of the synthesis is then interpreted via the low concentration of heavy water in our atmosphere.

Again, it should be stressed that synthesis (7.18) is *strictly prohibited* by quantum mechanics, e.g., because it does not admit the joint emission of neutrons. On the contrary, synthesis (7.18) is predicted and quantitatively treatable by hadronic mechanics. Oddly, the secondary emission of neutrons (or any other radiation) would *prohibit* synthesis (7.18) within the context of hadronic mechanics.

Again, the lack of existence of synthesis (7.18) is a *purely personal opinion and absolutely not a scientific truth*, for real science can only be conducted via experiments, particularly those that could disprove existing theories. Opposition against the test of synthesis (7.18) (as well as several others predicted by hadronic mechanics [5]) based on mere personal opinions are ascientific and asocial.

The potentially catastrophic environmental problems of our contemporary society demand the dismissal of personal opinions one way or another, and the conduction of serious science, that are based on direct experimental verifications or disproofs.

Intriguingly, stimulated nuclear transmutation (7.18) is among the simplest ones predicted by hadronic mechanics because *the carbon nucleus ${}_6C^{12}$ has zero spin, thus eliminating the need for complex equipment suitable to realize singlet couplings, as necessary for nuclei with spin.*

We have the following conservation of the angular momentum and parity for stimulated nuclear transmutation (7.18) (under the selection of a trigger hereon tacitly implied):

$$1^+ + 0^+ \rightarrow 1^+, \tag{7.19}$$

which is always verified for all possible triggers with spin zero. For the identification and verification of all other hadronic laws and geometries, we refer the interested reader to memoir [5] for brevity.

The energy output of synthesis (7.18) is impressive. The mass of the deuterium nucleus is

$$M_D = 1,876.122 \text{ MeV}/c_0^2. \tag{7.20}$$

The mass of the carbon nucleus is

$$M_C = (12.000 - 6 \times m_e - E) \text{ a.m.u.} \simeq 11,174.865 \text{ MeV}/c_0^2, \tag{7.21}$$

while the mass of the nitrogen nucleus is

$$M_N = 13,040.137 \text{ MeV}/c_0^2. \tag{7.22}$$

Therefore, the energy released in the synthesis of the nitrogen as per reaction (7.18) is

$$\Delta E = (M_D + M_C) - M_N = 14.850 \text{ MeV}/c_0^2. \tag{7.23}$$

By remembering that

$$1 \text{ MeV} = 1.6021 \times 10^{-13} \text{ Joule}, \tag{7.24}$$

and that in one mole we have 6.022×10^{23} atoms, the extremely low efficiency of one reaction (7.18) per 10^{10} nuclei per minute yields the sizable energy output of synthesis (7.18)

$$\begin{array}{c} 14.8 \times 10^6 \times 1.6 \times 10^{-19} \text{ Joule}) \times (6 \times 10^{23}) \times \\ \times (10^6 \text{ reaction per min. per mole}) = 1.42 \times 10^6 \text{ Joule/min.} \end{array} \tag{7.25}$$

Note that the instantaneous availability of such a large amount of energy provides the sole known, scientific explanation of *thunder*. Following studies that date back to the beginning of physics, the origin of thunder has never been *scientifically*, that is, numerically, interpreted, due to the magnitude of the needed energy and its instantaneous character. All "conventional" interpretations of thunder, those intended to preserve old doctrines, such as conventional chemical reactions, can be easily ruled out on numerical grounds because grossly insufficient to produce the needed large amount of energy in the extremely small volume and duration of lighting.

It is proposed that *the thunder is experimental evidence of the existence of stimulated nuclear transmutations at low energy without the emission of harmful radiations.*

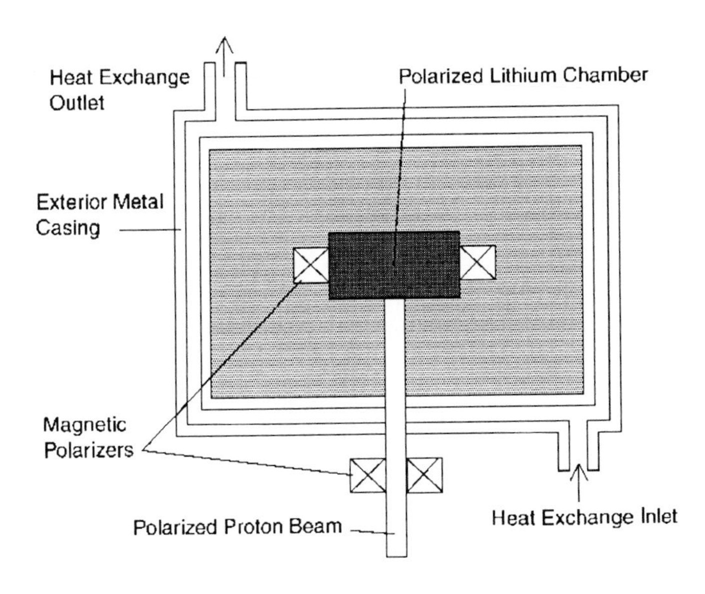

Figure 7.6. [5] Another example of industrial application of hadronic mechanics currently under study at *Clean Energy Technologies, Inc.*, of Palm Harbor, Florida. *The lithium enhanced combustion* (patents pending). Anomalous energy releases in the combustion of fuels enriched with lithium are well known. Hadronic mechanics has identified the physical laws, geometries, and conditions at the ultimate particle foundations of this technology. In particular, the reactions releasing the anomalous energy can only occur for singlet coupling (antiparallel spin), and other conditions currently occurring at random. The embodiment depicted in this figure is conceived to optimize said conditions, by therefore increasing significantly the energy output.

In closing, it should be indicated that hadronic mechanics provides the only known *invariant* representation of all other low energy nuclear transmutations predicted via *nonunitary* effects.

3) New clean energies of Class III, those occurring at the molecular level with contributions from energies of Classes I and II.

Hadronic reactors of Class I are intended to tap the energy within hadrons, e.g., via the stimulated decay of the neutron into the lighter proton and then the use of the released energy difference.

Hadronic reactors of Class II are intended to tap energy within nuclei, e.g., via the stimulated transmutation of selected nuclei into others of lower mass, with the consequential use of the energy difference.

Hadronic reactors of Class III are intended to tap energy within molecules, e.g., via stimulating the transition from given molecules to struc-

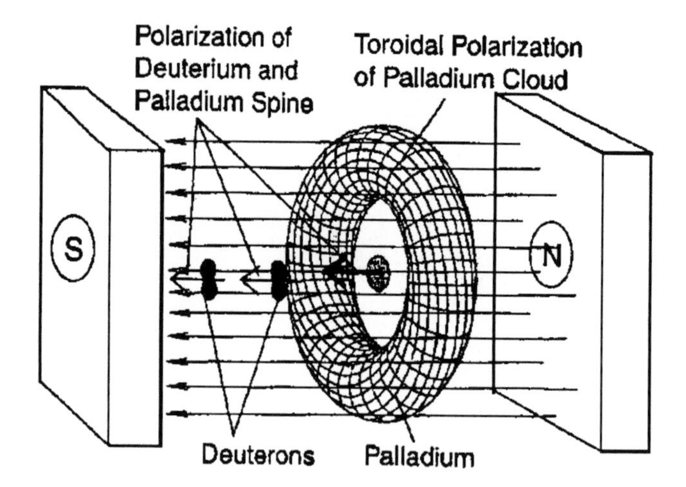

Figure 7.7. [5] Another industrial application of hadronic mechanics currently under study at *Clean Energies Technologies, Inc.* of Palm Harbor, Florida, *the enhancement of clean energy released in electrochemical reactions* (patents pending). Anomalous heat in the electrolytical compression of deuteron in palladium electrodes are well known since the early part of the 20-th century, although the effect is generally small. Hadronic mechanics has identified the physical laws, geometries and conditions under which heat is released, one of them being the singlet coupling of palladium and deuteron nuclei. These conditions occur at random in preceding tests, thus yielding minimal energy. The embodiment depicted in this figure is conceived to maximize the realization of these conditions, thus predicting an increase of heat production proportional to the efficiency of the equipment.

tures with lower energy with the consequential utilization of the energy difference.

An example of the third class of new clean energies predicted by hadronic mechanics is given by the *hadronic reactors of molecular type*, also called *PlasmaArcFlow*^{TM} *Reactors*, now in industrial production and sale, which are described in the subsequent Sections.

The hadronic reactors of Classes I, II and III are all based on the principle of stimulating the decay of the considered bound states via resonating effects acting on the *nonpotential* component of binding forces, which action constitutes a main novelty of approach over pre-existing quantum research.

In other words, the main physical principles underlying the new energies of Classes I, II and III are the same. This essentially implies that the verification of a novel energy of one given class supports the existence of the remaining two energies.

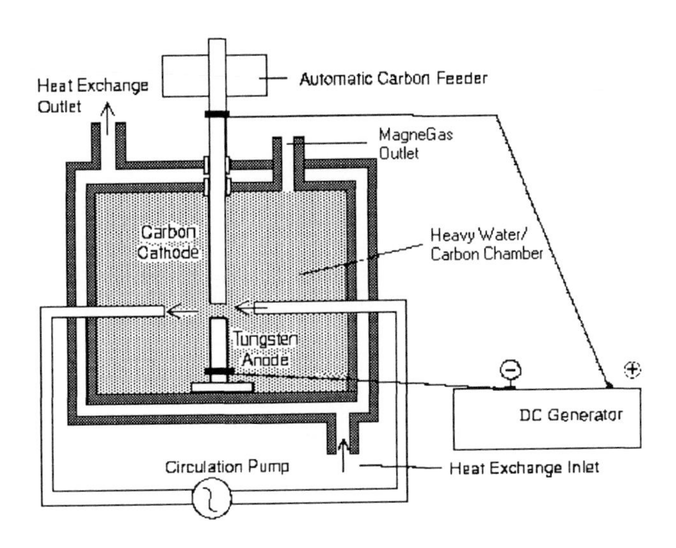

Figure 7.8. [5] Another new clean energy of Class II under study at *Clean Energies Technologies, Inc.* of Palm Harbor, Florida. It originates from the study of lighting and thunder and is called the Hadronic Nitrogen Reactor (patents pending). The embodiment here represented is essentially conceived to optimize the geometry and physical laws of the event according to stimulated nuclear transmutation (7.18) as described in the text, resulting in the production of the large amount of heat (7.25) which is clean because it does not emit harmful radiations and does not leave harmful waste.

8. PlasmaArcFlow Reactor for the Conversion of Liquid Waste into the Clean Burning Magnegas

Hadronic reactors of molecular type (Class III), also known as PlasmaArcFlow^{TM} Reactors (covered by patent with additional international patents pending), were first built by Santilli in 1998 at *Toups Technology Licensing, Inc.*, a public company in Largo, Florida, under the presidency of Leon Toups. Industrial development of the reactors is now under way at *EarthFirst Technologies, Inc.* (the new denomination of Toups Technology Licensing, Inc.) under the presidency of John Stanton, via three subsidiaries, *USMagnegas*, *EuroMagnegas* and *AsiaMagnegas*. The latter companies own the exclusive rights of the new technology for the corresponding geographical areas [8].

PlasmaArcFlow Reactors use a submerged DC electric arc to achieve the complete recycling of nonradioactive liquid waste into a clean burning combustible gas called *magnegas*^{TM}, heat usable via exchangers, and solid precipitates. The reactors are ideally suited to recycle antifreeze waste, oil waste, sewage, and other contaminated liquids, although they can also process ordinary fresh or sea water. However, the best liquid for these hadronic reactors is *crude oil*, which is processed into a combustible gas much cleaner than gasoline, at a cost smaller than that of refineries (see below).

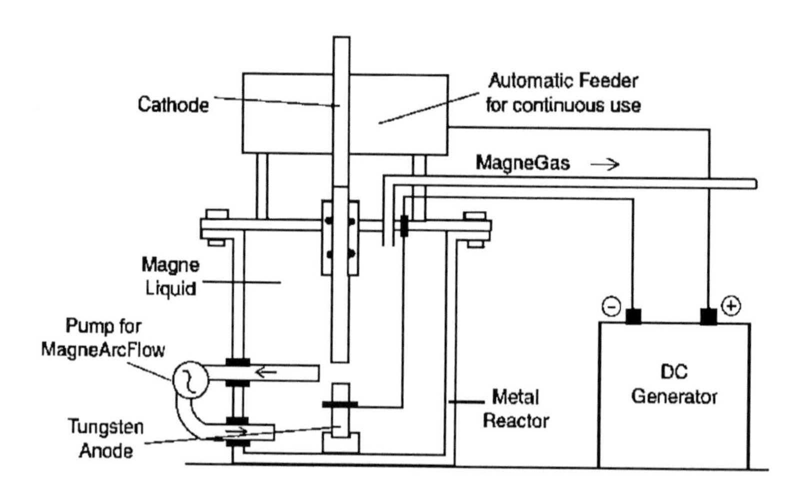

Figure 7.9. [8] A schematic view of Santilli's Hadronic Reactors. This figure provides a schematic view of: the DC power unit; the all encompassing metal vessel filled up with the liquid to be recycled; the submerged electrodes; the flow of the liquid through the electric arc; the bubbling of the combustible gas to the surface where it is collected. The heat acquired by the liquid is utilized via a heat exchanger not shown in this figure.

The new PlasmaArcFlow technology is essentially based on flowing liquids through a submerged DC arc with at least one consumable carbon electrode (see Fig. 7.9). The arc decomposes the liquid molecules and the carbon electrode into a plasma at about 3,500^o K, which plasma is composed of mostly ionized H, O and C atoms. Other solid substances generally precipitate at the bottom of the reactor where they are periodically collected. The technology moves the plasma away from the electric arc immediately following its formation, and controls the recombination of H, O and C into magnegas, which bubbles to the surface where it is collected with various means.

The known affinity of C and O permit the removal of oxygen from the plasma, resulting in the combustible CO. In turn, the removal of CO from the plasma immediately following its formation prevents the oxidation of CO into CO_2, thus reducing the CO_2 content of the gas to traces. Hydrogen then essentially recombines into H_2 and other species (see next Chapter).

By comparison, one should note that in other methods based on underwater arcs the stationary character of the plasma within the arc implies the creation of large percentages of CO_2 resulting in a CO_2 content of the exhaust much greater than that of gasoline and natural gas. The resulting fuel is then environmentally unacceptable.

Recall that the primary source of the large glow created by underwater arcs is the recombination of H and O into H_2O following its separation. This recombination is the reason for the low efficiency of underwater arcs and consequential lack of industrial development until recently.

By comparison, the PlasmaArcFlow causes the removal of H and O from the arc immediately following their creation, thus preventing their recombination into H_2O, with consequential dramatic increase of the efficiency, that is, of the volume of combustible gas produced per kWh.

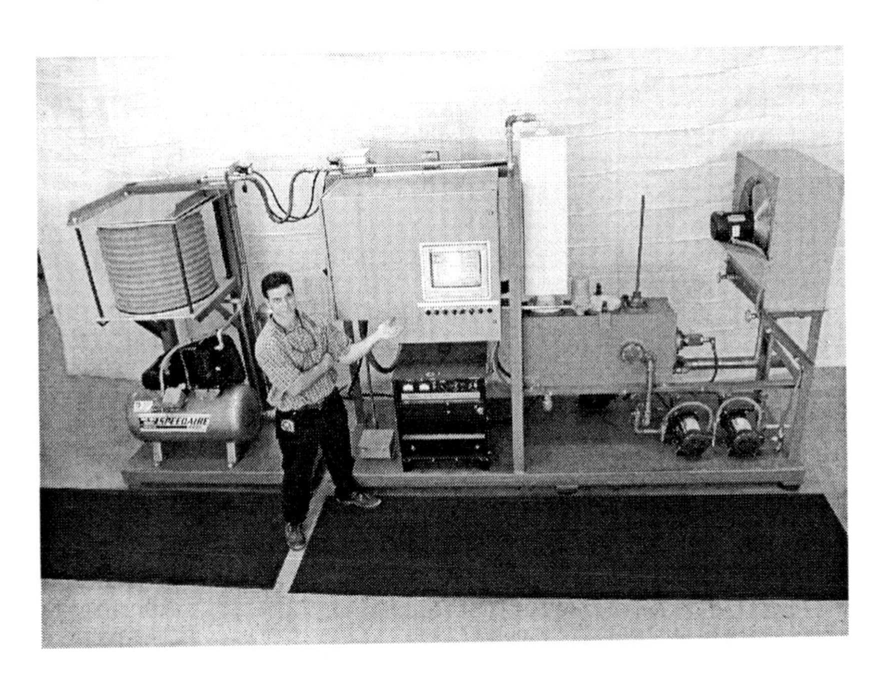

Figure 7.10. [8] A picture of a *PlasmaArcFlow^{TM} Reactor* with 75 kW DC power unit which produces about 1,000 cf of magnegas per hour possessing about 870 BTU/cf plus about 500,000 BTU of heat per hour.

A first important feature of the PlasmaArcFlow Reactors is that of having a *large efficiency, thus permitting the production of a combustible gas at a cost competitive with that of fossil fuels*, of course, when con-

Figure 7.11. [8] A picture of a Ferrari 308 GTSi 1980 and two Honda Civic cars converted by *USMagnegas, Inc.*, to operate with the new clean burning magnegas without catalytic converter, yet surpassing all EPA exhaust requirements, having no carcinogenic or other toxic substance in the exhaust, reducing of about 50% the CO_2 emission due to gasoline combustion, and emitting in the exhaust 10% to 14% breathable oxygen.

sidering comparatively large productions, including the income from the recycling of liquid wastes and the utilization of the heat.

A Ferrari 308 GTSi and two Honda Civic automobiles have been converted by *USMagnegas, Inc.*, to operate on magnegas. One of these vehicles has been subjected to intensive tests at an EPA certified automotive laboratory in Long Island, New York reviewed in details in Sect. 7.9, which tests have established that magnegas exhaust surpasses all EPA requirements *without catalytic converter*, emits no carcinogenic, CO or other toxic substances in the exhaust, reduce the CO_2 emission due to gasoline combustion by about 40%, and emits in the exhaust 12% to 15% breathable oxygen.

Figure 7.12. [8] A picture of the *Magnegas Production and Refilling Station*, consisting of a reactor as per Fig. 7.10, plus a standard compressor as used for natural gas. The simplicity of this station should be compared with the extreme complexity of corresponding station for liquid hydrogen. Note that the station depicted in this figure allows current *distributors* of fuel, such as gas stations, to become *fuel producers*.

In addition to the production of magnegas as a fuel, *the PlasmaArcFlow Reactors can be viewed as the most efficient means for producing a new form of hydrogen, called MagH^{TM}, a carbon-free version of magnegas, with energy content and output greater than the conventional hydrogen, and at a cost smaller than that of the latter.*

In reading this chapter, the reader should, therefore, keep in mind that, in the words of John Stanton, President of *EarthFirst Technologies, Inc.*, "the new technology of PlasmaArcFlow Reactors is *evolutionary*, rather than revolutionary, because conceived to be primarily beneficial for crude oil, piston engines and hydrogen industries".

9. Surpassing by Magnegas Exhaust of EPA Requirements without Catalytic Converter

While the chemical composition of magnegas is unknown at this writing because it is anomalous (see the remaining sections), the chemical composition of magnegas combustion exhaust is fully conventional, and it has been measured to sufficient accuracy.

The tests were conducted on a Honda Civic Natural Gas Vehicle (NGV) VIN number 1HGEN1649WL000160 (the white car of Fig. 7.11), produced in 1998 to operate with Compressed Natural Gas (CNG). The car was purchased new in 1999 by *USMagnegas, Inc.*, of Largo, Florida, and converted to operate on Compressed MagneGas (CMG) in early 2000. All tests reported in this section were done with magnegas produced by recycling antifreeze waste. The conversion from CNG to CMG was done via:

1) the replacement of CNG with CMG in a 100 liter tank at 3,600 psi which contains about 1,000 cf of magnegas;

2) the disabling of the oxygen sensor because magnegas has about 20 times more oxygen in the exhaust than natural gas, thus causing erroneous readings by the computer set for natural gas; and

3) installing a multiple spark system to improve magnegas combustion.

The rest of the vehicle was left unchanged, including its computer.

Comparative tests on performance (acceleration, full load, etc.) have established that *the output power of the vehicle operating on compressed magnegas is fully equivalent to that of the same car operating on compressed natural gas.*

Comparative tests on consumption also indicate similar results. In fact, measurements of magnegas consumption per hour in ordinary city driving were conducted with the following results:

TANK CAPACITY:	1,096 cf at 3,500 psi,	
TOTAL DURATION:	about 2.5 hours,	(7.26)
CONSUMPTION:	about 7 cf/minute.	

As one can see, a magnegas pressure tank of 1,500 cf at 5,000 psi would provide a range of about 4 hours, which is amply sufficient for all ordinary commuting and travel needs. Measurements of magnegas consumption rate per mile on highway are under way, and they are expected to yield essentially the same results holding for natural gas, namely,

$$\text{Gasoline gallon equivalent: 120 cf of magnegas.} \tag{7.27}$$

Preliminary measurements of magnegas combustion exhaust were conducted by the laboratory *National Technical Systems, Inc.*, of Largo, Florida, resulting in the following exhaust composition under proper combustion:

WATER VAPOR:	50% - 60%,	
OXYGEN:	10% - 12%,	
CARBON DIOXIDE:	6% - 7%,	(7.28)
BALANCE:	atmospheric gases,	
HYDROCARBONS, CARBON MONOXIDE, NITROGEN OXIDES:	in parts per million (ppm).	

Detailed magnegas exhaust measurements were then conducted at the EPA Certified, Vehicle Certification Laboratory *Liphardt & Associates* of Long Island, New York, under the Directorship of *Peter di Bernardi*, via the Varied Test Procedure (VTP) as per EPA Regulation 40-CFR, Part 86.

These EPA tests consisted of three separate and sequential tests conducted on a computerized dynamometer, the first and the third tests using the vehicle at its maximal possible capability to simulate an up-hill travel at 60 mph, while the second test consisted in simulating normal city driving.

Three corresponding bags with the exhaust residues were collected, jointly with a fourth bag containing atmospheric contaminants. The final measurements expressed in grams/mile are given by the average of the measurements on the three EPA test bags, less the measurements of atmospheric pollutants in the fourth bag.

The following three measurements were released by Liphardt & Associates:

1) Magnegas exhaust measurements with catalytic converter:

HYDROCARBONS:	0.026 grams/mile, which is 0.063 of the EPA standard of 0.41 grams/mile;	
CARBON MONOXIDE:	0.262 grams/mile, which is 0.077 of the EPA standard of 3.40 grams/mile;	
NITROGEN OXIDES:	0.281 grams/mile, which is 0.28 of the EPA standard of 1.00 grams/mile;	(7.29)
CARBON DIOXIDE:	235 grams/mile, corresponding to about 6%; there is no EPA standard on CO_2 at this time;	
OXYGEN:	9.5% to 10%; there is no EPA standard for oxygen at this time.	

The above tests have established the important feature that *magnegas exhaust with catalytic converter imply a reduction of about 1/15 of current EPA requirement.*

2) Magnegas exhaust measurements without catalytic converter in the same car and under the same conditions as (1):

HYDROCARBONS:	0.199 grams/mile, which is 0.485 of the EPA standard of 0.41 grams/mile;	
CARBON MONOXIDE:	2.750 grams/mile, which is 0.808 of the EPA standard of 3.40 grams/mile;	
NITROGEN OXIDE:	0.642 grams/mile, which is 0.64 of the EPA standard of 1.00 grams/mile;	(7.30)
CARBON DIOXIDE:	266 grams/mile, corresponding to about 6%;	
OXYGEN:	9.5% to 10%.	

As a result of the latter tests, the laboratory *Liphardt & Associates* released the statement that *magnegas exhaust surpasses the EPA requirements* without *the catalytic converter.* As such, magnegas can be used in *old cars without catalytic converter while meeting, and actually surpassing EPA emission standards.*

3) Natural gas exhaust measurements without catalytic converter in the same car and under the same conditions as (1):

HYDROCARBONS:	0.380 grams/mile, which is 0.926 of the EPA standard of 0.41 grams/mile;	
CARBON MONOXIDE:	5.494 gram/mile, which is 1.615 of the EPA standard of 3.40 grams/mile;	
NITROGEN OXIDES:	0.732 grams/mile, which is 0.73 the EPA standard of 1.00 grams/mile;	(7.31)
CARBON DIOXIDE:	646.503 grams/mile, corresponding to about 9%;	
OXYGEN:	0.5% to 0.7%.	

The latter tests established the important property that *the combustion of natural gas emits about 2.5 times the CO_2 emitted by magnegas without catalytic converter. Note that, as well known, natural gas exhaust without catalytic converter does not meet EPA requirements.*

As an additional comparison for the above measurements, a similar Honda car running on indolene (a version of gasoline) was tested in the same laboratory with the same EPA procedure, resulting in the following data:

4) Gasoline (indolene) exhaust measurements conducted on a two liter Honda KIA:

HYDROCARBONS:	0.234 grams/mile equal to 9 times the corresponding magnegas emission;	
CARBON MONOXIDE:	1.965 grams/mile equal to 7.5 times the corresponding magnegas emission;	
NITROGEN OXIDES:	0.247 grams/mile equal to 0.86 times the corresponding magnegas emission;	(7.32)
CARBON DIOXIDE:	458.655 grams/mile equal to 1.95 times the corresponding of magnegas emission,	
OXYGEN:	No measurement available.	

The above data establish the environmental superiority of magnegas over natural gas and gasoline. The following comments are now in order:

1) Magnegas does not contain (heavy) hydrocarbons since it is created at 3,500^o K. Therefore, the measured hydrocarbons are expected to be due to combustion of oil, either originating from magnegas compression pumps (thus contaminating the gas), or from engine oil.

2) Carbon monoxide is fuel for magnegas (while being a combustion product for gasoline and natural gas). Therefore, any presence of CO in the exhaust is evidence of insufficient combustion.

3) The great majority of measurements originate from the first and third parts of the EPA test at extreme performance, because, during ordinary city traffic, magnegas exhaust is essentially pollutant free, as shown in Fig. 7.13.

4) Nitrogen oxides are not due, in general, to the fuel (whether magnegas or other fuels), but to the temperature of the engine and other factors, thus being an indication of the quality of its cooling system. Therefore, for each given fuel, including magnegas, NOx's can be decreased by improving the cooling system and via other means.

5) The reported measurements of magnegas exhaust do not refer to the best possible combustion of magnegas, but only to the combustion of magnegas in a vehicle whose carburization was developed for natural gas. Alternatively, the test was primarily intended to prove that magnegas is interchangeable with natural gas without any major automotive changes, while keeping essentially the same performance and consumption. The measurements for combustion specifically conceived for magnegas are under way.

We should also indicate considerable research efforts under way to further reduce the CO_2 content of magnegas exhaust via disposable cartridges of CO_2-absorbing chemical sponges placed in the exhaust system (patent pending). Additional research is under way via *liquefied magnegas* obtained via *catalytic* (and *not* conventional) liquefaction, which liquid is expected to have an anomalous energy content with respect to

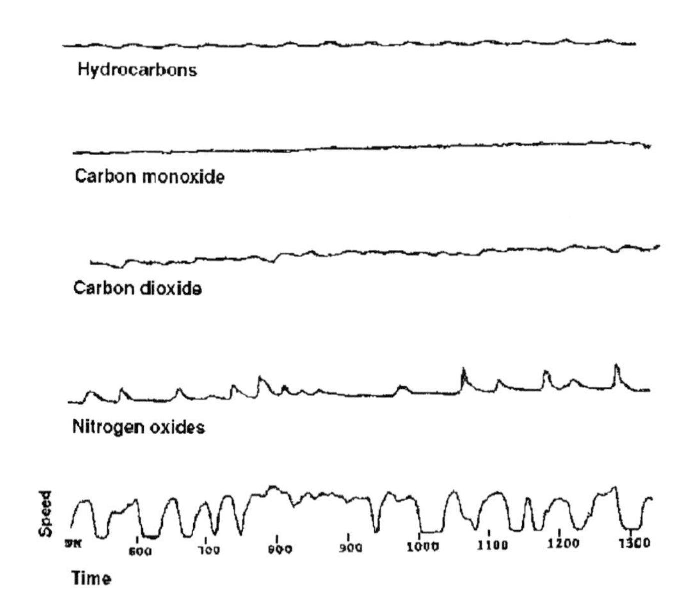

Figure 7.13. An illustration of the city part of the reported EPA test according to Regulation 40-CFR, Part 86, conducted at the Vehicle Certification Laboratory *Liphardt & Associates* of Long Island, New York on a Honda Civic Natural Gas Vehicle converted to magnegas by *USMagnegas, Inc.*, of Largo, Florida. The first three diagrams illustrate the very low combustion emission of magnegas in city driving, by keeping in mind that most of measured emission is due to the heavy duty, hill climbing part of the EPA test. The fourth diagram on nitrogen oxides is an indication of insufficient cooling of the engine. The bottom diagram indicates the simulated speed of the car versus time, where flat tracts simulate idle portions at traffic lights.

other liquid fuels, and an expected consequential decrease of pollutants. As a result of these efforts, the achievement of an exhaust essentially free of pollutants and CO_2, yet rich in oxygen, appears to be within technological reach.

10. Anomalous Chemical Composition of Magnegas

As studied in the preceding section, the chemical composition of the magnegas exhaust is fully known. However, the chemical composition of magnegas itself is unknown at this writing (spring of 2001). This unexpected occurrence, particularly for a *light gas* such as magnegas, is due to a variety of facts.

To begin, numerous tests in various analytic laboratories reviewed in Chapter 8 have established that the chemical composition of magnegas cannot be identified via conventional Gas Chromatographic Mass Spectrometric (GC-MS) measurements, since the gas results in being characterized by large peaks in macroscopic percentage all the way to 1,000 a.m.u. in molecular weight, which peaks remain individually unidentified by the MS computer after scanning all known molecules.

As also reviewed in Chapter 8, the chemical structure of magnegas is equally unidentifiable via InfraRed Detectors (IRD), because the new peaks composing magnegas have no IR signature at all, thus establishing the presence of bonds of non-valence type (because these large clusters cannot possibly be all symmetric).

Moreover, the IR signature of conventional molecules such as CO results in being *mutated* (in the language of hadronic mechanics) with the appearance of new peaks, which evidently indicate *new* internal bonds in *conventional* molecules.

In addition to all the above, dramatic differences between the prediction of quantum chemictry and physical reality exist for the energy content of magnegas. For instance, when produced with PlasmaArcFlow Reactors operating an electric arc between at least one consumable electrode within pure water, quantum chemistry predicts that magnegas should be a mixture of 50% H_2 and 50% CO, with traces of O_2 and CO_2.

This prediction is dramatically disproved by the fact that *both the CO and the CO_2 peaks do not appear in the MS scan in the predicted percentages, while they appear in the IR scan although in a mutated form.*

Moreover, quantum chemistry predicts that the indicated composition consisting of 50% H_2 and 50% CO should have an energy content of about 315 BTU/cf, namely, an energy content insufficient to cut metal. This prediction is also disproved by the experimental evidence that *magnegas cuts metal at least 50% faster than acetylene (which has 2,300 BTU/cf).*

Such a performance in metal cutting is more indicative of a *plasma cutting* feature, such as the metal cutting via a plasma of ionized hydrogen atoms which recombine into H_2 when cooling in the metal surface, thus releasing the energy needed for metal cutting. The problem is that magnegas is at room temperature when used for metal cutting, and it is subjected to ordinary combustion, thus requiring basically new approaches for its correct interpretation.

Nevertheless, the plasma cutting feature is indicative of the presence of isolated atoms and dimers in the magnegas structure which recombine

Figure 7.14. A view of metal cutting via magnegas. Independent certifications by various users have established that: 1) magnegas has a pre-heat time at least half that by acetylene (which is currently used for metal cutting and has an energy content of 2,300 BTU/cf); 2) magnegas cuts metal at least 50% faster than acetylene; 3) the cut produced by magnegas is much smoother without edges as compared to that by acetylene; 4) magnegas exhaust does not contain carcinogenic or other toxic substances, while that of acetylene is perhaps the most carcinogenic and toxic of all fuels; 5) magnegas cutting does not produce the "flash-back" (local explosion of paint over metal) typical of acetylene; 6) magnegas is dramatically safer than acetylene, which is unstable and one of the most dangerous fuels currently used; and 7) magnegas cost about 1/2 that of acetylene.

under combustion, thus yielding a behavior and a performance similar to that of plasma cutters.

In fact, as also shown later on, GC-MS scans have indicated the presence in the anomalous peaks of *individual atoms of hydrogen, oxygen, and carbon* evidently in addition to individual molecules.

To conclude, the composition of magnegas in H, C and O *atoms* can be easily identified from the liquid used in the reactors. For instance, when magnegas is produced from water, it is composed of 50% H, 25% O, and 25% C, with corresponding percentages for other liquids such as antifreeze, crude oil, etc.

However, all attempts to reduce the chemical composition of magnegas to conventional molecules conducted by the author as well as independent chemists, have been disproved by a variety of experimental evidence.

In particular, any belief that magnegas is entirely composed by ordinary molecules, such as H_2 and CO, is disproved by experimental evidence via GC-MS and IRD detectors.

The only possible scientific conclusion at this writing is that *magnegas is composed of a new chemical species* studied in Chapter 8.

11. Anomalous Energy Balance of Hadronic Molecular Reactors

As is well known, the *scientific efficiency* of any equipment is *under-unity* in the sense that, from the principle of conservation of the energy and the unavoidable energy losses, *the ratio between the total energy produced and the total energy used for its production is smaller than one.*

For the case of magnegas production, the total energy produced is the sum of the energy contained in magnegas plus the heat acquired by the liquid, while the total energy available is the sum of the electric energy used for the production of magnegas plus the energy contained in the liquid recycled. Therefore, from the principle of conservation of the energy we have the scientific energy balance

$$\frac{\text{Total energy produced}}{\text{Total energy available}} = \frac{E_{mg} + E_{heat}}{E_{electr} + E_{liq}} < 1. \tag{7.33}$$

An important feature of hadronic reactors is that they are *commercially over-unity*, namely, the ratio between the total energy produced and only the electric energy used for its production, is bigger than one,

$$\frac{E_{mg} + E_{liq}}{E_{electr}} > 1. \tag{7.34}$$

In this commercial calculation the energy contained in the liquid is not considered because liquid wastes imply an income, rather than costing money.

As a result, Santilli's hadronic molecular reactors can be viewed as reactors capable of tapping energy from liquid molecules, in much of the same way as nuclear reactors can tap energy from nuclei. An important difference is that the former reactors release no harmful radiation and leave no harmful waste, while the latter reactors do release harmful radiations and leave harmful waste.

The energy used for the production of the carbon rod, the steel of the reactors, etc. is ignored in commercial over-unity (7.34) because its numerical value per cubic foot of magnegas produced is insignificant.

The commercial over-unity of hadronic reactors is evidently important for the production of the combustible magnegas or magnetically polarized hydrogen ($MagH^{TM}$) at a price competitive over conventional fossil fuels.

A first certification of the commercial over-unity (7.34) was done on September 18 and 19, 1998, for the very first, manually operated prototype of hadronic reactors by the independent laboratory *Motorfuelers, Inc.*, of Largo, Florida, and included (see [8]):

1) Calibrating the cumulative wattmeter provided by *WattWatchers, Inc.*, of Manchester, New Hampshire, which was used to measure the electric energy drawn from the power lines per each cubic foot of magnegas produced;

2) The verification of all dimensions, including the volume of the column used for gas production, the volume of the liquid used in the process, etc.;

3) Repetition of numerous measurements in the production of magnegas and its energy content, calculation of the average values, identification of the errors, etc.

During the two days of tests, *Motorfuelers* technicians activated the electric DC generator and produced magnegas, which was transferred via a hose to a transparent plexyglass tower filled up with tap water, with marks indicating the displacement of one cubic foot of water due to magnegas production.

After the production of each cubic foot, the gas was pumped out of the tower, the tower was replenished with water, and another cubic foot of magnegas was produced. The procedure was repeated several times to have sufficient statistics. The electric energy from the electric panel required to produce each cubic foot of magnegas was measured via the previously calibrated cumulative wattmeter.

As a result of several measurements, *Motorfuelers, Inc.* certified [8] that the production of one cubic foot of magnegas with the first prototype required an average electric energy of

$$E_{electr} = 122 \text{ W/cf} = 416 \text{ BTU/cf } \pm 5\%. \tag{7.35}$$

It should be stressed that this is the electric energy from the electric panel, thus including the internal losses of the DC rectifier. Alternatively, we can say that the arc is served by only 65% of the measured electric energy, corresponding to

$$E_{electr} = 79.3 \text{ W/cf} = 270 \text{ BTU/cf}. \tag{7.36}$$

The energy content of magnegas was measured on a comparative basis with the BTU content of natural gas (1,050 BTU/cf). For this purpose, technicians of *Motorfuelers, Inc.*, used two identical tanks, one of natural gas and one of magnegas, at the same initial pressure of 110 psi. Both tanks were used for 5 psi pressure decreases, under the same gas flow, to increase the temperature of the same amount of water in the same pot at the same initial temperature. The ratio of the two temperature increases is evidently proportional to the ratio of the respective BTU contents.

Following several measurements, *Motorfuelers, Inc.* certified [8] that magnegas produced from the antifreeze waste used in the reactor has about 80% of the BTU content of natural gas, corresponding to

$$E_{mg} = 871 \text{ BTU/cf } \pm 5\%. \tag{7.37}$$

All other more scientific tests of BTU content of magnegas conducted at various academic and industrial laboratories failed to yield meaningful results due to the energy content of magnegas for various reasons. Despite their empirical character, the measurement of BTU content done by *Motorfuelers, Inc.*, remains the most credible one.

It should be noted that the value of 871 BTU/cf is a lower bound. In fact, automotive tests reviewed in Sect. 7.9 have established that the energy output of internal combustion engines powered by magnegas is fully equivalent to that of natural gas, thus yielding a realistic value of about

$$E_{mg} = 1,000 \text{ BTU/cf}. \tag{7.38}$$

During the tests, it was evident that the temperature of the liquid waste in the reactor experienced a rapid increase, to such an extent that the tests had to be stopped periodically to cool down the equipment, in order to prevent the boiling of the liquid with consequential damage to the seals.

Following conservative estimates, technicians of *Motorfuelers, Inc.*, certified [8] that, jointly with the production of 1 cf of magnegas, there was the production of heat in the liquid of 285 BTU/cf plus 23 BTU/cf of heat acquired by the metal of the reactor itself, yielding

$$E_{heat} = 308 \text{ BTU/cf}. \tag{7.39}$$

In summary, the average electric energy of 122 W = 416 BTU calibrated from the electric panel produced one cf of magnegas with 871 BTU/cf, plus heat in the liquid conservatively estimated to be 308 BTU/cf. These independent certifications established the following *commercial over-unity* of the first, manually operated hadronic reactor within ±5% error:

$$\frac{871 \text{ BTU/cf} + 308 \text{ BTU/cf}}{416 \text{ BTU/cf}} = 2.83. \tag{7.40}$$

Note that, if one considers the electric energy used by the arc itself corresponding to 79.3 W/cf = 270 BTU/cf), we have the following commercial over-unity:

$$\frac{871 \text{ BTU/cf} + 285 \text{ BTU/cf}}{270 \text{ BTU/cf}} = 4.36. \tag{7.41}$$

In releasing the above certification, *Motorfuelers, Inc.*, noted that the arc had a poor efficiency, because it was manually operated, thus resulting in large variation of voltage, at times with complete disconnection of the process and need for its reactivation.

Motorfuelers technicians also noted that the BTU content of magnegas, Eq. (7.37), is a minimum value, because measured in comparison to natural gas, not with a specially built burner, but with a commercially available burner that had large carbon residues, thus showing poor combustion, while the burner of natural gas was completely clean.

Immediately after the above certification of commercial over-unity, a number of safety and health measurements were conducted on hadronic molecular reactors, including measurements on the possible emission of neutrons, hard photons, and other radiation.

David A. Hernandez, Director of the *Radiation Protection Associates*, in Dade City, Florida conducted comprehensive measurements via a number of radiation detectors placed around the reactor, with particular reference to the only radiations that can possibly escape outside the heavy gauge metal walls, low or high energy neutrons and hard photons.

Under the presence of eyewitnesses, none of the various counters placed in the immediate vicinity of the reactor showed any measurement of any radiation at all. As a result, Radiation Protection Associates released an official Certificate stating that:

> "Santilli's PlasmaArcFlow^{TM} Reactors met and exceed the regulatory regulations set forth in Florida Administrative Code, Chapter 64-E. Accordingly, the reactors are declared free of radiation leakage."

Subsequent certifications of more recent hadronic reactors operating at atmospheric pressure with 50 kW and used to recycled antifreeze waste, this time done on fully automated reactors, have produced the following measurements:

$$E_{mg} = 871 \text{ BTU/cf}, \tag{7.42a}$$

$$E_{liq} = 326 \text{ BTU/cf}, \tag{7.42b}$$

$$E_{electr} = 100 \text{ W/cf} = 342 \text{ BTU/cf}, \tag{7.42c}$$

resulting in the following commercial over-unity of automatic reactors recycling antifreeze with about 50 kW and at atmospheric pressure:

$$\frac{871 \text{ BTU/cf} + 326 \text{ BTU/cf}}{342 \text{ BTU/cf}} = 3.5. \tag{7.43}$$

When ordinary tap water is used in the reactors, various measurements have established a commercial over-unity of about 2.78.

It should be indicated that the commercial over-unity of the hadronic reactors increases nonlinearly with the increase of the kiloWatts, pressure and temperature. Hadronic reactors with 250 kW are under construction for operation at 250 psi and 400° F. The latter reactors have a commercial over-unity considerably bigger than (7.43).

The origin of the commercial over-unity (7.43) is quite intriguing and not completely known at this writing. In fact, conventional chemical structures and reactions have been studied by Aringazin and Santilli [9] and shown not to be sufficient for a quantitative explanation, thus requiring a new chemistry.

Following Aringazin and Santilli [9], our first task is to compute the electric energy needed to create one cubic foot of plasma in the PlasmaArcFlow reactors as predicted by conventional quantum chemistry. Only after identifying the deviations of the experimental data from these predictions, the need for the covering hadronic chemistry can be properly appraised.

For these objectives we make the following assumptions. First, we consider PlasmaArcFlow reactor processing distilled water with the DC arc occurring between a consumable pure graphite cathode and a non-consumable tungsten anode. As indicated earlier, said reactors yield a commercial over-unity also when used with pure water. Therefore, quantum chemical predictions can be more effective studied in this setting without un-necessary ambiguities. We also assume that water and the solid graphite rod are initially at 300° K and that the plasma created by the DC electric arc is at $3{,}300^o$ K.

The electric energy needed to create one cubic foot of plasma must perform the following transitions (see Appendix 7.A for basic units and their conversions):

1) Evaporation of water according to the known reaction

$$H_2O(\text{liquid}) \rightarrow H_2O(\text{vapor}) - 10.4 \text{ Kcal/mole}, \tag{7.44}$$

2) Separation of the water molecule,

$$H_2O \rightarrow H_2 + \frac{1}{2}O_2 - 57 \text{ Kcal/mole}, \tag{7.45}$$

3) Separation of the hydrogen molecule,

$$H_2 \rightarrow H + H - 104 \text{ Kcal/mole}, \tag{7.46}$$

4) Ionization of H and O, yielding a total of 1,197 Kcal.

We then have the evaporation and ionization of the carbon rod,

$$\text{C(solid)} \rightarrow \text{C(plasma)} - 437 \text{ Kcal/mole}, \tag{7.47}$$

resulting in the total 1,634 Kcal for 4 moles of plasma, i.e.

$$\begin{aligned} 408.5 \text{ Kcal/mol} = 0.475 \text{ kWh/mol} = 1621 \text{ BTU/mol} = \\ = 515.8 \text{ Kcal/cf} = 0.600 \text{ kWh/cf} = 2,047 \text{ BTU/cf}, \end{aligned} \tag{7.48}$$

to which we have to add the electric energy needed to heat up the non-consumable tungsten anode which is estimated to be 220 BTU/cf, resulting in the total of 2,267 BTU/cf. This total, however, holds at the electric arc itself without any loss for the creation of the DC current from conventional alternative current. By assuming that rectifiers, such as the welders used in PlasmaArcFlow reactors have an efficiency of 70%, we reach the total electric energy from the source needed to produce one cubic foot of plasma

$$\text{Total Electric Energy } = 3,238 \text{ BTU/cf} = 949 \text{ W/cf}. \tag{7.49}$$

We now compute the total energy produced by PlasmaArcFlow reactors according to quantum chemistry. For this purpose we assume that the gas produced is composed of 50% hydrogen and 50% carbon monoxide with ignorable traces of carbon dioxide. The latter is indeed essentially absent in PlasmaArcFlow reactors, as indicated earlier. In addition, CO_2 is not combustible. Therefore, the assumption of ignorable CO_2 in the gas maximizes the prediction of energy output according to quantum chemistry, as desired.

Recall that the glow of underwater arcs is mostly due to the combustion of hydrogen and oxygen back into water which is absorbed by the water surrounding the arc and it is not present in appreciable amount in the combustible gas bubbling to the surface. Therefore, any calculation of the total energy produced must make an assumption of the percentage of the original H and O which recombine into H_2O (the evidence of this recombination is established by the production of water during the recycling of any type of oil by the hadronic reactors).

In summary, the calculation of the energy produced by the PlasmaArcFlow reactors requires: the consideration of the cooling down of the plasma from 3,300^o K to 300° K with consequential release of energy; the familiar reactions

$$H + H \rightarrow H_2 + 104 \text{ Kcal/mole}, \tag{7.50a}$$

$$H_2 + \frac{1}{2}O_2 \rightarrow H_2O + 57 \text{ Kcal/mole}, \tag{7.50b}$$

$$C + O \rightarrow CO + 255 \text{ Kcal/mole}. \tag{7.50c}$$

Under the assumption of 100% efficiency (that is, no recombination of water), the total energy produced is given by

$$398 \text{ Kcal/mole} = 1,994 \text{ BTU/cf}. \tag{7.51}$$

By assuming that the entire energy needed to heat up the non-consumable tungsten is absorbed by the liquid surrounding the electric arc in view of its continuous cooling due to the PlasmaArcFlow, we have the total heat energy of 2,254 BTU/cf.

In addition, we have the energy content of the combustible gas produced. For this purpose we recall the following known reactions:

$$H_2(mg) + \frac{1}{2}O_2(atm) \rightarrow H_2O + 57 \text{ Kcal/mole}, \tag{7.52a}$$

$$CO(mg) + \frac{1}{2}O_2(atm) \rightarrow CO_2 + 68.7 \text{ Kcal/mole}. \tag{7.52b}$$

Consequently, the 50%-50% mixture of conventional gases H_2 and CO has the following

$$\text{Conventional energy content of magnegas produced from water} =$$
$$= 62.8 \text{ Kcal/mole} = 249.19 \text{ BTU/mole} = 315 \text{ BTU/cf}. \tag{7.53}$$

Therefore, the total energy output of the PlasmaArcFlow Reactors is given by

$$E(mg) + E(heat) = 315 \text{ BTU/cf} + 2,254 \text{ BTU/cf} = 2,569 \text{ BTU/cf}. \tag{7.54}$$

It then follows that the energy efficiency of the PlasmaArcFlow reactors is under-unity for the case of maximal possible efficiency,

$$\text{Energy efficiency predicted by quantum chemistry} =$$
$$= \frac{\text{Total energy out}}{\text{Electric energy in}} = \frac{E_{mg} + E_{heat}}{E_{electr}} = \frac{2,569 \text{ BTU/cf}}{3,238 \text{ BTU/cf}} = 0.79. \tag{7.55}$$

It is possible to show that, for the case of 50% efficiency (i.e., when 50% of the original H and O recombine into water) the total energy output evidently decreases. For detail, we refer the interested reader to Aringazin and Santilli [9].

12. Concluding Remarks

The most important experimental evidence presented in this chapter is the independent certification of hadronic reactors of molecular type as being commercially over-unity, that is, the ratio between the total energy produced and the electric energy needed for its production is much larger than one.

The principle of conservation of the energy then establishes that the missing energy originates from the recycled liquids, namely, that said hadronic reactors can tap energy from liquids beginning with water, and then having bigger energy efficiencies for liquids rich in H, C and O, such as those of fossil origin.

The above experimental reality has been released by the U.S. public company *EarthFirst Technologies, Inc.*, following numerous verifications by independent laboratories. As an example, the measurement of the crucial electric energy used for the production of one cubic foot of magnegas was released following the intervention of: the company producing the cumulative wattmeters used in the measurements, *WattWatchers, Inc.*, from New Hampshire; an independent company which verified the calibration of said wattmeters, *Tampa Transformers, Inc.*; the company producing the welders used to power the reactors, Miller Corporation, to verify that the selected measurements of Watts absorption was accurate for their power source; an independent laboratory which actually conducted all measurements, *Motorfuelers, Inc.* of Clearwater, Florida; and scientists of the *Institute for Basic Research*, Palm Harbor, Florida, which supervised the organization of the measurements and verified all results.

The most important information presented in the preceding section is that quantum mechanics and chemistry prohibit the production of energy from the recycling of liquids, whether water or not, thus resulting in a dramatic disagreement with experimental evidence on an issue of such a societal relevance as the need for new clean energy.

Additional dramatic disagreements exist between the predictions of quantum mechanics and chemistry and experimental data on the energy efficiency. In fact, as illustrated in detail in the preceding section, quantum mechanics and chemistry predict the need of 3,238 BTU/cf = 949 W/cf to yield one cf of H_2 and CO in one-to-one ratio. The same disciplines predict that the total energy produced is given by 2,569 BTU/cf.

Comparison of these theoretical predictions with experimental data on PlasmaArcFlow reactors operating with water establishes the following dramatic departures:

1) The electric energy predicted to produce one cubic foot of magnegas, 949 W/cf, is almost ten times the electric energy actually measured at the source, yielding a *ten-fold error in excess*;

2) the heat energy predicted to be released during the formation of magnegas, 1,994 BTU/cf, is also about ten times that actually measured in the reactor, yielding a *ten-fold error in defect*; and

3) the energy content predicted for magnegas produced from water, 315 BTU/cf, is at least half the value actually measured, yielding a *two-fold error in defect.*

It is evident that all these deviations are excessively large to permit credible accommodations of experimental data with established doctrines. Therefore, hadronic reactors of molecular type constitute the strongest evidence of the need for covering theories so as to adapt the theories to reality.

In particular, the experimental evidence establishes the need for:

1) A new mechanics, superconductivity and chemistry for quantitative interpretation of the separation of water and other liquids with a positive energy balance. The line of research here adopted is that via hadronic theories as outlined in Sect. 7.7, thus centrally dependent on contact, nonpotential, nonhamiltonian, and nonunitary effects since, as recalled earlier, the indicated positive energy balance is impossible for Hamiltonian theories. Studies along the indicated lines are in progress and will be reported at some future time.

2) A new chemical species capable or representing the anomalous energy content and behavior of magnegas. These studies are outlined in the next chapter.

3) New technologies for the combustion, compression and liquefaction of magnegas, which are outlined in Ref. [8].

It is evident that the achievement of an in depth knowledge of the above aspect will permit their optimization thus resulting in significant advances toward the societal needs for new clean energies and fuels.

APPENDIX 7.A

Table 7.A.1. Basic units and their conversions.

1 kWh	860 Kcal = 3413 BTU	1 cf	28.3 liters
1 Kcal	3.97 BTU	1 cf[a]	1.263 mol
1 eV	3.83×10^{-23} Kcal	N_A	6.022×10^{23} mol^{-1}
1 cal	4.18 J	$N_A k/2$	1 cal/(mol·K)
1 mole[a]	22.4 liters = 0.792 cf	R	8.314 J/(mol·K) = 1.986 cal/(mol·K)

[a] An ideal gas, at normal conditions.

Table 7.A.2. Specific heat capacities. $p = 1$ atm, $T = 25^o$ C.

H_2(gas)	29.83 J/(mol·K)	7 cal/(mol·K)	
H_2O (liquid)	4.18 J/(gram·K)	1 cal/(gram·K)	18 cal/(mol·K)
Graphite (solid)	0.71 J/(gram·K)	0.17 cal/(gram·K)	2 cal/(mol·K)
O_2 (gas)	29.36 J/(gram·K)	7 cal/(gram·K)	
H (gas)	14.3 J/(gram·K)	3.42 cal/(gram·K)	
O (gas)	0.92 J/(gram·K)	0.22 cal/(gram·K)	
Fe (solid)	0.45 J/(gram·K)	0.11 cal/(gram·K)	6 cal/(mol·K).

Table 7.A.3. Average binding energies, at T=25° C.

	Kcal/mol		*Kcal/mol*		*Kcal/mol*
H–H	104.2[a]	C=O	192.0[d]	O=O	119.1[b]
C–C	82.6	O–H	110.6	C=C	145.8
C–O	85.5	C≡C	199.6	C=O	255.8[c]

[a] in H_2; [b] in O_2; [c] in carbon monoxide; [d] in carbon dioxide.

Table 7.A.4. Evaporation heats and first ionization potentials.

	Kcal/mol	*Atoms*	*eV*
Water	10.4	H	13.6
Graphite	171.7	C	11.26
		O	13.6

References

[1] Jena, P., Rao, B.K. and Khanna, S.N.: Physics and Chemistry of Small Clusters, in *NATO ASI Series, Series B: Physics*, Vol. **158**, Plenum Press (1986). Jena, P., Khanna, S.N., and Rao, B.K.: Physics and Chemistry of Finite Systems: From Clusters to Crystals, in *NATO ASI Series, Series C: Mathematical and Physical Sciences*, Vol. **374**, Vol. **1** and **2** (1991). Jena, P., Khanna, S.N., and Rao, B.K.: in *Proceedings of the Science and Technology of Atomically Engineered Materials*, World Scientific Press (1996).

[2] http://www.eia.doe.gov/emeu/international/energy.html.

[3] Spath, P. and Mann, M.: A Complete Look at the Overall Environmental Impact of Hydrogen Production, in *Proceedings of HY 2000 EFO Energy Forum*, GmbH, p. 523 (2000).

[4] Frank, D., Wolf, J., and Pehr, K.: Visions Come True: BMW Hydrogen Vehicles lead the Way, in *Proceedings of HY 2000 EFO Energy Forum*, GmbH, p. 181 (2000).

[5] Santilli, R.M.: The Physics of New Clean Energies and Fuels According to Hadronic Mechanics, *Journal of New Energy* **4**, Special Edition, No. 1 (1999), 318 pages.

[6] Santilli, R.M.: Nuovo Cimento Lettere **37**, 545 (1983) [6a]; J. Moscow Phys. Soc. **3**, 255 (1998) [6b]; and Intern. J. Modern Phys. **D7**, 351 (1998) [6c].

[7] Santilli, R.M.: Hadronic J. **13**, 513 (1990) [7a]. Santilli, R.M., JINR Comm. E4-93-352 (1993) [7b]. Santilli, R.M., Chinese J. Syst. Eng. and Electr. **6**, 177 (1995) [7c]. Santilli, R.M.: Hadronic J. **17**, 311 (1994) [7d]. Borghi, C., Giori, C., and Dall'Oilio, A.: (Russian) J. Nucl. Phys. **56**, 147 (1993) [7e]. Tsagas, N.F., Mystakidis, A., Bakos, G., and Seftelis, L.: Hadronic J. **19**, 87 (1996) [7f]. Smith, S.: in *Proceedings of the International Symposium on New Energies*, ed. by M. Shawe *et al.*, Association of New Energy, Denver, Colorado (1996) [7g].

[8] http://www.magnegas.com.

[9] Aringazin, A.K. and Santilli, R.M.: A study of the energy efficiency of hadronic reactors of molecular type (2001), *in preparation.*

Chapter 8

THE NEW CHEMICAL SPECIES OF MAGNECULES

1. Introduction

The only chemical species with a clearly identified bond which was known prior to the advent of hadronic chemistry was that of *molecules* and related *valence bonds*, whose identification dates back to the 19-th century, thanks to the work by Avogadro (1811), Canizzaro (1858), and several others, following the achievement of scientific measurements of atomic weights.

Various candidates for possible additional chemical species are also known, such as the delocalized electron bonds. However, none of them possess a clearly identified attractive force clearly distinct from the valence.

Also, as recalled in Chapter 7, various molecular clusters have been studied in more recent times, although they either are unstable or miss a precise identification of their internal attractive bond.

An example of unstable molecular cluster occurs when the internal bond is due to an *electric polarization* of atomic structures, that is, a deformation from a spherical charge distribution without a net electric charge to an ellipsoidical distribution in which there is the predominance of one electric charge at one end and the opposite charge at the other end, thus permitting atoms to attract each other with opposite electric polarities. The instability of these clusters then follows from the known property that the smallest perturbation causes nuclei and peripheral electrons to reacquire their natural configuration, with the consequential loss of the polarization and related attractive bond.

An example of molecular clusters without a clear identification of their internal attractive bond is given by *ionic clusters*. In fact, ionized molecules have the *same positive charge* and, therefore, they *repel*, rather than attract, each other. As a result, not only the internal attractive bond of ionic clusters is basically unknown at this writing, but, when identified, it must be so strong as to overcome the repulsive force among the ions constituting the clusters.

In 1998, R.M. Santilli [1] submitted the hypothesis of a new type of stable clusters composed of molecules, dimers and atoms under a new, clearly identified, attractive internal bond which permits their industrial and practical use. The new clusters were called **magnecules** (patents pending) because of the dominance of magnetic effects in their formation, as well as for pragmatic needs of differentiations with the ordinary molecules, with the understanding that a technically more appropriate name would be *electromagnecules*.

Thanks to an invaluable support by Leon Toups, President of *Toups Technologies Licensing, Inc.*, a public company in Largo, Florida, the new chemical species of magnecules received, also in 1998 [1], numerous experimental verifications. Additional experimental verifications and industrial developments have been more recently conducted at *EarthFirst Technologies, Inc.*, the new denomination of *Toups Technologies Licensing, Inc.* (EFTI [2]), under the presidency of John Stanton, which company controls the world wide patent rights via its subsidiaries for their respective geographical areas *USMagnegas, Inc.*, Largo, Florida, *EuroMagnegas, Ltd.*, London England, and *AsiaMagnegas, Lim.*, Hong Kong, under the presidency of Leon Toups.

This chapter is dedicated to the review of the basic notions underlying the new chemical species of magnecules, their experimental detection, their industrial implications, and the resulting *new technology of magnetically polarized fuels* [2].

The following terminology will be used herein:

1) The word *atom* is used in its conventional meaning as denoting a stable atomic structure, such as a hydrogen, carbon or oxygen, irrespective of whether the atom is ionized or not and paramagnetic or not.

2) The word *dimer* is used to denote part of a molecule under a valance bond, such as H–O, H–C, *etc.*, irrespective of whether the dimer is ionized or not, and whether it belongs to a paramagnetic molecule or not;

3) The word *molecule* is used in its internationally known meaning of denoting stable clusters of atoms under conventional, valence, electron bonds, such as H_2, H_2O, C_2H_2, *etc.*, irrespective of whether the molecule is ionized or not, and paramagnetic or not;

4) The word *magnecule* is used to denote stable clusters of two or more molecules, and/or dimers and/or atoms and any combination thereof formed by a new internal attractive bond of primarily magnetic type

identified in detail in this chapter; the word *magnecular* will be used in reference to substances with the structure or features of magnecules;

5) The words *chemical species* are used to denote an essentially pure population of stable clusters with the same internal bond, thus implying the conventional chemical species of molecules as well as that of magnecules, under the condition that each species admits an ignorable presence of the other species.

In this chapter we study the theoretical prediction permitted by hadronic mechanics and chemistry of the new chemical species of magnecules and its experimental verifications, which were apparently presented for the first time by Santilli in memoir [1] of 1998.

2. The Hypothesis of Magnecules

The main hypothesis, studied in details in the rest of this Chapter, can be formulated as follows:

DEFINITION 8.2.1 (patent pending) [1]: **Magnecules** *in gases, liquids, and solids consist of stable clusters composed of conventional molecules, and/or dimers, and/or individual atoms bonded together by opposing magnetic polarizations of the orbits of at least the peripheral atomic electrons when exposed to sufficiently strong external magnetic fields, as well as the polarization of the intrinsic magnetic moments of nuclei and electrons. A population of magnecules constitutes a chemical species when essentially pure, i.e., when molecules or other species are contained in very small percentages in a directly identifiable form. Magnecules are characterized by, or can be identified via the following main features:*

I) Magnecules primarily exist at large atomic weights where not expected, for instance, at atomic weights which are ten times or more the maximal atomic weight of conventional molecular constituents;

II) Magnecules are characterized by large peaks in macroscopic percentages in mass spectrography, which peaks remain unidentified following a search among all existing molecules;

III) Said peaks admit no currently detectable infrared signature for gases and no ultraviolet signature for liquids other than those of the conventional molecules and/or dimers constituting the magnecule;

IV) Said infrared and ultraviolet signatures are generally altered (a feature called "mutation") with respect to the conventional versions, thus indicating an alteration (called infrared or ultraviolet mutation) of the conventional structure of dimers generally occurring with additional peaks in the infrared or ultraviolet signatures not existing in conventional configurations;

V) Magnecules have an anomalous adhesion to other substances, which results in backgrounds (blank) following spectrographic tests which are often similar to the original scans, as well as implying the clogging of small feeding lines with consequential lack of admission into analytic instruments of the most important magnecules to be detected;

VI) Magnecules can break down into fragments under sufficiently energetic collisions, with subsequent recombination with other fragments and/or conventional molecules, resulting in variations in time of spectrographic peaks (called time mutations of magnecular weights);

VII) Magnecules can accrue or lose during collision individual atoms, dimers or molecules;

VIII) Magnecules have an anomalous penetration through other substances indicating a reduction of the average size of conventional molecules as expected under magnetic polarizations;

IX) Gas magnecules have an anomalous solution in liquids due to new magnetic bonds between gas and liquid molecules caused by magnetic induction;

X) Magnecules can be formed by molecules of liquids which are not necessarily solvable in each other;

XI) Magnecules have anomalous average atomic weights in the sense that they are bigger than that of any molecular constituent and any of their combinations;

XII) A gas with magnecular structure does not follow the perfect gas law because the number of its constituents (Avogadro number), or, equivalently, its average atomic weight, varies with a sufficient variation of the pressure;

XIII) Substances with magnecular structure have anomalous physical characteristics, such as anomalous specific density, viscosity, surface tension, etc., as compared to the characteristics of the conventional molecular constituents;

XIV) Magnecules release in thermochemical reactions more energy than that released by the same reactions among unpolarized molecular constituents;

XV) All the above characteristic features disappear when the magnecules are brought to a sufficiently high temperature, which varies from species to species, called Curie Magnecular Temperature; in particular, combustion eliminates all magnetic anomalies resulting in an exhaust without magnecular features.

Magnecules are also called: **elementary** *when only composed of two molecules;* **magneplexes** *when entirely composed of several identical molecules;* and **magneclusters** *when composed of several different molecules.*

The primary objective of this chapter is, first, to study the characteristic features of magnecules from a theoretical viewpoint, and then present at least two independent experimental verifications for each feature.

The reader should keep in mind that *magnegas*, the new, clean combustible gas developed by *USMagnegas, Inc.* [2], of Largo, Florida, has precisely a magnecular structure from which it derives its name. Nevertheless, we shall identify in this chapter other gases, liquids and solids with a magnecular structure.

By denoting the conventional valence bond with the symbol "–" and the new magnetic bond with the symbol "×", examples of *elementary magnecules* in gases and liquids are respectively given by

$$\{H-H\} \times \{H-H\}, \quad \{O-O\} \times \{O-C-O\}, \ etc., \tag{8.1a}$$

$$\{C_{15}-H_{20}-O\} \times \{C_{15}-H_{20}-O\}, \ etc.; \tag{8.1b}$$

examples of *magneplexes* in gases and liquids are respectively given by

$$\{H-H\} \times \{H-H\} \times \{H-H\} \times \ldots, \ etc., \tag{8.2a}$$

$$\{H-O-H\} \times \{H-O-H\} \times \ldots, \ etc.; \tag{8.2b}$$

and examples of *magneclusters* are given by

$$\{H-H\} \times \{C-O\} \times \{O-C-O\} \times \{C=O\} \times \ldots, \ etc., \tag{8.3a}$$

$$\{C_{13}-H_{18}-O\} \times \{C_{14}-H_{12}-O_3\} \times \{C_{15}-H_{20}-O\} \times \ldots, \ etc. \tag{8.3b}$$

A generic representation of a gas magnecules requires the presence of individual atoms and dimers, such as:

$$\{H-H\} \times \{C-O\} \times H \times \{H-O-H\} \times C \times \{H-O\} \times \ldots, \ etc. \tag{8.4}$$

One of the most important features of magnecules is their anomalous release of energy in thermochemical reactions (Feature XIV of Definition 8.2.1), in view of its evident importance for the industrial development of new clean fuels such as magnegas (Sects. 7.10 and 7.11).

As we shall see in detail later on, this feature is crucially dependent on the existence within the magnecules of individual atoms, such as H, C and O, and/or individual unpaired dimers, such as H–O and H–C. In fact, at the breakdown of the magnecules due to combustion, these individual atoms and dimers coupled themselves into conventional molecules via known exothermic reactions such as

$$\begin{aligned} H + H &\rightarrow H_2 + 105 \text{ Kcal/mole}, \\ C + O &\rightarrow CO + 255 \text{ Kcal/mole}, \\ H-O + H &\rightarrow H_2O + 28 \text{ Kcal/mole}, \ etc., \end{aligned} \tag{8.5}$$

with consequential release during combustion of a large amount of energy that does not exist in fuels with a conventional molecular structure.

In reading this chapter, the reader should keep in mind that, in view of the above important industrial, consumer and environmental implications, a primary emphasis of the presentation is the study of magnecules with the largest possible number of *unpaired atoms and dimers*, rather than molecules.

In inspecting the above representation of magnecules, the reader should also keep in mind that their linear formulation in a row is used mainly for practical purposes. In fact, the correct formulation should be via *columns*, rather than rows, since the bond occurs between one atom of a given molecule and an atom of another molecule, as we shall see in detail later on.

3. The Five Force Fields Existing in Atoms

The attractive bond responsible for the creation of magnecules originates within the structure of individual *atoms*. Therefore, it is recommendable to initiate our study via the identification of all force fields existing in a conventional atomic structure.

The sole fields in the atomic structure studied by chemists prior to Ref. [1] were the intrinsic electric and magnetic fields of electrons and nuclei (see Fig. 8.1.A). It was proved a century ago that these fields can only produce *valence bonds*, thus explaining the reason why molecules were the only form of atomic clustering with a clear bond admitted by chemistry until recently.

Santilli's [1] main contribution has been the identification of a *new force field in the atomic structure*, which is sufficiently strong to permit a new chemical species.

Since the inception of atomic physics, the electron of the hydrogen atom (but not necessarily peripheral electrons of more complex atoms) has been assumed to have a spherical distribution, which is indeed the case for isolated and unperturbed atomic structures (see also Fig. 8.1.A).

However, electrons are charged particles, and all charges rotating in a planar orbit create a magnetic field in the direction perpendicular to the orbital plane, and such to exhibit the North polarity in the semi-space seeing a counter-clockwise rotation (see Fig. 8.1.B).

A main point of Ref. [1] is that the distribution in space of electron orbits is altered by sufficiently strong external magnetic fields. In particular, the latter cause the transition from the conventional spherical distribution to a new distribution with the same cylindrical symmetry of the external field, and such to exhibit magnetic polarities opposite to the external ones (Fig. 8.1.C).

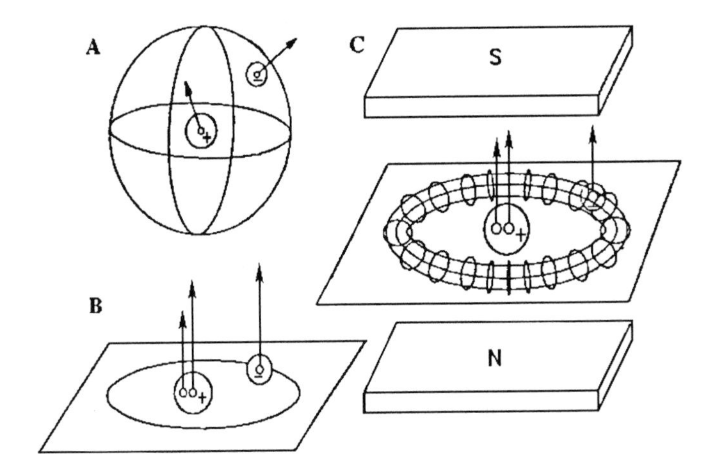

Figure 8.1. A schematic view of the force fields existing in the hydrogen atom. Fig. 8.1.A depicts an isolated hydrogen atom in its conventional spherical configuration when at absolute zero degree temperature, in which the sole force fields are given by the electric charges of the electron and of the proton, as well as by the intrinsic magnetic moments of the same particles. Fig. 8.1.B depicts the same hydrogen atom in which the orbit of the peripheral electron is polarized into a plane. In this case there is the emergence of a fifth field, the magnetic dipole moment caused by the rotation of the electron in its planar orbit. Fig. 8.1.C depicts the same hydrogen atom under an external magnetic field which causes the transition from the spherical distribution of the peripheral electron as in Fig. 8.1.A to a new distribution with the same cylindrical symmetry as that of the external field, and such to offer magnetic polarities opposite to the external ones. In the latter case, the polarization generally occurs within a toroid, and reaches the perfectly planar configuration of Fig. 8.1.B only at absolute zero degree temperature or under extremely strong magnetic fields.

Therefore, the magnetic fields of atoms *are not* solely given by the intrinsic magnetic fields of the peripheral electrons and of nuclei because, under the application of a sufficiently strong external magnetic field, atoms exhibit the additional magnetic moment caused by a polarization of the electron orbits. This third magnetic field was ignored by chemists until 1998 (although not by physicists) because nonexistent in a conventional atomic state.

As a matter of fact, it should be recalled that *orbits are naturally planar in nature, as established by planetary orbits, and they acquire a spherical distribution in atoms because of various quantum effects, e.g., uncertainties.* Therefore, in the absence of these, all atoms would naturally exhibit *five* force fields and not only the four fields currently assumed in chemistry.

On historical grounds it should be noted that theoretical and experimental studies in physics of the hydrogen atom subjected to an external (homogeneous) magnetic field date to Schrödinger's times.

4. Magnecules Internal Bonds

In the preceding section we have noted that a sufficiently strong external magnetic field polarizes the orbits of peripheral atomic electrons resulting in a magnetic field which does not exist in a conventional spherical distribution. Needless to say, the same external magnetic fields also polarize the intrinsic magnetic moments of the peripheral electrons and of nuclei, resulting into *three net magnetic polarities* available in an *atomic* structure for a new bond.

When considering molecules, the situation is different because valence electrons are bonded in singlet couplings to verify Pauli's exclusion principle, as per our hypothesis of the *isoelectronium* of Chapter 4. As a result, their net magnetic polarities can be assumed in first approximation as being null. In this case, only *two* magnetic polarities are available for new bonds, namely, the magnetic field created by the rotation of paired valence electrons in a polarized orbit plus the intrinsic magnetic field of nuclei.

It should be noted that the above results persist when the inter-electron distance of the isoelectronium assumes orbital values. In this case the total intrinsic magnetic moment of the two valence electrons is also approximately null in average due to the persistence of antiparallel spins and, therefore, antiparallel magnetic moments, in which absence there would be a violation of Pauli's exclusion principle.

The calculation of these *polarized magnetic moments at absolute zero degree temperature* is elementary [1]. By using rationalized units, the magnetic moment $M_{\text{e-orb.}}$ of a polarized orbit of one atomic electron is given by the general quantum mechanical law:

$$M_{\text{e-orb.}} = \frac{q}{2m} L\mu, \tag{8.6}$$

where L is the angular momentum, μ is the rationalized unit of the magnetic moment of the electron, $q = -e$, and $m = m_e$.

It is easy to see that *the magnetic moment of the polarized orbit of the isoelectronium with characteristics (4.25) coincides with that of one individual electron.* This is due to the fact that, in this case, in Eq. (8.6) the charge in the numerator assumes a double value $q = -2e$, while the mass in the denominator also assumes a double value, $m = 2m_e$, thus leaving value (8.6) unchanged.

By plotting the various numerical values for the ground state of the hydrogen atom, one obtains:

$$M_{\text{e-orb.}} = M_{\text{isoe-orb.}} = 1,859.59\mu. \qquad (8.7)$$

By recalling that in the assumed units the proton has the magnetic moment 1.4107 μ, we have the value [1]:

$$\frac{M_{\text{e-orb.}}}{M_{\text{p-intr.}}} = \frac{1,856.9590}{1.4107} = 1,316.33, \qquad (8.8)$$

namely, *the magnetic moment created by the orbiting in a plane of the electron in the hydrogen atom is 1,316 times bigger than the intrinsic magnetic moment of the nucleus*, thus being sufficiently strong to create a bond.

It is evident that the *polarized magnetic moments at ordinary temperature* are smaller than those at absolute zero degrees temperature. This is due to the fact that, at ordinary temperature, the perfect polarization of the orbit in a plane is no longer possible. In this case the polarization occurs in a *toroid*, as illustrated in Fig. 8.2, whose sectional area depends on the intensity of the external field.

As an illustrative example, under an external magnetic field of 10 Tesla, an *isolated hydrogen atom* has a total magnetic field of the following order of magnitude:

$$M_{\text{H-tot.}} = M_{\text{p-intr.}} + M_{\text{e-intr.}} + M_{\text{e-orb.}} \approx 3,000\mu, \qquad (8.9)$$

while the same hydrogen atom under the same conditions, when a component of a *hydrogen molecule* has the smaller value

$$M_{\text{H}_2\text{-tot.}} = M_{\text{p-intr.}} + M_{\text{isoe-orb.}} \approx 1,500\mu, \qquad (8.10)$$

again, because of the absence of the rather large contribution from the intrinsic magnetic moment of the electrons, while the orbital contribution remains unchanged.

The above feature is particularly important for the study of magnecules and their applications because it establishes the theoretical foundations for the presence of isolated atoms in the structure of magnecules since *the magnetic bonds of isolated atoms can be at least twice stronger than those of the same atoms when part of a molecule.*

An accurate independent verification of the above calculations was conducted by M.G. Kucherenko and A.K. Aringazin [3], who obtained the following value via the use of alternative models,

$$\frac{M_{\text{e-orb.}}}{M_{\text{p-intr.}}} \approx 1,315\mu. \qquad (8.11)$$

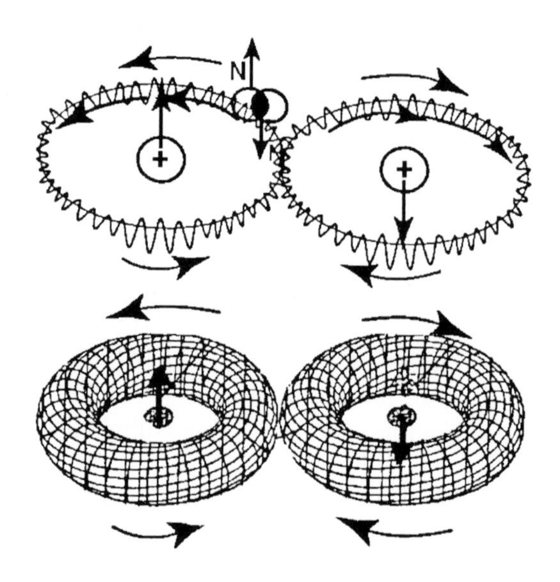

Figure 8.2. A schematic view of the magnetic fields of the isochemical model of the hydrogen molecule with isoelectronium assumed to be a stable quasi-particle. The top view represents the molecule at absolute zero degree temperature with polarization of the orbit in a plane, while the bottom view represents the molecule at ordinary temperature with a polarization of the orbit within a toroid. In both cases there is the disappearance of the *total intrinsic* magnetic moments of the electrons because they are coupled in the isoelectronium with antiparallel spin and magnetic moments due to Pauli's exclusion principle. The *lack* of contribution of the intrinsic magnetic moments of the electrons persists even when the isoelectronium has dimension much bigger than 1 fm, because the antiparallel character of the spins and magnetic moments persists, resulting in an average null total intrinsic magnetic moment of the electrons. Therefore, the biggest magnetic moment of the hydrogen molecule which can be obtained via polarizations is that of the electrons *orbits*. Note, as recalled in Sect. 4.2, the *oo*-shaped (also called figure eight) configuration has been recently proved in mathematics to be one of the most stable solutions of the N-body problem.

Needless to say, the quantized value of the angular momentum of the ground state of the conventional (unpolarized) hydrogen atom is null, $L = 0$, thus implying a null magnetic moment, $M = 0$. This occurrence confirms the well known feature that the magnetic moment of the orbit of the peripheral electron of a conventional (unpolarized) hydrogen atom is null.

Consequently, expressions (8.6)-(8.11) should be considered under a number of clarifications. First, said expressions refer to *the orbit of the peripheral electron under an external magnetic field* which implies an evident alteration of the value of the magnetic moment. Note that this external magnetic field can be either that of an electric discharge, as in the PlasmaArcFlow reactors, or that of another polarized hydrogen atom, as in a magnecule. This occurrence confirms a main aspect of the new chemical species of magnecules, namely, that the plane polarization of the orbits of the peripheral atomic electron is stable if and only if said polarization is coupled to another because, if isolated, the plane polarization is instantly lost due to rotations with recover the conventional spheroidal distribution of the orbits.

Moreover, expressions (8.6)-(8.11) refer to the angular momentum of the orbit of the peripheral electron *polarized in a plane*, rather than that with a spherical distribution as in the conventional ground state of the hydrogen atom. The latter condition, alone, is sufficient to provide a non-null quantized orbital magnetic moment.

Finally, the value $L = 1$ needed for expressions (8.6)-(8.11) can be obtained via *the direct quantization of the plane polarization of a classical orbit*. These aspects have been studied in detail by Kucherenko and Aringazin [2] and Aringazin [8] (see Appendix 8.A). These studies clarify a rather intriguing property mostly ignored throughout the 20-th century according to which, contrary to popular beliefs, *the quantized angular momentum of the ground state of the hydrogen atom is not necessarily zero, because its value depends on possible external fields.*

It is important to note that the magnetic polarizations herein considered are *physical notions*, thus being best expressed and understood via *actual orbits* as treated above rather than *chemical orbitals*. This is due to the fact that *orbits are physical entities* actually existing in nature, and schematically represented in the figures with standing waves, in semiclassical approximation. By contrasts, *orbitals are purely mathematical notions* given by probability density. As a result, magnetic fields can be more clearly associated with orbits rather than with orbitals.

Despite the above differences, it should be stressed that, magnetic polarizations can also be derived via the *orbitals* of conventional use in chemistry. For example, consider the description of an isolated atom via the conventional Schrödinger equation

$$H|\psi\rangle = \left(\frac{p^2}{2m} + V\right)|\psi\rangle = E|\psi\rangle, \tag{8.12}$$

where $|\psi\rangle$ is a state in a Hilbert space. Orbitals are expressed in terms of the probability density $|\langle\psi| \times |\psi\rangle|$. The probability density of the electron of a hydrogen atom has a spherical distribution, namely, the electron of an isolated hydrogen atom can be found at a given distance from the nucleus with the same probability in any direction in space.

Assume now that the same hydrogen atom is exposed to a strong external homogeneous and static magnetic field B. This case requires the new Schrödinger equation,

$$\left((p - \frac{e}{c}A)^2/2m + V\right)|\psi'\rangle = E'|\psi'\rangle, \tag{8.13}$$

where A is vector-potential of the magnetic field B. It is easy to prove that, in this case, the new probability density $|\langle\psi'| \times |\psi'\rangle|$ possesses a *cylindrical symmetry* precisely of the type indicated above, thus confirming the results obtained on physical grounds. A similar confirmation can be obtained via the use of Dirac's equation or other chemical methods.

An accurate recent review of the Schrödinger equation for the hydrogen atom under external magnetic fields is that by A.K. Aringazin [8], which study confirms the toroidal configuration of the electron orbits which is at the foundation of the new chemical species of magnecules. A review of Aringazin studies is presented in Appendix 8.A. As one can see, under an external, strong, homogeneous, and constant magnetic fields of the order of 10^{13} Gauss = 10^7 Tesla, the solutions of Schrödinger equation of type (8.13) imply the restriction of the electron orbits within a single, small-size toroidal configuration, while the excited states are represented by the double-splitted toroidal configuration due to parity.

Intriguingly, the binding energy of the ground state of the H atom is much higher than that in the absence of an external magnetic field, by therefore confirming another important feature of the new chemical species of magnecules, that of permitting new means of storing energy within conventional molecules and atoms, as discussed in Sect. 8.9.

For magnetic fields of the order of 10^9 Gauss, spherical symmetry begins to compete with the toroidal symmetry, and for magnetic fields of the order of 10^5 Gauss or less, spherical symmetry is almost completely restored by leaving only ordinary Zeeman effects. This latter result confirms that the creation of the new chemical species of magnecules in gases as per Definition 8.2.1 requires very strong magnetic fields. The situation for liquids is different, as shown in Sect. 8.10.

The magnetic polarization of atoms larger than hydrogen is easily derived from the above calculations. Consider, for example, the magnetic polarization of an isolated atom of oxygen. For simplicity, assume that an external magnetic field of 10 Tesla polarizes only the two peripheral valence electrons of the oxygen. Accordingly, its total polarized magnetic field of orbital type is of the order of twice value (8.9), *i.e.*, about 6,000 μ. However, when the same oxygen atom is bonded into the water or other molecules, the maximal polarized magnetic moment is about half the preceding value.

Note the dominance of the magnetic fields due to polarized electron *orbits* over the intrinsic *nuclear* magnetic fields. This is due not only to the fact that the former are 1,316 times the latter, but also to the fact that nuclei are at a relative great distance from peripheral electrons, thus providing a contribution to the bond even smaller than that indicated. This feature explains the essential novelty of magnecules with respect to established magnetic technologies, such as that based on *nuclear magnetic resonances.*

Note also that a main mechanism of polarization is dependent on an external magnetic field and the force actually providing the bond is of magnetic type. Nevertheless, the ultimate origin is that of charges rotating in an atomic orbit. This illustrates that, as indicated in Sect. 8.1, the name "magnecules" was suggested on the basis of the predominant magnetic origin, as well as for the pragmatic differentiation with molecules without using a long sentence, although a technically more appropriate name would be "electromagnecules."

Needless to say, the polarization of the orbits is not necessarily restricted to valence electrons because the polarization does not affect the quantum numbers of any given orbit, thus applying for all atomic electrons, including those of complete inner shells, of course, under a sufficiently strong external field. As a consequence, *the intensity of the magnetic polarization generally increases with the number of atomic electrons*, namely, the bigger is the atom, the bigger is, in general, its magnetic bond in a magnecule.

Ionizations do not affect the *existence* of magnetic polarizations, and they may at best affect their *intensity*. An ionized hydrogen atom is a naked proton, which acquires a polarization of the direction of its magnetic dipole moment when exposed to an external magnetic field. Therefore, an ionized hydrogen atom can indeed bond magnetically to other polarized structures. Similarly, when oxygen is ionized by the removal of one of its peripheral electrons, its remaining electrons are unchanged. Consequently, when exposed to a strong magnetic field, such an ionized oxygen atom acquires a magnetic polarization which is similar to that of an unpolarized oxygen atom, except that it lacks the contribution from the missing electron. Ionized molecules or dimers behave along similar lines. Accordingly, the issue as to whether individual atoms, dimers or molecules are ionized or not will not be addressed hereon.

The magnetic polarizations here considered are also independent as to whether the substance considered is paramagnetic or not. This is evidently due to the fact that the polarization deals with the individual orbits of individual peripheral electrons, irrespective of whether paired or unpaired, belonging to a saturate shell or not. Therefore, the issue as to whether a given substance is paramagnetic or not will be ignored hereon.

Similarly, the polarizations here considered do not require molecules to have a net total magnetic polarity, which would be possible only for paramagnetic substances, again, because they act on individual orbits of individual atomic electrons.

We should also indicate that another verification of our isochemical model of molecular structures is the resolution of the inconsistency of the conventional model in predicting that all substances are paramagnetic, as illustrated in Figs. 1.4 and 1.5.

Recall that the atoms preserve their individualities in the conventional molecular model, thus implying the *individual* acquisition of a magnetic polarization under an external field, with consequential net total magnetic polarities for all molecules which is in dramatic disagreement with experimental; evidence.

By comparison, in the isochemical molecular model the valence electrons are actually bonded to each other, with consequential *oo*-shaped orbit around the respective nuclei. This implies that the rotational directions of the *o*-branches are opposite to each other. In turn, this implies that magnetic polarizations are also opposite to each other, resulting in the lack of a net magnetic polarity under an external field, in agreement with nature (see Figs. 4.5 and 8.3 for more details).

5. Production of Magnecules in Gases, Liquids and Solids

At its simplest, the creation of magnecules can be understood via the old method of magnetization of a paramagnetic metal by induction. Consider a paramagnetic metal which, initially, has no magnetic field. When exposed to a constant external magnetic field, the paramagnetic metal acquires a permanent magnetic field that can only be destroyed at a sufficiently high temperature varying from metal to metal and called the *Curie Temperature.*

The mechanism of the above magnetization is well known. In its natural unperturbed state, the peripheral atomic electrons of a paramagnetic metal have a space distribution that results in the lack of a total magnetic field. However, when exposed to an external magnetic field, the orbits of one or more unpaired electrons are polarized into a toroidal shape with end polarities opposite to those of the external field.

This mechanism is called magnetic induction, and results in a stable chain of magnetically polarized orbits from the beginning of the metal to its end with polarities North-South/North-South/North-South/ ... This chain of polarizations is so stable that it can only be destroyed by high temperatures.

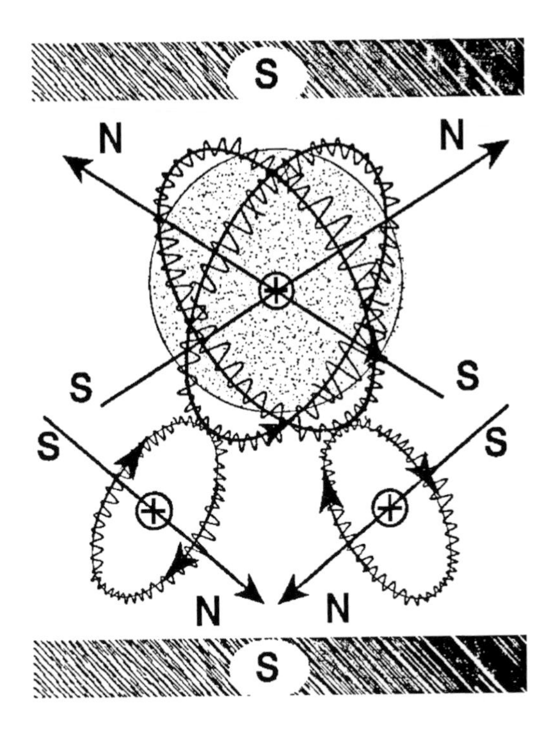

Figure 8.3. A schematic view of the resolution for the case of the water molecule of the inconsistent prediction of the conventional molecular model that water is paramagnetic (Fig. 1.5), as permitted by the Santilli-Shillady isochemical model of water molecule (Chapter 5). As one can see, the resolution is given by the impossibility for the water molecule to acquire a net magnetic polarity. Note the complexity of the geometry of the various magnetic fields which, according to ongoing research, apparently permits the first explanation on scientific record of the 105° angle between the two H–O dimers. The corresponding resolution for the case of the hydrogen is outlined in Fig. 4.5.

The creation of magnecules can be essentially understood with a similar polarization of the peripheral electron orbits, with the main differences that: no total magnetic polarization is necessary; the polarization generally apply to all electrons, and not necessarily to unpaired electrons only; and the substance need not to be paramagnetic.

To illustrate these differences, consider a diamagnetic substance, such as the hydrogen at its gaseous state at ordinary pressure and temperature. As well known, the hydrogen molecule is then a perfect sphere whose radius is equal to the diameter of a hydrogen atom, as illustrated in Fig. 8.5.A. The creation of the needed magnetic polarization requires the use of external magnetic fields capable, first, to remove the rotation of the atoms, as illustrated in Fig. 8.4.B, and then the removal of the

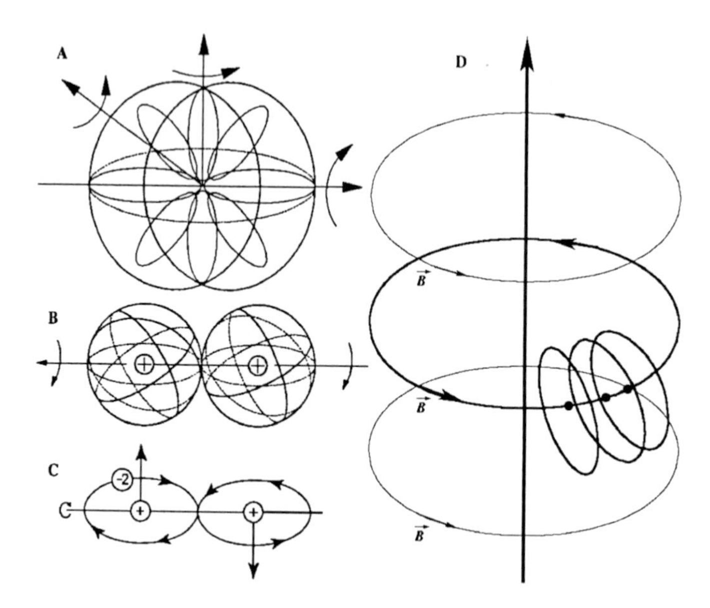

Figure 8.4. A schematic view of the main mechanism underlying the creation of magnecules, here illustrated for the case of the hydrogen molecule. It consists in the use of sufficiently strong external magnetic fields which can progressively eliminate all rotations, thus reducing the hydrogen molecule to a configuration which, at absolute zero degrees temperature, can be assumed to lie in a plane. The planar configuration of the electron orbits then implies the manifestation of their magnetic moment which would be otherwise absent. The r.h.s. of the above picture outlines the geometry of the magnetic field in the immediate vicinity of an electric arc as described in the text for the case of hadronic molecular reactors (Chapter 7). Note the *circular* configuration of the magnetic field lines around the electric discharge, the *tangential* nature of the symmetry axis of the magnetic polarization of the hydrogen atoms with respect to said circular magnetic lines, and the consideration of hydrogen atoms at *orbital distances* from the electric arc 10^{-8} cm, resulting in extremely strong magnetic fields proportional to $(10^{-8})^{-2} = 10^{16}$ Gauss, thus being ample sufficient to create the needed polarization (see Appendix 8.A for details).

internal rotations of the same, resulting in a planar configuration of the orbits as illustrated in Fig. 8.4.C.

Once the above polarization is created in two or more hydrogen molecules sufficiently near each other, they attract each other via opposite magnetic polarities, resulting in the elementary magnecules of Fig. 8.5.A. Additional elementary magnecules can then also bond to each other, resulting in clusters with a number of constituents depending on the conditions considered.

A most efficient industrial production of gas and liquid magnecules is that via the *PlasmaArcFlow Reactors* studied in the preceding Chap-

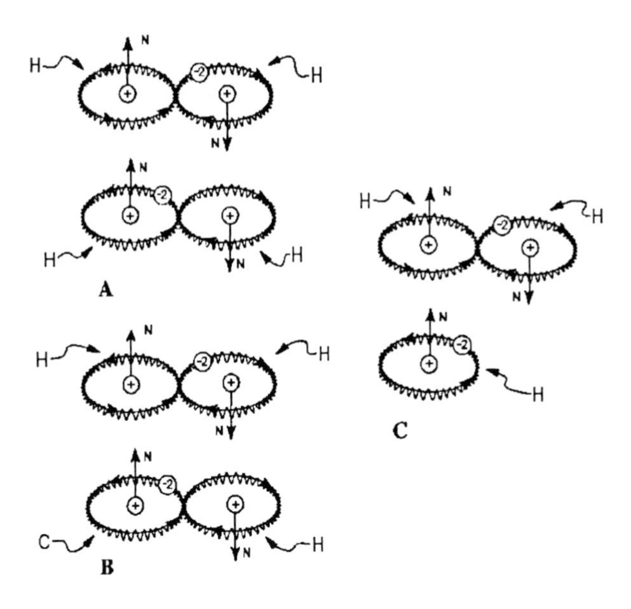

Figure 8.5. A schematic view of the simplest possible bonds due to the polarization of the orbits of peripheral electrons. Case A illustrates the *elementary hydrogen magnecule*, namely, that composed of two hydrogen molecule under said magnetic bond. Note that the magnecule has atomic weight very close to that of the helium. Therefore, the detection in a GC-MS scan of a peak with 4 a.m.u., by no means, necessarily identifies the helium because the peak could belong to the elementary hydrogen magnecule. Case B illustrates a magnecule composed by a molecule and a dimer. Case C illustrates the hypothesis submitted in this monograph that the structure with 3 a.m.u. generally interpreted as a conventional "molecule" H_3 may in reality be a magnecule between a hydrogen molecule and an isolated hydrogen atom. This is due to the fact that, once the two valence electrons of the hydrogen molecule are bonded-correlated, they cannot admit the same valence bond with a third electron for numerous physical reasons, such as: the bond cannot be stable because the former is a Boson while the latter is a Fermion; the former has charge $-2e$ while the latter has charge $-e$, thus resulting in a large repulsion; *etc.*

ter [2]. As we shall see via the experimental evidence presented below, said reactors can produce an essentially pure population of gas and liquid magnecules without appreciable percentages of molecules directly detectable in the GC- or LC-MS.

The reason for these results is the intrinsic geometry of the PlasmaArcFlow itself. With reference to Fig. 7.9, recall that this technology deals with a DC electric arc submerged within a liquid waste to be recycled. The arc decomposes the molecules of the liquid into its atomic constituents; ionizes the same; and creates a plasma of mostly ionized H, C and O atoms at about $3,500°$ K. The flow of the liquid through the arc then continuously removes the plasma from the arc following its formation. Said plasma then cools down in the surrounding liquid, and a number of chemical reactions take place resulting in the formation of magnegas which bubbles to the surface of the liquid where it is collected for industrial or consumer use.

To understand the creation of a *new chemical species* defined according to Sect. 8.1 as an essentially pure population of *gas magnecules*, recall that magnetic fields are inversely proportional to the square of the distance,

$$F_{\text{magnetic}} = \frac{m_1 m_2}{r^2}. \tag{8.14}$$

Therefore, an atom in the immediate vicinity of a DC electric arc with 1,000 A and 30 V, experiences a magnetic field which is inversely proportional to the square of the *orbital* distance $r = 10^{-8}$ cm, resulting in a magnetic field proportional to 10^{16} units.

No conventional space distribution of peripheral atomic electrons can exist under these extremely strong magnetic fields, which are such to generally cause the polarization of the orbits of *all* atomic electrons, and not only those of valence type, as well as their essential polarization in a plane, rather than a toroid.

As soon as two or more molecules near each other possessing such an extreme magnetic polarization are created, they bond to each other via opposing magnetic polarities, resulting in the elementary magnecule of Fig. 8.3.A.

Moreover, as shown earlier, isolated atoms have a magnetic field with an intensity double that of the same atom when belonging to a molecule. Therefore, as soon as created in the immediate vicinity of the electric arc, individual polarized atoms can bond to polarized molecules without any need to belong themselves to a molecule, as illustrated in Fig. 8.5.C.

Finally, recall that the PlasmaArcFlow is intended to destroy liquid molecules such as that of water. It then follows that the plasma can also contain individual highly polarized molecular fragments, such as the dimer H–O. The notion of gas magnecules as per Definition 8.2.1 then follows as referred to stable clusters of molecules, and/or dimers, and/or isolated atoms under an internal attractive bond among opposing polarities of the magnetic polarization of the orbits of peripheral electrons, nuclei and electrons when the latter are not coupled into valence bonds.

Effective means for the creation of an essentially pure population of *liquid magnecules* are given by the same PlasmaArcFlow Reactors. In fact, during its flow through the DC arc, the liquid itself is exposed to the same extreme magnetic fields as those of the electric arc indicated above. This causes the creation of an essentially pure population of liquid magnecules composed of highly polarized liquid molecules, dimers of the same liquid, and individual atoms, as established by LC-MS/UVD tests.

One way to create an essentially pure population of *solid magnecules* is given by freezing the new chemical species at the liquid level and then verifying that the latter persists after defrosting, as confirmed by various tests. Therefore, the case of solid magnecules is ignored hereon for simplicity.

By denoting with the arrow $\uparrow$ the vertical magnetic polarity North-South and with the arrow $\downarrow$ the vertical polarity South-North, and by keeping the study at the absolute zero degree temperature, when exposed to the above indicated extreme magnetic fields, the hydrogen molecule H–H can be polarized into such a form that the orbit of the isoelectronium is in a plane with resulting structure $H_\uparrow - H_\downarrow$ (Fig. 8.2).

The elementary hydrogen magnecule can then be written

$$\{H^a_\uparrow - H^b_\downarrow\} \times \{H^c_\uparrow - H^d_\downarrow\}, \tag{8.15}$$

where: a, b, c, d denote different atoms; the polarized hydrogen atom $H^a_\uparrow$ is bonded magnetically to the polarized atom $H^c_\uparrow$ with the South magnetic pole of atom a bonded to the North pole of atom c; and the North polarity of atom b is bonded to the South polarity of atom d (see, again, Fig. 8.5.A). This results in a strong bond due to the flat nature of the atoms, the corresponding mutual distance being very small and the magnetic force being consequently very large. Moreover, unlike the case of the unstable clusters due to electric polarization discussed in Sect. 8.1, the above magnetic bonds are very stable because motions due to temperature apply to the bonded couple (8.15) as a whole.

For other magnecules we can then write

$$\{H_\uparrow - H_\downarrow\} \times \{C_\uparrow - O_\downarrow\}; \tag{8.16}$$

or, more generally

$$\{H_\uparrow\text{–}H_\downarrow\} \times H_\downarrow \times \{C_\uparrow\text{–}O_\downarrow\} \times \{H_\uparrow\text{–}O_\downarrow\} \times \{H_\uparrow\text{–}C_\downarrow\text{–}A\text{–}B\text{–}C\ldots\} \times \ldots, \tag{8.17}$$

where A, B, and C are generic atoms in a conventional molecular chain and the atoms without an indicated magnetic polarity may indeed be polarized but are not necessarily bonded depending on the geometric distribution in space.

Magnecules can also be formed by means other than the use of external magnetic fields. For instance, magnecules can be produced by electromagnetic fields with a distribution having a cylindrical symmetry;

or by microwaves capable of removing the rotational degrees of freedom of molecules and atoms, resulting in magnetic polarizations. Similarly, magnecules can be formed by subjecting a material to a pressure that is sufficiently high to remove the orbital rotations. Magnecules can also be formed by friction or by any other means not necessarily possessing magnetic or electric fields, yet capable of removing the rotational degrees of freedom within individual atomic structures, resulting in consequential magnetic polarizations.

It is, therefore, expected that a number of substances which are today listed as of unknown chemical bond, may eventually result to have a magnecular structure.

Magnecules of type (8.15) may well have been detected in past mass spectrometric measurements, but believed to be the helium (because its molecular weight is very close to that of the helium). In fact, the same happens for the "molecule" H_3 which, in reality may be the magnecule of Fig. 8.5.C.

The destruction of magnecules is achieved by subjecting them to a temperature greater than the magnecules Curie Temperature which varies from magnecule to magnecule.

6. New Molecules Internal Bonds

As indicated in Sect. 8.2, and verified experimentally later on, the IR signatures of conventional molecules such as CO_2 are mutated due to the appearance of two new peaks which do not exist for the conventional molecule. By recalling that peaks in the IR signature generally represent bonds, this evidence indicates the capability by the CO_2 molecule to acquire new internal bonds in addition to those of conventional valence type.

The magnetic polarization at the foundations of magnecules predicts the existence of these new internal bonds and permits their quantitative study. Recall that external magnetic fields can polarize the orbit of valence electrons, but cannot possibly break or alter valence bonds. Recall that, consequently, sufficiently strong external magnetic fields can polarize the orbits of all atomic electrons, and not only those of the valence electrons.

Consider then a conventional molecule such as C=O. When exposed to the extreme magnetic fields as existing in the PlasmaArcFlow technology, the orbits of all internal electrons can be polarized, individually, for the carbon and the oxygen, in addition to the polarization of the two pairs of valence bonds. Note that the planes of these polarizations need not be necessarily parallel to each other, because their relative orientation dependents on the geometry at hand.

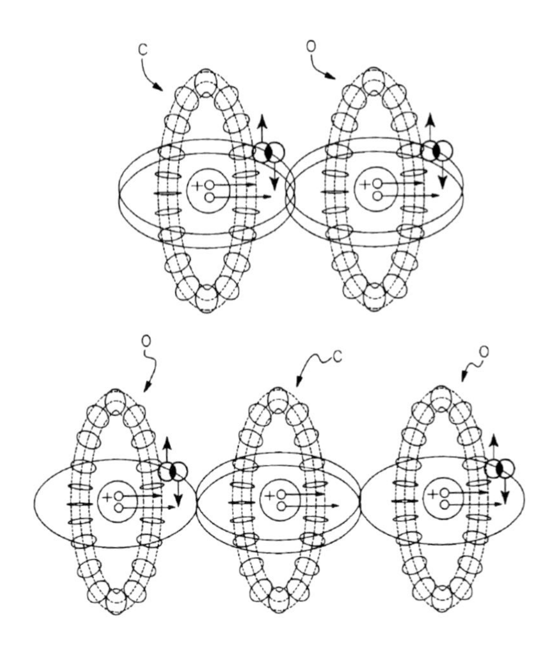

Figure 8.6. A schematic view for the cases of C=O and O–C–O of the polarization of internal atomic electrons, while preserving conventional valence bonds, and the consequential creation of new bonds in conventional molecules which are not of valence type, as later on verified experimentally via IR scans.

One of the various possible geometries is that in which the plane of the polarization of the internal electrons is perpendicular to that of the two pairs of valence bonds. In this case we have the birth of a *new bond of magnetic origin in the interior of a conventional molecule*, which is evidently given by the alignment of the two polarities North-South and North-South in the carbon and oxygen, and the consequential attraction of opposite polarities of different atoms, as illustrated in Fig. 8.6.A.

For the case of the O–C–O molecule we can evidently have two internal bonds of magnetic type in addition to the valence bonds, which are also given by the alignment of the magnetic polarities, resulting in one new bond for the O–C dimer and a second one for the C–O dimer, as illustrated in Fig. 8.6.B.

As we shall see later on, the above new internal molecular bonds have major industrial and consumer implications, inasmuch as they permit the production of fuels capable of releasing under combustion anomalous amounts of energy, with consequential reduction of pollutants in the exhaust, as already proved by magnegas.

Needless to say, the creation of new internal bonds is an extreme case of IR mutation. In reality, numerous other weaker forms of mutations without the appearance of new peaks are possible and their study is left to the interested reader.

7. Main Features of Magnecules to be Detected

The experimental detection of gas magnecules requires the verification of a number of characteristic features of magnecules identified in Definition 8.2.1. In the following we focus the reader's attention on the main features of gas magnecules which must be verified via GC-MS tests. The remaining features will be considered later on.

Feature 1: Appearance of unexpected heavy MS peaks.

Gas magnecules are generally heavier than the heaviest molecule in a given gas. Peaks in the GC-MS are, therefore, expected in macroscopic percentages with atomic weights bigger than the heaviest molecule. As a concrete example, the heaviest molecule in magnegas in macroscopic percentage is CO_2 with 44 a.m.u. Therefore, GC-MS scans should only show background noise if set for over 44 a.m.u. On the contrary, peaks in macroscopic percentages have been detected in magnegas all the way to 1,000 a.m.u.

Feature 2: Unknown character of the unexpected MS heavy peaks.

To provide the initial premises for the detection of magnecules, all MS peaks of feature 1 should result in being "unknown" following the computer search among all known molecules, usually including a minimum of 150,000 molecules. Evidently, this lack of identification of the peaks, *per se*, does not guarantee the presence of a new chemical species.

Feature 3: Lack of IR signature of the unknown MS peaks.

Another necessary condition to claim the detection of magnecules is that the unknown MS peaks of feature 1 should have no IR signature other than that of the molecules and/or dimers constituents. This feature guarantees that said heavy peaks cannot possibly represent molecules, thus establishing the occurrence of a new chemical species. In fact, only very few and very light molecules can have such a perfect spherical symmetry to avoid IR detection, while such a perfect spherical symmetry is manifestly impossible for large clusters. In regard to the constituents we are referring to IR signatures, *e.g.*, of the CO_2 at 44 a.m.u. in a cluster having 458 a.m.u.

Feature 4: Mutation of IR signatures.

The infrared signatures of conventional molecules constituting magnecules are expected to be *mutated*, in the sense that the shape of their peaks is not the conventional one. As indicated in the preceding section, the mutations most important for industrial applications are those due to the presence of *new IR peaks* representing new internal bonds. Nevertheless, various other forms of IR mutations are possible.

Feature 5: Mutation of magnecular weights.
While molecules preserve their structure and related atomic weight at conventional temperatures and pressures, this is not the case for gas magnecules, which can *mutate* in time, that is, change their atomic weight with consequential change of the shape and location of their MS peaks. Since we are referring to gases whose constituents notoriously collide with each other, magnecules can break-down during collisions into fragments which can then recombine with other fragments or other magnecules to form new clusters.

Feature 6: Accretion or emission of individual atoms, dimers or molecules.
Magnecules are expected to experience accretion or emission of individual atoms, dimer or molecules without necessarily breaking down into parts. It follows that the peaks of Feature 1 are not expected to remain the same over a sufficient period of time for the same gas under the same conditions.

Feature 7: Anomalous adhesion.
Magnetically polarized gases have anomalous adhesion to walls of disparate nature, not necessarily of paramagnetic character, as compared to the same unpolarized gas. This is due to the well-known property that magnetism can be propagated by induction, according to which a magnetically polarized molecule with a sufficiently intense magnetic moment can induce a corresponding polarization of valence and/or other electrons in the atoms constituting the wall surface. Once such a polarization is created by induction, magnecules can have strong magnetic bonds to the indicated walls. In turn, this implies that the background of GC-MS following scans and conventional flushing are often similar to the scan themselves. As a matter of fact, backgrounds following routine flushing are often used to identify the most dominant magnecules. Notice that the magnetic polarization here considered does not require that the walls of the instrument are of paramagnetic type, since the polarization occurs for the orbits of arbitrary atoms.

Magnetically polarized gases additionally have mutated physical characteristics and behavior because the very notion of polarization of electron orbits implies physical alterations of a variety of characteristics, such as average size. Mutations of other characteristics are then consequential.

We should finally recall that the above features are expected to disappear at a sufficiently high temperature, evidently varying from gas to gas (Curie Temperature), while the features are expected to be enhanced at lower temperature and at higher pressure, and survive liquefaction.

8. Necessary Conditions for the Correct Detection of Molecules and Magnecules

8.1. Selection of Analytic Instruments

Current technologies offer an impressive variety of analytic instruments (see, *e.g.*, Ref. [4]), which include: Gas Chromatography (GC), Liquid Chromatography (LC), Capillary Electrophoresis Chromatography (CEC), Supercritical Chromatography (SCC), Ion Chromatography (IC), Infrared Spectroscopy (IR), Raman Spectroscopy (RS), Nuclear Magnetic Resonance Spectroscopy (NMRS), X-Ray Spectroscopy (XRS), Atomic Absorption Spectroscopy (AAS), Mass Spectrometry (MS), Laser Mass Spectrometry (LMS), Flame Ionization Spectrometry (FIS), and others.

Only some of these instruments are suitable for the detection of magnecules and, when applicable, their set-up and use are considerably different than those routinely used with great success for molecules.

Among all available chromatographic equipment, that suitable for the detection of gas magnecules is the GC with column having ID of at least 0.32 mm operated according to certain criteria outlined in Sect. 8.8.3. By comparison, other chromatographs do not appear to permit the entrance of large magnecules, such as the CEC, or be potentially destructive of the magnecules to be detected, such as the IC.

Among all available spectroscopic equipment, that preferable is the IR, with the understanding that such an instrument is used in a *negative* way, that is, to verify that the magnecule considered has no IR signature. The RS may also result in being preferable in various cases, while other instruments, such as the NMRS do not appear to be capable of detecting magnecules despite their magnetic nature, evidently because NMRS are most effective for the detection of microscopic magnetic environment of H-nuclei rather than large structures. Other spectroscopic instruments have not been studied at this writing.

In regard to spectrometric equipment, the most recommendable one is the low ionization MS due to the fact that other instruments seemingly destroy magnecules at the time of their detection. The study of other spectrometric equipment is left to interested researchers. Chemical analytical methods (i.e. via chemical reactions) to *detect* gas magnecules are probably not very effective since they necessarily destroy the magnecules in reaction.

As it is well known, when used individually, the above suggested instruments have considerable limitations. For instance, the GC has a great resolution of a substance into its constituent, but it has very limited capabilities to identify them. By comparison, the MS has great capabilities to identify individual species, although it lacks the ability to separate them.

For these reasons, some of the best analytic instruments are given by the combination of two different instruments. Among them, the most recommendable one is the GC combined with the MS, and denoted GC-MS. A similar occurrence holds for the IR combined to the GC-MS. As indicated since the early parts of this Chapter, the best instrument for the detection of both molecules and magnecules in gases is the GC-MS equipped with the IRD denoted GC-MS/IRD while that for liquids is the LC-MS equipped with UVD and denoted LC-MS/UVD.

Among a large variety of GC-MS instruments, only a few are truly effective for the detection of gas magnecules for certain technical reasons identified below. The instrument which has permitted the first identification of magnecules and remains the most effective at this writing (despite its considerable age for contemporary standards) is the GC Hewlett-Packard (HP) model 5890 combined with the MS HP model 5972 equipped with a large ID column and feeding line operated at the lowest temperature permitted by the instrument (about 10° C) and the longest elusion time (about 25 min).

A secondary function of the IRD is that of identifying the *dimers constituting a magnecule*, a task which can be fulfilled by various IRD. That which was used for the original discovery of magnecules and still remains effective (again, despite its age by current standards) is the IRD HP model 5965, when operated with certain criteria identified below.

A most insidious aspect in the detection of magnecule is the protracted use of any given instrument with great success in the detection of conventional molecules, and the consequential expectation that the same instrument should work equally well for the detection of magnecules, resulting in an analysis without any real scientific value because:

i) the species to be detected may not even have entered the instrument, as it is routinely the case for small syringes and feeding lines particularly for liquid magnecules (which can be so big as to be visible to the naked eye, as shown in Sect. 8.10);

ii) the species to be detected may have been destroyed by the measurement itself, as it is routinely the case for instruments operated at very high temperature, or flame ionization instruments which, when used for combustible gases with magnecular structure, cause the combustion of magnecules at the very time of their detection; or

iii) the detection itself may create magnecules which do not exist in the original species, as it is the case of peaks with 3 a.m.u. discussed in Fig. 8.4.

In conclusion, the separation between a true scientific measurement and a personal experimental belief requires extreme scientific caution in the selection of the analytic instrument, its use, and the interpretation of the results.

8.2. Unambiguous Detection of Molecules

As it is well known, a *gas molecule* is identifiable by unique and unambiguous GC-MS peaks, which are distinctly different from those of any other gas molecule. In addition, this GC-MS identification can be confirmed by IRD peaks and related resonating frequencies, which are also distinctly different for different gas molecule. Additional confirmations are possible using other analytic methods, such as those based on average molecular weight, chemical reactions and other procedures.

The advent of the new chemical species of magnecules suggests a reexamination of these analytic methods and procedures so as to separate personal opinions from actual scientific identifications. Such a reexamination is warranted by the fact that, due to extended use, claims of specific molecular identifications are nowadays generally voiced via the use of only one analytic detector.

As an illustration, most contemporary analytic laboratories conduct chemical analyses on gases via the sole use of the IRD. However, *infrared detectors do not identify complete molecules, since they can only identify the bond in their dimers.* For instance, for the case of H_2O, the IRD does not identify the complete molecule, but only its dimer H–O.

This method of identification of molecules is certainly acceptable for gases whose lack of magnetic polarization has been verified by the analysts. However, the same method is highly questionable for gases of unknown origin. In fact, we shall soon show experimental evidence of clear IR signatures for molecules which have no MS identification at all, in which case the claim of such a molecule evidently has no scientific value.

The inverse occurrence is equally questionable, namely, the claim of a given molecule from its sole identification in the MS without a confirmation of exactly the same peak in the IRD. In fact, there are several MS peaks in magnetically polarized gases which may be easily identified with one or another molecule, but which have no IR signature at all at the MS value of the atomic weight, in which case the claim of molecular identification evidently has no scientific value.

Note that the great ambiguities in the separate use of disjoint GC-MS and IRD. In fact, in this case there is no guarantee or visible evidence that exactly the same peak is jointly inspected under the MS and, separately, the IRD. In fact, a given molecule can be tentatively identified in the MS at a given a.m.u., while the same molecule may indeed appear in the IRD, although at a different value of a.m.u., in which case, again, the claim to have detected a given molecule is a personal experimental belief, rather than a scientific truth.

In conclusion, *a serious scientific identification of any given molecule requires the joint use of at least two different analytic methods, both giving exactly the same result for exactly the same peak in a unique and unambiguous way, such as the detection via MS scans with unequivocal computer identifications, confirmed by IR scans without ambiguities, thus requiring the use of GC-MS equipped with IRD.*

Additional ambiguities result from the rather widespread belief that molecules are the only possible chemical species in nature, in which case small deviations from exact identifications are generally ignored for the specific intent of adapting experimental evidence to pre-existing knowledge, rather than modifying old interpretations to fit new experimental evidence. This widespread tendency is also a reason why magnecules have not been identified until now.

As an illustration, suppose that: a GC-MS equipped with IRD detects a peak with 19 a.m.u.; said peak is identified by the MS search as the water molecule with 18 a.m.u.; and the IRD confirms the presence of the HO-dimer. Under these conditions, it is almost universally accepted in contemporary analytic laboratories that said peak with 19 a.m.u. represents the water molecule, and the spurious single a.m.u. is just an "impurity" or something to be ignored, in which case, however, we do not have a true scientific identification of the species.

In fact, it is well possible that the peak at 19 a.m.u. is constituted by a highly polarized water molecule magnetically bonded to one isolated hydrogen atom with structure

$$\{H_{\downarrow} - O_{\uparrow\uparrow} - H_{\downarrow}\} \times H_{\downarrow}. \tag{8.18}$$

In this case, according to our terminology, the peak at 19 a.m.u. is a magnecule and *not* a molecule, even though the MS search gives 99.99% confidence and the IR search gives 100% confidence that the species is the ordinary water molecule. After all, the magnecular bond is transparent to current IR detection, then, the latter confirms an erroneous belief.

At any rate, no claim on the peak with 19 a.m.u. can be truly scientific or otherwise credibly, unless it explains in a specific and numerical way, without vague nomenclatures, how the single a.m.u. entity is attached to the water molecule.

Recall that the valence bond requires singlet couplings to verify Pauli's exclusion principle. As a consequence, coupled pairs of valence electrons are *Bosonic states with zero spin.* Under these conditions, no nomenclature suggesting one or another type of valence can credible explain the bonding of one single H atom to the H–O–H molecule because it would imply the bond of a *Fermion with spin* 1/2 (the valence electron of the hydrogen) with a *Boson* (the coupled valence electron pair of the water), which bond is an impossibility well known in particle physics. By comparison, the magnecular hypothesis identifies the *attractive* character of the bond in a clear and unambiguous way, and then its *numerical value* (8.9).

The detection of *liquid molecules* has problems greater than those for gas molecules, because liquid magnecules can be so big to be visible by the naked eye, in which case only their conventional molecular constituents are generally permitted to enter current instruments, resulting again in a lack of real detection.

In conclusion, the separation in the identification of molecules between a true scientific process and a personal experimental belief requires extreme care before claiming that a certain peak characterizes a molecule, since possible ambiguities exist in all cases, from small to large atomic weights. In the final analysis, as stressed above, the difference between a molecule and a magnecule may be given by what is generally considered noise, or instrument malfunction.

The most unreassuring occurrence is that all GC-MS equipped with IRD identified by this author in the USA following a laborious search belong to military, governmental, or law enforcement institutions, and none of them was identified in commercial or academic laboratories. Therefore, the great majority of analytic laboratories lack the very instrument necessary for a final and unequivocal identification of a conventional *molecules*, let alone that of magnecules.

8.3. Unambiguous Detection of Magnecules

Since magnecules have properties very different from those of conventional molecules, the experimental detection of magnecules requires a special care. In particular, methods which have been conceived and developed for the detection of molecules are not necessarily effective for the detection of the different chemical species of magnecules precisely in view of the indicated differences.

The first indication of a possible *gas magnecule* is given by MS peaks with large atomic weight which cannot be explained via conventional molecular hypotheses. The second indication of a gas magnecule is given by the lack of identification of said heavy peaks in the MS following a search among all known molecules. A third indication of a gas magnecule then occurs when said unknown MS peak has no IR signature, except those of its constituents with much smaller atomic weight, which occurrence establishes the lack of a valence bond. Final identification of a gas magnecule requires the knowledge of the method used in the production of the gas and other evidence.

As it is the case also for molecules, a serious spectrographic analysis of magnecules requires GC-MS detectors necessarily equipped with IRD, because only such an instrument permits the direct test of the *same peaks* under both the MS and IR scan. Again, if the IRD operates separately from the GC-MS, the indicated joint inspection is not possible; the IRD can only detect ordinary molecular dimers; the experimental belief that the MS peak must be a molecule is then consequential.

As a concrete example verified later on with actual tests, consider the spectrographic analysis of magnegas. This is a light gas whose heaviest molecule in macroscopic percentages should be the CO_2 at 44 a.m.u. Consider now an MS peak of magnegas at 481 a.m.u. It is evident that, while small deviations could be adapted to quantum chemistry, large deviations of such an order of magnitude cannot be reconciled with established knowledge in a credible way, thus permitting the hypothesis that the MS peak in a *light* gas with 481 a.m.u. can be a magnecule. The MS scan of the peak soon establishes the impossibility for the computer to identify the peak among all existing molecules. When the GC-MS is equipped with IRD, the analyst can scan the same peak with 481 a.m.u. under the IRD and detect no signature at the 481 a.m.u. value, the only IR signature being that at 44 a.m.u. of the CO_2 as well as those of smaller molecules. The production of the gas under intense magnetic fields then confirm that the peak here considered at 481 a.m.u. is indeed a magnecule composed of a large number of ordinary light molecules, dimers and individual atoms, in accordance with Definition 8.2.1.

Note that the IRD scan in the above test has solely identified conventional molecules without any additional unknown. Yet, the conclusion that the gas considered is solely composed of molecule would be nonscientific for numerous reasons, such as: 1) magnetic bonds are transparent to IR scans with available frequencies; 2) there is no IR detection, specifically, at 481 a.m.u.; and 3) IRD do not detect molecules, but only dimers.

Therefore, even though the IRD has detected CO_2 in the above test, the actual detection was for the C–O dimer, in which case the claim of the presence of the full CO_2 molecule is a personal opinion, and not an experimental fact.

The anomalous energy content, weigh and other features of magnegas confirm the above conclusions, because the latter can only be explained by assuming that a certain percentage of IR counts is indeed due to complete molecules, while the remaining percentage is due to unpaired dimers trapped in the magnecules. The freeing of these dimers and atoms at the time of the combustion, and their re-combination into molecules as in Eqs. (8.5) then explains the anomalous energy content.

In addition to the above basic requirements, numerous other precautions in the use of the GC-MS equipped with IRD are necessary for the detection of magnecules, such as:

i) the MS equipment should permit measurements of peaks at ordinary temperature, and avoid the high temperatures of the GC-MS column successfully used for molecules;

ii) the feeding lines should be cryogenically cooled;

iii) the GC-MS/IRD should be equipped with feeding lines of at least 0.5 mm ID;

iv) the GC-MS should be set to detect peaks at large atomic weights usually not expected; and

v) the ramp time should be the longest allowed by the instrument, *e.g.*, of at least 25 minutes.

It should be stressed that *the lack of verification of any one of the above conditions generally implies the impossibility to detect magnecules.* For instance, the use of a feeding line with 0.5 mm ID is unnecessarily large for a conventional light gas, while it is necessary for a gas with magnecular structure such as magnegas. This is due to the unique adhesion of the magnecules against the walls of the feeding line, resulting in occluded lines which prevent the passage of the most important magnecules to be detected, those with large atomic weight.

Similarly, it is customary for tests of conventional gases to use GC-MS with columns at high temperature to obtain readings in the shortest possible time, since conventional molecules are perfectly stable under the temperatures here considered. The use of such method would equally prevent the test of the very species to be detected, because, as indicated earlier, they have a characteristic Curie Temperature at which all magnetic features are lost. Magnecules are stable at ordinary temperatures and, consequently, they should be measured at ordinary temperatures.

Along similar lines, recall that GC-MS with a short ramp time are generally used for rapidity of results. Again, the use of such a practice, which has been proven by extensive evidence to be effective for molecules, prevents clear detection of magnecules. If the ramp time is not of the order of 25 minutes, *e.g.*, it is of the order of one minute, all the peaks of magnecules generally combine into one single large peak, as described below. In this case the analyst is generally lead to inspect an individual section of said large peak. However, in so doing, the analyst identifies conventional molecules constituting the magnecule, and not the magnecule itself.

When these detectors with short ramp times are equipped with IRD, the latter identify the infrared signatures of individual conventional molecules constituting said large unique peak, and do not identify the possible IR signature of the single large peak itself. Therefore, a GC-MS with short ramp time is basically unsuited for the detection of magnecules because it cannot separate all existing species into individual peaks.

In conclusion, the experimental evidence of the above occurrences establishes the need in the detection of gas magnecules of *avoiding, rather than using, techniques and equipment with a proved efficiency for molecules*, thus avoiding the use of GC-MS without IRD, with short ramp time, high column temperatures, microscopic feeding lines, and other techniques. On the contrary, new techniques specifically conceived for the detection of magnecules should be worked out.

The conditions for scientific measurements of *liquid magnecules* via LC-MS/UVD are more stringent than those for gases, because of the great increase, in general, of the atomic weight of liquid magnecules which are generally much larger than the IR of conventionally used feeding lines, as shown below.

This implies the possible erroneous claim that magnecules do not exist because they are not detected by the LC-MS, while in reality the magnecule to be detected could not enter into at all into the instrument.

8.4. Apparent Magnecular Structure of H_3 and O_3

As it is well known, chemistry has identified in GC-MS tests clusters with 3 a.m.u., which can only be constituted of three H atoms, H_3, while the familiar ozone O_3 has been known since quite some time. These structures are generally assumed to be molecules, that is, to have a valence bond according to one nomenclature or another, although this author is aware of no in depth theoretical or experimental identification of the attractive force necessary to bond the third atom to a conventional molecule.

There are serious doubts as to whether such a conventional molecular interpretation will resist the test of time as well as of scientific evidence. To begin, a fundamental property of valence bonds is that *valence electrons correlate in pairs.* Since the H_3 and O_3 structures contain the molecules H_2 and O_2 in which all available valence electrons are already bonded in pairs, the belief that an additional third valence electron could be correlated to the preceding ones violates basic chemical knowledge on valence.

Moreover, we have stressed earlier that the assumption of a third valence electron bonded to a valence pair is in violation of basic physical knowledge, because it would require the bond of a Fermion (the third electron with spin 1/2) with a Boson (the singlet valence pair with spin 0) both possessing the same negative charge. Such a hypothetical bond under molecular conditions would violate various laws in particle physics, e.g., it would imply a necessary violation of Pauli-s exclusion principle since the assumed "triplet" of electrons would have *two* identical electrons in the same structure with the same energy.

In view of the above (as well as other) inconsistencies, we here assume that *the familiar H_3 and O_3 clusters are magnecules consisting of a third H and O atom magnetically bonded to the conventional H_2 and O_2 molecules, respectively, along the structure of Fig. 8.5.C.* Note that this assumption is fully in line with Definition 8.2.1 according to which a magnecule also occurs when one single atom is magnetically bonded to a fully conventional molecule.

The plausibility of the above structure is easily illustrated for the case of O_3. In fact, the oxygen is known to be paramagnetic, and the ozone is known to be best created under an electric discharge. These are the ideal conditions for the creation of a magnetic polarization of the orbits of (at least) the paramagnetic electrons. The attraction of opposing magnetic polarities is then consequential, and so is the magnetic bond of the third oxygen to the oxygen molecule, resulting in the magnecule $O_2 \times O$.

The above magnecular interpretation of O_3 is confirmed by various GC-MS detections of peaks with 32 a.m.u. in a magnetically treated gas originally composed of pure oxygen, in which case the sole possible interpretation is that of two magnetically bonded oxygen molecules, resulting in the magnecule $O_2 \times O_2$.

The plausibility of the magnecular interpretation is less trivial for the H_3 structure since hydrogen is diamagnetic. Nevertheless, the assumption remains equally plausible by recalling that a central feature of the new chemical species of magnecules is that *the magnetic polarization occurs at the level of each individual atom, and not at the level of a diamagnetic molecule, whose total magnetic moment remains null as illustrated in Fig. 8.2.*

In particular, the magnecular interpretation of the MS peaks at 3 a.m.u. is numerical and without ambiguities. Recall that GC equipment works by ionizing molecules. When testing a hydrogen gas, a number of H_2 molecules are separated into individual H atoms by the ionization itself. Moreover, the ionization occurs via the emission of electrons from a filament carrying current, which is very similar to that of the Plasma-ArcFlow Reactors producing magnecules. Under these conditions, the filament of the GC can not only separate H-molecules but also polarize them when sufficiently close to the filament. Once such polarizations are created, their bond is a known physical law, resulting in the magnecule of Fig. 8.5.C, *i.e.*

$$\{H_{\downarrow}-H_{\uparrow}\} \times H_{\downarrow}. \tag{8.19}$$

As one can see, under the magnecular structure the bond is manifestly attractive, very strong, and numerically identified in Eq. (8.9). Other interpretations of the peak at 3 a.m.u. are here solicited, provided that, to be credible, they are not of valence type and the internal bond is identified in a clear, unambiguous, and numerical way.

The magnecular interpretation of H_3 is confirmed by numerous GC-MS detections of a cluster with 4 a.m.u. in a magnetically treated gas which originally was composed by pure hydrogen, under which conditions such a peak can only be constituted by two hydrogen molecules resulting in the magnecule $H_2 \times H_2$ illustrated in Fig. 8.5.A.

It is an easy prediction that numerous peaks detected in contemporary GC-MS or LC-MS equipment may need a magnecular re-interpretation since, as indicated earlier, the method of detection itself can create magnecules. This is typically the case when the comparison of a given MS cluster with the actual peak of a given molecules contains additional lines.

As a specific example, when the peak representing a hexanal molecule (whose heaviest constituent has 100 a.m.u.) contains additional lines at 133 a.m.u., 166 a.m.u., and 207 a.m.u., it is evident that the latter lines cannot cluster with the hexanal molecule via valence bond. The plausibility of the magnecular interpretation is then evident.

For copies of the GC-MS scans mentioned in this section, which are not reproduced here for brevity, we suggest the interested reader to contact the author.

8.5. Need for New Analytic Methods

In closing, we should stress that the methods for the detection and identification of magnecules are at their infancy and numerous issues remain open at this writing (spring 2001). One of the open issues relates to several detections in magnegas of IR signatures apparently belonging to complex molecules, such as light hydrocarbons, while such molecules have not been identified in the MS scans. This occurrence creates the realistic possibility that certain complex magnecules may indeed have an IR signature in view of their size. More specifically, as indicated earlier, magnecules are assumed to be transparent to currently available IRD because their inter-atomic distance is expected to be 10^4 times smaller than the inter-atomic distance in molecules, thus requiring test frequencies which simply do not exist in currently available IRD.

However, such an argument solely applies for magnecules with small atomic weight, such as the elementary magnecules of Fig. 8.5. On the contrary, magnecules with heavy atomic weight may well have an IR signature and, in any case, the issue requires specific study.

This possibility is confirmed by the fact that magnegas is created via underliquid electric arcs whose plasma can reach up to $10{,}000^o$ K. The insistence that light hydrocarbons could survive in these conditions, let alone be created, is not entirely clear. This direct observation is confirmed by the fact that no hydrocarbon has been detected in the combustion of magnegas. In fact, the cars running on magnegas (as reviewed in the preceding Chapter) operate without catalytic converter. Direct analysis of the combustion exhaust show a *negative count* of hydrocarbons, that is, the exhaust contains less hydrocarbons than the local atmosphere which is used for basic calibration of the instrument.

In summary, we have a case in which light hydrocarbons are seemingly indicated by IR scans to exist in small percentages in magnegas, while no hydrocarbon has ever been identified in the MS scans, no hydrocarbon is expected to survive at the extreme temperatures of the electric arcs used for their production, and no hydrocarbon has been detected in the combustion exhaust.

These occurrences illustrate again that the identification of conventional molecules via the sole use of IR scans or, equivalently, the sole use of MS scans, is, in general, a mere personal opinion without scientific foundations.

9. Experimental Evidence of Gas Magnecules

9.1. Conventional Chemical Composition of Magnegas Used in the Tests

As it is well known, the underwater arc was developed in the mid of the 19-th century. The combustible character of the gas produced by underwater arcs was discovered by sailors of the same period who used to ignite the bubbles of gas surfacing from undersea arcs.

Numerous patents were subsequently obtained on the production of a combustible gas from underwater arcs (see, *e.g.*, U.S. patents [5]). Their main differences relate to the efficiency of the production, the CO_2 and energy content, and other aspects. The technology achieved industrial maturity only recently via the PlasmaArcFlow (patent and patents pending), as reviewed in the preceding chapter.

As indicated earlier in this chapter, an efficient way to produce an essentially pure population of gas magnecules is the production of a combustible gas via an electric arc between carbon electrodes submerged within a liquid.

All these gases are called "magnegas" because they have a magnecular structure irrespective of the liquid and the method used for their production.

The version of magnegas used in the tests reviewed in this section was produced via an electric arc between carbon electrodes submerged within ordinary tap water. According to several chemical analyses conducted at NASA and other laboratories, such a gas has the following *conventional chemical composition*:

$$50\%\ H_2, \quad 40\%\ CO, \quad 9\%\ CO_2, \text{ traces of } O_2 \text{ and } H_2O. \tag{8.20}$$

Therefore, the maximal chemical constituent in macroscopic percentage of this particular type of magnegas is given by CO_2 with 44 a.m.u.

In regard to preceding publications of direct relevance for the detection of magnecules, we recall that, as indicated earlier, the "H_3" structure is routinely detected in GC-MS analyses of hydrogen base gases. Some of its recent chemical study can be found in Ref. [6] and quoted literature. However, the physical reasons for the impossibility to have a valence bond and the consequential new structure as per Fig. 8.5.C do not appear to be identified in Ref. [6].

Similarly, there exist numerous theoretical and experimental studies in the chemical literature on molecular magnetic properties, such as those in Refs. [7] and quoted literature. These studies refer to "global" magnetic properties of conventional "molecule" and none of them appears to identify a "new chemical species." By contrast, the studies present in this Chapter refers to the magnetic fields of "individual atoms" and their capability to form a "new chemical species" beyond that of molecules.

The author (a physicist) would gratefully appreciate receiving from colleagues copies of papers more directly related to the use of magnetic fields of the "orbits of peripheral atomic electrons" for the creation of a "new chemical species" beyond that of molecules for proper quotation of the related literature in possible new edition of this monograph.

9.2. GC-MS/IRD Measurements of Magnegas at the McClellan Air Force Base

Santilli [1] had predicted that gases produced from underwater electric arcs had the new chemical structure of magnecules as clusters of molecules, dimers and individual atoms as per Definition 8.2.1, in which case conventional chemical structure (8.20) is valid only in first approximation.

Following a laborious search, Santilli [*loc. cit.*] located a GC-MS equipped with IRD suitable to measure magnecules at the *McClellan Air Force Base* in North Highland, near Sacramento, California. Thanks to the invaluable assistance and financial support by Leon Toups, President, of *Toups Technologies Licensing, Inc.*, of Largo, Florida, GC-MS/IRD measurements were authorized at that facility on magnegas with conventional chemical structure (8.20).

On June 19, 1998, Santilli visited the analytic laboratory of *National Technical Systems* (NTS) located at said *McClellan Air Force Base* and using instruments belonging to that base. The measurements on magnegas were conducted by analysts Louis A. Dee, Branch Manager, and Norman Wade who operated an *HP GC model 5890, an HP MS model 5972, equipped with an HP IRD model 5965.* Upon inspection at arrival, the instrument met all conditions indicated in the preceding sections then, and only then, measurements were permitted.

Thanks to a professional cooperation by the NTS analysts, the equipment was set at all the unusual conditions indicated in Sect. 8.8.3. In particular, the equipment was set for the analytic method VOC IRMS.M utilizing an HP Ultra 2 column 25 m long with a 0.32 mm ID and a film thickness of 0.52 μm. It was also requested to conduct the analysis from 40 a.m.u. to the instrument limit of 500 a.m.u. This condition was necessary to avoid the expected large CO peak of magnegas at 28 a.m.u.

Moreover, the GC-MS/IRD was set at the low temperature of 10°C; the biggest possible feeding line with an ID of 0.5 mm was installed; the feeding line itself was cryogenically cooled; the equipment was set at the longest possible ramp time of 26 minutes; and a linear flow velocity of

50 cm/sec was selected. A number of other technical requirements are available in the complete documentation of the measurements.

The analysts first secured a documentation of the *background* of the instrument prior to any injection of magnegas (also called *blank*). Following a final control that *all* requested conditions were implemented, the tests were initiated. The results, reported in part via the representative scans of Figs. 8.7 to 8.12, constitute the first direct experimental evidence of the existence of magnecules in gases.

After waiting for 26 minutes, sixteen large peaks appeared on the MS screen between 40 and 500 a.m.u. as shown in Fig. 8.7. Each of these sixteen MS peaks resulted to be "unknown," following a computer search of database on all known molecules available at *McClellan Air Force Base*, as illustrated in Fig. 8.8 No identifiable CO_2 peak was detected at all in the MS spectrum between 40 and 500 a.m.u., contrary to the presence of 9% of such a molecule in magnegas as per conventional analyses (8.20).

Upon the completion of the MS measurements, exactly the same range of 40 to 500 a.m.u. was subjected to IR detection. As expected, none of the sixteen peaks had any infrared signature at all, as shown in Fig. 8.9. Furthermore, the IR scan for these MS peaks shows only one peak, that belonging to CO_2, with additional small peaks possibly denoting traces of other substances.

Note that the IR signature of the other components, such as CO or O_2 *cannot* be detectable in this IR test because their atomic weights are below the left margin of the scan. In addition, the IR peak of CO_2 is itself mutated from that of the unpolarized molecule, as shown in Fig. 8.10. Note that the mutation is due to the appearance of *two new peaks* which are absent in the conventional IR signature of CO_2, exactly as expected, thus confirming the hypothesis of new internal bonds as submitted in Fig. 8.6.

Note also in Fig. 8.10 that the computer interprets the IR signature as that belonging to CO which interpretation is evidently erroneous because CO is outside of the selected range of a.m.u.

All remaining small peaks of the IR scan resulted to be "unknown," thus being possible magnecules, following computer search in the database of IR signatures of all known molecules available at the *McClellan Air Force Base*, as illustrated in Fig. 8.11.

Following the removal of magnegas from the GC-MS/IRD, the background continued to show the same anomalous peaks of Fig. 8.7, and reached the configuration of Fig. 8.12 only after a weekend bakeout with an inert gas. Note that the latter background is itself anomalous because the slope should have been the opposite of that shown. The background

```
Information from Data File:
File        : C:\HPCHEM\1\DATA\0618004.D
Operator    : NAW
Acquired    : 18 Jun 98   3:01 pm using AcqMethod VOC_IRMS
Sample Name: TOUP'S TECH
Misc Info   : 1ML LOOP; 10C ● ULTRA COLUMN
Vial Number: 1
CurrentMeth: C:\HPCHEM\1\METHODS\DEFAULT.M
```

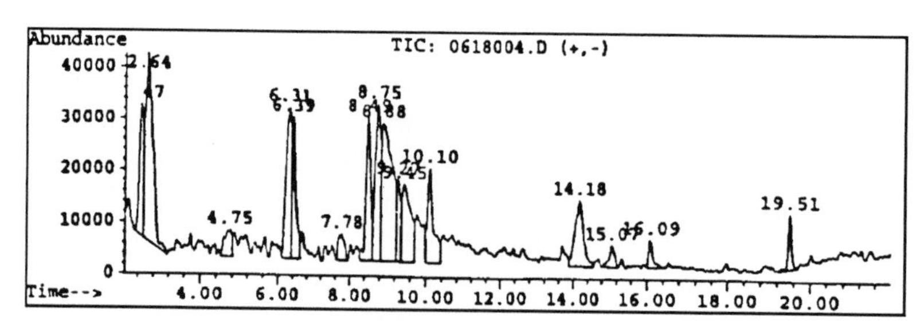

Retention Time	Area	Area %	Ratio %
Total Ion Chromatogram			
2.474	1753306	5.386	32.724
2.644	5091514	15.641	95.030
4.754	641528	1.971	11.974
6.307	2737749	8.411	51.098
6.390	2211258	6.793	41.272
7.782	592472	1.820	11.058
8.490	2357396	7.242	43.999
8.754	2784829	8.555	51.977
8.882	5357812	16.460	100.000
9.265	1123809	3.452	20.975
9.448	2421234	7.438	45.191
10.098	1946292	5.979	36.326
14.177	2129791	6.543	39.751
15.073	435208	1.337	8.123
16.085	389822	1.198	7.276
19.509	577433	1.774	10.777

Figure 8.7. A reproduction of the MS peaks providing the *first experimental evidence of the existence of magnecules* identified on June 19, 1998, by analysts *Louis A. Dee* and *Norman Wade* of the branch of *National Technical; Systems* (NTS) located at the *McClellan Air Force Base* in North Highland, near Sacramento, California, with support from *Toups Technologies Licensing, Inc.* (TTL) of Largo, Florida. The scan is restricted from 40 a.m.u to 500 a.m.u. The peaks refer to magnegas produced via an electric arc between consumable carbon electrodes within ordinary tap water with conventional chemical composition (8.20). Therefore, only the CO_2 peak was expected to appear in the scan with any macroscopic percentage, while no CO_2 was detected at all in the MS scan.

finally recovered the conventional shape only after flushing the instrument with an inert gas at high temperature.

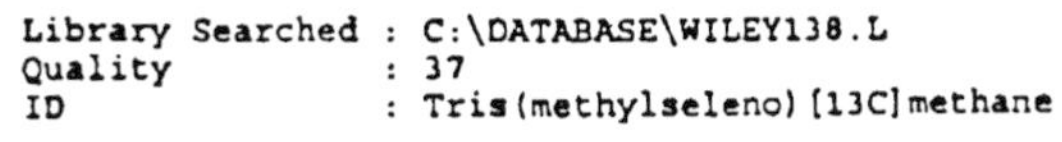

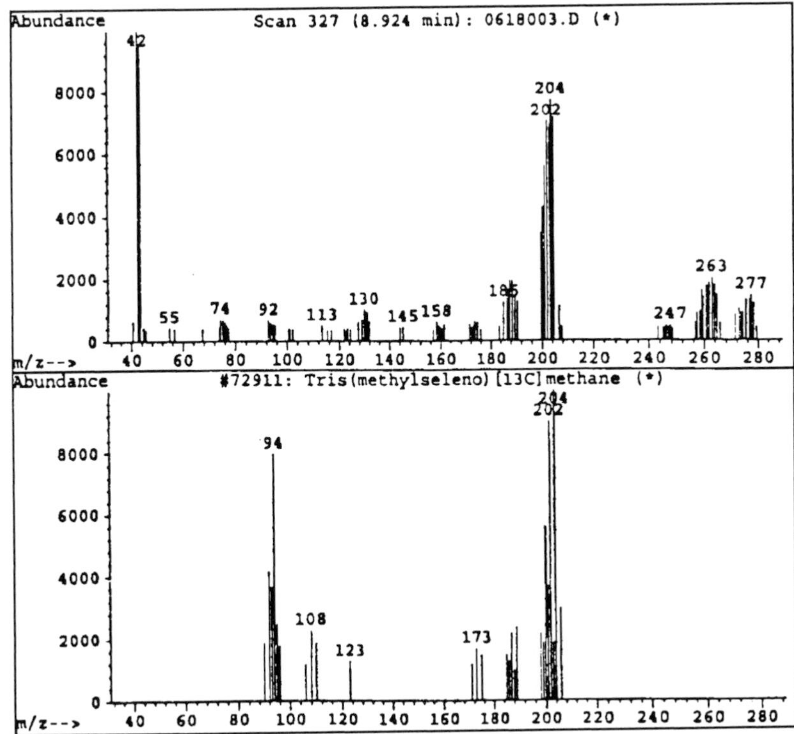

Figure 8.8. A representation of *the first experimental evidence at NTS that the peaks of Fig. 8.7 are "unknown."* The peak at the top is at 8.924 minutes, and that at the bottom shows the lack of its identification by the computer search. Note that the best fit identified by the computer does not match the peak considered. Moreover, the identified substance (methylseleno) cannot possibly exist in magnegas because of the impossible presence of the necessary elements. The same situation occurred for all remaining fifteen peaks of Fig. 8.7.

9.3. GC-MS/IRD Tests of Magnegas at Pinellas County Forensic Laboratory

Measurements on the same sample of magnegas tested at NTS were repeated on July 25, 1998, via a GC-MS/IRD located at the *Pinellas County Forensic Laboratory* (PCFL) of Largo, Florida, with support from *Toups Technologies Licensing, Inc.*

The equipment consisted of a *HP GC model 5890 Series II, an HP MS model 5970 and an HP IRD model 5965B.* Even though similar to the equipment used at NTS, the PCFL equipment was significantly different

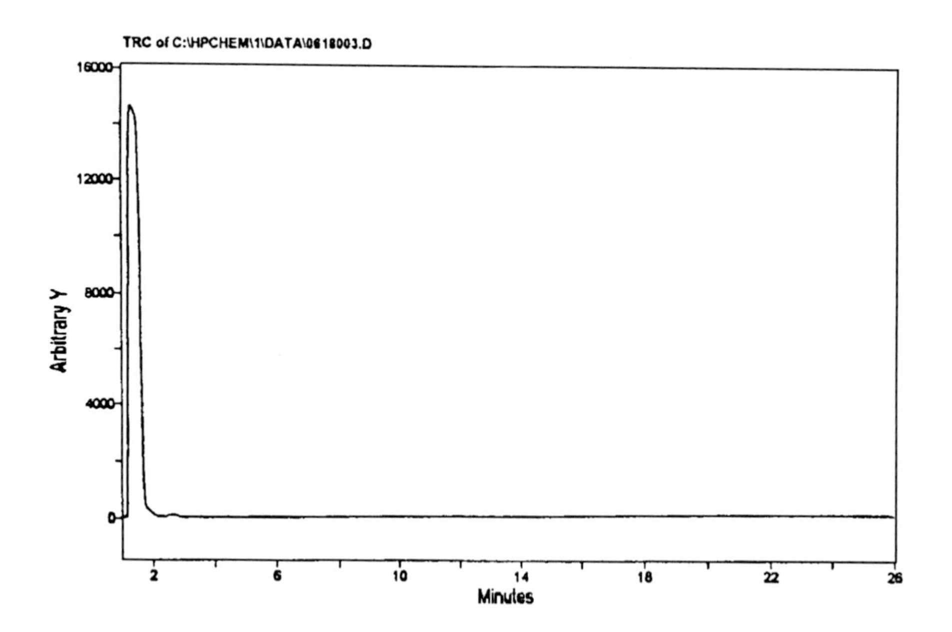

Figure 8.9. The first experimental evidence at NTS of the lack of IR signature of MS peaks. The evidence establishes the existence of large peaks in the MS that have no IR signature at all. The only identified IR signature, that for CO_2, refers to the *constituents* of the peaks of Fig. 8.7. In the above figure only the IR signature of CO_2 appears because the scan was from 40 a.m.u. to 500 a.m.u. and, as such, could not include the IR signatures of other molecules such as O_2 and CO (H_2 has no IR signature).

inasmuch as the temperature had to be increased from 10°C to 55°C and the ramp time reduced from 26 to 1 minute. The latter reduction implied the cramping of all peaks of Fig. 8.7 into one single large peak, a feature confirmed by all subsequent GC-MS tests with short ramp time.

Despite these differences, the test at PCFL, reported in part via the representative scans of Figs. 8.13 to 8.18, confirmed *all* features of magnecules first detected at NTS. In addition, the tests provided the experimental evidence of additional features.

Following Santilli's request [1], the analysts conducted *two* MS tests of the *same* magnegas at *different times* about 30 minutes apart. As one can see in Figs. 8.13 and 8.14, *the test at PCFL provided the first experimental evidence of mutation in time of the atomic weight of magnecules.* In fact, the peak of Fig. 8.13 is macroscopically different than that of Fig. 8.14.

This difference provides evidence that, when colliding, magnecules can break down into ordinary molecules, atoms, and fragments of mag-

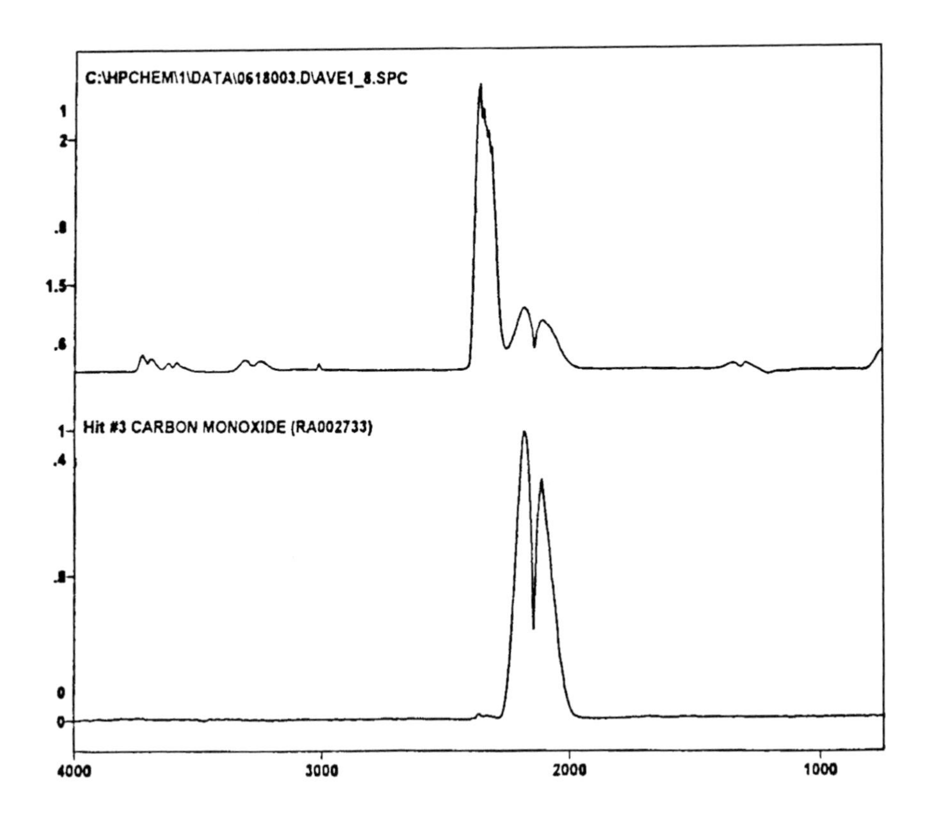

Figure 8.10. The first experimental evidence at NTS on the mutation of the IR signature of magnetically polarized conventional molecules, here referring to the CO_2 (top) compared to the result of the computer search (bottom). Note that the known, double-lobe peak of CO_2 persists in the detected peak with the correct energy, and only with decreased intensity. Jointly, there is the appearance of *two new peaks*, which are evidence of *new internal bonds within the conventional CO_2 molecule*. This evidently implies an increased energy content, thus establishing experimental foundations for the new technology of magnetically polarized fuels such as magnegas [2].*Note that the computer interprets the IR signature as that of CO, which is erroneous since CO is out of the selected range of detection.*

neclusters, which then recombine with other molecules, atoms, and/or magnecules to form new clusters. The same scan provides first experimental evidence of the accretion or loss by magnecules of individual atoms, dimers and molecules, as discussed later on.

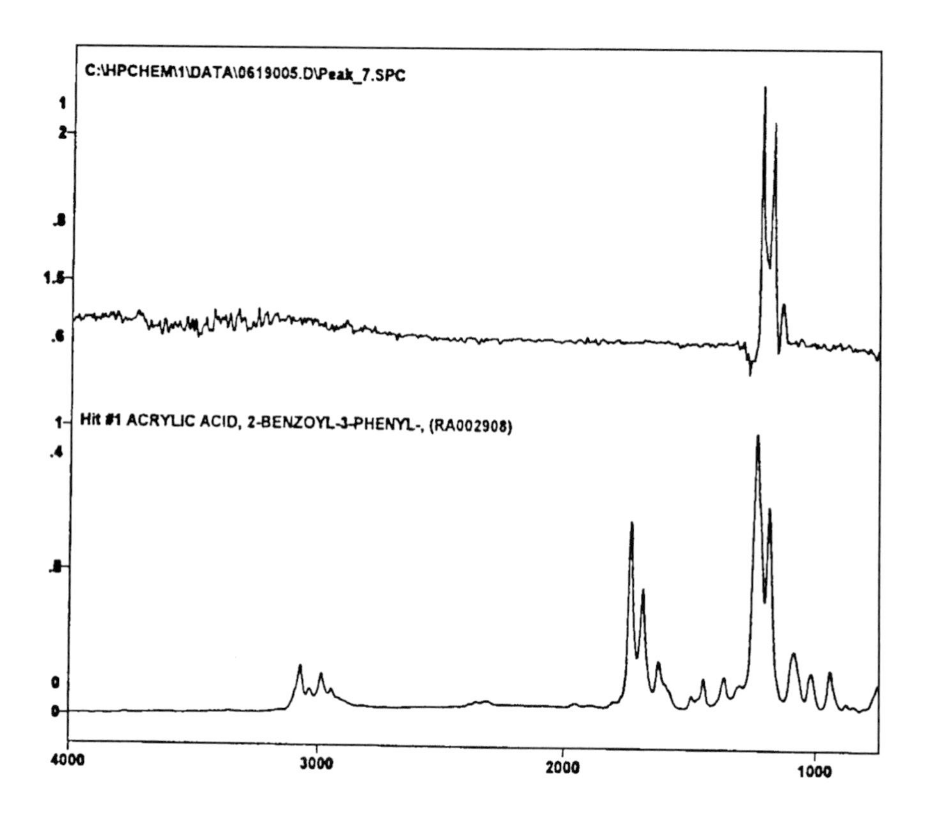

Figure 8.11. A reproduction of the lack of identification in the computer search of small peaks in the IR scan, which can therefore be additional magnecules, or IR signatures of the magnecules appearing in the MS scan.

Figure 8.15 depicts the failure by the GC-MS/IRD to identify the peaks of Figs. 8.13 and 8.14 following a search in the database among all known molecules.

Figure 8.16 provides an independent confirmation that the IR scan of Fig. 8.9, namely, that the MS peaks, this time of Figs. 8.13 and 8.14, have no IR signature except for the single signature of the CO_2. However, the latter was not detected at all in said MS scans. Therefore, the CO_2 detected in said IR scan is a *constituent* of the new species detected in Figs. 8.13 and 8.14. The lack of IR signature of the MS peaks confirms that said peaks *do not* represent molecules.

```
File        : C:\HPCHEM\1\DATA\0622005.D
Operator    : NAW
Acquired    : 22 Jun 98   1:16 pm using AcqMethod VOC_MS
Instrument  :    5972A
Sample Name: BLANK
Misc Info   : AFTER WEEKEND BAKEOUT
Vial Number: 1
```

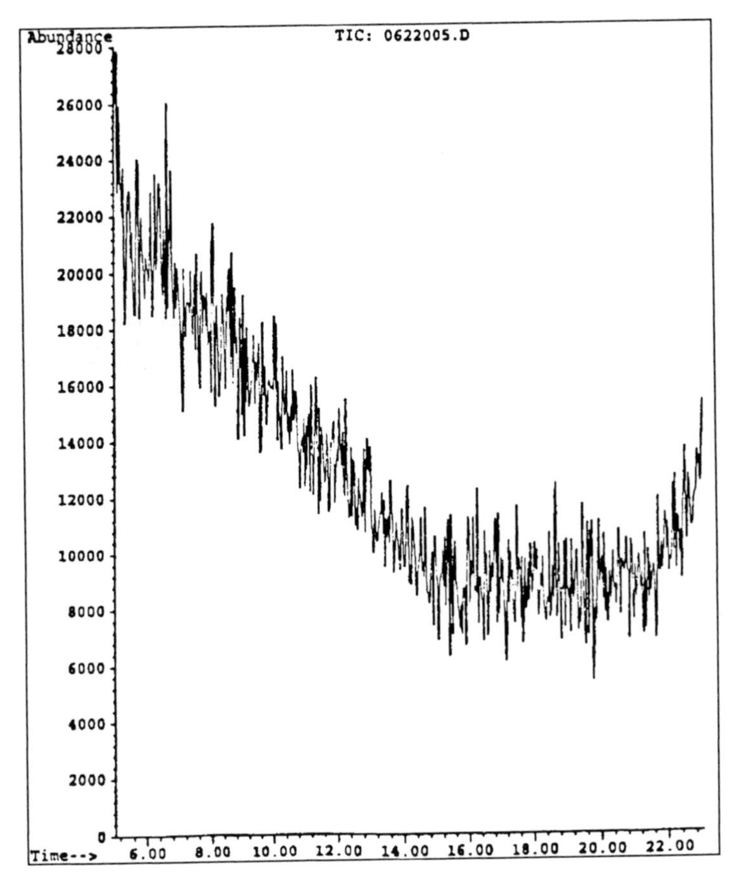

Figure 8.12. A view of the background of the preceding tests following a weekend bakeout.

Figure 8.17 confirms in full the mutated IR signature of CO_2 previously identified in Fig. 8.10, including the important presence of two new peaks, with the sole difference that, this time, the computer correctly identifies the IR signature as that of carbon dioxide.

Figure 8.18 presents the background of the instrument after routine flushing with an inert gas, which background essentially preserves the

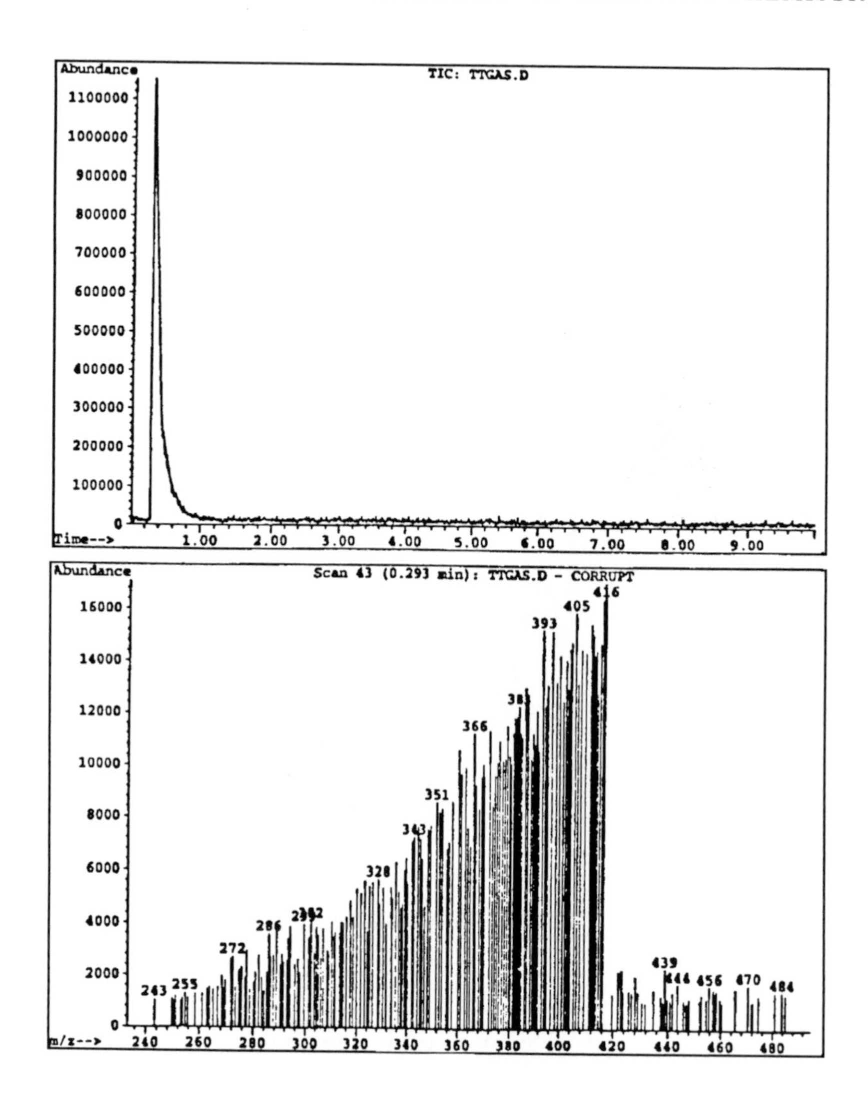

Figure 8.13. A view of the Total Ion Count (top) and MS spectrum (below) of magnegas conducted on July 25, 1998, via a HP GC-MS/IRD at the *Pinellas County Forensic Laboratory* (PCFL) of Largo, Florida, under support from *Toups Technologies Licensing, Inc.* (TTL) also of Largo, Florida. The scan is restricted to the range 40 a.m.u to 500 a.m.u. and confirm all results of the preceding NTS tests.

peaks of the MS scans, thus confirming the unique adhesion of magnecules to the instrument walls.

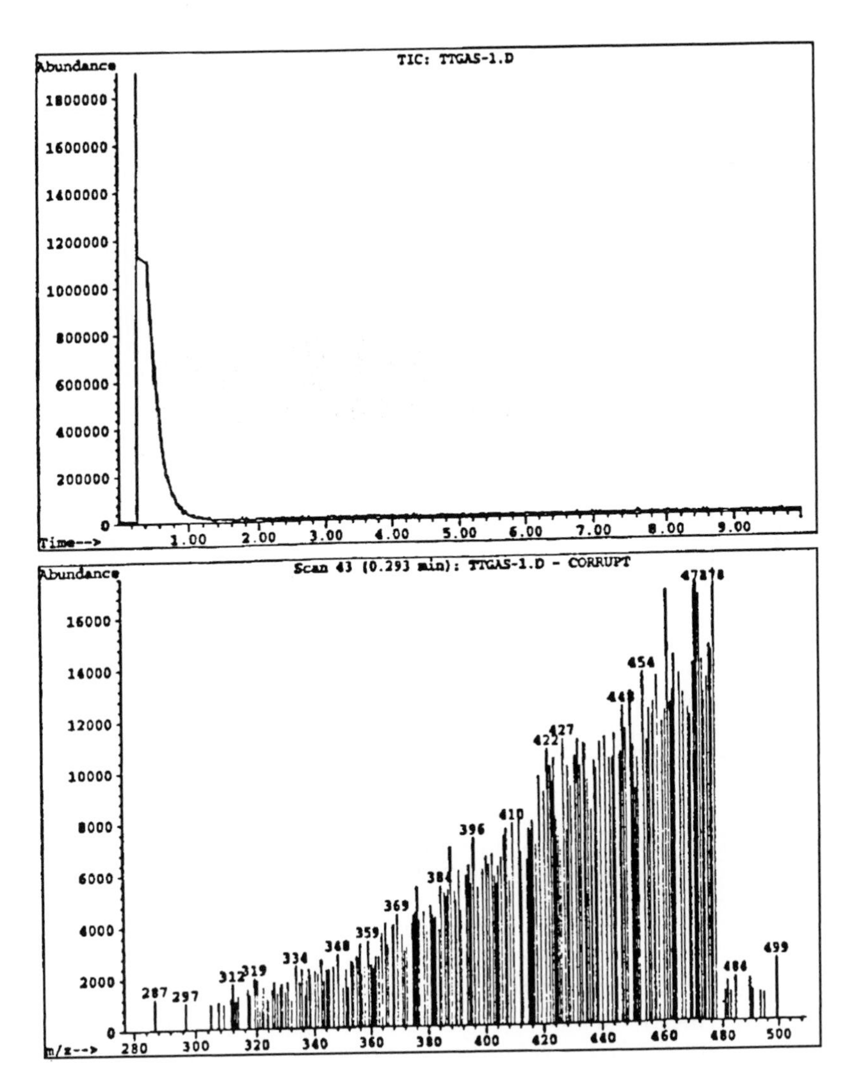

Figure 8.14. A repetition of the scan of the preceding figure conducted at PCFL in the same sample of magnegas on the same instrument and under the same conditions, but 30 minutes later. The scan provides *the first experimental evidence of the mutation of atomic weight of magnecules,* as one can see from the variation of the peaks of this figure compared with that of the preceding figure.

9.4. Interpretations of the Results

A few comments are now in order for the correct interpretation of the results. First, note in the GC-MS/IRD scans that the CO_2 detected in the IRD has no counterpart in the MS scans, while none of the peaks in the MS have a counterpart in the IR scans. Alternatively, the CO_2 peak

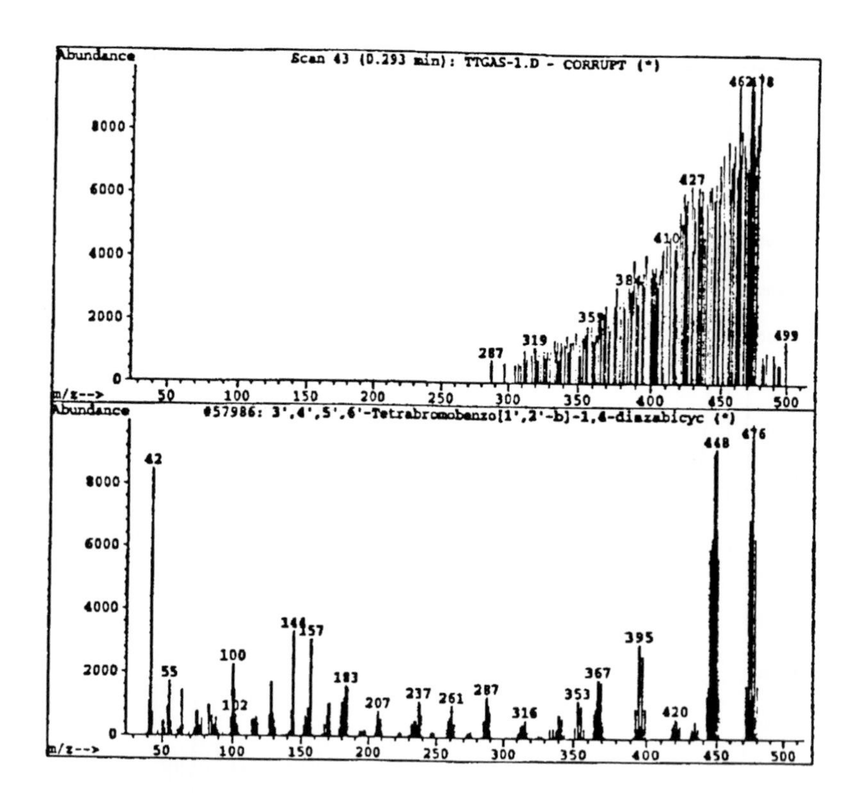

Figure 8.15. Lack of identification by the computer of the GC-MS/IRD at PCFL of the MS peaks of the preceding two scans following search among the database on all available molecules.

detected in the IR scans of Figs. 8.10 and 8.17 *does not* correspond to any peak in the MS scans in Figs. 8.7, 8.13 and 8.14. Therefore, said IR peak identifies a *constituent* of the MS clusters, and not an isolated molecule.

Moreover, the IR scan was done for the entire range of 40 to 500 a.m.u., thus establishing that said IR peak is the sole conventional constituent in macroscopic percentage in said a.m.u. range of *all* MS peaks, namely, the single constituent identified by the IRD is a constituent of all MS peaks.

It should also be noted that, as recalled earlier, the IR only detects *dimers* such as C–O, H–O, *etc.*, and does not detect complete molecules. Therefore, the peak detected by the IRD *is not* sufficient to establish the presence of the complete molecule CO_2 unless the latter is independently identified in the MS. Yet the MS scan does not identify any peak for the CO_2 molecule, as indicated earlier. Despite that, the presence in the

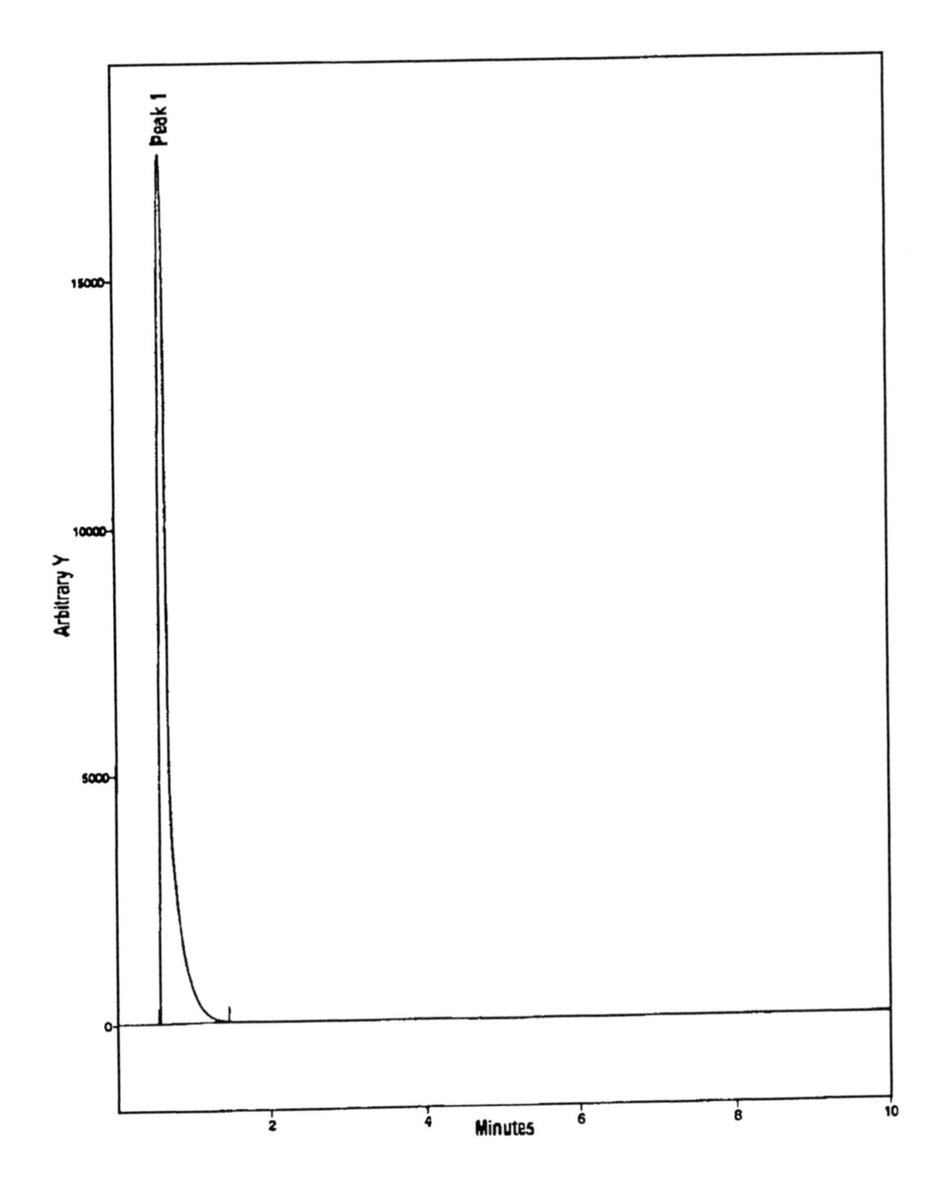

Figure 8.16. A confirmation of the lack of IR signature of the peaks of Figs. 8.13 and 8.14, as occurred for Fig. 8.9, which establishes that the MS peaks of Figs. 8.13 and 8.14 cannot have a valence bond, thus constituting a new chemical species.

MS peaks of complete molecules CO_2 cannot be ruled out. Therefore, the most plausible conclusion is that the MS peaks represent clusters composed of a percentage of C–O dimers and another percentage of CO_2 molecules, plus other dimers, and/or molecules, and/or atoms with

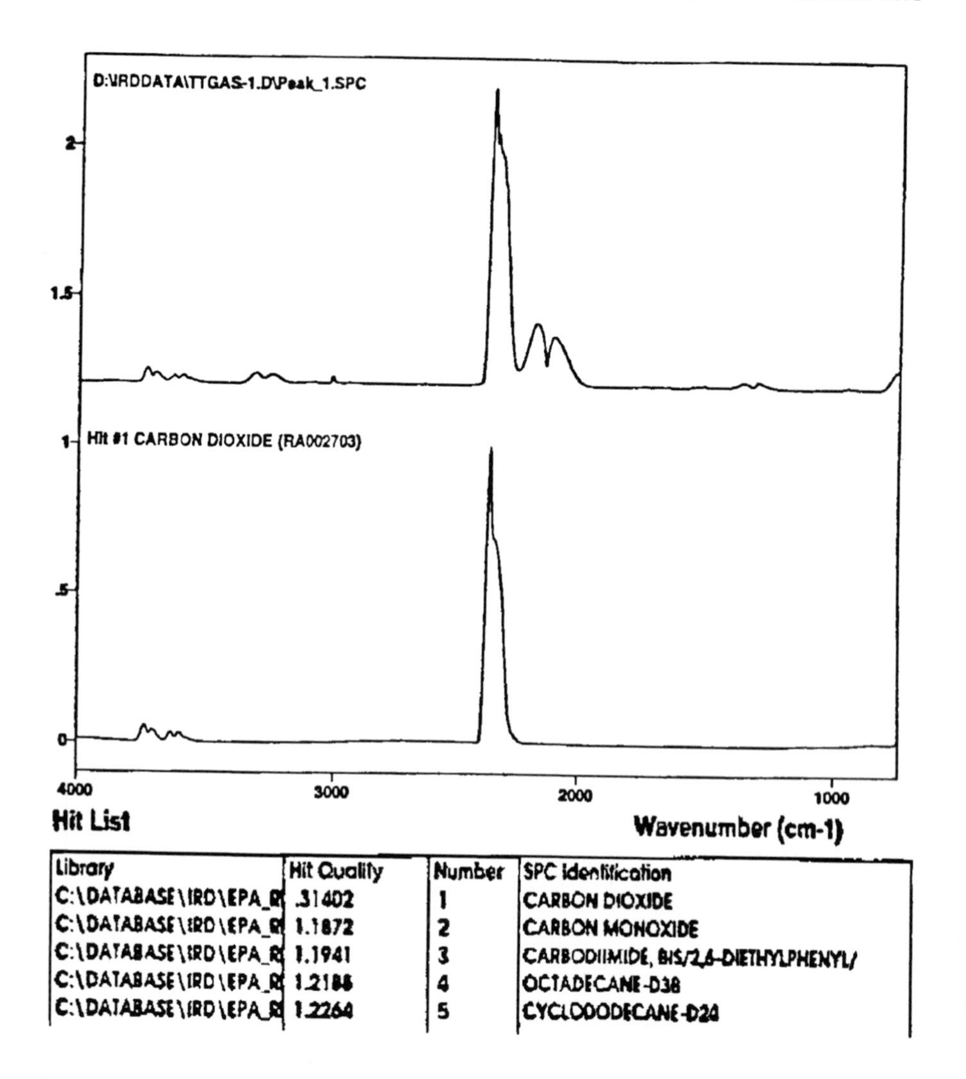

Library	Hit Quality	Number	SPC Identification
C:\DATABASE\IRD\EPA_R	.31402	1	CARBON DIOXIDE
C:\DATABASE\IRD\EPA_R	1.1872	2	CARBON MONOXIDE
C:\DATABASE\IRD\EPA_R	1.1941	3	CARBODIIMIDE, BIS/2,6-DIETHYLPHENYL/
C:\DATABASE\IRD\EPA_R	1.2186	4	OCTADECANE-D38
C:\DATABASE\IRD\EPA_R	1.2264	5	CYCLODODECANE-D24

Figure 8.17. The independent confirmation at the PCFL of the NTS finding of Fig. 8.10 regarding the mutated IR signature of the CO_2 in magnegas. Note the identical shapes of the mutated IR peak in the top of the above figure, and that in Fig. 8.10 obtained via a different instrument. Note also the appearance again of two new peaks in the IR signature of CO_2, which indicate the presence of *new internal bonds* not present in the *conventional* molecule. Note finally that the instrument now correctly identifies the signature as that of the CO_2.

atomic weight smaller than 40 a.m.u., thus outside the range of the considered scans.

As indicated earlier, the presence of dimers and individual atoms in magnegas is essential for a quantitative interpretation of the large excess

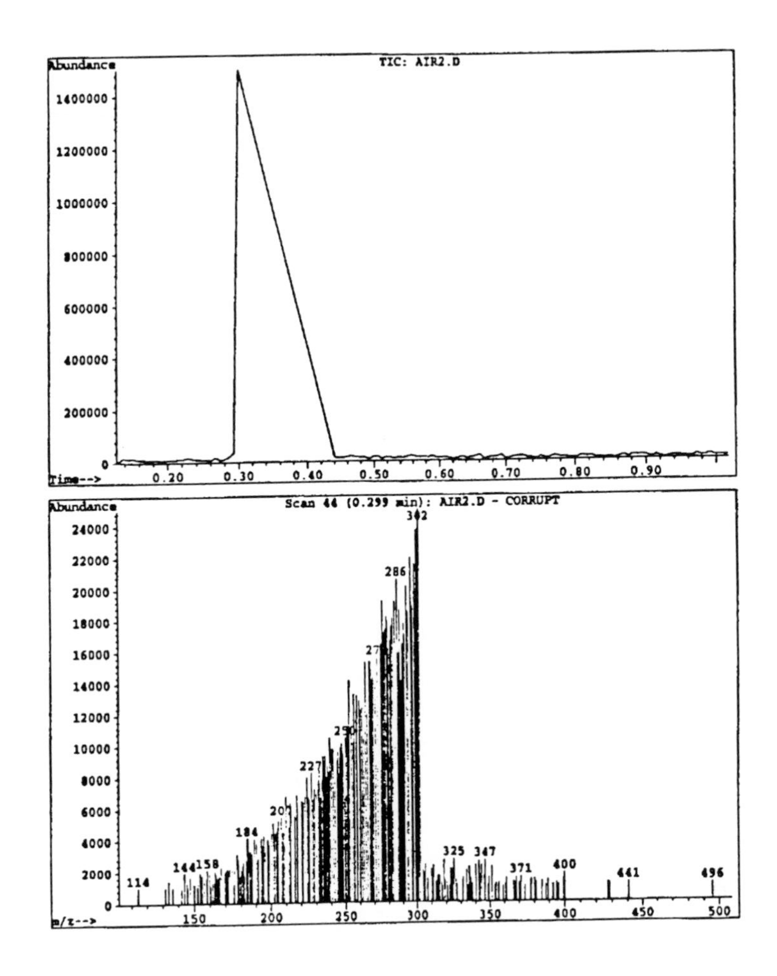

Figure 8.18. The first direct experimental verification at PCFL of the anomalous adhesion of magnecules. The figure reproduces the background of the instrument upon completion of the measurements, removal of magnegas, and conventional flushing. As one can see, the background results in being very similar to the MS scan during the tests, thus establishing that part of the gas had remained in the interior of the instrument. This behavior can only be explained via the induced magnetic polarization of the atoms in the walls of the instrument, with consequential adhesion of magnecules. It should be noted that this anomalous adhesion has been confirmed by all subsequent tests for both the gaseous and liquid states. The removal of magnecules in the instrument after tests required flushing with an inert gas at high temperature.

of energy contained in this new fuel, the order of at least three times the value predicted by quantum chemistry, which energy is released during combustion. The admission of dimers and atoms as constituents of magnecules readily explains this anomalous energy content because said dimers and atoms are released at the time of the combustion, thus being able at that time to form molecules with exothermic reactions of type (8.5). In the event magnecules would not contain dimers and atoms, their only possible constituents are conventional molecules, in which case no excess energy is possible during combustion.

The large differences of MS peaks in the two tests at NTS and at PCFL of exactly the same gas in exactly the same range from 40 to 500 a.m.u., even though done with different GC-MS/IRD equipment, illustrates the importance of having a ramp time of the order of 26 minutes. In fact, sixteen different peaks appear in the MS scan following a ramp time of 26 minutes, as illustrated by Fig. 8.7, while all these peaks collapsed into one single peak in the MS scan of Figs. 8.13 and 8.14, because the latter were done with a ramp time of about 1 minute. Therefore, *the collapse of the sixteen peaks of Fig. 8.7 into the single large peak of Figs. 8.13 and 8.14 is not a feature of magnecules, but rather it is due to the insufficient ramp time of the instrument.*

9.5. Concluding Remarks

The tests on magnegas conducted at the *McClellan Air Force Base* and at *Pinellas County Forensic Laboratory*, as per representative scans of Figs. 8.7 to 8.18, constitute two independent verifications of all characteristics of gas magnecules according to Definition 8.2.1 which can be detected via GC-MS equipped with IRD, the remaining characteristics requiring separate measurements presented in the next section.

The first and perhaps most important evidence achieved for the first time in these tests is that *the scans of Figs. 8.7, 8.13 and 8.14 constitute independent experimental evidence of the new chemical species of magnecules in the terminology of Sect. 8.1 because they constitute not only an essentially pure, but actually a pure population of magnecules due to the complete absence of molecules in a directly detectable percentage.*

More generally, said independent tests, plus numerous additional tests not reported here for brevity, confirm the following features of Definition 8.2.1:

I) Magnecules have been detected in MS scans at high atomic weights where no molecules are expected for the gas considered. In fact, the biggest molecule in macroscopic percentages of the magnegas tested, that produced from tap water with conventional chemical composition (8.20), is CO_2 with 44 a.m.u., while peaks in macroscopic percentages have been detected with *ten times* such an atomic weight and more.

II) The MS peaks characterizing magnecules remain unidentified following a computer search among all known molecules. This feature has been independently verified for *each* of the sixteen peaks of Fig. 8.7, for all peaks of Figs. 8.13 and 8.14, as partially illustrated in Figs. 8.8 and 8.15, as well as for all additional MS scans not reported here for brevity.

III) The above MS peaks characterizing magnecules admit no IR signature, thus confirming that they do not have a valence bond. In fact, none of the peaks here considered had any IR signature as partially illustrated in Figs. 8.9 and 8.16, thus confirming the achievement of an essentially pure population of magnecules.

IV) The IR signature of the only molecule detected in macroscopic percentage, that of the CO_2, is mutated precisely with the appearance of two additional peaks, as shown in Fig. 8.10 and independently confirmed in Fig. 8.17. Since any peak in the IR signature represents an internal bond, the mutation here considered confirms the creation by the PlasmaArcFlow technology of new internal magnetic bonds within conventional molecules, as per Fig. 8.6.

V) The anomalous adhesion of magnecules is confirmed in both tests from the evidence that the background (blank) at the end of the tests following conventional flushing continued to show the presence of essentially the same magnecules detected during the tests, as illustrated in Figs. 8.12 and 8.18.

VI) The atomic weight of magnecules mutates in time because magnecules can break down into fragments due to collisions, and then form new magnecules with other fragments. This feature is clearly illustrated by the macroscopic differences of the two scans of Figs. 8.13 and 8.14 via the same instrument on the same gas under the same conditions, only taken 30 minutes apart.

VII) Magnecules can accrue or lose individual atoms, dimers or molecules. This additional feature is proved in the scans of Figs. 8.13 and 8.14 in which one can see that: the peak at 286 a.m.u. of the former becoming 287 a.m.u. in the latter, thus establishing the accretion of one hydrogen *atom*; the peak at 302 a.m.u. in the former becomes 319 a.m.u. in the latter, thus establishing the accretion of the H–O dimer; the peak at 328 a.m.u. in the former becomes 334 a.m.u. in the latter, thus establishing the accretion of one O_2 molecule; the peak at 299 a.m.u. in the former become 297 a.m.u. in the latter, thus exhibiting the loss of one H_2 molecule; *etc.* It should be indicated that these features have been confirmed by all subsequent GC-MS/IRD scans not reported here for brevity.

The other features of Definition 8.2.1 require measurements other than those via GC-MS/IRD and, as such, they will be discussed in the next section.

Notice, as illustrated in Figs. 4.5 and 8.3, that there is no need for a total magnetic polarity in the clusters composing magnegas. As a matter of fact, its *absence* is necessary to avoid the inconsistent prediction that all clusters are paramagnetic. The general expectation that a cluster can have a magnetic moment if and only if exhibited by the cluster as a whole is, perhaps, a reason for the lack of prediction and detection of magnecules until now.

Similarly, ionic clusters must be excluded for any credible interpretation of the peaks of Figs. 8.7 to 8.18 due to the fact that ions have the same charge and, therefore, they repel, rather than attract each other.

Since the experimental evidence eliminates the possible valence or electric origin of the bond, the sole remaining possibility is that the attractive force responsible for the peaks is of magnetic character, namely, that said peaks constitute magnecules.

In closing the author would like to stress that the above findings, even though independently confirmed numerous times, should be considered preliminary and in need of additional independent verifications, which are here solicited under the suggestion that:

1) Only peaks with macroscopic percentages should be initially considered to avoid shifting issues of primary relevance into other of comparatively marginal importance at this time;

2) The internal *attractive* force necessary for the very existence of cluster is identified in clear numerical terms without vague nomenclatures deprived of an actual physical reality, or prohibited by physical laws; and

3) The adopted terminology is identified with care. The word "magnecule" is a mere name intended to denote a chemical species possessing the specifically identified characteristics I) to XV) of Definition 8.2.1 which are distinctly different than the corresponding characteristics of molecules. Therefore, the new species can not be correctly called molecules. The important features are these distinctly new characteristics, and not the name selected for their unified referral.

10. Experimental Evidence of Liquid Magnecules

10.1. Preparation of Liquid Magnecules used in the Tests

In early 1998 Santilli [1] obtained a number of samples of *fragrance oils* from *Givaudan-Roure Corporation* (GR) with headquarters in Teaneck, New Jersey. About 50 cc of various samples of perfectly transparent fragrance oils were placed in individual glass containers. One polarity of an alnico permanent magnet with 12,000 G and dimension $1/2'' \times 1'' \times 2''$ was immersed within said oils.

Starting with a perfect transparency, after a few days a darkening of the oils became visible, jointly with a visible increase of the viscosity, with changes evidently varying from oil to oil. Subsequently, there was the appearance of granules of dark complexes in the interior of the oil which were visible to the naked eye. Both the darkening and the viscosity increased progressively in subsequent days, to reach in certain cases a dark brown color completely opaque to light. The viscosity increased to such an extent that the oil lost all its fluidity.

It should be stressed that the above visible effects are of pure magnetic origin because of the lack of any other contribution, *e.g.*, the complete absence of any additives. After the immersion of the permanent magnets, all samples were left undisturbed at ordinary room conditions. The indicated effects remain unchanged to this day, thus showing that the changes were stable at ordinary conditions of temperature and pressure.

Santilli's [1] main hypothesis on the darkening of the oils is that their molecules acquire a magnetic polarization in the orbits of at least some of their atomic electrons (called in chemistry *cyclotron resonance orbits*), by therefore bonding to each other according to Definition 8.2.1 in a way similar to the corresponding occurrence for gases.

It should also be indicate that the immersion of one polarity of a permanent magnet in fragrance oils is, evidently, a rudimentary way to create magnecules in detectable percentage although not an essentially pure population of magnecules as requested for a new chemical species (see Sect. 8.1). A number of more sophisticated magnetic polarization techniques are now available with rather complex geometries. Also, as indicated in Sect. 8.5, an essentially pure population of liquid magnecules can be reached via the PlasmaArcFlow reactors described in the preceding chapter.

10.2. Photographic Evidence of Magnecules in Liquids

The above alteration of the structure of fragrance oils was confirmed by photographs taken by the GR Research Laboratory in Dubendorf, Switzerland, via a microscope with minimal magnification, as illustrated in the pictures of Figs. 8.19 and 8.20.

The pictures of Fig. 8.19 refer to the GR fragrance oil received under the code "ING258AIN, Text 2" subjected to the rudimentary magnetic polarization indicated in the preceding section under the respective magnification 10X and 100X.

As one can see, these photographs establish that, under the indicated magnetic treatment, the oil has acquired a structure of the type of "brick layering" which is visible under only 10X magnification, and is per se

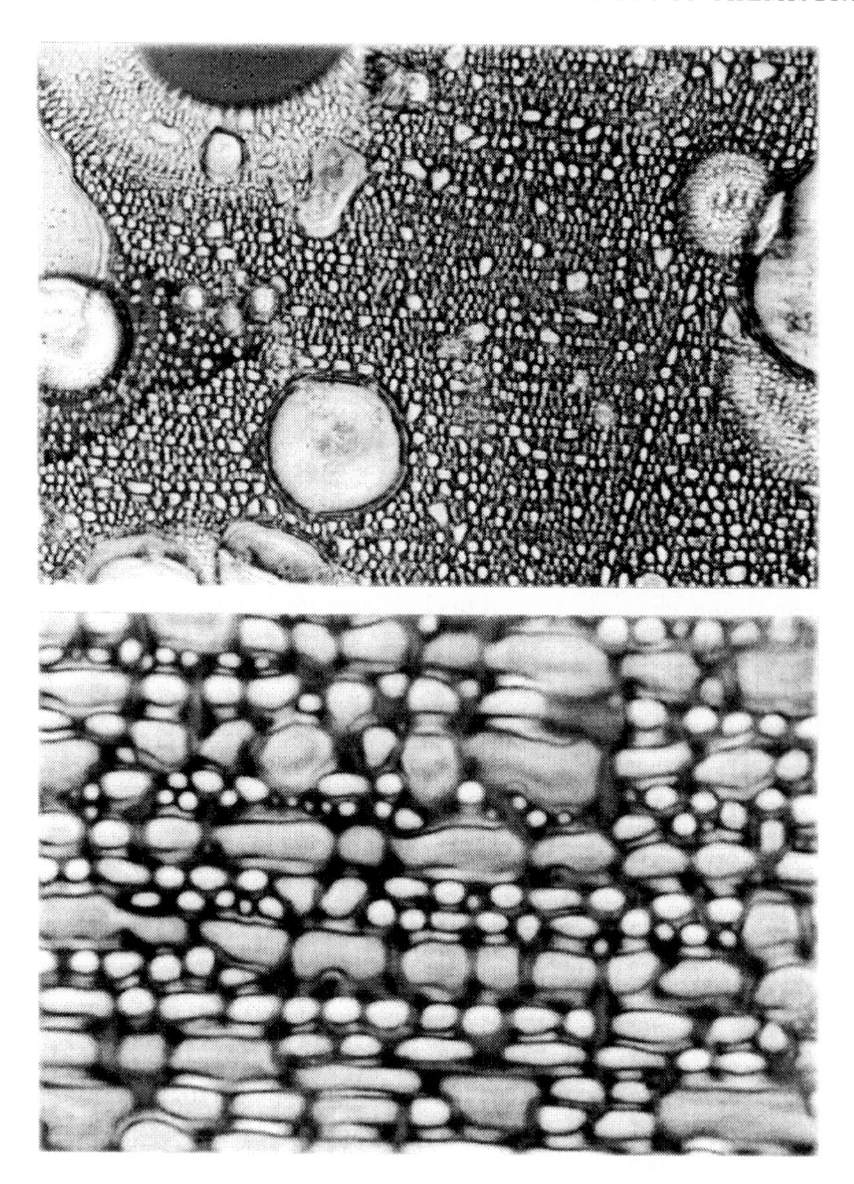

Figure 8.19. A photographic evidence of magnecules in liquids obtained at the *Givaudan-Roure Research Laboratory* in Dubendorf, Switzerland, in the GR fragrance oil "ING258IN Test 2" under magnifications 10X and 100X [1].

highly anomalous for a liquid that was originally fully transparent. Note that the magnecules are not constituted by the individual "bricks," but rather by the dark substance which interlock said "bricks." This point

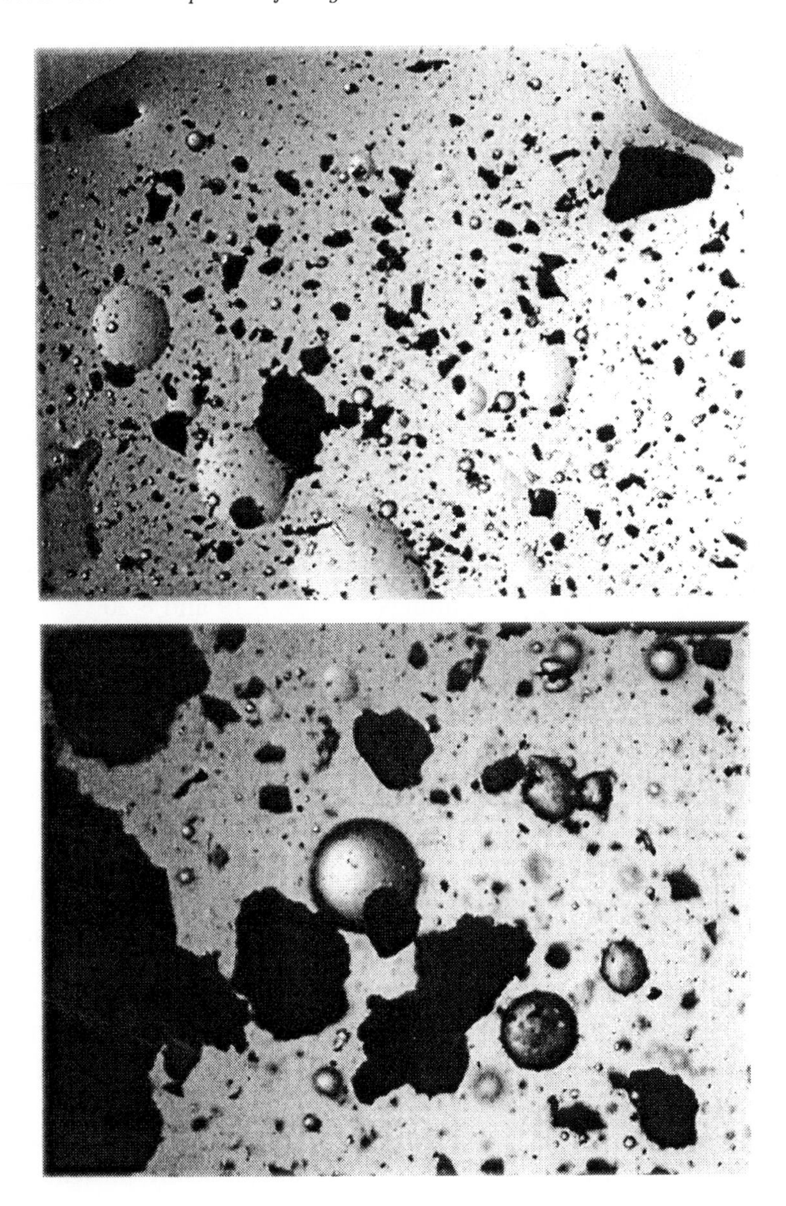

Figure 8.20. Confirmation of magnecules in GR fragrance oil "Mixture 2" under 10X and 100X obtained at the University of South Florida in St. Petersburg. Note the difference in sizes of the magnecules and their difference with those of Fig. 8.19 [1].

is important to understand the size of the magnecule here considered which covers the entire 50 cc of the liquid.

The photographs in Figs. 8.20 were taken at the University of South Florida in St. Petersburg via a microscope with the same magnifications 10X and 100X, but refer to a different GR fragrance oil received under the code "Mixture 2" and magnetically treated to such a point of completely losing transparency and fluidity. As one can see, the latter picture provides confirmation that, following exposure to a 12,000 G magnetic field, fragrance oil molecules bond together into rather large clusters estimated to be well in excess of 10,000 a.m.u., that is, with an atomic weight which is dramatically bigger than that of the largest molecule composing the oil, as per Feature I) of Definition 8.2.1.

Inspection of the various photographs shows a variety of sizes of magnecules, thus establishing their lack of unique characteristics for any given liquid. This evidently confirms the *lack* of a valence bond. Inspection of the samples also show the magnecules capability of increasing their size via the accretion of further oil molecules.

Other photographic documentations of various magnecules in liquids were done, by confirming the findings of Figs. 8.19 and 8.20.

10.3. Spectroscopic Evidence of Liquid Magnecules at the Tekmar-Dohrmann Corporation

The first experimental evidence of magnecules in liquids was established on May 5, 1998, by analysts *Brian Wallace* and *Mia Burnett* at *Tekmar-Dohrmann Corporation* (TDC) in Cincinnati, Ohio, operating a *Tekmar 7000 HT Static Headspacer Autosampler* equipped with a Flame Ionization Detector (FID). The tests were repeated on May 8 and 11, 1998, by confirming the preceding results. It should be noted that the Tekmar equipment lacks the computer search as well as the UV scan. Also, the instrument had limited capability in atomic weight. Finally, the FID was permitted in this case because the liquids were not combustible.

The measurements were done on: Sample 1, pure (magnetically untreated) GR "Fragrance Oil 2"; Sample 2, magnetically untreated tap water; and Sample 3, a magnetically treated mixture of the two.

Despite these limitations, *the results of the Tekmar tests provided the first direct spectroscopic evidence of the existence of magnecules in liquids, including the first direct experimental evidence of water magneplexes* as per Definition 8.2.1. In particular, these tests established that magnecules in liquids have the same main features of the magnecules in gases.

To avoid a prohibitive length we reproduce only a few representative scans in Figs. 8.21 to 8.25 [1]. Figure 8.21 reproduces the origin test of

```
Software Version: 4.0<4J28>
Date: 5/5/98  08:18 AM
Sample Name   : 500ul Perfume Oil ID#1
D-ta File     : C:\TC4\HP210\MY04003.RAW   Date: 5/4/98  04:44 PM
 uence File:    C:\TC4OLD\HP210\MY04.SEQ   Cycle: 3   Channel : B
Instrument    : 772_-_2   Rack/Vial: 0/0   Operator: mb
Sample Amount   : 1.0000                   Dilution Factor  : 1.00
```

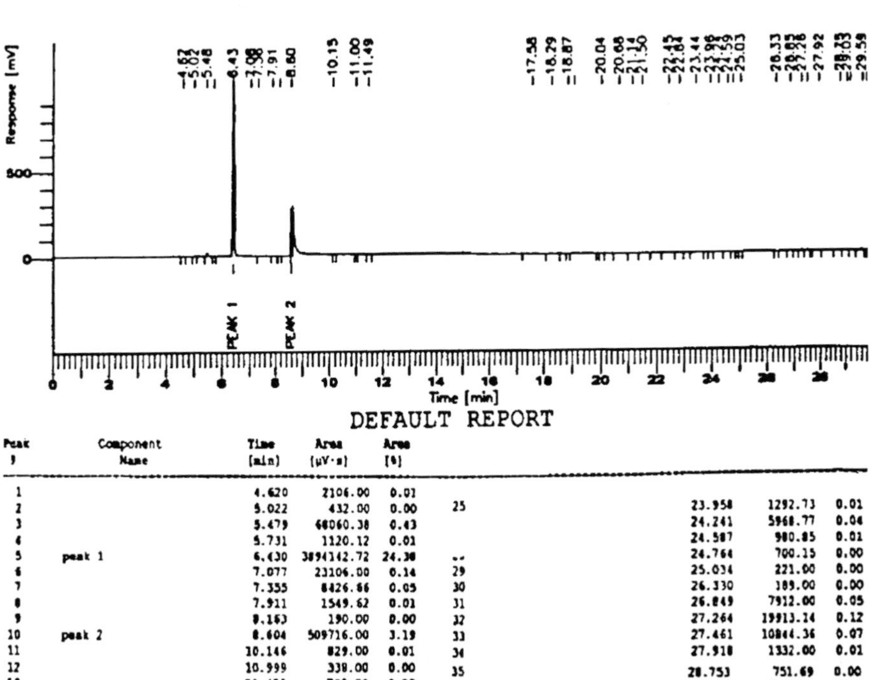

DEFAULT REPORT

Peak #	Component Name	Time [min]	Area [µV·s]	Area [%]
1		4.620	2106.00	0.01
2		5.022	432.00	0.00
3		5.479	68060.38	0.43
4		5.731	1120.12	0.01
5	peak 1	6.430	3894142.72	24.38
6		7.077	23106.00	0.14
7		7.355	8426.66	0.05
8		7.911	1549.62	0.01
9		8.163	190.00	0.00
10	peak 2	8.604	509716.00	3.19
11		10.146	829.00	0.01
12		10.999	338.00	0.00
13		11.485	798.00	0.00
14		17.582	9646.00	0.06
15		18.294	814.50	0.01
16		18.871	748.51	0.00
17		19.082	12390.99	0.08
18		20.043	582.50	0.00
19		20.679	4000.46	0.03
20		21.139	654.15	0.00
21		21.500	716.38	0.00
22		22.452	4196.12	0.03
23		22.837	1128.08	0.01
24		23.437	10546.00	0.07
25		23.958	1292.73	0.01
		24.241	5968.77	0.04
		24.587	980.85	0.01
..		24.764	700.15	0.00
29		25.034	221.00	0.00
30		26.330	189.00	0.00
31		26.849	7912.00	0.05
32		27.264	19913.14	0.12
33		27.461	10844.36	0.07
34		27.918	1332.00	0.01
35		28.753	751.69	0.00
..		29.026	3094.14	0.02
		29.163	2312.68	0.01
		29.589	30846.00	0.19
39		29.750	383.00	0.00
40		30.320	6254.00	0.04
41		31.370	3617.37	0.02
42		31.720	51605.63	0.32
43		32.296	268.00	0.00
44		32.519	87913.29	0.55
45	Peak 3	32.742	11181133.21	70.00
			15973772.00	100.00

Figure 8.21. A first scan done on May 5, 1998, 8.18 a.m. at *Tekmar Dohrmann Company* (TDC) in Cincinnati, Ohio, via a Tekmar 7000 HT Static Headspacer Autosampler with a Flame Ionization Detector (FID).

the fragrance oil without magnetic treatment. Note the dominance of three molecules denoted "Peak 1" with 24.28%, "Peak 2" with 3.19% and "Peak 3" with 70.00%. Figure 8.22 depicts the background which is shown to be correct. Figure 8.23 represent the scan of magnetically treated water with a large "unknown 1" with 64.24% and "unknown 2" with 33.53% totaling 97.78%. This is evidence of the creation of mag-

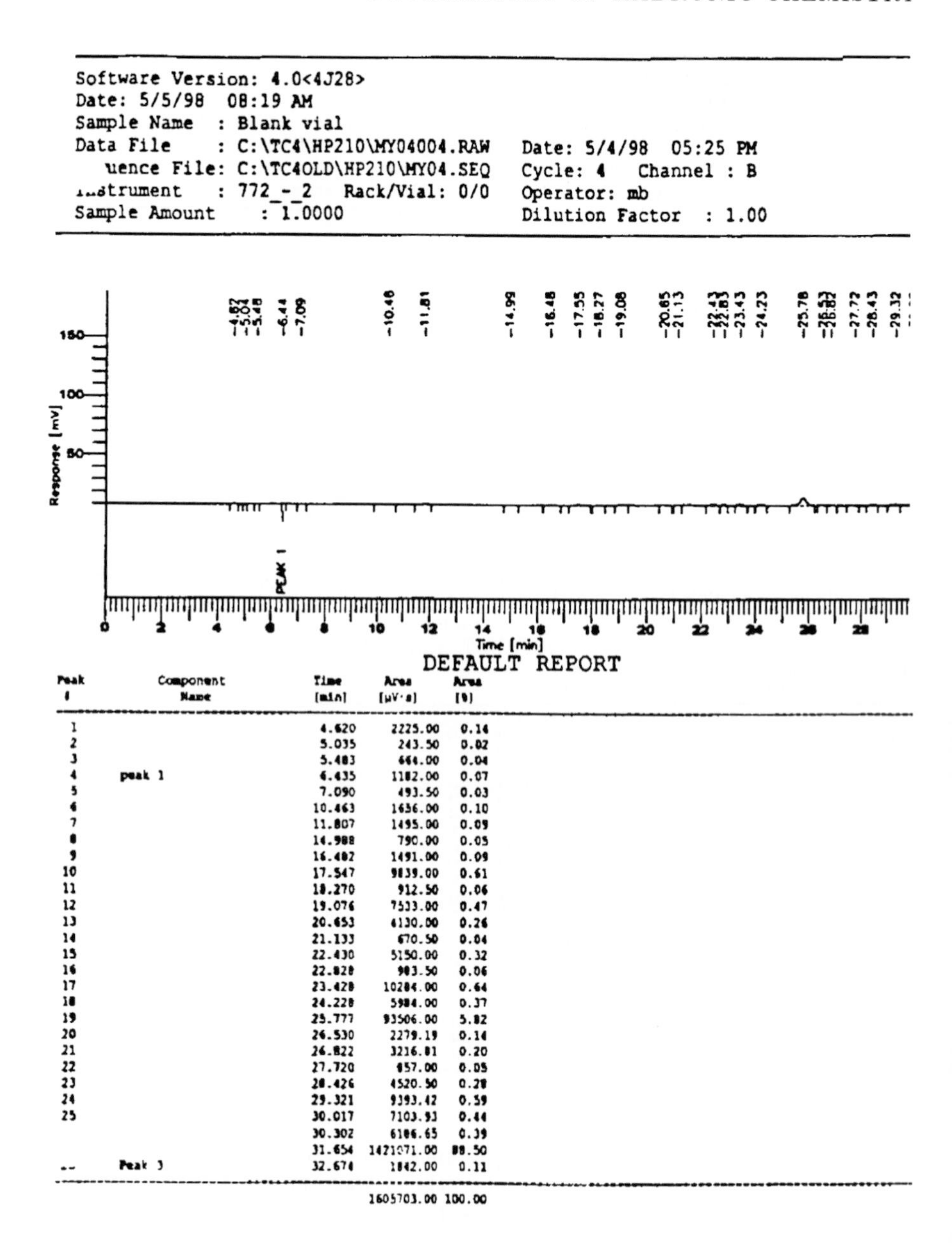

Software Version: 4.0<4J28>
Date: 5/5/98 08:19 AM
Sample Name : Blank vial
Data File : C:\TC4\HP210\MY04004.RAW Date: 5/4/98 05:25 PM
uence File: C:\TC4OLD\HP210\MY04.SEQ Cycle: 4 Channel : B
..strument : 772_-_2 Rack/Vial: 0/0 Operator: mb
Sample Amount : 1.0000 Dilution Factor : 1.00

DEFAULT REPORT

Peak #	Component Name	Time [min]	Area [µV·s]	Area [%]
1		4.620	2225.00	0.14
2		5.035	243.50	0.02
3		5.483	664.00	0.04
4	peak 1	6.435	1182.00	0.07
5		7.090	493.50	0.03
6		10.463	1636.00	0.10
7		11.807	1495.00	0.09
8		14.988	790.00	0.05
9		16.482	1491.00	0.09
10		17.547	9839.00	0.61
11		18.270	912.50	0.06
12		19.076	7533.00	0.47
13		20.653	4130.00	0.26
14		21.133	670.50	0.04
15		22.430	5150.00	0.32
16		22.828	983.50	0.06
17		23.428	10284.00	0.64
18		24.228	5984.00	0.37
19		25.777	93506.00	5.82
20		26.530	2279.19	0.14
21		26.822	3216.81	0.20
22		27.720	857.00	0.05
23		28.426	4520.50	0.28
24		29.321	9393.42	0.59
25		30.017	7103.93	0.44
		30.302	6186.65	0.39
		31.654	1421071.00	88.50
--	Peak 3	32.674	1842.00	0.11
			1605703.00	100.00

Figure 8.22. The scan at TDC on 5/5/98 at 8.19 a.m. to check that the background is correct.

necules in water, also called magneplexes according to Definition 8.2.1. Figure 8.24 represents a scan of the magnetically treated combination of water and fragrance oil with "unknown 1" 1.75% and "unknown 2"

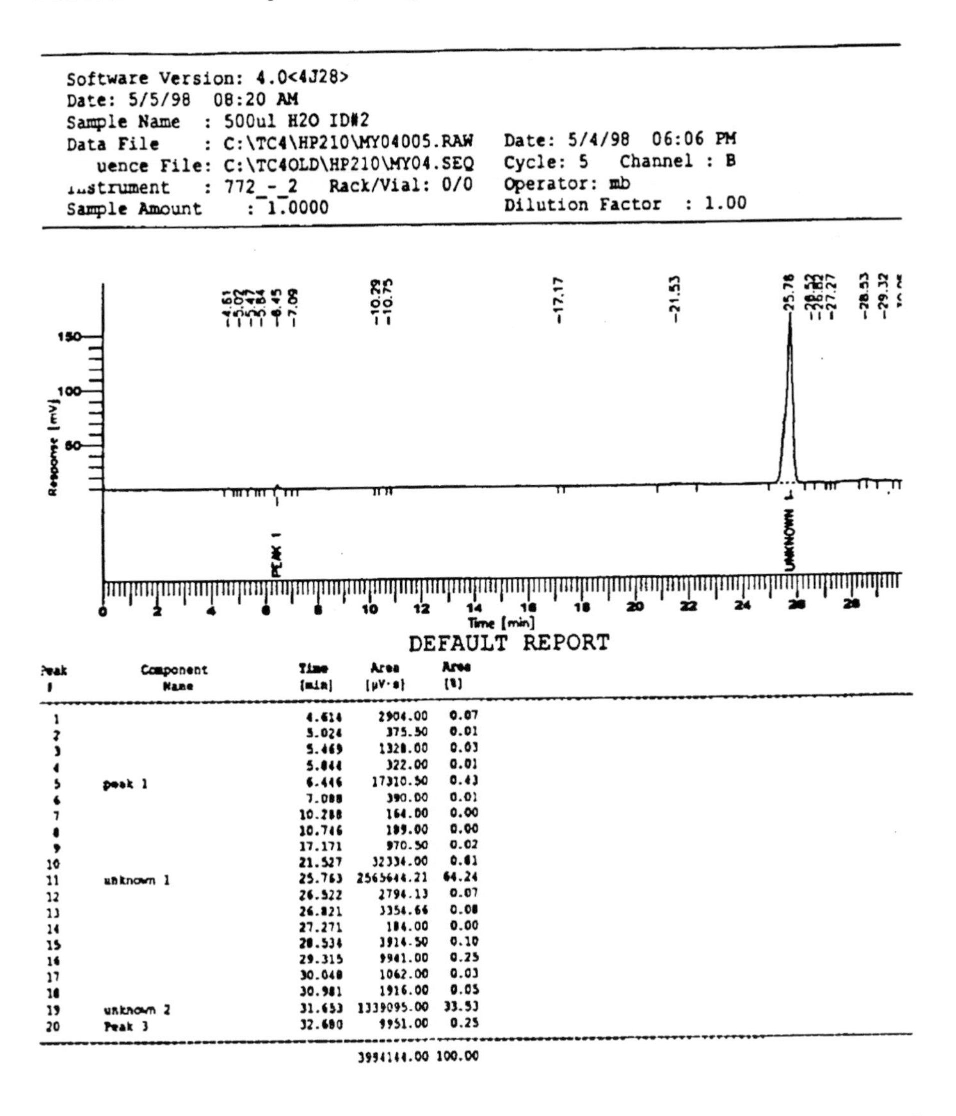

Peak #	Component Name	Time [min]	Area [µV·s]	Area [%]
1		4.614	2904.00	0.07
2		5.024	375.50	0.01
3		5.469	1328.00	0.03
4		5.844	322.00	0.01
5	peak 1	6.446	17310.50	0.43
6		7.088	390.00	0.01
7		10.288	164.00	0.00
8		10.746	189.00	0.00
9		17.171	970.50	0.02
10		21.527	32334.00	0.81
11	unknown 1	25.763	2565644.21	64.24
12		26.522	2794.13	0.07
13		26.821	3354.66	0.08
14		27.271	184.00	0.00
15		28.534	3914.50	0.10
16		29.315	9941.00	0.25
17		30.048	1062.00	0.03
18		30.981	1916.00	0.05
19	unknown 2	31.653	1339095.00	33.53
20	Peak 3	32.680	9951.00	0.25
			3994144.00	100.00

Figure 8.23. The scan at TDC on 5/5/98 at 8.19 a.m. on the magnetically treated water which constitutes experimental evidence of magnecules in water given by the large unknown peak.

with 0.45%. An important information of this scan is that the original Peak 1 of Fig. 8.21 with 24.28% and Peak 3 with 70.00% have been decreased to the values 5.33% and 68.71%, respectively. This is evidence that the missing percentages of these molecules have been used in the formation of magnecules. Figure 8.25 reproduces the background following the tests and routine flushing. As one can see, the scan pre-

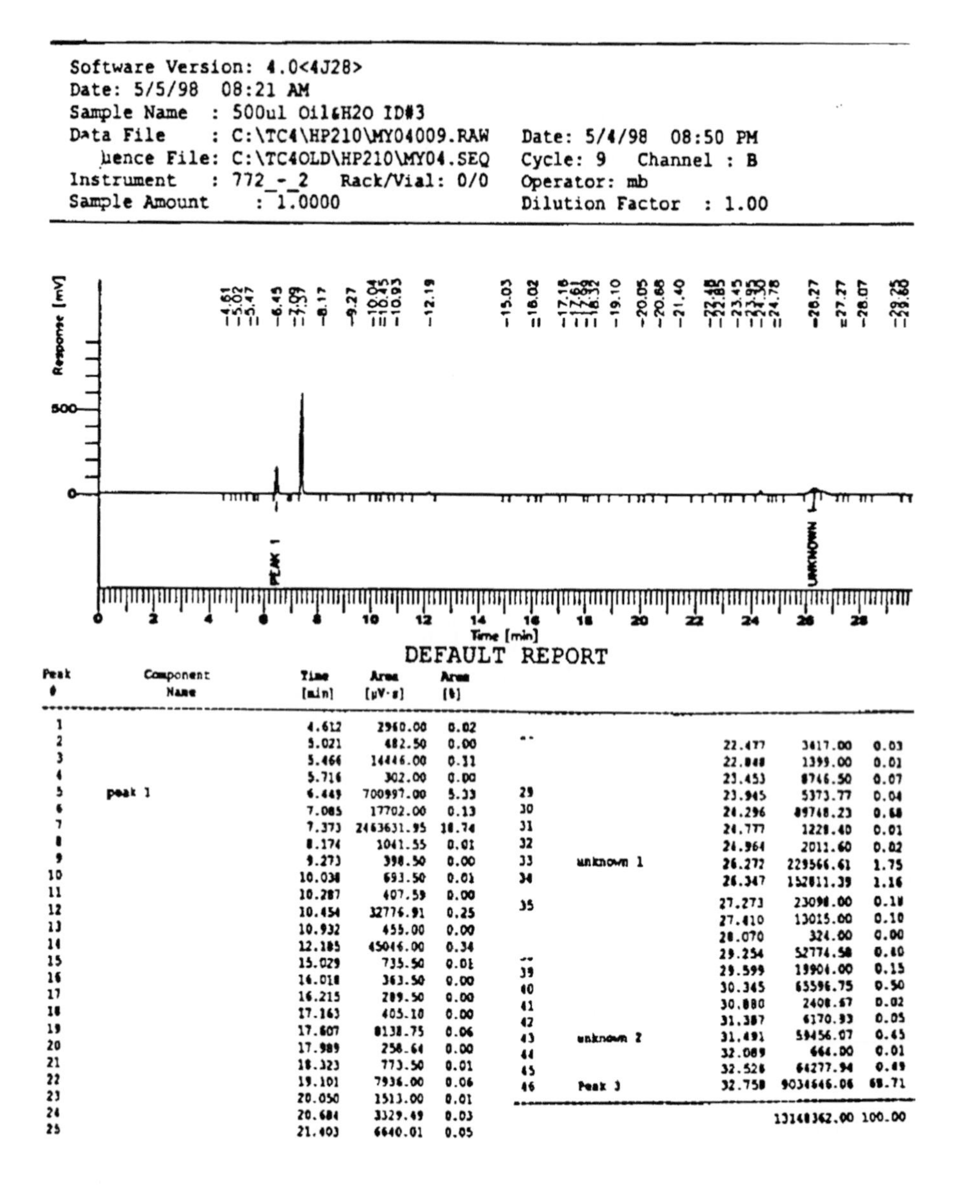

Software Version: 4.0<4J28>
Date: 5/5/98 08:21 AM
Sample Name : 500ul Oil&H2O ID#3
Data File : C:\TC4\HP210\MY04009.RAW Date: 5/4/98 08:50 PM
…uence File: C:\TC4OLD\HP210\MY04.SEQ Cycle: 9 Channel : B
Instrument : 772_-_2 Rack/Vial: 0/0 Operator: mb
Sample Amount : 1.0000 Dilution Factor : 1.00

DEFAULT REPORT

Peak #	Component Name	Time [min]	Area [µV·s]	Area [%]
1		4.612	2960.00	0.02
2		5.021	482.50	0.00
3		5.466	14446.00	0.11
4		5.716	302.00	0.00
5	peak 1	6.449	700997.00	5.33
6		7.085	17702.00	0.13
7		7.373	2463631.95	18.74
8		8.174	1041.55	0.01
9		9.273	398.50	0.00
10		10.038	693.50	0.01
11		10.287	407.59	0.00
12		10.454	32776.91	0.25
13		10.932	455.00	0.00
14		12.185	45046.00	0.34
15		15.029	735.50	0.01
16		16.018	363.50	0.00
17		16.215	289.50	0.00
18		17.163	405.10	0.00
19		17.607	8138.75	0.06
20		17.989	258.64	0.00
21		18.323	773.50	0.01
22		19.101	7936.00	0.06
23		20.050	1513.00	0.01
24		20.684	3329.49	0.03
25		21.403	6640.01	0.05
[illegible]		22.477	3417.00	0.03
[illegible]		22.848	1399.00	0.01
[illegible]		23.453	8746.50	0.07
29		23.945	5373.77	0.04
30		24.296	89748.23	0.68
31		24.777	1228.40	0.01
32		24.964	2011.60	0.02
33	unknown 1	26.272	229566.61	1.75
34		26.347	152811.39	1.16
35		27.273	23098.00	0.18
[illegible]		27.410	13015.00	0.10
[illegible]		28.070	324.00	0.00
[illegible]		29.254	52774.58	0.40
39		29.599	19904.00	0.15
40		30.345	65596.75	0.50
41		30.880	2408.67	0.02
42		31.387	6170.93	0.05
43	unknown 2	31.491	59456.07	0.45
44		32.089	664.00	0.01
45		32.528	64277.94	0.49
46	Peak 3	32.758	9034646.06	68.71
			13148362.00	100.00

Figure 8.24. The scan on 5/5/98 at 8.21 a.m. on the magnetically treated mixture of water and fragrance oil of scan 8.21 which constitutes evidence of magnecules given by two unknown peaks.

serves macroscopic percentages of the preceding scans, thus confirming the anomalous adhesion also existing in gas magnecules.

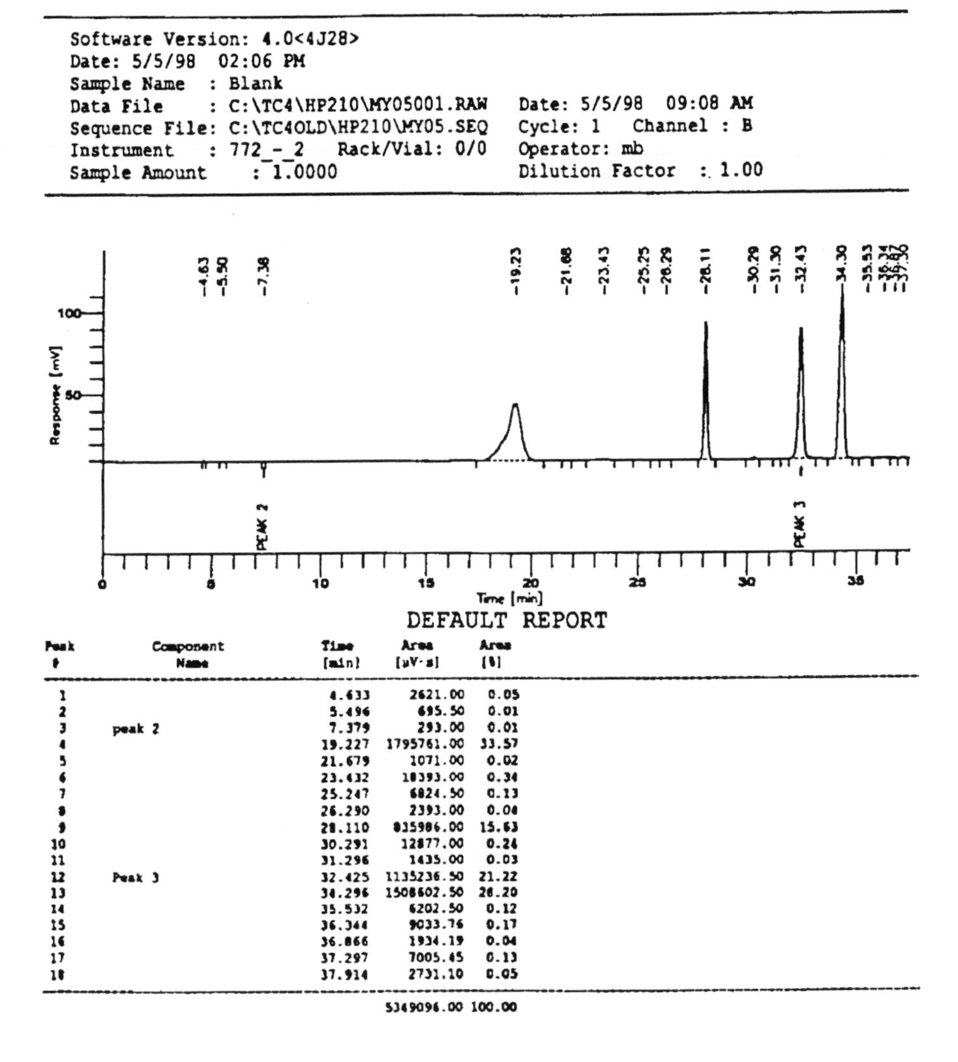

Software Version: 4.0<4J28>
Date: 5/5/98 02:06 PM
Sample Name : Blank
Data File : C:\TC4\HP210\MY05001.RAW Date: 5/5/98 09:08 AM
Sequence File: C:\TC4OLD\HP210\MY05.SEQ Cycle: 1 Channel : B
Instrument : 772_-_2 Rack/Vial: 0/0 Operator: mb
Sample Amount : 1.0000 Dilution Factor : 1.00

DEFAULT REPORT

Peak #	Component Name	Time [min]	Area [µV·s]	Area [%]
1		4.633	2621.00	0.05
2		5.496	695.50	0.01
3	peak 2	7.379	293.00	0.01
4		19.227	1795761.00	33.57
5		21.679	1071.00	0.02
6		23.432	18393.00	0.34
7		25.247	6824.50	0.13
8		26.290	2393.00	0.04
9		28.110	835986.00	15.63
10		30.291	12877.00	0.24
11		31.296	1435.00	0.03
12	Peak 3	32.425	1135236.50	21.22
13		34.296	1508602.50	28.20
14		35.532	6202.50	0.12
15		36.344	9033.76	0.17
16		36.866	1934.19	0.04
17		37.297	7005.45	0.13
18		37.914	2731.10	0.05
			5349096.00	100.00

Figure 8.25. The scan at TDC on 5/5/98 at 2.26 p.m. on the background with anomalous adhesion confirming the corresponding anomalous background for gas magnecules.

10.4. Spectroscopic Evidence of Liquid Magnecules at Florida International University

Additional comprehensive tests via a modern equipment for LC-MS equipped with UVD were conducted on the GR fragrance oil "ING258IN Test 2" of Figs. 8.19 on December 1, 1998, at the chemistry laboratory of *Florida International University* (FIU) in Miami, Florida. The tests were then repeated on December 17 and 18 by confirming the preceding results.

The tests were conducted under a number of technical characterizations specifically selected to detect magnecules, among which include:

1) Total Ion Chromatogram (TIC) collected under the positive ion atmospheric pressure electrospray ionization (ESI+) mode;

2) Integrated TIC with retention times and areas for the most abundant peaks;

3) Raw mass spectra for all peaks identified in item 2;

4) HP LC chromatograms collected at fixed wavelength of 254 cm; and

5) UV-visible spectra form the HPLC diode array detector with 230–700 mm.

The tests were conducted on the following samples:

I) Sample GR331, the magnetically untreated, fully transparent GR fragrance oil "ING258IN Test 2";

II) Sample GR332, magnetically treated "ING258IN Test 2" with 10% Dipropylene Glycol (DPG);

III) Sample GR332S, bottom layer of the preceding sample;

IV) Sample GR335, magnetically treated mixture 4% GR fragrance oil "ING258IN Test 2", 0.4% DPG and 95% tap water; and

V) Sample GR335O, visible dark clusters in the preceding sample.

To avoid a prohibitive length of this presentation, only representative scans are reproduced in Figs. 8.26 to 8.30 [1]. As one can see, these scans provide a second experimental evidence of magnecules in liquids as evident in comparing the peaks of the untreated liquid with those of the treated one.

A few comments are in order. To understand the FIU measurements the reader should keep in mind that the liquid is that of Fig. 8.18. Consequently, *the magnecules to be tested are visible to the naked eye. Therefore, only minute fragments entered the capillary feeding lines of the LC-MS/UVD instrument.*

Finally, the reader should keep in mind that the magnetic polarization of the test has been minimal, and *the liquid does not constitute a pure population of liquid magnecules.* The latter case is available from the PlasmaArcFlow reactors of Chapter 7 whose study is here omitted.

11. Experimental Verification of Mutated Physical Characteristics

In addition to the preceding *chemical* features, the existence of magnecules implies the mutation of *physical* characteristics, such as increase of the specific density and viscosity. This is due to the fact that magnetic

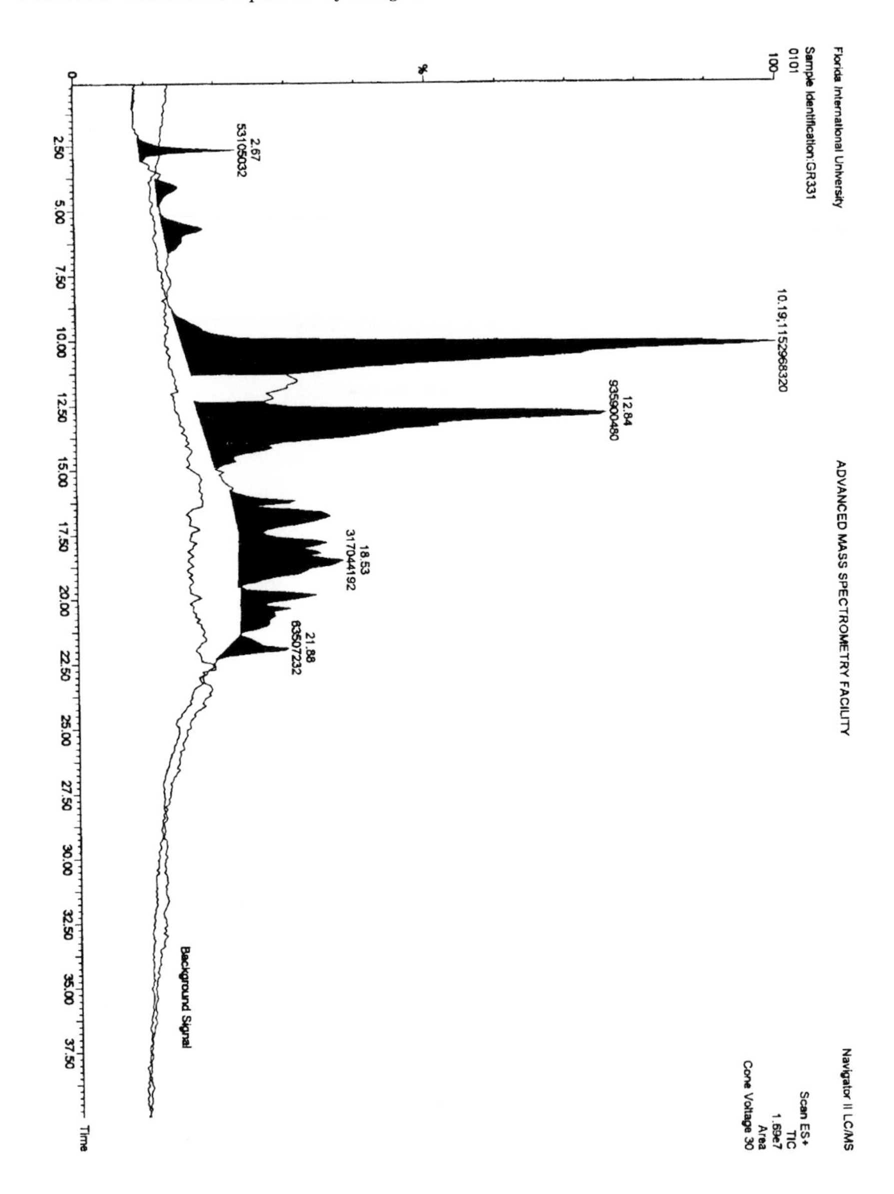

Figure 8.26. Scan on the untreated GR oil "ING258IN Test 2" of Fig. 8.18 (GR331 of the text) conducted at *Florida International University* (FIU).

bonds among ordinary molecules imply an evident reduction of intermolecular distances, thus resulting in more molecules per unit volume, as compared to the magnetically untreated substance. The increases in density and viscosity are then consequential.

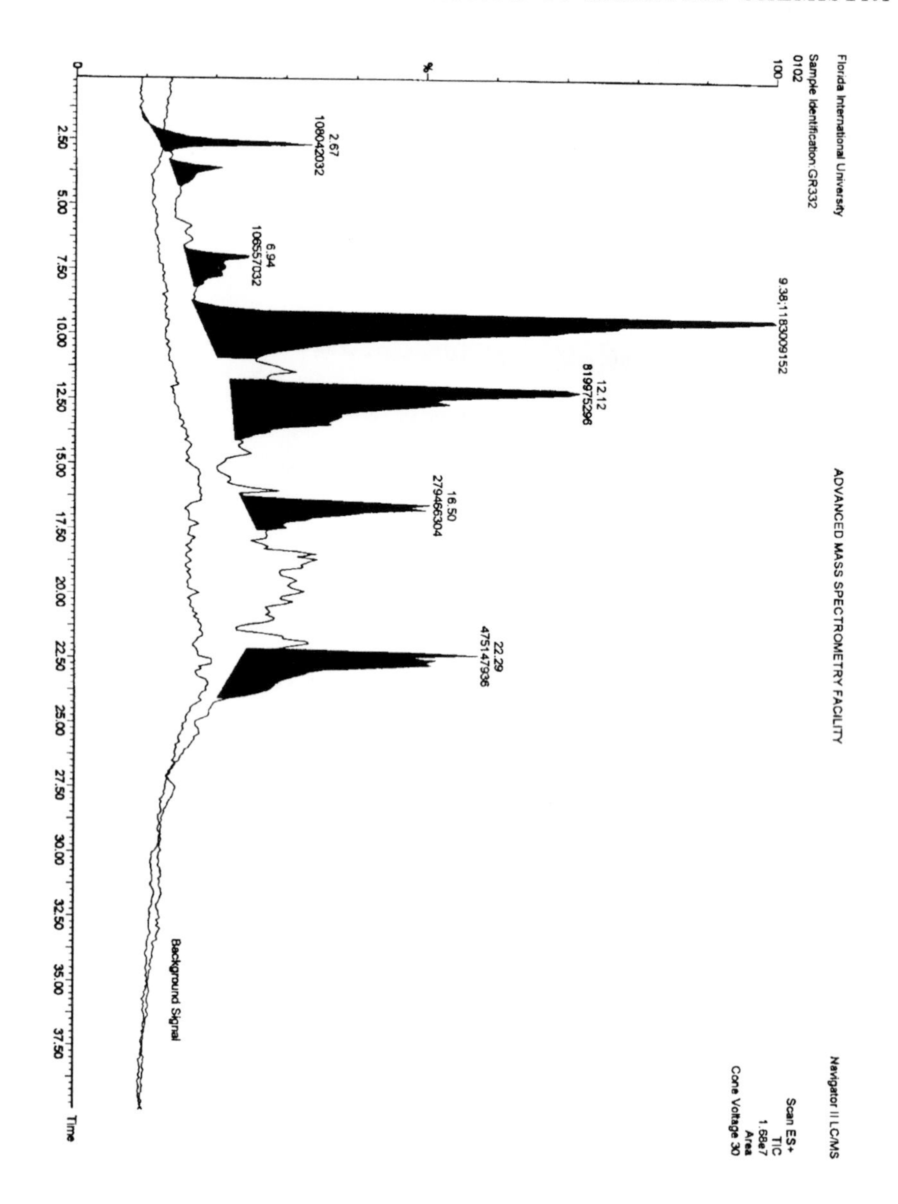

Figure 8.27. Scan at FIU of Sample GR332.

A most intriguing feature of gas magnecules with important scientific and industrial implications is that *the Avogadro number of a gas with magnecular structure is not constant, or, equivalently, the so-called "gas constant" R of a gas with magnecular structure is an (expectedly non-linear) function of P, V, T, $R = R(P, V, T)$*, resulting in the generalized

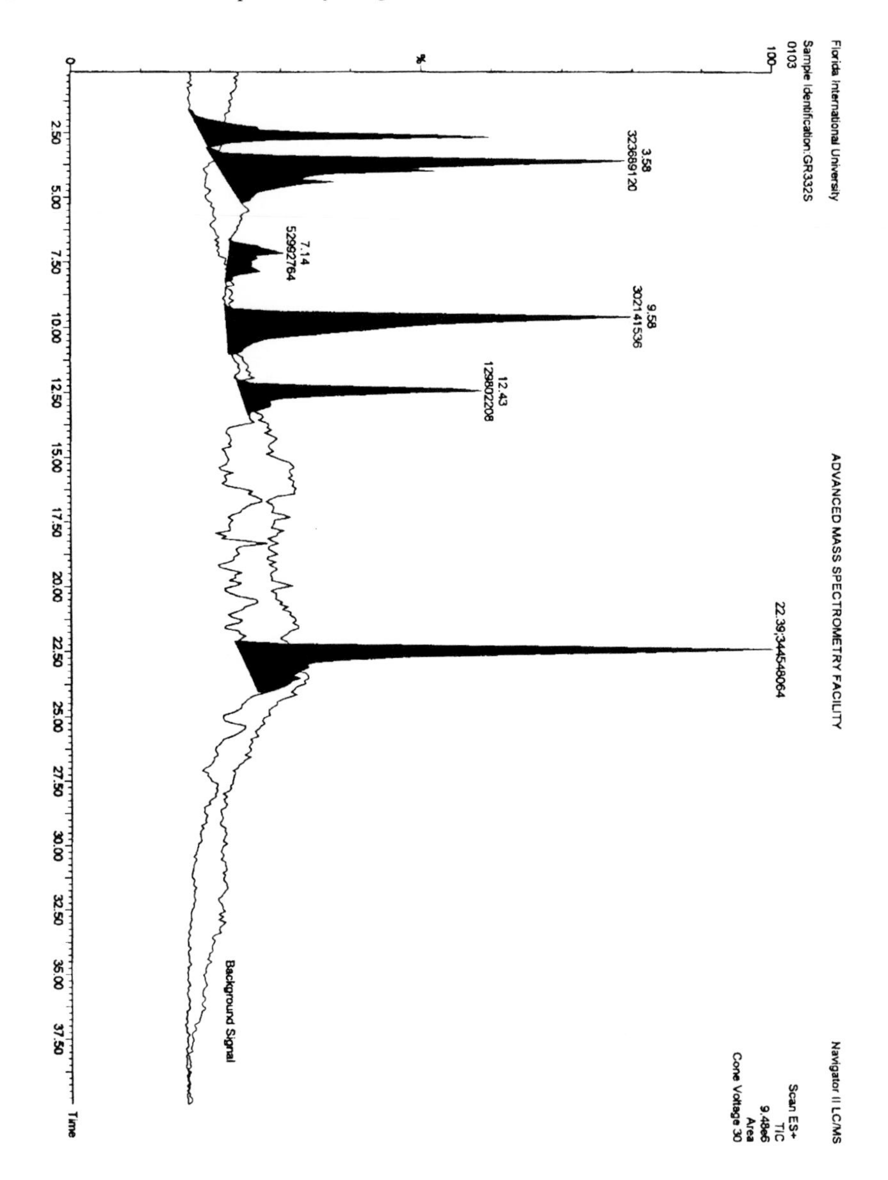

Figure 8.28. Scan at FIU of Sample GR332S.

gas law

$$\frac{PV}{T} = nR(P, V, T), \tag{8.21}$$

where the explicit dependence of R on P, V, and T depends on the magnecular gas considered.

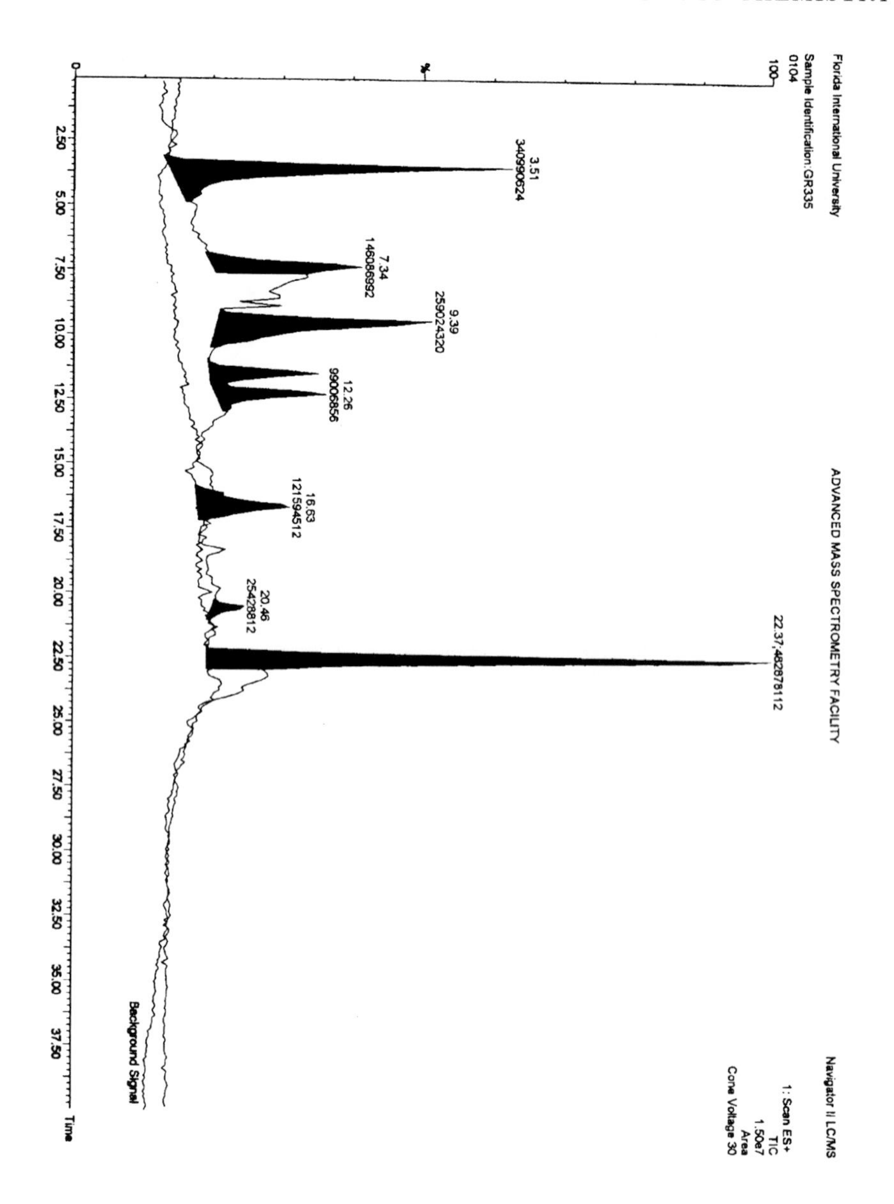

Figure 8.29. Scan at FIU of Sample GR335.

The variation of the Avogadro number for gas with magnecular structure has been proved by routine tests at *USMagnegas, Inc.,* Largo, Florida, establishing that:

1) The number of constituents of a gas with magnecular structure decreases with a sufficient increase of the pressure;

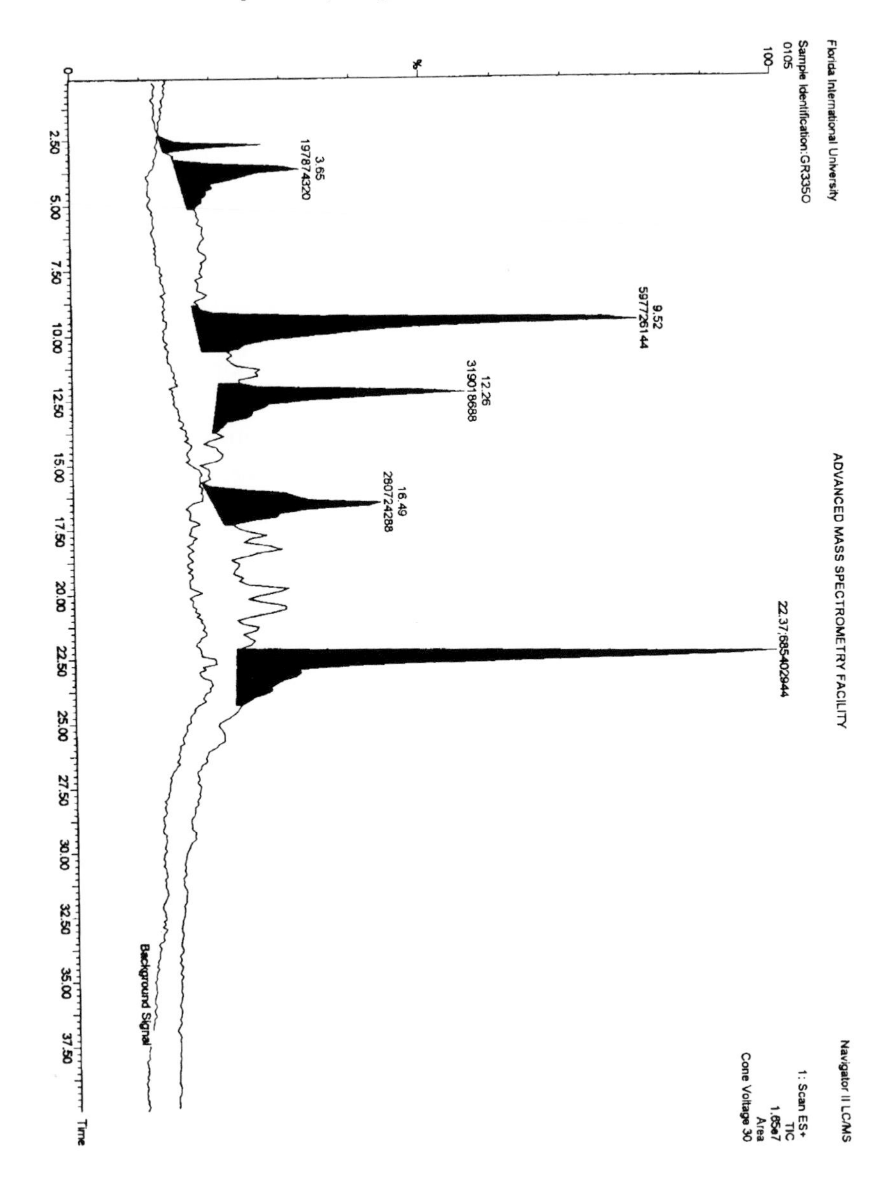

Figure 8.30. Scan at FIU of Sample GR335O.

2) Given a fixed and sealed tank with volume V of a gas with magnecular structure at given pressure P and temperature T, after bringing this tank to a sufficiently higher temperature $T' > T$, and then returning it to the original temperature T, the pressure of the tank is not the original pressure P but a generally bigger pressure $P' > P$;

3) The increase of pressure of a gas with magnecular structure requires a volume which generally increases with the pressure itself, that is, if the increase of pressure in a given tank from 100 psi to 200 psi requires V cf of magnecular gas, the same increase of pressure in the same tank via the same gas, this time from 4,000 psi to 4,100 psi at the same temperature does not require the same volume V but a volume V' of the magnecular gas bigger than the original volume, $V' > V$.

The above deviations from the conventional gas law are easily explained by the fact that *the increase of pressure in a gas with magnecular structure generally implies the aggregation of magnecules into bigger clusters, with consequential decrease of the number of constituents.* Similarly, *the increase of temperature generally implies the breaking down of magnecules into smaller clusters, with consequential increase of the number of constituents and resulting anomalous increase of pressure.* It then follows that, if the increase of temperature of a given fixed volume is beyond the Curie Magnecular Point (Definition 8.2.1), all magnetic polarizations are terminated with consequential increase of the number of constituents due to the reduction of magnecules to molecules. This implies that the return of the gas to the original temperature does not restore the original magnecules, and, consequently, the return to the original temperature generally occurs at an increased pressure due to the increased number of constituents.

We now report measurements of specific density, viscosity and other characteristics of fluids with magnecular structure which confirm the above GC-MS/IRD and LC-MS/UVD tests, by providing final evidence on the existence of magnecules as per Definition 8.2.1.

All tests were done via the use of ordinary tap water and a number of GR fragrance oils. All samples here considered were prepared by conventionally mixing tap water and one fragrant oil, and then submitting that mixture to rather weak permanent magnets of 200 G (much weaker than those used for the fragrance oils of Figs. 8.18 and 8.19). All samples resulted in being very stable without any measurable change over a period of about one year, and survived freezing followed by defrosting. The various samples were numbered from 1 to 25.

The measurements of the specific density were conducted on March 9, 1998 by the *U.S. Testing Company, Inc.* (USTC) of Fairfield, New Jersey. The results of the tests are presented in Figs. 8.31 and 8.32.

Sample 1 is ordinary untreated tap water. Sample 2 is ordinary tap water magnetically treated for about 5 minutes. Samples 3 and 4 were tap water treated with other magnetic equipment. Sample 5 was ordinary untreated GR fragrance oil "APC Fragrance." Sample 6 was a mixture of fragrance oil 5 with tap water magnetically treated for about

REPORT OF TEST

SGS SGS U.S. Testing Company Inc.

291 Fairfield Avenue
Fairfield, NJ 07004-3833
Tel: 973-575-5252
Fax: 973-244-1694

Report Number: 103947
Date: 03/09/98
Page: 1 of 1

Millennium Results

Density of	g/mL	% Change Density vs Ordinary Water
Sample #1	0.9805	0
Sample #2	0.9889	+0.86
Sample #3	0.9804	0
Sample #4	0.9853	+0.49
Fragrant #5	0.9720	NA
Mixture #6	0.9967	+1.65
Mixture #7	0.9982	+1.80
Mixture #8	0.9902	0.99
Treated Water #16	0.9893	0.89
Frag Treated # 17	0.9453	NA
Mixture #18	0.9902	0.99
Mixture #19	0.9929	1.28

Samples were transferred to a separatory funnel. The layers were allowed to separate. The water layer was withdrawn into a funnel with Whatman #4 filter paper. The filtrate was transferred to a preweighed 10 mL volumetric flask. The sample was weighed to 0.0001 grams and the density calculated.

When the samples were pure substances, they were transferred directly to preweighed 10 mL volumetric flasks.

Calculations:

Weight flash with sample - weight flask + volume of flask = g/mL

Arlyn Sibille, Ph.D.

Figure 8.31. USTC measurements of specific density on magnetically treated liquids.

REPORT OF TEST

SGS SGS U.S. Testing Company Inc.

291 Fairfield Avenue
Fairfield, NJ 07004-3833
Tel: 973-575-5252
Fax: 973-244-1694

Report Number: 103947
Date: 03/09/98
Page: 1 of 1

SUBJECT: Three (3) samples received on 02/09/98 and identified by the client as:

PURPOSE: Determine the density and viscosity of the three samples.

TEST DATE: 02/25/98

PROCEDURE: Three 10 milliliter volumetrtic flasks were pre-weighed. One of the samples was transferred to each of the volumetric flasks with a pipet. The samples were weighed again. The density of each sample was calculated.

The three oil samples were measured for viscosity using a Kinematic viscometer (ASTM D-445).

RESULTS:

Sample Identification	Density, g/mL)	Viscosity (cps)	Increase Viscosity, %
1) Motor Oil, "as is"	0.8682	199.8	0
2) Motor Oil, Treatment Type A	0.8714	288.7	44.5
3) Motor Oil, Treatment Type B	0.8689	302.0	51.2

SIGNED FOR THE COMPANY BY:

James R. Tyminski
Laboratory Supervisor

Arlyn Sibille, Ph.D.
Laboratory Director

/mo

Member of the SGS Group

ANALYTICAL SERVICES · PERFORMANCE TESTING · STANDARDS EVALUATION · CERTIFICATION SERVICES

Figure 8.32. USTC measurements of viscosity on magnetically treated liquids.

5 minutes. Mixtures 7 and 8 were the same mixture 5 although treated with other equipment. Sample 17 was a magnetically treated GR oil "Air Freshener 1." Mixture 19 was Fragrance 17 with tap water 16 magnetically treated for 5 minutes. Note that all measurements were done to an accuracy of the fourth digit. Therefore, numerical results up to the third digit can be considered accurate.

In the transition from Sample 1 (untreated water) to Sample 2 (magnetically treated water) there is an increase in the specific density in the macroscopic amount of 0.86%, thus confirming the indicated mutation of the specific density of water under a magnetic treatment. In turn, the increase in density supports the existence of magneplexes in magnetically treated water as per the scan of Fig. 8.27.

As well known, fragrance oils are (generally) *lighter* than water, *i.e.*, the specific density of the untreated fragrance in Sample 5 is *smaller* than that of the untreated water in Sample 1. According to quantum chemistry, the specific density of any mixture of the above two liquids, whether solution, suspension or dispersion, should be *in between* the lighter and heavier specific densities.

On the contrary, as one can see, *the specific density of the magnetically treated mixture of GR fragrance with tap water, Sample 6, resulted in being bigger than that of the densest liquid, the water.* This measurement constitutes additional, rather strong, direct experimental verification of the mutation of physical characteristics in liquids under magnetic fields.

A remarkable point is that the *magnetic mutations of density are macroscopically large.* In fact, they were called by an analyst "UPS-type anomalies", meaning that the shipment via UPS of a given volume of a magnetically treated liquid may require an increase of the shipping cost of the same volume of untreated liquid due to the macroscopic increase in the weight.

A further prediction of magnetically polarized liquids is the increase of its viscosity. This is evidently due to the arbitrary size of an individual magnecule, as well as the tendency of the same to bond to near-by molecules, resulting in accretions, not to mention the anomalous adhesion to the walls of the container, which has been systematically detected for all magnetically polarized liquids.

As indicated earlier, in certain cases the increase of viscosity is so large as to be first visible to the naked eye, and, when the treatment is sufficiently protracted, the increase in viscosity is such as to lose the customary liquid mobility.

Ordinary engine oils are particularly suited for magnetic treatment because, when properly treated, their increase in viscosity is so dramatic as to be visible to the naked eye jointly with a visible change in visual appearance (color, texture, opacity, *etc.*).

The measurements on viscosity are reported in Fig. 8.32. The selected engine oil was an ordinarily available 30-40 Castrol Motor Oil subjected to a particular type of magnetic treatments via two different kinds of equipment called of Type A and B. All treatments were done at ordinary conditions without any additive or change of any type. As one can see, *measurement 2 shows a dramatic increase in the viscosity in the magnetically treated oil of 44.5%.*

The above experimental results evidently provide additional support for the existence of magnecules.

The tests also provide evidence of the anomalous adhesion of liquids with magnecules, which is established in this case by a dramatic, macroscopic increase of adhesion of the oil to the walls of the glass container.

The same macroscopic anomaly is confirmed at the microscopic level. During the measurement of viscosity there was such an anomalous adhesion of the magnetically treated oils to the walls of the instrument that said oil could not be removed via routine cleaning with acetone and required the use of strong acids.

This anomalous adhesion is further experimental evidence of the existence of magnecules, because of their predicted capability to induce the polarization of the orbits of the valence electrons of the atoms in the walls of the container, thus resulting in anomalous adhesion via magnetic bonds due to induction.

It is evident that the mutations of density and viscosity implies the expected mutation of *all* other physical characteristics of the liquid considered. These measurements are left to the interested researchers.

The existence of mutation of *physical* characteristics then implies the mutation of *chemical* features. At this moment, we can only indicate the visual evidence reported by the analysts of USTC according to whom the reaction of magnetically treated oils with acetone is dramatically different from that with untreated oil, including mutations in color, texture and other appearances.

12. Concluding Remarks

The theoretical and experimental evidence presented in this Chapter establishes that the chemical species of molecules, defined as stable clusters of atoms under a valance bond, does not exhaust all possible chemical species existing in nature.

This conclusion is proved beyond scientific doubt, for instance, by macroscopic percentage of stable clusters, with atomic weight of several hundreds a.m.u., in light gases without an infrared signature where heaviest possible detected molecule is the CO_2 with 44 a.m.u.; the mutation of transparent oils into a completely opaque substance without fluidity; the joint increase of the specific density for both gaseous and liquid cases; and other evidence.

Needless to say, the final *characterization* and *detection* of the new chemical species submitted in Ref. [1] and reviewed in this chapter will require a considerable collegial effort, since the methods presented in this chapter are manifestly preliminary, with the understanding that, again, the *existence* of the new chemical species is outside scientific doubts.

As a matter of fact, the proposed new chemical species of magnecules, which, according to Definition 8.2.1 includes that of molecules, cannot be considered itself as the final chemical species in nature as it is the fate proved by history for all scientific discoveries.

As an example, the reformulation of magnecules via the hyperstructural branch of hadronic chemistry implies the prediction of the broader chemical species of *hypermagnecules* which is apparently more suitable to represent living organisms due to its inherent irreversibility, multidimensional structure compatible with our three-dimensional sensory perception, and other features needed for a more adequate representation of the complexities of living organisms. The *novelty* of this possible species is then an evident consequence of its novel features. Its *need* is established by the fact that current attempts to decipher the DNA code via the numbers used for molecules and magnecules dating back to biblical times have little chance of success, thus mandating the use of broader numbers, such as the hypernumbers and related multi-dimensional structures.

All in all, we can safely conclude that science is a discipline which will never admit final theories.

APPENDIX 8.A

Aringazin's Studies on Toroidal Orbits of the Hydrogen Atom under an External Magnetic Field

In the main text of this chapter we have presented the theoretical and experimental foundations of the new chemical species of magnecules which is centrally dependent on individual atoms acquiring a generally toroidal configuration of the orbits of at least the peripheral electrons when exposed to sufficiently intense external magnetic fields, as originally proposed by Santilli [1] and reviewed in the main text of this Chapter.

In this Appendix we outline the studies by Aringazin [8] on the Schrödinger equation of the hydrogen atom under a strong, external, static and uniform magnetic field which studies have confirmed the toroidal configuration of the electron orbits so crucial for the existence of the new chemical species of magnecules.

It should be stressed that when considered at orbital distances (i.e., of the order of 10^{-8} cm), atoms and molecules near the electric arc of hadronic reactors (Chapter 7), and in the plasma region, are exposed to a strong magnetic field, whose intensity may be high enough to cause the needed magnetic polarization (see Fig. 8.4.D).

A weak, external, static, and uniform magnetic field B causes an anomalous Zeeman splitting of the energy levels of the hydrogen atom, with ignorably small effects on the electron charge distribution. In the case of a more intense magnetic field which is strong enough to cause decoupling of a spin-orbital interaction (in atoms), $e\hbar B/2mc > \Delta E_{jj'} \simeq 10^{-3}$ eV, i.e., for $B \simeq 10^5$ Gauss, a normal Zeeman effect is observed, again, with ignorably small deformation of the electron orbits.

More particularly, in the case of a *weak* external magnetic field B, one can ignore the quadratic term in the field B because its contribution is small in comparison with that of the other terms in Schrödinger equation, so that the *linear* approximation in the field B can be used. In such a linear approximation, the wave function of electron remains unperturbed, with the only effect being the well known Zeeman splitting of the energy levels of the H atom. In both Zeeman effects, the interaction energy of the electron with the the magnetic field is assumed to be much smaller than the binding energy of the hydrogen atom, $e\hbar B/2mc \ll me^4/2\hbar^2 = 13.6$ eV, i.e., the intensity of the magnetic field is much smaller than some characteristic value, $B \ll B_0 = 2.4 \cdot 10^9$ Gauss = 240000 Tesla (recall that 1 Tesla = 10^4 Gauss). Thus, the action of a weak magnetic field can be treated as a small perturbation of the hydrogen atom.

In the case of a *very strong* magnetic field, $B \gg B_0$, the quadratic term in the field B makes a great contribution and cannot be ignored. Calculations show that, in this case, a considerable deformation of the electron charge distribution in the hydrogen atom occurs. More specifically, under the influence of a very strong external magnetic field a magnetic confinement takes place, i.e., in the plane perpendicular to the direction of magnetic field (see Fig. 8.4.D), the electron dynamics is determined mainly by the action of the magnetic field, while the Coulomb interaction of the electron with the nucleus can be viewed as a small perturbation. This adiabatic approximation allows one to separate variables in the associated Schrödinger equation [9]. At the same time, in the direction of the magnetic field the motion of electron is governed both by the magnetic field and the Coulomb interaction of the electron with the nucleus.

The highest intensities of magnetic fields maintained macroscopically at large distances in modern magnet laboratories are of the order of $10^5 - 10^6$ Gauss (~ 50 Tesla), i.e., they are much below $B_0 = 2.4 \cdot 10^9$ Gauss ($\sim 10^5$ Tesla). Extremely intense external magnetic fields, $B \geq B_c = B_0/\alpha^2 = 4.4 \cdot 10^{13}$ Gauss, correspond to the interaction energy of the order of the mass of electron, $mc^2 = 0.5$ MeV, where $\alpha = e^2/\hbar c$ is the fine structure constant. In this case, despite the fact that the extremely strong magnetic field does characterize a stable vacuum in respect to creation of electron-positron pairs, one should account for relativistic and quantum electrodynamics (QED) effects, and invoke Dirac or Bethe-Salpeter equation. These contributions are of interest in astrophysics, for example, in studying the atmosphere of neutron stars and white dwarfs which are characterized by $B \simeq 10^9 \ldots 10^{13}$ Gauss.

Aringazin [8] has focused his studies on magnetic fields with intensities of the order of $2.4 \cdot 10^{10} \leq B \leq 2.4 \cdot 10^{13}$ Gauss, at which value nonrelativistic studies via the Schrödinger equation can be used to a very good accuracy, and the adiabatic approximations can be made.

Relativistic and QED effects (loop contributions), as well as effects related to finite mass, size, and magnetic moment of the nucleus, and the finite electromagnetic radius of electron, reveal themselves even at low magnetic field intensities, and can be accounted for as very small perturbations. Additional effects are related to the apparent deviation from QED of strongly correlated valence bonds as studies in Chapter 4. These effects are beyond the scope of the presented study, while being important for high precision studies, such as those on stringent tests of the Lamb shift.

It should be noted that locally high-intensity magnetic fields may arise in plasma as the result of nonlinear effects, which can lead to the creation of stable self-confined structures having nontrivial topology with knots [10]. More particularly, Faddeev and Niemi [10] recently argued that the static equilibrium configurations within the plasma are topologically stable solitons describing knotted and linked fluxtubes of helical magnetic fields. In the region close to such fluxtubes, we suppose the magnetic field intensity may be as high as B_0. In view of this, a study of the action of strong magnetic field and the fluxtubes of magnetic fields on atoms and molecules becomes of great interest in theoretical and applicational *plasmachemistry*. Possible applications are conceivable for the new chemical species of magnecules.

As a result of the action of a very strong magnetic field, atoms attain a great binding energy as compared to the case of zero magnetic field. Even at intermediate $B \simeq B_0$, the binding energy of atoms greatly deviates from that of the zero-field case, and even lower field intensities may essentially affect chemical properties of molecules of heavy atoms. This occurrence permits the creation of various other bound states in molecules, clusters and bulk matter [9, 11, 12].

The paper by Lai [12] is focused on very strong magnetic fields, $B \gg B_0$, motivated by astrophysical applications, and provides a good survey of the early and recent studies in the field, including studies on the intermediate range, $B \simeq B_0$, multi-electron atoms, and H_2 molecule. Several papers using variational/numerical and/or analytical approaches to the problem of light and heavy atoms, ions, and H_2 molecule in strong magnetic field, have been published within the last years (see, e.g., references in [12]). However, highly magnetized molecules of heavy atoms have not been systematically investigated until Santilli's proposal for the new species of magnecules [1]. One of the surprising implications is that for some diatomic molecules of heavy atoms, the molecular binding energy is predicted to be several times bigger than the ground state energy of individual atom [13].

To estimate the intensity of the magnetic field which causes considerable deformation of the ground state electron orbit of the H atom, one can formally compare Bohr radius of the H atom in the ground state, in zero external magnetic field, $a_0 = \hbar^2/me^2 \simeq 0.53 \cdot 10^{-8}$ cm $= 1$ a.u., with the radius of orbit of a single electron moving in the external static uniform magnetic field $\vec{B}$.

The mean radius of the orbital of a single electron moving in a static uniform magnetic field can be calculated exactly by using Schrödinger's equation, and it is given by

$$R_n = \sqrt{\frac{n+1/2}{\gamma}}, \tag{8.A.1}$$

where $\gamma = eB/2\hbar c$, B is intensity of the magnetic field pointed along the z axis, $\vec{B} = (0,0,B)$, $\vec{r} = (r,\varphi,z)$ in cylindrical coordinates, and $n = 0,1,\ldots$ is the principal quantum number. Thus, the radius of the orbit takes *discrete* set of values (8.A.1), and is referred to as Landau radius. This is in contrast to well known *classical* motion of electrons in an external magnetic field, with the radius of the orbit being of a continuous set of values.

The energy levels E_n of a single electron moving in said external magnetic field are referred to as Landau energy levels,

$$E_n = E_n^{\perp} + E_{k_z}^{\parallel} = \hbar\Omega(n+\frac{1}{2}) + \frac{\hbar^2 k_z^2}{2m}, \tag{8.A.2}$$

where $\Omega = eB/mc$ is so called cyclotron frequency, and $\hbar k_z$ is a projection of the electron momentum $\hbar\vec{k}$ on the direction of the magnetic field, $-\infty < k_z < \infty$, m is mass of electron, and $-e$ is charge of electron.

Landau's energy levels $E_n^{\perp}$ correspond to a discrete set of round orbits of the electron which are projected to the transverse plane. The energy $E_{k_z}^{\parallel}$ corresponds to a free motion of the electron in parallel to the magnetic field (*continuous* spectrum), with a conserved momentum $\hbar k_z$ along the magnetic field.

In regard to the above review of Landau's results, we recall that in the general case of a *uniform* external magnetic field the coordinate and spin components of the total wave function of the electron can always be separated.

The corresponding coordinate component of the total wave function of the electron, obtained as an exact solution of Schrödinger equation for a single electron moving in the external magnetic field with vector-potential chosen as $A_r = A_z = 0$, $A_\varphi = rB/2$,

$$-\frac{\hbar^2}{2m}\left(\partial_r^2 + \frac{1}{r}\partial_r + \frac{1}{r^2}\partial_\varphi^2 + \partial_z^2 - \gamma^2 r^2 + 2i\gamma\partial_\varphi\right)\psi = E\psi, \tag{8.A.3}$$

is of the following form [9]:

$$\psi_{n,s,k_z}(r,\varphi,z) = \sqrt{2\gamma}I_{ns}(\gamma r^2)\frac{e^{il\varphi}}{\sqrt{2\pi}}\frac{e^{ik_z z}}{\sqrt{L}}, \tag{8.A.4}$$

where $I_{ns}(\rho)$ is Laguerre function,

$$I_{ns}(\rho) = \frac{1}{\sqrt{n!s!}}e^{-\rho/2}\rho^{(n-s)/2}Q_s^{n-s}(\rho); \tag{8.A.5}$$

Q_s^{n-s} is Laguerre polynomial, L is normalization constant, $l = 0,\pm1,\pm2,\ldots$ is azimuthal quantum number, $s = n - l$ is radial quantum number, and $\rho = \gamma r^2$.

Spin components of the total wave function are trivially given by

$$\psi(\frac{1}{2}) = \begin{pmatrix} 1 \\ 0 \end{pmatrix}, \quad \psi(-\frac{1}{2}) = \begin{pmatrix} 0 \\ 1 \end{pmatrix}, \tag{8.A.6}$$

with the corresponding energies $E_{spin} = \pm\mu_0 B$, to be added to the energy (8.A.2); $\mu_0 = e\hbar/2mc$ is Bohr magneton.

For the *ground* Landau level, i.e. at $n = 0$ and $s = 0$, and zero momentum of electron in the z-direction, i.e. $\hbar k_z = 0$, we have from (8.A.2)

$$E_0^{\perp} = \frac{e\hbar B}{2mc}, \tag{8.A.7}$$

and due to Eq. (8.A.4) the corresponding normalized ground state wave function is

$$\psi_{000}(r,\varphi,z) = \psi_{000}(r) = \sqrt{\frac{\gamma}{\pi}}\, e^{-\gamma r^2/2}, \tag{8.A.8}$$

$\int_0^\infty \int_0^{2\pi} r dr d\varphi\ |\psi_{000}|^2 = 1.$

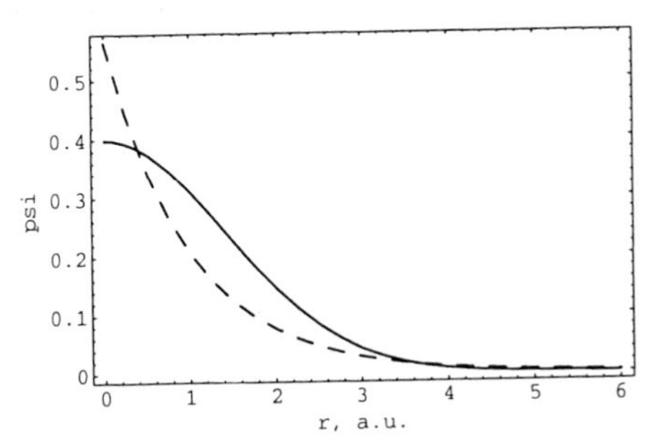

Figure 8.A.1. Landau's ground state wave function of a single electron, ψ_{000} (solid curve), Eq. (8.A.8), in a strong external magnetic field $B = B_0 = 2.4 \cdot 10^9$ Gauss, as function of the distance r in cylindrical coordinates, and (for a comparison) the hydrogen ground state wave function (at *zero* external magnetic field), $(1/\sqrt{\pi})e^{-r/a_0}$ (dashed curve), as function of the distance r in spherical coordinates. The associated probability densities are shown in Fig. 8.A.2; 1 a.u. $= a_0 = 0.53 \cdot 10^{-8}$ cm.

The corresponding (smallest) Landau's radius of the orbit of electron is

$$R_0 = \sqrt{\frac{\hbar c}{eB}} \equiv \sqrt{\frac{1}{2\gamma}}, \tag{8.A.9}$$

in terms of which ψ_{000} reads

$$\psi_{000} = \sqrt{\frac{1}{2\pi R_0^2}}\ e^{-\frac{r^2}{4R_0^2}}. \tag{8.A.10}$$

Figure 8.A.1 depicts Landau's ground state wave function of a single electron, ψ_{000}, in the strong external magnetic field $B = B_0 = 2.4 \cdot 10^9$ Gauss ($R_0 = 1$ a.u.), and (for a comparison) of the hydrogen ground state wave function, at *zero* external magnetic field, $(1/\sqrt{\pi})e^{-r/a_0}$. Figures 8.A.2 and 8.A.3 display the associated probability density of the electron as a function of the distance r from the center of the orbit, the radius of which is about 1 a.u.

The condition that Landau's radius is smaller than Bohr's radius, $R_0 < a_0$ (which is adopted here as the condition of a considerable "deformation" of the electron orbit

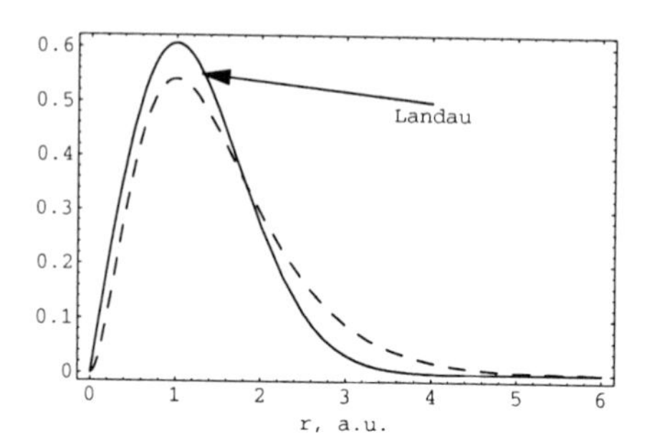

Figure 8.A.2. Probability density for the case of Landau's ground state of a single electron, $2\pi r|\psi_{000}|^2$ (solid curve), Eq. (8.A.8), in a strong external magnetic field $B = B_0 = 2.4 \cdot 10^9$ Gauss, as a function of the distance r in cylindrical coordinates, and (for a comparison) the probability density of the hydrogen atom ground state (at *zero* external magnetic field), $4\pi r^2|(1/\sqrt{\pi})e^{-r/a_0}|^2$ (dashed curve), as function of the distance r in spherical coordinates. The associated wave functions are shown in Fig. 8.A.1; 1 a.u. $= 0.53 \cdot 10^{-8}$ cm.

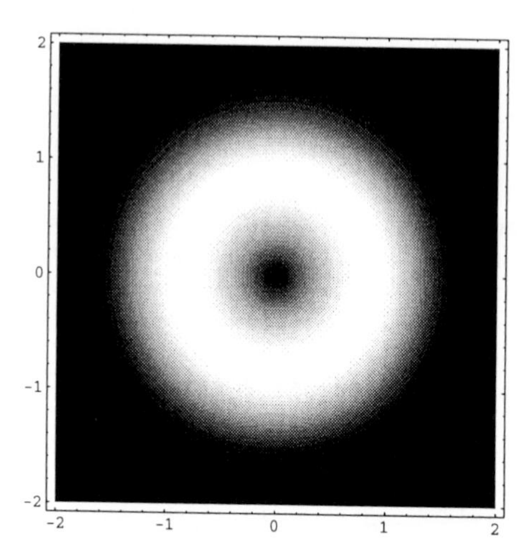

Figure 8.A.3. Contour plot of the (r, φ) probability density for the case of Landau's ground state of a single electron, $2\pi r|\psi_{000}|^2$, Eq. (8.A.8), in strong external magnetic field $B = B_0 = 2.4 \cdot 10^9$ Gauss, as a function of the distance in a.u. (1 a.u. $=$ $0.53 \cdot 10^{-8}$ cm). The lighter area corresponds to a bigger probability of finding the electron. The set of maximal values of the probability density is referred to as an "orbit".

of the H atom) then implies

$$B > B_0 = \frac{m^2 c e^3}{\hbar^3} = 2.351 \cdot 10^9 \text{ Gauss}, \tag{8.A.11}$$

where m is mass of electron. Equivalently, this deformation condition corresponds to the case when the binding energy of the H atom, $|E_0^{Bohr}| = |-me^4/2\hbar^2| = 0.5$ a.u. $= 13.6$ eV, is smaller than the ground Landau energy $E_0^{\perp}$.

The above critical value of the magnetic field, B_0, is naturally taken as an *atomic unit* for the strength of the magnetic field, and corresponds to the case when the pure Coulomb interaction energy of the electron with nucleus is equal to the interaction energy of the single electron with the external magnetic field, $|E_0^{Bohr}| = E_0^{\perp} = 13.6$ eV, or equivalently, when Bohr radius is equal to Landau radius, $a_0 = R_0 = 0.53 \cdot 10^{-8}$ cm.

It should be stressed here that the characteristic parameters, Bohr's energy $|E_0^{Bohr}|$ and Bohr's radius a_0, of the H atom have the purpose to establish a criterium for the critical strength of the external magnetic field of the hydrogen atom under the conditions here considered. For other atoms the critical value of the magnetic field may be evidently different.

After outlining the quantum dynamics of a single electron in an external magnetic field, Aringazin [8] turns to the consideration of the H atom under an external static uniform magnetic field.

In the cylindrical coordinate system (r, φ, z), in which the external magnetic field is $\vec{B} = (0, 0, B)$, i.e., the magnetic field is directed along the z-axis, Schrödinger's equation for an electron moving around a fixed proton (Born-Oppenheimer approximation) in the presence of the external magnetic field is given by

$$-\frac{\hbar^2}{2m}\left(\partial_r^2 + \frac{1}{r}\partial_r + \frac{1}{r^2}\partial_\varphi^2 + \partial_z^2 + \frac{2me^2}{\hbar^2\sqrt{r^2+z^2}} - \gamma^2 r^2 + 2i\gamma\partial_\varphi\right)\psi = E\psi, \quad (8.A.12)$$

where $\gamma = eB/2\hbar c$.

The main problem in the nonrelativistic study of the hydrogen atom in an external magnetic field is to solve the above Schrödinger equation and find the energy spectrum. This equation is not analytically tractable so that one is led to use approximations.

In the approximation of a *very* strong magnetic field, $B \gg B_0 = 2.4 \cdot 10^9$ Gauss, Coulomb interaction of the electron with the nucleus is not important, in the transverse plane, in comparison to the interaction of the electron with external magnetic field. Therefore, in accord to the exact solution (8.A.4) for a single electron, one can look for an approximate ground state solution of Eq. (8.A.12) in the form of factorized transverse and longitudinal parts,

$$\psi = e^{-\gamma r^2/2}\chi(z), \quad (8.A.13)$$

where $\chi(z)$ is the longitudinal wave function to be found. This is so called *adiabatic approximation*. In general, the adiabatic approximation corresponds to the case when the transverse motion of electron is totally determined by the intense magnetic field, which makes the electron "dance" at its cyclotron frequency. Specifically, the radius of the orbit is then *much smaller* than Bohr radius, $R_0 \ll a_0$. The remaining problem is thus to find longitudinal energy spectrum, in the z direction.

Inserting tehe wave function (8.A.13) into the Schrödinger equation (8.A.12), multiplying it by ψ^*, and integrating over variables r and φ in cylindrical coordinate system, one gets the following equation characterizing the z dependence of the wave function:

$$\left(-\frac{\hbar^2}{2m}\frac{d^2}{dz^2} + \frac{\hbar^2\gamma}{m} + C(z)\right)\chi(z) = E\chi(z), \quad (8.A.14)$$

where

$$C(z) = -\sqrt{\gamma}\, e^2 \int_0^\infty \frac{e^{-\rho}}{\sqrt{\rho + \gamma z^2}}\, d\rho = -e^2 \sqrt{\pi\gamma}\, e^{\gamma z^2} [1 - \mathrm{erf}(\sqrt{\gamma}|z|)], \qquad (8.A.15)$$

where $\mathrm{erf}(x)$ is the error function.

The arising effective potential $C(z)$ is of a nontrivial form, which does not allow to solve Eq. (8.A.14) analytically, so one can approximate it by simple potentials, to make an estimation on the ground state energy and wave function of the H atom.

At high intensity of the magnetic field, $\gamma \gg 1$ so that under the condition $\gamma\langle z^2\rangle \gg 1$ one can ignore ρ in the square root in the integrand in Eq. (8.A.15). Then, one can perform the simplified integral and obtain the result

$$C(z) \simeq V(z) = -\frac{e^2}{|z|}, \quad \text{at } \gamma\langle z^2\rangle \gg 1, \qquad (8.A.16)$$

which appears to be a pure Coulomb interaction of electron with the nucleus, in the z direction. Due to the exact result (8.A.15), $C(z)$ tends to zero as $z \to \infty$. However, a remarkable implication of the exact result is that $C(z)$ is *finite* at $z = 0$, namely, $C(0) = -\sqrt{\pi\gamma}\, e^2$, so that the effective potential $C(z)$ can *not* be well approximated by the Coulomb potential.

The exact potential $C(z)$ can be well approximated by the *modified* Coulomb potential,

$$C(z) \simeq V(z) = -\frac{e^2}{|z| + z_0}, \qquad (8.A.17)$$

where z_0 is a parameter, $z_0 \neq 0$, which depends on the field intensity B due to

$$z_0 = -\frac{e^2}{C(0)} = \frac{1}{\sqrt{\pi\gamma}} = \sqrt{\frac{2\hbar c}{\pi e B}}. \qquad (8.A.18)$$

The analytic advantage of this approximation is that $V(z)$ is *finite* at $z = 0$, being of Coulomb-type form. Therefore, Eq. (8.A.14) reduces to *one-dimensional* Schrödinger equation for the Coulomb-like potential,

$$\left(\frac{\hbar^2}{2m}\frac{d^2}{dz^2} + \frac{e^2}{|z| + z_0} + \frac{\hbar^2\gamma}{m} + E\right)\chi(z) = 0. \qquad (8.A.19)$$

In the atomic units ($e = \hbar = m = 1$), using the notation

$$E' = \frac{\hbar^2\gamma}{m} + E, \quad n^2 = \frac{1}{-2E'}, \qquad (8.A.20)$$

introducing the new variable $x = 2z/n$, and dropping $x_0 = 2z_0/n$, to simplify representation, the above equation can be rewritten as

$$\left[\frac{d}{dx^2} + \left(-\frac{1}{4} + \frac{n}{x}\right)\right]\chi(x) = 0, \qquad (8.A.21)$$

where $x > 0$ is assumed. Introducing new function $v(x)$ defined as $\chi(x) = xe^{-x/2}v(x)$, one gets the final form of the equation,

$$xv'' + (2 - x)v' - (1 - n)v = 0. \qquad (8.A.22)$$

Noting that it is a particular case of Cummer's equation,

$$xv'' + (b - x)v' - av = 0, \tag{8.A.23}$$

the general solution is given by

$$v(x) = C_1\,{}_1F_1(a, b, x) + C_2 U(a, b, x), \tag{8.A.24}$$

where

$${}_1F_1(a, b, x) = \frac{\Gamma(b)}{\Gamma(b-a)\Gamma(a)} \int_0^1 e^{xt} t^{a-1} (1-t)^{b-a-1}\, dt \tag{8.A.25}$$

and

$$U(a, b, x) = \frac{1}{\Gamma(a)} \int_0^\infty e^{-xt} t^{a-1} (1+t)^{b-a-1}\, dt \tag{8.A.26}$$

are the confluent hypergeometric functions, and $C_{1,2}$ are constants; $a = 1 - n$ and $b = 2$. Hence, for $\chi(x)$ one has

$$\chi(x) = (|x| + x_0) e^{-(|x|+x_0)/2} \left[C_1^{\pm}\,{}_1F_1(1-n, 2, |x| + x_0) + C_2^{\pm} U(1-n, 2, |x| + x_0) \right], \tag{8.A.27}$$

where the parameter x_0 has been restored, and the "$\pm$" sign in $C_{1,2}^{\pm}$ corresponds to the positive and negative values of x, respectively (the modulus sign is used for brevity).

Let us consider first the $x_0 = 0$ case. The first hypergeometric function ${}_1F_1(1 - n, 2, x)$ is finite at $x = 0$ for any n. At big x, it diverges exponentially, unless n is an integer number, $n = 1, 2, \ldots$, at which case it diverges polynomially. The second hypergeometric function $U(1-n, 2, x)$ behaves differently, somewhat as a mirror image of the first one. In the limit $x \to 0$, it is finite for integer $n = 1, 2, 3, \ldots$, and diverges as $1/x$ for noninteger $n > 1$ and for $0 \leq n < 1$. In the limit $x \to \infty$, it diverges polynomially for integer n, tends to zero for noninteger $n > 1$ and for $n = 0$, and diverges for noninteger $0 < n < 1$.

In general, because of the prefactor $xe^{-x/2}$ in the solution (8.A.27) which cancels some of the divergencies arising from the hypergeometric functions, we should take into account *both* of the two linearly independent solutions, to get the most general form of normalizable wave functions.

As a consequence, for $x_0 \neq 0$ the eigenvalues may *differ* from those corresponding to $n = 1, 2, \ldots$ (which is a counterpart of the principal quantum number in the ordinary hydrogen atom problem) so that n is allowed to take some *non-integer* values from 0 to ∞, provided that the wave function is normalizable.

For even states, in accord to the symmetry of wave function under the inversion $z \to -z$, one has

$$C_1^+ = C_1^-, \quad C_2^+ = C_2^-, \quad \chi'(0) = 0. \tag{8.A.28}$$

Also, since $n = 1$ gives $E' = -1/(2n^2) = -1/2$ a.u., one should seek normalizable wave function for n in the interval $0 < n < 1$, in order to achieve lower energy value. If successful, $n = 1$ indeed does not characterize the ground state. Instead, it may correspond to some excited state.

Analysis shows that *normalizable* wave functions, as a combination of *two* linearly independent solutions, for the modified Coulomb potential *does exist* for various *non-integer* n. Focusing on the ground state solution, Aringazin considers values of n ranging from 0 to 1. Remind that $E' = -1/2n^2$ so that for $n < 1$ the energy lower than $E' = -0.5$ a.u.

For $n < 1$, the first hypergeometric function is not suppressed by the prefactor $xe^{-x/2}$ in the solution (8.A.27) at large x so we are led to discard it as an unphysical solution by putting $C_1 = 0$. A normalizable ground state wave function for $n < 1$ is thus may be given by the second term in the solution (8.A.27). Indeed, the condition $\chi'(x)_{|x=0} = 0$ implies

$$\begin{aligned} &\tfrac{1}{2}e^{-(x+x_0)/2}C_2[(2-x-x_0)U(1-n,2,x+x_0)- \\ &\qquad -2(1-n)(x+x_0)U(2-n,3,x+x_0))]_{|x=0} = 0. \end{aligned} \tag{8.A.29}$$

The l.h.s of this equation depends on n and x_0, so one can select some field intensity B, calculate associated $x_0 = x_0(B)$ and find n, from which one obtains the ground state energy E'. On the other hand, for the ground state this condition can be viewed, *vice versa*, as an equation to find x_0 at some selected n.

For example, taking the noninteger value $n = 1/\sqrt{15.58} \simeq 0.253 < 1$ Aringazin found $x_0 = 0.140841$. This value is in confirmation with the result $x_0 = 0.141$ obtained by Heyl and Hernquist [14]. On the other hand, x_0 is related in accord to Eq. (8.A.18) to the intensity of the magnetic field, $x_0 = 2z_0/n$, from which one obtains $B \simeq 4.7 \cdot 10^{12}$ Gauss. Hence, at this field intensity the ground state energy of the hydrogen atom is determined by $n = 1/\sqrt{15.58}$.

The total ground state wave function is given by

$$\psi(r,\varphi,x) \simeq \sqrt{\frac{1}{2\pi R_0^2}}e^{-\frac{r^2}{4R_0^2}}(|x|+x_0)e^{(|x|+x_0)/2}U(1-n,2,|x|+x_0), \tag{8.A.30}$$

where n is determined due the above procedure, and the associated three-dimensional probability density is schematically depicted in Fig. 8.A.4.

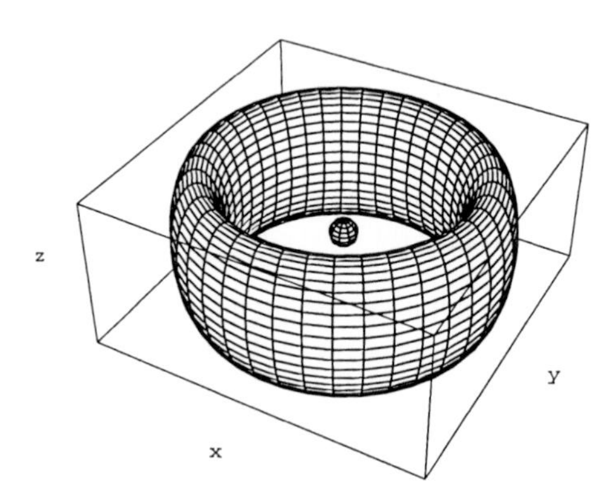

Figure 8.A.4. A schematic view on the H atom in the ground state under a very strong external magnetic field $\vec{B} = (0,0,B)$, $B \gg B_0 = 2.4 \cdot 10^9$ Gauss, due to the *modified* Coulomb approximation studied in the text. The electron moves on the Landau orbit of small radius $R_0 \ll 0.53 \cdot 10^{-8}$ cm resulting in the toroidal structure used for the new chemical species of magnecules. The vertical size of the atom is comparable to R_0. The spin of the electron is antiparallel to the magnetic field.

One can see that the problem remarkably difference than the ordinary three-dimensional problem of the hydrogen atom, for which the principal quantum number

n must be integer to get normalizable wave functions, and the value $n = 1$ corresponds to the lowest energy.

The modified Coulomb potential approach provides qualitatively correct behavior, and suggests a *single* Landau-type orbit shown in Fig. 8.A.4 for the *ground* state charge distribution of the hydrogen atom. This is in full agreement with Santilli's study [1, 11] of the hydrogen atom in a strong magnetic field.

Accurate analytic calculation of the ground and excited hydrogen wave functions made by Heyl and Hernquist [14] in the adiabatic approximation leads to the longitudinal parts of the wave functions shown in Fig. 8.A.5, which reproduces the original Fig. 3 of their work; $\zeta = 2\pi\alpha z/\lambda_e$; $B = 4.7 \cdot 10^{12}$ Gauss. They used the modified Coulomb potential of the type (8.A.17), and the additional set of linearly independent solutions of the one-dimensional modified Coulomb problem in the form

$$(|x| + x_m)e^{-(|x|+x_m)/2}{}_1F_1(1-n, 2, |x| + x_m) \int^{|x|+x_m} \frac{e^t}{(t\,{}_1F_1(1-n, 2, t))^2}\,dt, \quad (8.A.31)$$

where $m = 0$ corresponds to the ground state. For the ground state with $n = 1/\sqrt{15.58}$, they found $x_0 = 0.141$, which corresponds to $B = 4.7 \cdot 10^{12}$ Gauss. This result is in agreement with the study made above.

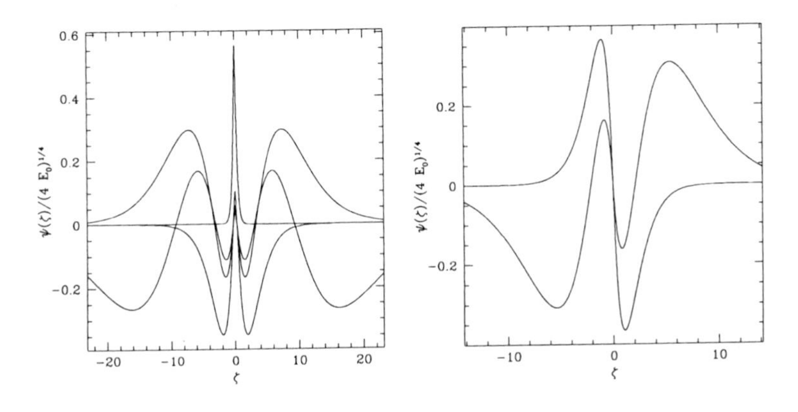

Figure 8.A.5. The axial wavefunctions of hydrogen in an intense magnetic field (analytic calculation) for $B = 4.7 \cdot 10^{12}$ Gauss. The first four even states with axial excitations, $|000\rangle$ (ground state), $|002\rangle$, $|004\rangle$, and $|006\rangle$ (left panel), and odd states $|001\rangle$ and $|003\rangle$ (right panel) are depicted; $n = 1/\sqrt{15.58}$, $\zeta = 2z/n$ corresponds to x in the used notation; z in a.u., 1 a.u. $= 0.53 \cdot 10^{-8}$ cm (reproduction of Figure 3 by Heyl and Hernquist [14]).

One can see from Fig. 8.A.5 that the peak of the ground state wave function $|000\rangle$ is at the point $z = 0$, while the largest peaks of the excited wave functions are away from the point $z = 0$ (as it was expected to be). Consequently, the associated longitudinal probability distributions (square modules of the wave functions multiplied by the volume factor of the chosen coordinate system) are symmetric with respect to $z \to -z$, and their maxima are placed in the center $z = 0$ for the ground state, and away from the center for the excited states. The computed ground state $|000\rangle$ binding energy of the hydrogen atom for different field intensities are [14]:

Magnetic field B (Gauss)	Binding energy, $\|000\rangle$ state (Rydberg)
4.7×10^{12}	15.58
9.4×10^{12}	18.80
23.5×10^{12}	23.81
4.7×10^{13}	28.22
9.4×10^{13}	33.21
23.5×10^{13}	40.75
4.7×10^{14}	47.20

Heyl and Hernquist calculated the first-order perturbative corrections to the above energies and obtained the values, which are in a good agreement with the results by Ruder *et al.* [9] and Lai [12].

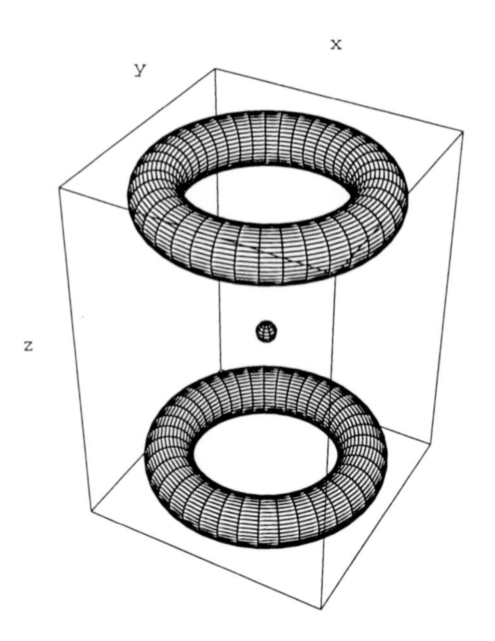

Figure 8.A.6. A schematic view on the H atom in an excited state under a very strong external magnetic field $\vec{B} = (0, 0, B)$, $B \gg B_0 = 2.4 \cdot 10^9$ Gauss. One electron moves simultaneously on two toroidal orbits of radius R_0 which are shown schematically as torii in the different (x, y) planes, one torus at the level $z = -L_z$ and the other at the level $z = +L_z$, with the nucleus shown in the center at $z = 0$. Each torus represents the (x, y) probability distribution as shown in Fig. 8.A.3 but with small Landau radius, $R_0 \ll a_0$. The spin of electron is aligned antiparallel to the magnetic field.

The associated probability density of the above *excited* states is evidently of a cylindrical (axial) symmetry and can be described as *two* Landau orbits of radius R_0 in different (r, φ) planes, one at the level $z = -L_z$, and the other at the level $z = +L_z$, with the nucleus at $z = 0$, as schematically depicted in Fig. 8.A.6. Presence of two

Landau orbits occurs in accord to the excited wave functions, which is symmetrical with respect to the inversion, $z \rightarrow -z$, and the largest peaks of which are away from the center $x = 0$. The electron moves simultaneously on these two Landau orbits.

A review of approximate, variational, and numerical solutions can be found in the paper by Lai [12]. The accuracy of numerical solutions is about 3%, for the external magnetic field in the range from 10^{11} to 10^{15} Gauss. Particularly, due to the variational results [12], the z-size of the hydrogen atom in the ground state is well approximated by the formula $L_z \simeq [\ln(B/B_0)]^{-1}$ a.u.; the transverse (Landau) size is $L_\perp \simeq (B/B_0)^{-1/2}$ a.u.; and the ground state energy $E \simeq -0.16[\ln(B/B_0)]^2$ a.u., with the accuracy of few percents, for $b \equiv B/B_0$ in the range from 10^2 to 10^6. One can see for $B = 100B_0$, that the variational study predicts the ground state energy $E = -3.4$ a.u. $= -92.5$ eV, the transverse size $L_\perp$ of about 0.1 a.u. $= 0.53 \cdot 10^{-9}$ cm, and the z-size L_z of about 0.22 a.u. This confirms the result of the modified Coulomb analytic approach.

Since a zero-field ground state case is characterized by perfect spherically symmetric electron charge distribution in the H atom, intermediate intensities of the magnetic field are naturally expected to imply a distorted spherical distribution. However, a deeper analysis is required for the intermediate magnetic field intensities because the adiabatic approximation is not longer valid in this case.

As to the multi-electron atoms, an interesting problem is to study action of very strong external magnetic field on He atom (see. e.g., Refs. [12, 14]) and on the multi-electrons heavy atoms, with outer electrons characterized by a *nonspherical* charge distribution, such as the p-electrons in Carbon atom, orbitals of which penetrate the orbitals of inner electrons. In fact, a very intense magnetic field would force such outer electrons to follow *small round* toroidal orbits. In addition to the effect of a direct action of the magnetic field on the inner electrons, a series of essential rearrangements of the whole electron structure of the atom seems to occur with the variation of the field strength. The magnetic field competes with the Coulomb energy, which is different for different states of electrons, and with the electron-electron interactions, including spin pairings. However, it is evident that at sufficiently strong fields, all the electron spins are aligned antiparallel to the magnetic field — fully spin polarized configuration — while at lower field intensities various partial spin polarized configurations are possible.

In accord to the numerical calculations based on the density matrix theory by Johnsen and Yngvason [13], which is in good agreement with the Hartree-Fock treatment of a very strong magnetic field, the inner domain in iron atom (26 electrons) is characterized by a slightly distorted spherically symmetric distribution, even at the intensities as high as $B = 100B_0 \dots 1000B_0$. The outer domain appears to be of specific, highly elongated distribution along the direction of the magnetic field as shown in Fig. 8.A.7. The possible interpretation that the inner electrons remain to have a spherical distribution while outer electrons undergo the squeeze seems to be not correct unless the spin state of the iron atom is verified to be partially polarized. So, we can conclude that all the electrons are in the highly magnetically polarized state (Landau state mixed a little by Coulomb interaction), and the electronic structure is a kind of *Landau multi-electron cylindrical shell*, with the spins of all the electrons being aligned antiparallel to the magnetic field (fully spin polarized configuration).

Another remark regarding Fig. 8.A.7 is that the contours indicating a nearly spherical distribution will always appear since the Coulomb center (nucleus) is not totally eliminated from the consideration (non-adiabatic approximation), and it forces a spherical distribution to some degree, which evidently depends on the distance

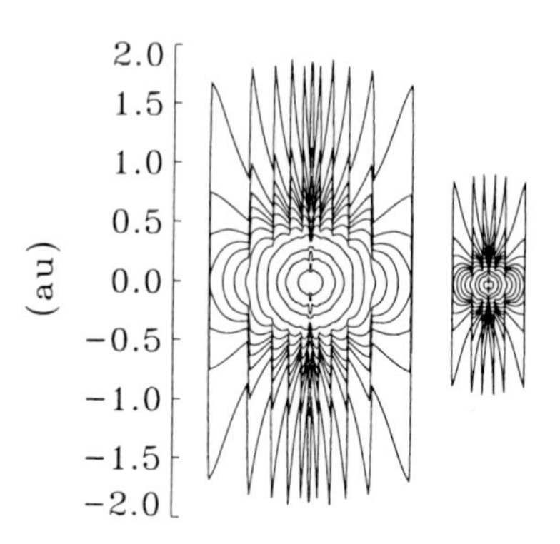

Figure 8.A.7. Contour plots of the (r, z) plane electronic density of iron atom according to the density matrix theory at two different magnetic field strengths, 10^{11} Gauss (left) and 10^{12} Gauss (right). The outermost contour encloses 99% of the negative charge, the next 90%, then 80% etc., and the two innermost 5% and 1% respectively (reproduction of Fig. 5 by Johnsen and Yngvason [13]).

from the center (closer to the center, more sphericity). We note that outer contours in Fig. 8.A.7 is in qualitative agreement with Fig. 8.A.6 in the sense that the predicted charge distribution reveals symmetry under the inversion $z \to -z$, with the characteristic z-elongated Landau-type orbits.

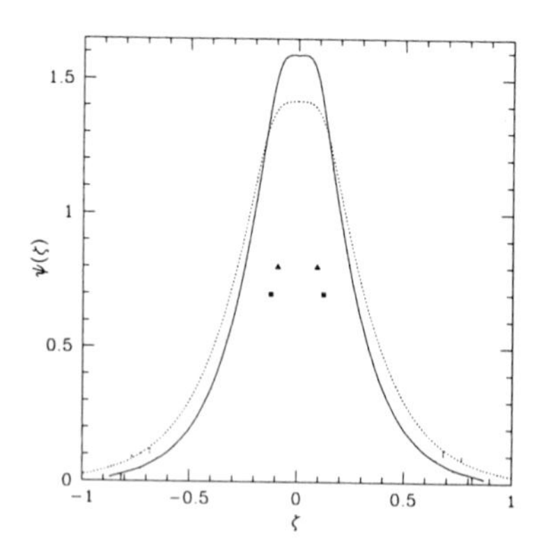

Figure 8.A.8. A schematic view of the ground and first-excited state of H_2^+ ion. The solid line traces $|000\rangle$, and the dashed line follows $|0\text{-}1\,0\rangle$. The triangles give the positions of the protons for the ground state and the squares for the excited state. The magnetic field $B = 4.7 \cdot 10^{12}$ Gauss is pointed along the internuclear axis; $\zeta = 2\pi\alpha z/\lambda_e$ denotes z in a.u.; 1 a.u. $= 0.53 \cdot 10^{-8}$ cm (reproduction of Figure 5 by Heyl and Hernquist [14]).

An interesting problem is to study H_2 molecule under the action of a strong external static uniform magnetic field using Schrödinger's equation. However, prior to that study, it would be useful to investigate the simpler two-center H_2^+ ion, since it can give valuable information on the features of the full hydrogen molecule under the action of a strong magnetic field. We refer the interested reader to Refs. [12, 14, 15] for studies on H_2^+ ion and H_2 molecule in strong magnetic field. Figure 8.A.8 displays the ground and first excited state wave functions of H_2^+ [14].

References

[1] Santilli, R.M.: Hadronic J. **21**, 789 (1998).

[2] http://www.magnegas.com.

[3] Kucherenko, M.G. and Aringazin, A.K.: Hadronic J. **21**, 895 (1998).

[4] Settle, F.A.: Editor, *Handbook of Instrumental Techniques for Analytic Chemistry*, Prentice Hall, Upper Saddle River, New Jersey (1997).

[5] Eldridge, H.: U.S. Patent no. 603,058, Spril [5a]; Dammann, W.A. and Wallman, D.: U.S. Patents nos. 5,159,900 (1992) and 5,417,817 (1995) [5b]; Richardson, W.H. jr.: U.S. Patents nos. 5,435,274 (1995), 5,692,459 (1997), 5,792,325 (1999) [5c]; See also for the Brown Gas Web Site http://www.eagle-research.com [5d],

and for the SkyGas Web Site http://www.mpmtech.com [5e].

[6] Sachse, T.I. and Kleinekathöfer, U.: "Generalized Heitler–London Theory for H_3: A Comparison of the Surface Integral Method with Perturbation Theory", e-print arXiv: physics/0011058 (November 2000).

[7] Kadomtsev, B.B. and Kudryavtsev, V.S.: Pis'ma ZhETF **13**, 15, 61 (1971); Sov. Phys. JETP Lett. **13**, 9, 42 (1971) (English Translation); ZhETF **62**, 144 (1972); Sov. Phys. JETP **35**, 76 (1972) (English Translation). Ruderman, M.: Phys. Rev. Lett. **27**, 1306 (1971); in: IAU Symposium 53, *Physics of Dense Matter*, C.J. Hansen (ed.), Dordrecht, Reidel (1974). Lai, D., Salpeter, E. and Shapiro, S.L.: Phys. Rev. **A45**, 4832 (1992). Lai, D. and Salpeter, E.: Phys. Rev. **A52**, 2611 (1995); Phys. Rev. **A53**, 152 (1996); Astrophys. J. **491**, 270 (1997).

[8] Aringazin, A.K.: Toroidal configuration of the orbit of the electron of the hydrogen atom under strong external magnetic fields, Hadronic J. **24** (2001), in press.

[9] Sokolov A.A., Ternov I.M., and Zhukovskii, V.Ch.: *Quantum mechanics*, Nauka, Moscow, 1979 (in Russian). Landau, L.D. and Lifshitz E.M.: *Quantum Mechanics: Non-Relativistic Theory*, 3rd ed., Pergamon, Oxford, 1989. Ruder, H., Wunner, G., Herold, H. Geyer, F.: *Atoms in Strong Magnetic Fields*, Springer, Berlin-Heidelberg-New York, 1994. Kadomtsev, B.B.: Soviet Phys. JETP **31** 945 (1970).

Kadomtsev, B.B. and Kudryavtsev, V.S.: JETP **13** 42 (1971); JETP Lett. **13** 9 (1971).

[10] Faddeev, L. and Niemi, A.J.: Magnetic geometry and the confinement of electrically conducting plasmas, physics/0003083, April 2000. R. Battye, R., Sutcliffe, P.: Phys. Rev. Lett. **81**, 4798 (1998); and Proc. R. Soc. Lond. **A455**, 4305 (1999). Hietarinta, J, Salo, P.: Phys. Lett. **B451**, 60 (1999); and The ground state in the Faddeev-Skyrme model, University of Turku preprint, 1999; for video animations, see `http://users.utu.fi/hietarin/knots/index.html`

[11] Santilli, R.M.: *The Physics of New Clean Energies and Fuels According to Hadronic Mechanics*, Journal of New Energy **4**, Special Edition, No. 1 (1999), 318 pages. Santilli, R.M. and Shillady, D.D.: Intern. J. Hydrogen Energy **24**, 943 (1999); Intern. J. Hadrogen Energy **25**, 173 (2000).

[12] Lai, D.: Matter in strong magnetic fields, chem-ph/0009333, September 2000.

[13] Johnsen, K. and Yngvason, J.: Density Matrix Functional Calculations for Matter in Strong Magnetic Fields: I. Atomic Properties, chem-ph/9603005, March 1996.

[14] Heyl, J.S. and Hernquist, L.: Hydrogen and Helium Atoms and Molecules in an Intense Magnetic Field, chem-ph/9806040, June 1998. Jones, M.D., Ortiz, G., and Ceperley, D.M.: Spectrum of Neutral Helium in Strong Magnetic Fields, chem-ph/9811041, November 1998.

[15] Lopez, J.C., Hess, P., and Turbiner, A.: H_2^+ ion in strong magnetic field: a variational study, Preprint ICN-UNAM 97-06, chem-ph/9707050. Turbiner, A., Lopez, J.C., and Solis, U.: H_3^{++} molecular ions can exist in strong magnetic fields, Preprint ICN-UNAM 98-05, chem-ph/9809298. Lopez, J.C. and Turbiner, A.: One-electron linear systems in a strong magnetic field, Preprint ICN-UNAM 99-03, chem-ph/9911535.

Index